AF800402

Smart Nanomaterials Technology

Series Editors

Azamal Husen, Wolaita Sodo University, Wolaita, Ethiopia

Mohammad Jawaid, Laboratory of Biocomposite Technology, Universiti Putra Malaysia, INTROP, Serdang, Selangor, Malaysia

Nanotechnology is a rapidly growing scientific field and has attracted a great interest over the last few years because of its abundant applications in different fields like biology, physics and chemistry. This science deals with the production of minute particles called nanomaterials having dimensions between 1 and 100 nm which may serve as building blocks for various physical and biological systems. On the other hand, there is the class of smart materials where the material that can stimuli by external factors and results a new kind of functional properties. The combination of these two classes forms a new class of smart nanomaterials, which produces unique functional material properties and a great opportunity to larger span of application. Smart nanomaterials have been employed by researchers to use it effectively in agricultural production, soil improvement, disease management, energy and environment, medical science, pharmaceuticals, engineering, food, animal husbandry and forestry sectors.

This book series in Smart Nanomaterials Technology aims to comprehensively cover topics in the fabrication, synthesis and application of these materials for applications in the following fields:

- Energy Systems—Renewable energy, energy storage (supercapacitors and electrochemical cells), hydrogen storage, photocatalytic water splitting for hydrogen production
- Biomedical—controlled release of drugs, treatment of various diseases, biosensors,
- Agricultural—agricultural production, soil improvement, disease management, animal feed, egg, milk and meat production/processing,
- Forestry—wood preservation, protection, disease management
- Environment—wastewater treatment, separation of hazardous contaminants from wastewater, indoor air filters.

Akil Ahmad · Mohammad Jawaid ·
Mohamad Nasir Mohamad Ibrahim ·
Asim Ali Yaqoob · Mohammed B. Alshammari
Editors

Nanohybrid Materials for Treatment of Textiles Dyes

Springer

Editors
Akil Ahmad
Department of Chemistry, College of Science and Humanities
Prince Sattam bin Abdulaziz University
Al-Kharj, Saudi Arabia

Mohamad Nasir Mohamad Ibrahim
School of Chemical Sciences
Universiti Sains Malaysia
Gelugor, Malaysia

Mohammed B. Alshammari
Department of Chemistry, College of Science and Humanities
Prince Sattam bin Abdulaziz University
Al-Kharj, Saudi Arabia

Mohammad Jawaid
Laboratory of Biocomposite Technology
Universiti Putra Malaysia
Serdang, Malaysia

Asim Ali Yaqoob
School of Chemical Sciences
Universiti Sains Malaysia
Gelugor, Malaysia

ISSN 3004-8273 ISSN 3004-8281 (electronic)
Smart Nanomaterials Technology
ISBN 978-981-99-3900-8 ISBN 978-981-99-3901-5 (eBook)
https://doi.org/10.1007/978-981-99-3901-5

© The Editor(s) (if applicable) and The Author(s), under exclusive license to Springer Nature Singapore Pte Ltd. 2023

This work is subject to copyright. All rights are solely and exclusively licensed by the Publisher, whether the whole or part of the material is concerned, specifically the rights of translation, reprinting, reuse of illustrations, recitation, broadcasting, reproduction on microfilms or in any other physical way, and transmission or information storage and retrieval, electronic adaptation, computer software, or by similar or dissimilar methodology now known or hereafter developed.
The use of general descriptive names, registered names, trademarks, service marks, etc. in this publication does not imply, even in the absence of a specific statement, that such names are exempt from the relevant protective laws and regulations and therefore free for general use.
The publisher, the authors, and the editors are safe to assume that the advice and information in this book are believed to be true and accurate at the date of publication. Neither the publisher nor the authors or the editors give a warranty, expressed or implied, with respect to the material contained herein or for any errors or omissions that may have been made. The publisher remains neutral with regard to jurisdictional claims in published maps and institutional affiliations.

This Springer imprint is published by the registered company Springer Nature Singapore Pte Ltd.
The registered company address is: 152 Beach Road, #21-01/04 Gateway East, Singapore 189721, Singapore

The editors are honored to dedicate this book to Late Jamil Ahmad and Mrs. Sajada Begum (Beloved Father and Mother of Dr. Akil Ahmad).

Preface

Due to rapid growth of various colorant (dyes) industries, dyes are discharged into the water bodies which led to the degrading of water quality. This may cause adverse effect to living organisms as well as human being. To treat these dyes at the outlet of coloring industries, various physical and chemical methods can be used such as adsorption, solvent extraction, ion exchange, precipitation, filtration and photocatalytic degradation. Given the increasing number of scientific publications dealing with various techniques and the huge progress in the processing possibility to produce an efficient and effective materials based on novel techniques, a book meeting and apprising all of these innovations is of great interest for academicians, researchers, industrialists and students who are concerned with the production and use of these techniques as innovative process for the treatment of textile dye from water and wastewater.

The main objective of this book is about the current demand of the nanohybrid materials and their application for dye removal. The specialism in nanohybrid materials inaugurated by its outstanding properties like high surface area, pore size, mechanical strength and toughness and electrical and thermal conductivity. Unlike other materials, this nanohybrid is easily synthesized and fabricated and doped with various metal oxide and other nanoparticles. Nanohybrid currently considered as an efficient material for dyes removal as compared to other available materials. This may be due to its unique characteristics such as low cost, high regenerability, high adsorption capacity, environmentally friendly and sustainability. The unique feature of this book is it presents a unified knowledge of this efficient nanohybrid on the basis of wide application, easy to synthesize and fabricate with metal oxide and other nanoparticles.

We are greatly thankful to all qualified researcher, scholar and leading experts to contribute their valuable work. The chapters provided cutting-edge up-to-date research findings on nanohybrid materials for dyes removal. We collected all the information given by eminent authors on nanohybrid materials and related research from Mexico, Australia, Turkey, India, Malaysia, Saudi Arabia, Nigeria, Sudan,

Iran, Ghana, Slovakia, Pakistan and Bangladesh and finally compiled this project in a fruitful way.

Al-Kharj, Saudi Arabia — Akil Ahmad
Serdang, Malaysia — Mohammad Jawaid
Gelugor, Malaysia — Mohamad Nasir Mohamad Ibrahim
Gelugor, Malaysia — Asim Ali Yaqoob
Al-Kharj, Saudi Arabia — Mohammed B. Alshammari

Acknowledgements Editors wish to appreciate the support of Prince Sattam bin Abdulaziz University, Saudi Arabia, Universiti Sains Malaysia and Universiti Putra Malaysia, Malaysia.

Contents

Introduction of Nanohybrid Materials 1
Thiruppathi Krithika, Thiruppathi Iswarya, and Thiruppathi Sowndarya

Survey of Nanohybrid Materials in Textile Dyes Removal 19
V. Mahalakshmi and Lali Growther

Textile Dyes and Their Effect on Human Beings 41
N. Hemashenpagam and S. Selvajeyanthi

Synthesis and Characterization of Nanohybrid Materials 61
Mustapha Omenesa Idris, Mohamad Nasir Mohamad Ibrahim, Akil Ahmad, and Mohammed B. Alshammari

Graphene-Supported Nanohybrid Materials for Removal of Textile Dyes 75
Mustapha Omenesa Idris, Najwa Najihah Mohamad Daud, Mohamad Nasir Mohamad Ibrahim, and Abdulmumuni Sumaila

Synthesis and Characterization of Nanohybrid Materials for Anionic Dye Removal 91
Alain R. Picos-Benítez, María M. Ramírez-Alaniz, Pablo Emilio Escamilla-García, and Blanca L. Martínez-Vargas

Decolourization of Textile Dyes Using CNT-Based Hybrid Materials 119
Rania Edrees Adam Mohammad, Abdullahi Haruna Birniwa, Shehu Sa'ad Abdullahi, and Ahmad Hussaini Jagaba

Synthesis and Catalytic Applications of PANI-Based Hybrid Materials for the Catalytic Removal of Organic Dyes from Wastewaters 157
Sandeep Kaushal and Karina Bano

Recovery and Removal of Textile Dyes Through Adsorption Process 179
Growther Lali, V. Mahalakshmi, M. Seenuvasan, and G. Sarojini

Photocatalytic Degradation of Textile Dyes Using Nanohybrid Materials 203
A. A. A. Mutalib and N. F. Jaafar

Montmorillonite (MMt) Clay-Based Hybrid Materials for Textile Dyes' Removal 223
Babak Jaleh, Ensiye Shabanlou, Atefeh Nasri, and Mahtab Eslamipanah

Metal-Decorated Nanohybrid Materials for Textile Dyes' Removal from Wastewater 247
Babak Jaleh, Ensiye Shabanlou, Mahtab Eslamipanah, and Atefeh Nasri

Nano-engineered Hybrid Materials for Cationic Dye Removal 273
Nana Aboagye Acheampong, Emmanuel Okoampah,
Nana Kobea Bonso, and Abubakari Zarouk Imoro

Textile Dyes Removal Using Silica-Dendrimer Hybrid Materials 303
Akbar Esmaeili

Nanohybrid-Based Catalysts for Degradation of Dyes from Aqueous Solution 343
Burcu Palas

Dye Removal Using Magnetized Nanohybrid Adsorbent 381
Akansha Mehta

The Impact of Textile Dyes on the Environment 401
Tanzina Akter, Anica Tasnim Protity, Modhusudon Shaha,
Mohammad Al Mamun, and Abu Hashem

Photoactive Titanium Dioxide Nanoparticles Hybrid for Dye Removal Under Light Irradiation 433
Mustaffa Ali Azhar Taib, Mohd Azam Mohd Adnan,
Mohd Fadhil Majnis, and Nurhidayatullaili Muhd Julkapli

Alginate-Based Hybrid Materials for the Treatment of Textile Dyes 471
Muhammad Alamzeb, Behramand Khan, and Haroon Subhani

About the Editors

Akil Ahmad is currently working at Prince Sattam bin Abdulaziz University, Saudi Arabia, as Assistant Professor in Chemistry and having the experience of seven years as Research Fellow, Teaching Fellow, and Postdoc and Visiting Researcher from Universiti Teknologi Malaysia, Universiti Sains Malaysia, University of KwaZulu-Natal, South Africa, and Universiti Kebangsaan Malaysia, Malaysia. He has completed Ph.D. in Analytical Chemistry (2011) from Aligarh Muslim University (AMU), India. He is Guest Editor of journals, namely *Adsorption Science and Technology*, Hindawi, *Polymers MDPI*, *Frontiers in Environmental Chemistry*, and *Journal of Chemistry*, Hindawi. He has edited more than 10 books, published in Elsevier, Springer Nature, IOP publisher, etc., and worked as Research Assistant at Ekahala Resources Sdn Bhd, Malaysia.

Dr. Mohammad Jawaid is currently working as Senior Fellow (Professor) at Biocomposite Technology Laboratory, Institute of Tropical Forestry and Forest Products (INTROP), Universiti Putra Malaysia (UPM), Serdang, Selangor, Malaysia, and also has been Visiting Professor at the Department of Chemical Engineering, College of Engineering, King Saud University, Riyadh, Saudi Arabia, since June 2013. He also obtained three patents and six copyrights. Recently, he joined as Chief Executive Editor of Pertanika UPM Journals. He is founding Series Editor of *Composite Science and Technology* Book Series from Springer Nature, and also Series Editor of *Springer Proceedings in Materials*

and Springer Nature and also International Advisory Board Member of *Springer Series on Polymer and Composite Materials*. Besides that, he is also Reviewer of several high-impact international peer-reviewed journals of Elsevier, Springer, Wiley, Saga, ACS, RSC, Frontiers, etc. Presently, he is supervising 12 Ph.D. and six master's students.

Dr. Mohamad Nasir Mohamad Ibrahim obtained his B.Sc. (1994), M.Sc. (1997), and Ph.D. (1999) from Missouri S&T (formerly known as University of Missouri-Rolla, USA). He is currently serving as Associate Professor in the School of Chemical Sciences, Universiti Sains Malaysia (USM). He had received fourteen international awards for his research outputs and currently served as a Guest Editor for Frontiers in Chemistry. He had supervised more than 20 graduate students. Recently, he is busy working in the microbial fuel cells topic especially in developing a novel electrode and had published several papers in well-reputed journals such as *Chemical Engineering Journal* and *Journal of Cleaner Production*. At the moment, he is Pioneer on MFCs research topic in the School of Chemical Sciences, USM.

Dr. Asim Ali Yaqoob obtained his B.Sc. (2010), M.Sc. (2014), and M.Phil. (2018) from Mirpur University of Science and Technology Mirpur, Mirpur, 10250, Ajk-Pakistan. Recently, he completed his Ph.D. (2021) from School of Chemical Sciences, Universiti Sains Malaysia (USM), under the supervision of Dr. Mohamad Nasir Mohamad Ibrahim. His area of interest is energy generation coupled with wastewater treatment, and currently, his project is on Microbial Fuel Cells. He has published 22 articles in high reputed journals and several book chapters. He is also Co-author for an academic book (*Graphene: A Versatile Advanced Material*) published by USM Press, Malaysia. He has completed M.Phil. in Material Chemistry (2018 from Mirpur University of Science and Technology (MUST), AJK Pakistan. He is also Invited Reviewer of many reputed international journals such as *Nanotechnology Reviews*, *International Journal of Energy Research*, *Journal of Chemistry*, *PLOS ONE*, and many others.

Prof. Dr. Mohammed B. Alshammari is currently working at Prince Sattam bin Abdulaziz University in Al-Kharj, Saudi Arabia, as Professor in Organic Chemistry. He received his B.Sc. and M.Sc. degrees in Chemistry from King Saud University, Riyadh, KSA, in 2007 under supervision of Prof. Abdullah Almajed and Prof. Hassan Alhazmi. He worked in chemistry department in KSU for nine years as Researcher. He received his Ph.D. degree from Cardiff University, UK, in 2013 under supervision of Prof. Keith Smith. His research focused in using of organometallic intermediates in organic synthesis. He has published more than 60 International peer-reviewed journal publications and 20 conference proceedings in the area of organic chemistry and chemistry.

Introduction of Nanohybrid Materials

Thiruppathi Krithika, Thiruppathi Iswarya, and Thiruppathi Sowndarya

Abstract Nanohybrid materials get their name from the fact that they are composed of synthetic organic and inorganic parts that are linked at the nanoscale in either a covalent or non-covalent fashion. The chemical reactivity of the organic base material is increased as a result of the contribution of the inorganic groups, which serve as functional groups. The development of these materials is currently at the forefront of a cutting-edge field that merges nanotechnology, material science, and life sciences. This is a multidisciplinary field that is also on the cutting edge of technological advancement. The nanoscale size, structure, form, and surface chemistry of the hybrid material all contribute to an increase in its already impressive multifunctionality. This is due to the fact that the ratio of the surface area of the material to its volume is increasing, which in turn causes the atoms on the material's surface to have a greater influence on the performance of the material. Because they are so much smaller than the bulk material, nanoparticles have a surface area-to-volume ratio that is significantly higher than that of the bulk material. This is because surface area contributes more to the total volume of a substance. Because of this quality of nanohybrid materials, it is possible for them to have unexpected optical, physical, and chemical properties. This property is made possible due to the fact that nanohybrid materials are small enough to produce quantum effects and trap their electrons within their own structures. Intelligent, integrated nanohybrid agents are of particular interest because of the increased functionality, target selectivity, and applicability that they offer. In recent times, there has been a lot of focus placed on the development of innovative strategies for the synthesis and characterization of nanohybrid materials. This chapter provides a concise overview of the concept of nanohybrid materials, as well as a discussion of the numerous types of these materials, the benefits they offer, and the applications they can have in a wide variety of industries.

T. Krithika (✉)
Department of Microbiology, Hindusthan College of Arts and Science, Coimbatore, Tamil Nadu 641028, India
e-mail: krithikat2294@gmail.com

T. Iswarya · T. Sowndarya
Department of Environmental Sciences, Bharathiar University, Coimbatore, Tamil Nadu 641046, India

© The Author(s), under exclusive license to Springer Nature Singapore Pte Ltd. 2023

A. Ahmad et al. (eds.), *Nanohybrid Materials for Treatment of Textiles Dyes*,
Smart Nanomaterials Technology, https://doi.org/10.1007/978-981-99-3901-5_1

Keywords Nanoparticles · Hybrid · Nanocomposites · Nanohybrid materials

1 Introduction

Norio Taniguchi of the Tokyo University of Science was the first person to use the term "nanotechnology" in 1974. He was referring to the processes that are used in the manufacturing of semiconductors. The study of functional materials at length scales on the nanometer (1 nm = 1×10^{-9} m) is the focus of the multidisciplinary branch of research known as nanotechnology. The study of nanoscale functional materials is within the purview of this inter-disciplinary field. The enormous leaps in scientific and technical advancement that led to the development of this cutting-edge technology had a significant influence on the progression of civilization as a whole [1].

As a result of the use of nanotechnology, a wide variety of new kinds of materials have been fabricated at the nanoscale. Nanoparticles (NPs), also known as particulate materials having at least one dimension less than 100 nm, are a diverse family of materials. They are commonly defined as particulate materials with at least one dimension less than 100 nm. Nanoparticles (NPs) are considered to be a kind of nanomaterial. Depending on the overall shape, each of these materials may take the form of a 0D, 1D, 2D, or 3D object. When researchers realized that the size of a substance might have an effect on its physiochemical properties, it became abundantly evident to them the relevance of the substances that were in question [2].

Nanoscale particles have better attributes like as particle size distribution and shape, in contrast to larger particles of bulk material, which do not possess these properties. Nanoscale particles also have unique qualities. Nanoparticle characteristics may be seen to reflect both the solute and the distinct particle phase. Nanoparticles have a surface-to-volume ratio that is 35–45% greater than that of atoms or large particles, depending on the size of the particle. Large particles have a surface area that is equal to their volume in a ratio of 1–1. One of the many intrinsic properties of a nanoparticle that is influenced by its particular surface area, an extrinsic property that is particular to the nanoparticle, is strong surface reactivity, which is size dependent and contributes to the high value of the particle. The specific surface area of the nanoparticle is one of the nanoparticle's extrinsic characteristics that make it unique. When considering a nanoparticle's specific surface area, this is just one of many properties that are impacted. Overall, the unique characteristics of nanoparticles are what confer upon them their multifunctional qualities and contribute to the expanding interest in using them in fields such as medical, the manufacturing of food and beverages, and the field of energy [3].

The term "metallic nanoparticles," which is more often referred to as simply "metal nanoparticles," is a relatively recent addition to the lexicon around nanoparticles. Noble metals like gold, silver, and platinum are used in the production of nanoparticles not only due to the fact that they are employed to generate metallic nanoparticles, but also due to the fact that they have favorable effects on the health of humans. Researchers are currently focusing their attention on metal nanoparticles, nanostructures, and the synthesis of nanomaterials because of the unique properties that make them useful for a variety of applications, including catalysis, composite preparations like polymers, disease diagnosis and treatment, sensor technology, and labeling of optoelectronic recorded media. These applications are made possible by the nanoparticles' small size and high surface area [4].

Traditional nanomaterials that consist of a single component provide considerable support for a wide variety of features, but they can only carry out a limited number of functions. The concept of composites, which involves the fusion of two or more fundamental materials, was developed in order to increase performance and find new applications. Composites are made by fusing together two or more basic materials. This concept is shown in a more tangible form by hybrid materials, which consist of a combination of organic and inorganic components. Inorganic materials provide a variety of desirable features, such as high strength and stiffness as well as resistance to the effects of ageing, despite the fact that it may be difficult to create these types of materials. Organic materials, on the other hand, provide the possibility of adding flexibility and make it possible to process the material while it is exposed to its natural environment. The integration of organic and inorganic components will make it easier to create one-of-a-kind chemical, biological, electrical, and optical properties that are not possible with single-component materials. These properties cannot be imagined with single-component materials [5].

Dendrimers are a subclass of repetitively branching monodisperse molecules. These macromolecules take the form of trees and are noted for their tree-like structures. This structure is composed of three components: the repeating unit, the surface functional groups on the exterior, and the inner initiator core. The repeating unit is located in the centre of the structure. Using the procedures that have been described, a wide array of dendrimers that have diverse chemical functions and interior chambers for encasing visiting molecules have been manufactured. It serves as an organic building block for the production of hybrid materials by encapsulating material inside intramolecular cavities, interacting electrostatically with end groups, and chemically conjugating end groups. All of these features contribute to its versatility. However, in order to create high generation dendrimers with dense end groups, one must first employ challenging synthesis and separation procedures. This, in turn, typically results in a low yield as well as defects in the structure of the final product. In recent years, there has been an increase in the amount of focus placed on the use of multifunctional inorganic cores as the building blocks for the development of dendrimer nanohybrid materials with high density end groups. This is mostly attributable to the fact that the production of these goods is uncomplicated and dependable [5]. The creation of bio-nanohybrid materials begins with the assembly of molecular or polymeric species that have a biological origin and inorganic substrates through interactions that take place on the nanometric scale. This lays the groundwork for the development of bio-nanohybrid materials.

This chapter will provide an introduction to nanohybrid materials, along with details on their development, history, and advantages of using them in a different industries.

2 Nanotechnology

"Nanotechnology" refers to the process of changing matter at a scale that is very near to that of atoms in order to generate unique structures, materials, and devices. This process may be used to manufacture a variety of products. The method makes it possible for scientific advances to be made that may be used to a diverse variety of sectors, such as the production of consumer products, the energy business, the medical industry, and many others. The term "nanotechnology" refers to a wide category that includes the aforementioned engineered systems, tools, and structures. This term is also used to describe these things. Nanomaterials may have dimensions ranging from one nanometer to one hundred nanometers in length. At this size, materials begin to display numerous features, which impact how they operate physically, chemically, and physiologically. These qualities may be broken down into two categories: macroscopic and microscale. The pursuit of new information, the production of novel items, and the practical implementation of these technologies are the driving forces behind the rapid development of current technology (The National Institute for Occupational Safety and Health [NIOSH]).

Richard Feynman, a physicist, delivered a talk with the title "There's Plenty of Room at the Bottom" on the 29th of December, 1959, at the California Institute of Technology, which was hosting a conference of the American Physical Society. Before the term "nanotechnology" was coined, this era lasted for a very considerable amount of time (CalTech). During his talk, Richard Feynman presented a strategy for how scientists could one day be able to exercise control over particular atoms and molecules and make modifications to them. The concept was explained in detail. Professor Norio Taniguchi, who was doing research on ultraprecision machining at the time, is credited with coming up with the word "nanotechnology" more than 10 years after the phrase was first used. The development of the scanning tunneling microscope in 1981, which made it possible to "see" individual atoms, is generally regarded as the moment that marked the beginning of modern nanotechnology.

The capacity to observe and manipulate individual atoms and molecules is an essential part of both nanoscience and nanotechnology. Atoms are the building blocks of everything that exists on our planet, including the food we eat, the clothes we wear, the houses and other buildings we live in, and even our own bodies. This includes everything from the clothing we wear to the food we consume. However, the human eye is incapable of seeing anything that is any smaller than an atom. It may be somewhat difficult to make observations with the microscopes that are typically used in science classes taught at high schools. In the early 1980s, the first commercially available microscopes with the resolution to see objects on the nanoscale appeared on the market.

The acquisition of the essential equipment by scientists, such as the scanning tunneling microscope (STM) and the atomic force microscope, marked the beginning of a new era in technology that is known as nanotechnology. Despite the fact that the fields of nanoscience and nanotechnology are still in their infancy, materials on the nanoscale have been used for millennia. The use of gold and silver granules of varying sizes was necessary in order to achieve the desired effect of recreating the colors of the stained glass windows that were common in medieval cathedrals. These cathedrals have been standing in this location for quite some time. To put it more simply, the painters of that era were not aware that the methods they used to create such exquisite works of art actually changed the chemical composition of the materials they were using to create the works of art they were creating. This is because the painters of that era were not aware of the connection between the techniques they used to create the works of art they were creating and the materials they were using.

To take advantage of the improved properties of materials at the nanoscale, such as higher strength, lighter weight, increased control over the light spectrum, and greater chemical reactivity than their larger-scale counterparts, scientists and engineers are currently developing a wide variety of deliberate manufacturing processes. This is being done in order to take advantage of the improved properties of materials at the nanoscale. These processes include the following steps.

3 Nanoparticles

Nanoparticles, also known as NPs, are a broad class of materials that are particulate and have at least one dimension that is less than 100 nm. NPs are classified as a subcategory of nanomaterials. Depending on the shape as a whole, these materials may either have a 0D, 1D, 2D, or 3D form. When researchers realized that the size of a substance may have an effect on its physiochemical properties, such as its optical capabilities, they came to an understanding of the relevance of these materials. Wine red, yellowish gray, black, and dark black are the typical hues of 20-nm gold (Au), platinum (Pt), and silver (Ag) nanoparticles, respectively. Palladium (Pd) nanoparticles have the darkest black hue [6].

The NPs' surface layer is the first of three levels that make up the object. This layer may be functionalized with a variety of small molecules, metal ions, surfactants, and polymers. The object also has two other layers.

(a) The core, which is the fundamentally central region of the NP and frequently corresponds to the NP itself;
(b) The shell layer, which is composed of material that is chemically distinct from the core in all regards;
(c) The core, which is essentially the central portion of the NP and usually refers the NP itself [7].

Because of the exceptional features that these materials had, researchers from a wide variety of fields showed a significant amount of interest in studying them. Mesoporousness provides nanoparticles additional characteristics. The NPs have potential applications in the capture of carbon dioxide [8], the delivery of medications [9], chemical and biological sensing [10], and other applications related to these fields [11].

4 Types, Synthesis, and Characterization of Nanoparticles

It is possible to categorize nanoparticles (NPs) into a variety of different groups dependent on their dimensions, their three-dimensional shapes, and their chemical composition. There are many different types of nanoparticles (NPs) that can be categorized according to the physical and chemical properties of the nanoparticles themselves. Some of the more common types of NPs include lipid-based nanoparticles, metal nanoparticles, ceramic nanoparticles, semi-conductor nanoparticles, polymeric nanoparticles, and carbon-based nanoparticles [2].

There are many different kinds of nanoparticles, some of which include semiconductor nanoparticles, polymeric nanoparticles, and nanoparticles formed of carbon. Carbon, polymeric, and semiconductor nanoparticles are a few examples of the many different kinds of nanoparticles that can be found in the world. As can be seen in Fig. 1, there are numerous ways to synthesize NPs, each of which can be classified into one of two major categories: either an approach that works from the bottom up or an approach that works from the top down [12].

It is possible to further categorize these strategies into a variety of different subclasses by thinking about the circumstances in which they are used, the results they produce, and the methods they employ. A destructive method is required in order to carry out top-down synthesis successfully. These units were originally a part of a larger starting molecule that needed to be broken down into their component parts in order to be converted into the appropriate NPs. This process required a disassembly step. NPs can be manufactured through the use of a process that is also known as the

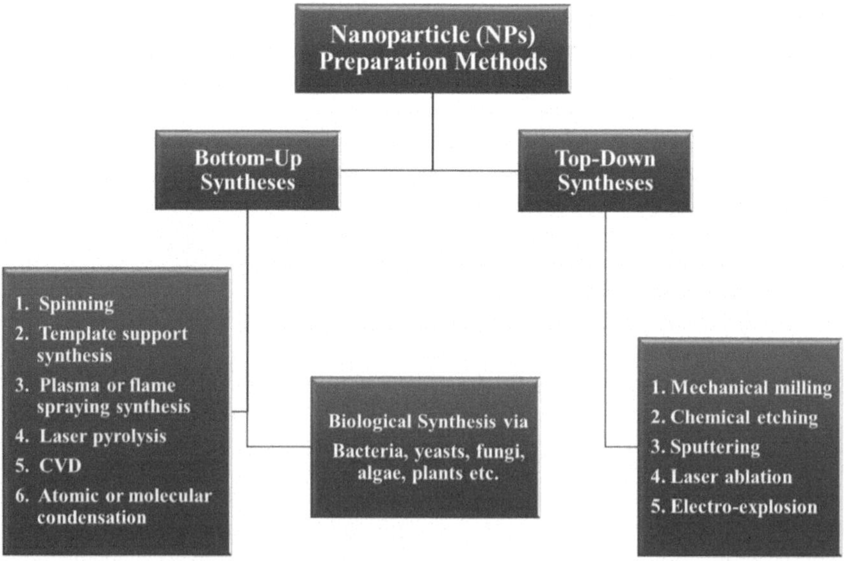

Fig. 1 Typical synthetic methods for NPs for the [12]

bottom-up syntheses procedure and is referred to as the building up strategy. This is due to the fact that the molecules used to make NPs are of an extremely fundamental nature.

In order to investigate the numerous physicochemical characteristics of NPs, researchers have employed a wide variety of different methods that can be categorized as falling under the heading of characterization. X-ray diffraction, X-ray photoelectron spectroscopy, infrared, scanning electron microscopy (SEM), transmission electron microscopy (TEM), Brunauer–Emmett–Teller (BET), and particle size analysis are some of the techniques that are utilized in this field. Infrared is yet another method that has the potential to be utilized in this region.

4.1 Hybrid Materials

There is a good chance that a hybrid material will have at least one inorganic component; but, there is also the possibility that there may be two or more inorganic components, all of which will be of quite different types. At the micrometric and sub-micrometric sizes, the components often mix together and/or interact with one another; this behavior persists all the way down to the nanometric and molecular levels [13]. Synergistic effects are commonly connected with the interactions between the different components, which result in the production of hybrid materials as a consequence of the interactions. These materials are capable of displaying either a diverse range of capabilities or distinguishing traits. Synergistic effects are often linked with interactions between numerous different components [14].

Both nacre, which is a crystallized compacted lamellar structure made of aragonite and conchiolin, and the natural pigment known as Blue Maya, which is created by mixing natural dyes (derived from molecules of an indigo-type with lamellar clays), are examples of hybrid materials that are created by nature. Blue Maya is a natural pigment, and nacre is an example of a hybrid material. When natural dyes are mixed together, a natural color called Blue Maya is produced. Blue Maya is a pigment that is produced by combining natural colors with lamellar clay. In contrast, nacre is a material that is manufactured by combining aragonite and conchiolin. Gemstones may include blue maya.

In accordance with the characteristics of the interfacial interactions between the phases or components, hybrid materials may be divided into one of the following two primary categories: (1) Class I structures have both organic and inorganic components contained inside of one another, and they are kept together by hydrogen, van der Waals, or electrostatic interactions. (2) Class II does not include any organic or inorganic components at all. The elements that make up Class II are held together in part by covalent bonds, which are chemical connections that are very long-lasting [15].

Bottom-up techniques, such as those that begin with precisely specified "nano-objects" and chemical precursors, might potentially be employed to generate these materials. Bottom-up tactics include: Casting, electrospinning, dip-, spin-, and spray coating, soft/hard lithography, and spray drying are just some of the many processes that may be used to manufacture a broad variety of materials. Some of these materials include monoliths, foams, fibers, membranes, films, patterns, and particles. There are many other methods, such as electrospinning, spray drying, and soft/hard lithography. There is also the possibility of using the process of spray drying while making these materials [16]. Because there has already been a substantial amount of in-depth research done on the topic, there is no need to perform a major study here to investigate the key chemical routes for the synthesis of functional hybrid nanomaterials [17, 18].

Because of their unique mechanical, optical, electrical, and thermal characteristics, organic–inorganic hybrid materials have attracted the attention of a significant number of researchers. Because of these features, organic–inorganic hybrid materials are appropriate for a wide variety of applications. These applications include those in the disciplines of energy, biology, medicine, optics, electronics, mechanics, and the environment. Over the course of the last five years, a broad variety of hybrid organic–inorganic materials have been developed, and many of these materials have the potential to be used in applications that will be immensely useful to society as a whole [19].

The vast adaptability of synthetic processes and the almost uncountable permutations that may be put to use in the production of organic–inorganic hybrid structures are the primary factors that have contributed to this progression. There is a direct correlation between this development and the production of hybrid materials. Throughout the course of its existence, the evolution of the discipline has been profoundly impacted by these two components [20].

4.2 Nanohybrid Materials

The process of fusing, merging, or mixing properties on the molecular level to create a hybrid material that possesses the advantages of several components while doing away with the disadvantages of each of those components is referred to as "hybridization." This process creates a hybrid material that possesses the advantages of several components while doing away with the disadvantages of each of those components [21].

Because of this, a one-of-a-kind medication that incorporates both the positive and negative qualities of its constituent parts has been developed. When we speak of "hybrid nanomaterials," we are referring to substances that include two or more organic or inorganic parts that are linked at the nanoscale. These parts can be connected to one another in a variety of ways. There are many different ways that these parts can be put together to form the whole. These molecules are made up of components such as cellulose and starch, which are examples of organic–organic substances; cellulose and TiO_2, which are examples of inorganic-inorganic substances; and TiO_2-Ag, which is an example of an inorganic–inorganic component. Compounds such as TiO_2-Ag are examples of molecules that have properties of both inorganic and organic matter [22].

5 Classification of Hybrid Nanomaterials

Generally, hybrid materials are classified into two categories depending on the intra- and intermolecular interactions among the organic matrix and cross-linking agent [23].

Class I (organic and inorganic exhibiting weaker interactions such as non-covalent interactions; van der Waals and hydrogen bonding).

Class II (organic and inorganic exhibiting strong interactions such as covalent, ionic, ionocovalent, and coordinative bonding).

Table 1 represents some selected examples of bio-nanohybrid materials [24].

Table 1 Examples of nanohybrid materials

Inorganic moiety	Biological species	Bio-nanohybrid features
Silica nanoparticles	Poly-L-lysine (PLL)	Biomimetic nanocomposites with controlled morphology
Siloxane networks	Living bacteria	Encapsulation by sol–gel
Calcium carbonate	Chitosan and poly(aspartate)	Biomimetic preparation towards artificial nacre
Hydroxyapatite (HAP)	Collagen	Biomimetic porous scaffolds for bone regeneration
Layered clay minerals (montmorillonite)	Chitosan	Functional bio-nanocomposite for ion-sensing applications
Fibrous clay minerals (sepiolite)	Caramel	Bio-nanocomposite as precursor of multifunctional carbon–clay nanostructured materials
Organoclays	PLA	Green nanocomposites as biodegradable bioplastics
Layered double hydroxides (LDHs)	Deoxyribonucleic acid (DNA)	Bio-nanocomposite as non-viral vector for gene transfection
Gold nanoparticles	Chitosan	Bio-nanohybrid processable as self-supporting films for biosensor applications
Magnetite nanoparticles	Phosphatidylcholine	Magnetocerasomes for targeted drug delivery
Carbon nanotubes (CNTs)	Galactose	Modified CNTs able to capture pathogens by protein binding
Layered perovskites (CsCa2Nb3O10)	Gelatin	Bio-nanocomposite thin films with dielectric properties

5.1 Metal and Metal Oxide Nanohybrid Materials

Magnetic, semiconducting, or metallic nanoparticles may be incorporated into polymeric matrices to create a variety of nanocomposite materials that are beneficial in a number of settings. In the present world, the usage of biopolymers for the aforementioned purpose is becoming more and more common since these materials provide an option that is more environmentally friendly than the standard polymers currently in use, especially for applications in the area of medicine. For instance, bio-nanocomposites containing silver oxide nanoparticles have been created using gelatin. The luminous intensity of this novel bio-nanohybrid material is substantially greater than that of femtosecond laser stimulated untreated nanoparticles. This characteristic, together with the required mechanical qualities, allows the gelatin-Ag_2O nanoparticle nanocomposite to be employed in light-emitting materials for all-optical logic circuits or data storage media [25].

5.2 Polymer-Based Nanohybrid

It is possible for hybrid organic–inorganic composites, also known as polymer-based composites or nanocomposites, to include either the nanoscale form of the organic or inorganic component. In order to produce them, a very minute amount of an inorganic component is incorporated into either an organic or a polymeric matrix. Utilizing this approach results in the production of the nanoparticles. This is done in order to develop an entirely new component that has functionality that is superior to that which was previously available [26].

The term "bio-nanocomposites" refers to a kind of material that combines biologically produced materials, such as biopolymers, with particles that have at least one dimension that is between 1 and 100 nm. The term "bio-nanocomposites" was coined by researchers specifically for the purpose of describing the aforementioned types of materials. It is a term used to describe entities such as biopolymers and particles with at least one dimension ranging from 1 to 100 nm that include components of biological origin. Numerous researches on polymer nanocomposites have been published; nevertheless, the vast majority of these materials were manufactured artificially and cannot be broken down by natural processes. This is despite the fact that the bulk of these materials have been studied. There have only been a few of works that make use of bionanocomposites, such as those that are based on chitin or cellulose nanocomposites [27].

The development of research based on lignin paves the way for the creation of novel opportunities for the eco-friendly synthesis of nanocomposites, which does not need the use of potentially harmful chemicals or abrasive processing methods. Lignin is an excellent component for use in polymer matrices not just because of its low price, but also because it can be easily renewed and is readily available. Several examples of polymer matrices include rubber, epoxy, and polyvinyl alcohol, which is often referred to by its abbreviation PVA [28].

The greatest level of adaptability that may be reasonably achieved Rubber is a material that is purposefully created to be hazardous, and it has a broad variety of applications, including the manufacture of tires and seals, among other things. However, unrefined rubber is almost never used in industrial applications due to the fact that it possesses a low mechanical strength in addition to other qualities that are not desirable. The incorporation of a wide range of different reinforcing chemicals into the rubber matrix is one potential solution to these issues. Products made of high-tech rubber that are not only lightweight but also have the potential to contain fillers with improved thermal, mechanical, and dielectric properties The field of bio-nanocomposites is garnering an increasing amount of interest as more and more people become aware of the numerous benefits that it will bring to the economy as well as the environment once it has been developed. This composite, in contrast to others, contains a significant amount of more conventional fillers [29].

6 Advantages of Polymer-Based Nanohybrid Materials

Natural polymers, also known as biopolymers, have lately attracted a lot of interest due to the fact that they are readily available, have a low level of toxicity, have a low cost, are biodegradable, are biocompatible, and have a wide range of applications [30]. As an environmentally friendly alternative to cellulose, chitin, chitosan, pectin, starch, dextran, xanthan, guar gum, fucoidan, heparin, hyaluronan, and pullulan, numerous biopolymers have been used. These include proteins (albumin, casein, collagen, fibrinogen, and gelatin), polylactic acid (PLA), and nucleic acids [31]. Polysaccharide-based hybrid nanocomposites have emerged as increasingly significant materials over the course of the last several decades [32].

Studies have demonstrated that natural polymers, which are more widely known as polysaccharide-based nanocomposites, are used in a range of industries. Some of these industries include food, biotechnology, environmentally responsible and sustainable food packaging, environmental pollution management, and environmental pollution remediation [33–35].

The purpose of combining synthetic polymers or nanomaterials with natural polymers, also known as biopolymers, is to improve the characteristics and uses of the synthetic or nanomaterials, respectively. Because natural polymers do not naturally possess sufficient levels of mechanical, insulating, and processing characteristics, this step is necessary [36]. When it comes to biopolymers, scientists are most interested in polysaccharides like starch, cellulose, chitin, and chitosan. Chitin and chitosan are also included in this category. These polysaccharides are then combined with metal nanoparticles (such as Au, Ag, Cu, and Pd), metal oxide nanoparticles (such as TiO_2, ZnO, CuO, SiO_2, Fe_2O_3, and Fe_3O_4), and carbon nanomaterials to produce biodegradable nanocomposite [30].

6.1 Carbon-Based Bio-Nanohybrid Materials

Since Iijima [37] made the discovery of carbon nanotubes, they have piqued the interest of scientists working in a wide variety of scientific fields. Carbon nanotubes are also referred to by their acronym, CNT. Structures that have the appearance of tubes but are actually rolled up from sheets of graphene are called carbon nanotubes (also abbreviated as CNTs). Single-walled carbon nanotubes (SWCNTs) have a diameter between 0.7 and 1.5 nm, whereas multiwalled carbon nanotubes (MWCNTs) have a diameter between 2 and 30 concentric tubes with an inner tubular layer that has a diameter between 2 and 10 nm and an additional thickness of about 0.7 nm for every additional wall. MWCNTs also have an inner tubular layer that has an additional thickness of about 0.7 nm for every SWCNTs have an additional thickness of approximately 0.7 nm for every additional wall that is added to the structure. Carbon nanotubes, also known as CNTs, can be separated into two distinct categories according to the number of tubular walls that each one possesses. These are groupings of lonely persons that have gathered together [38].

Variations in the chiral vectors, which describe how graphene sheets are rolled up, as well as topological defects that are present on the surface of the tube are responsible for controlling the various electronic topologies and, as a result, the semiconducting or metallic behavior of CNTs. This is the case because of the interplay between these two factors. These variables are controlled by topological imperfections on the surface of the tube. Chiral vectors provide an explanation for how graphene sheets could be rolled up. The surface of the tube exhibits a number of topological abnormalities. Inaccuracies in the topology of the circuit have the potential to have an effect on a variety of different electrical designs. Because of their exceptionally high aspect ratio, carbon nanotubes, which are also known as CNTs and are more commonly referred to by their abbreviated form, exhibit exceptional mechanical qualities in addition to their remarkable electrical properties. This is due to the fact that carbon nanotubes are made up of extremely small tubes made of carbon atoms. As a consequence of this, they are gaining an increasing amount of notoriety and are in extremely high demand for application as nanofillers in a wide variety of polymers [39].

7 Synthesis of Nanohybrid Materials

To develop these hybrid materials, a number of processes, including but not limited to electrostatic binding, covalent immobilization, and polymerization approaches, are required. It is possible for the structure of a hybrid material, the interfaces between the hybrid's various component components, the constituent parts themselves, and the interfaces between the hybrid's constituent parts to all have an effect on the hybrid's qualities. In addition to the ability of the material to withstand extremes in temperature while still retaining its mechanical strength, the ideal mixture may also be able to control the material's optical, anticorrosive, magnetic, electrical, thermal, and fire-retardant properties. These capabilities are in addition to the material's ability to maintain its mechanical strength. In addition to this, it has the potential to improve the substance's chemistry in addition to its thermal stability [40].

Because of their outstanding mechanical, physical, and tribological properties, hybrid nanomaterials are finding growing usage in a wide variety of applications across the agri-food, environmental, and biotechnology industries. In addition to the purposes listed below, further applications include electrochemistry, the protection of plants, and the packaging of food. This is due to the fact that hybrid nanomaterials may be used in a broad number of applications [41].

Within the discipline of materials chemistry, the study of functional nanoscale hybrid materials is widely recognized as one of the most intriguing and forward-thinking subfields of research now being conducted. At the nanometer scale, it is possible to assemble synthetic materials composed of both organic and inorganic components using either covalent (Class II, connected by hydrogen bonds, electrostatic forces, or van der Waals forces) or non-covalent (Class I, connected by hydrogen bonds, electrostatic forces, or van der Waals forces) bonding. Covalent bonding is classified as Class II, while non-covalent bonding is classified as Class I.

The formation of non-covalent bonds is placed in Class I, whereas the formation of covalent bonds is placed in Class II. The formation of non-covalent bonds falls under the category of Class I, while the development of covalent bonds falls under the category of Class II. The formation of non-covalent bonds falls into the category of Class I, while the formation of covalent bonds is categorized as falling into the category of Class II. A lot of attention has been paid to the many molecular and nanoscale combinations of inorganic, organic, and even bioactive components that may exist in a single material. These combinations may be found in a single material. These different combinations may all be found in the same chemical.

The prospect of locating components of this sort in a single material served as the impetus for the development of this area of study. Producing a broad variety of complicated and sophisticated materials, such as structures that have been subjected to stringent maintenance and are capable of a wide range of functions, is made possible with the assistance of this technology. The distinctive properties of advanced hybrid nanomaterials could be of use in a wide variety of industries, including those dealing with optical and electrical materials, biomaterials, catalysis, sensing, coating, and energy storage, among a great number of other potential applications. These industries could also benefit from the unique properties of advanced hybrid nanomaterials.

Hybrid materials have the potential to simultaneously produce cutting-edge materials and pique public interest, they are becoming increasingly popular. The articles that make up this issue cover a variety of topics, one of which is hybrid materials, which has contributed to the rise in popularity of these materials. This is without a doubt one of the most important problems that needs to be fixed in order for hybrid materials to gain widespread acceptance [42].

8 Advantages of Nanohybrid

When using the nanohybrid material, it will be of particular interest to upgrade and diversify polyolefin compositions that are produced by transition metal catalyzed polymerizations. The production of materials made of polyolefin is beneficial to both the environment and the economy, in addition to a number of different applications:

- In terms of the economy, it is a significant industrial sector that is responsible for the production of more than fifty percent of the two hundred million tons of plastics that are manufactured each year. The use of highly efficient catalytic processes that only require a small amount of energy can help to keep the energy content of polyolefins comparable to that of oil without having a negative impact on the environment. At temperatures higher than 300 °C, the polyolefins begin to degrade, which ultimately results in the release of synthetic oil and natural gas.
- Applications in the Industrial Sector: There is a diverse selection of materials that can be modified to fulfill the requirements of contemporary technologies, such as those for the field of communication technology, engineering materials with a low weight, and packaging. Adjustments of this kind can be made in order to fulfill the prerequisites of a wide variety of applications. This is dependent on the kind of catalyst that is utilized, the conditions of the process, the method that is utilized to process the polymer, and, most importantly, the type of olefin copolymer and functional fillers that are utilized.
- Improvements in properties applicable to applications that are made possible by the use of nanostructured hybrid polyolefins Self-assembly of block copolymers will, in the medium to long term, offer special advantages in addition to opportunities for the creation of new products and applications. These advantages will come in addition to the opportunities. This can be accomplished through the combination of block copolymer self-assembly and catalytic polymerization. A few of the intended improvements include scratch resistance, improved modulus without a loss in stiffness, heat distortion temperature, flame retardancy, and barrier qualities (super hydrophobic/hydrophilic, UV radiation, conductivity, antibacterial, and so on). The automotive, telecommunications, microelectromechanical systems (MEMS), packaging, and textile industries are the ones that are being investigated for this particular case.

9 Conclusion

Nanohybrid materials have a great deal of promise given that environmentally friendly materials of the future will be formed of organic ingredients. This is particularly the case when considering the fact that biopolymers and the composites made from them are more affordable than other forms of polymers. Bionanocomposites are an example of a material that may be used for food packaging that is both environmentally friendly and long-lasting. In conclusion, the use of bionanocomposites for environmental applications still remains a challenge to environmental researchers and scientists. Despite this, the relevance of these materials is only going to grow as more and more goods based on these materials make their way into the marketplace.

References

1. Saba N, Alothman OY, Almutairi Z, Jawaid M, Asad M (2019) Introduction of graphene-based nanotechnologies. Graphene-Based Nanotechnol Energy Environ Appl 3–21.https://doi.org/10.1016/B978-0-12-815811-1.00001-6
2. Khan I, Saeed K, Khan I (2019) Nanoparticles: properties, applications and toxicities. Arab J Chem 12(7):908–931. https://doi.org/10.1016/j.arabjc.2017.05.011
3. Jamkhande PG, Ghule NW, Bamer AH, Kalaskar MG (2019) Metal nanoparticles synthesis: an overview on methods of preparation, advantages and disadvantages, and applications. J Drug Deliv Sci Technol 53:101174. https://doi.org/10.1016/j.jddst.2019.101174
4. Shaikh SF, Shinde NM, Lee D, Al-Enizi AM, Kim KH, Mane RS (2019) Nanostructures in dye-sensitized and perovskite solar cells. Nanostructures. IntechOpen. https://doi.org/10.5772/intechopen.85699
5. Li Z, Hu J, Yang L, Zhang X, Liu X, Wang Z, Li Y (2020) Integrated POSS-dendrimer nanohybrid materials: current status and future perspective. Nanoscale 12(21):11395–11415. https://doi.org/10.1039/D0NR02394A
6. Tiwari JN, Tiwari RN, Kim KS (2012) Zero-dimensional, one-dimensional, two-dimensional and three-dimensional nanostructured materials for advanced electrochemical energy devices. Prog Mater Sci 57(4):724–803. https://doi.org/10.1016/j.pmatsci.2011.08.003
7. Shin WK, Cho J, Kannan AG, Lee YS, Kim DW (2016) Cross-linked composite gel polymer electrolyte using mesoporous methacrylate-functionalized SiO_2 nanoparticles for lithium-ion polymer batteries. Sci Rep 6(1):1–10. https://doi.org/10.1038/srep26332
8. Ganesh M, Hemalatha P, Peng MM, Jang HT (2017) One pot synthesized Li, Zr doped porous silica nanoparticle for low temperature CO_2 adsorption. Arab J Chem 10:S1501–S1505. https://doi.org/10.1016/j.arabjc.2013.04.031
9. Lee JE, Lee N, Kim T, Kim J, Hyeon T (2011) Multifunctional mesoporous silica nanocomposite nanoparticles for theranostic applications. Acc Chem Res 44(10):893–902. https://doi.org/10.1021/ar2000259
10. Khan I, Yamani ZH, Qurashi A (2017) Sonochemical-driven ultrafast facile synthesis of SnO_2 nanoparticles: growth mechanism structural electrical and hydrogen gas sensing properties. Ultrason Sonochem 34:484–490. https://doi.org/10.1016/j.ultsonch.2016.06.025
11. Shaalan M, Saleh M, El-Mahdy M, El-Matbouli M (2016) Recent progress in applications of nanoparticles in fish medicine: a review. Nanomed: Nanotechnol Biol Med 12(3):701–710. https://doi.org/10.1016/j.nano.2015.11.005
12. Iravani S (2011) Green synthesis of metal nanoparticles using plants. Green Chem 13(10):2638–2650. https://doi.org/10.1039/c1gc15386b
13. Nicole L, Laberty-Robert C, Rozes L, Sanchez C (2014) Hybrid materials science: a promised land for the integrative design of multifunctional materials. Nanoscale 6(12):6267–6292. https://doi.org/10.1039/C4NR01788A
14. Fahmi A, Pietsch T, Mendoza C, Cheval N (2009) Functional hybrid materials. Mater Today 12(5):44–50. https://doi.org/10.1016/S1369-7021(09)70159-2
15. Sanchez C, Rozes L, Ribot F, Laberty-Robert C, Grosso D, Sassoye C, Nicole L (2010) "Chimie douce": a land of opportunities for the designed construction of functional inorganic and hybrid organic-inorganic nanomaterials. C R Chim 13(1–2):3–39. https://doi.org/10.1016/j.crci.2009.06.001

16. Sanchez C, Boissiere C, Cassaignon S, Chanéac C, Durupthy O, Faustini M, Sassoye C (2014) Molecular engineering of functional inorganic and hybrid materials. Chem Mater 26(1):221–238. https://doi.org/10.1021/cm402528b
17. Kickelbick G (2014) Hybrid materials–past, present and future. Hybrid Mater 1(1):39–51. https://doi.org/10.2478/hyma-2014-0001
18. Hood MA, Mari M, Muñoz-Espí R (2014) Synthetic strategies in the preparation of polymer/inorganic hybrid nanoparticles. Materials 7(5):4057–4087. https://doi.org/10.3390/ma7054057
19. Sanchez C, Julián B, Belleville P, Popall M (2005) Applications of hybrid organic–inorganic nanocomposites. J Mater Chem 15(35–36):3559–3592. https://doi.org/10.1039/B509097K
20. Cui J, Jia S (2017) Organic–inorganic hybrid nanoflowers: a novel host platform for immobilizing biomolecules. Coord Chem Rev 352:249–263. https://doi.org/10.1016/j.ccr.2017.09.008
21. Chauhan BP (ed) (2011) Hybrid nanomaterials: synthesis, characterization, and applications. Wiley
22. Meroni D, Ardizzone S (2018) Preparation and application of hybrid nanomaterials. Nanomaterials 8(11). https://doi.org/10.3390/nano8110891
23. Gamage A, Punniamoorthy T, Madhujith T (2022) Starch-based hybrid nanomaterials for environmental remediation. In: Starch: evolution and recent advances, vol 4, no 171. https://doi.org/10.5772/intechopen.101697
24. Ruiz-Hitzky E, Darder M, Aranda P (2008) An introduction to bio-nanohybrid materials. Bio-inorg Hybrid Nanomaterials 1–40.https://doi.org/10.1039/b509097k
25. Gleitsmann T, Bernhardt TM, Wöste L (2006) Luminescence properties of femtosecond-laser-activated silver oxide nanoparticles embedded in a biopolymer matrix. Appl Phys A 82(1):125–130. https://doi.org/10.1007/s00339-005-3372-4
26. Nguyen TP, Yang SH (2018) Hybrid materials based on polymer nanocomposites for environmental applications. In: Polymer-based nanocomposites for energy and environmental applications. Woodhead Publishing, pp 507–551. https://doi.org/10.1016/B978-0-08-102262-7.00019-2
27. Gopalan Nair K, Dufresne A, Gandini A, Belgacem MN (2003) Crab shell chitin whiskers reinforced natural rubber nanocomposites. 3. Effect of chemical modification of chitin whiskers. Biomacromolecules 4(6):1835–1842. https://doi.org/10.1021/bm030058g
28. Saito T, Brown RH, Hunt MA, Pickel DL, Pickel JM, Messman JM, Baker FS, Keller M, Naskar AK (2012) Turning renewable resources into value-added polymer: development of lignin-based thermoplastic. Green Chem 14(12):3295–3303. https://doi.org/10.1039/C2GC35933B
29. Parvathy G, Sethulekshmi AS, Jayan JS, Raman A, Saritha A (2021) Lignin based nanocomposites: synthesis and applications. Process Saf Environ Prot 145:395–410. https://doi.org/10.1016/j.psep.2020.11.017
30. Vilela C, Pinto RJB, Pinto S, Marques P, Silvestre A, Barros CSDRF (2018) Polysaccharide based hybrid materials: metals and metal oxides, graphene and carbon nanotubes. Springer. https://doi.org/10.1007/978-3-030-00347-0_1
31. Russo T, Fucile P, Giacometti R, Sannino F (2021) Sustainable removal of contaminants by biopolymers: a novel approach for wastewater treatment. Current state and future perspectives. Processes 9(4):719. https://doi.org/10.3390/pr9040719
32. Zheng Y, Monty J, Linhardt RJ (2015) Polysaccharide-based nanocomposites and their applications. Carbohydr Res 405:23–32. https://doi.org/10.1016/j.carres.2014.07.016
33. Arora B, Bhatia R, Attri P (2018) Bionanocomposites: green materials for a sustainable future. In: New polymer nanocomposites for environmental remediation. Elsevier, pp 699–712. https://doi.org/10.1016/B978-0-12-811033-1.00027-5
34. Bilal M, Gul I, Basharat A, Qamar SA (2021) Polysaccharides-based bio-nanostructures and their potential food applications. Int J Biol Macromol 176:540–557. https://doi.org/10.1016/j.ijbiomac.2021.02.107
35. Wen Y, Oh JK (2014) Recent strategies to develop polysaccharide-based nanomaterials for biomedical applications. Macromol Rapid Commun 35(21):1819–1832. https://doi.org/10.1002/marc.201400406

36. Kotharangannagari VK, Krishnan K (2016) Biodegradable hybrid nanocomposites of starch/lysine and ZnO nanoparticles with shape memory properties. Mater Des 109:590–595. https://doi.org/10.1016/j.matdes.2016.07.046
37. Iijima S (1991) Helical microtubules of graphitic carbon. Nature 354(6348):56–58. https://doi.org/10.1038/354056a0
38. Saito R, Dresselhaus G, Dresselhaus MS (1999) Physical properties of carbon nanotubes. Imperial College, Singapore, 1998 (C. Google Scholar C. Dekker, Phys. Today)
39. Tjong SC (2006) Structural and mechanical properties of polymer nanocomposites. Mater Sci Eng: R: Rep 53(3–4):73–197. https://doi.org/10.1016/j.mser.2006.06.001
40. Rejab MRBM, Hamdan MHBM, Quanjin M, Siregar JP, Bachtiar D, Muchlis Y (2020) Historical development of hybrid materials. Mat Sci Mat Eng 4:445–455. https://doi.org/10.1016/B978-0-12-803581-8.10546-6
41. Abd-Elsalam KA (2020) Multifunctional hybrid nanomaterials for sustainable agri-food and ecosystems: a note from the editor. In: Multifunctional hybrid nanomaterials for sustainable agri-food and ecosystems. Elsevier, pp 1–19. https://doi.org/10.1016/B978-0-12-821354-4.00001-7
42. Liu LH, Métivier R, Wang S, Wang H (2012) Advanced nanohybrid materials: surface modification and applications modification and applications.https://doi.org/10.1155/2012/536405

Survey of Nanohybrid Materials in Textile Dyes Removal

V. Mahalakshmi and Lali Growther

Abstract Industrial revolution has undoubtedly upgraded the quality of life but on the other side, it has culminated in severe environmental pollution due to indiscreet discharge of toxic wastes. Diverse industries use dyes, viz., textiles, paint, pulp, paper, food, plastic, cosmetics, rubber, tannery, etc., and thereby polluting the aquatic ecosystems with these toxic dyes. Since conventional treatment methods have proved inefficient in the shifting of toxic dyes, in particular, originating at fabrics manufacturing industries, novel approaches to use nanohybrid materials for dye removal is currently in practice. Designing of nanohybrid materials is carried out by combining nanoscience with technology thereby constructing high-quality substances for effective withdrawal of pollutants. A collection of nanohybrid materials, especially adsorbents, are being used for their removal efficiency, recyclability, economic value and easy operation. These include different nanostructures like nanotubes, nanorods, nanocrystals, nanomembrane, metal oxides framework, etc. A plethora of nanohybrid materials are being produced and used, namely amorphous carbon nanotubes, chitosan magnetic iron oxides, graphene metal oxides, carrageenan-based hybrid with silica, biochar, etc. Though the protocols of synthesis of these nanohybrid materials differ, they are found to play a vital role in remediating polluted water bodies with some nanohybrid material (e.g. microspheres) which are discovered to even exhibit a dye removal efficiency of 95.4%. Thus, nanohybrid materials are found to have the potential scavenging ability to cleanse water pollutants.

Keywords Nanohybrid materials · Nanomaterials · Toxic dyes · Remediation

V. Mahalakshmi (✉)
Department of Microbiology, Madras Christian College, Chennai, India
e-mail: mahalakshmi@mcc.edu.in

L. Growther
Department of Microbiology, Hindusthan College of Arts and Science, Coimbatore, India
e-mail: laligrowther@hicas.ac.in

© The Author(s), under exclusive license to Springer Nature Singapore Pte Ltd. 2023
A. Ahmad et al. (eds.), *Nanohybrid Materials for Treatment of Textiles Dyes*,
Smart Nanomaterials Technology, https://doi.org/10.1007/978-981-99-3901-5_2

1 Introduction

Fabrics manufacturing industries gobble huge quantities of water for clarification throughout the production process and hence considered to be the biggest polluter of water, due to the release of diverse dyes and chemicals. Dyes are chemical substances that impart colour to the fabrics and are not easily removed by heat, light, water and any other factors to which the textile materials are exposed. Even a small percentage of dyes released into the effluent will result in water pollution due to their conspicuous nature. Zollinger [1] in the year 1987 has estimated that nearly 100,000 commercial dyes are available with over 7×10^5 dyes synthesized annually. Their compounded chemical structures make them resistant to decolourisation. In addition to their resistance towards decolourization, many of the dyes exhibited toxicity. The Ecological and Toxicological Association of Dyestuff manufacturing industry (ETAD), conducted a survey, which showed that over 90% of 4000 dyes had LD_{50} values more than 2×10^3 mg/kg, of which the highest toxicity was reported for basic and diazo direct dyes [2]. Azo dyes are largely mutagenic and carcinogenic. One such chemical is the reactive red-35, the potent mutant azo dye, but has increased use in textile manufacturing for its energetic shade, endurance strength to washing and application methods [3].

The problem with the textile effluent is the extremely dense coloured waste water with the combination of varied dyes, which further spoil the of elegance of aquatic bodies by drastically turn down illumination diffusion and hence affect the photosynthetic activity of aquatic plants. Furthermore, they are destructive to the aquatic life forms too [4]. Biomagnification of dyes through food chain causes severe health impacts in human such as dermatitis, cancer and kidney-related ailment. Especially those people, who man handle sensitive colourants will be at the risk of showing hypersensitive responses such as sneezing, itching of the nose, eyes or roof of the mouth, runny, stuffy nose, watery, red or swollen eyes [5]. Stringent laws are being enforced in well-developed countries to control the release of dyes in the effluent. Since 1997, UK Environmental policy has enforced zero per cent release of colourants into the aquatic ecosystems and therefore fabrics manufacturing companies are liable to treat the dyes containing effluent to the said standards. Similarly, European Community (EU) legislation is turning rigorous [6].

Remediation of textile dye removal is being attempted by researchers using strategies like decomposition by chemical means, agglomeration, precipitation, adsorption, clarification, ion exchange and biological methods such as waste activated biosolids, bioreactors employing membrane, bio-film activator and artificial wetlands. Each technique has its own merits and demerits. Among the many techniques, high dye removal efficiency is found be possible with adsorption methods. Though a plethora of compounds are used as conventional and non-conventional adsorbents, nanomaterials are predominantly used in dye removal because of their build up exterior surface and maximum adsorption to mass correlation. A blend of nanomaterials, termed as hybrid nanomaterials, due to their polymathic nature are used nowadays for efficient dye removal from textile effluents [7].

2 Conventional Methods for Dye Removal and Their Limitations

The last stage in the dyeing procedure in textile industries involves gushing in water, which in turn discharges surplus dyes into water that persists for ages. Therefore, to develop a sustained process is crucial in colour removal technology. Depending on the method adapted, the dye removal techniques could be grouped as physical methods, synthetic or chemical methods and biotic methods or the combination of methods may be utilized [8].

2.1 Physical Methods for Dye Removal

The physical methods depending on the mechanism of mass transfer employ adsorption, ion exchange and membrane filtration, with a removal efficiency of 85–99% according to a research done by Samsami et al. [9], in the year 2020. The advantages of these methods are a basic model, easy handling, economically feasible, little need for consumables and no inhibitory effect undesirable products present in the waste water [10]. But generation of unsafe end products is slurry, increased mercury levels, increased oxygen level due to chemical and biological methods, pH, colour and death metals often impede their utilization in dye removal from fabrics manufacturing industries [8].

2.1.1 Adsorption

The procedure by which ions or molecules are attached to a solid surface with porous structure, by physical forces (physisorption) or by chemical interactions (chemisorption) is termed as adsorption. The extensively used adsorbents to achieve efficient dye removal are zeolites, alumina, stiffened silicon oxide and triggered charcoal. Jadhav and Jadhav [11] in the year 2021, reported the usage of triggered carbonaceous compounds by industries, due to their adsorptive nature. The decolourization percentages of treating wastewater with dyes using the biopolymer zeolite material were found to be more than 75% as experimented by Brião et al. [12]. Employing zeolite composite with zinc oxide material by Madan et al. [13] attained 90% Congo red decolourization in waste water.

2.1.2 Ion Exchange

The system of dye removal in industrial effluents by ion exchange is done effectively by the creation of strong bonds between the functional groups of the adhesives present in a packed bed activator and the charges on the dissolved substances. Martin et al.

[14] in the year 2019 executed 97% removal of Orange dye, using an Amberlite IRA 400 as an anion-exchange adhesive from wastewater. However, a cationic-exchanger adhesive achieved a 91.7% of dye removal such as Disperse Violet 28, as revealed by Bayramoglu et al. [15].

2.1.3 Membrane Filtration

The next colour removal technology is by the usage of membranes with pores that could entrap dyes from the source, though this protocol is advantageous, membranes need to be changed on a regular basis. Three different of filtration process are there, namely (i) Microfiltration, with membrane pore sizes of 0.1–10 μm, (ii) Nanofiltration with membrane pore sizes of 0.5–0.2 nm and (iii) Ultrafiltration, with membrane pore sizes fluctuating between 0.1 and 0.001 μm.

2.2 Chemical Methods for Dye Removal

Agglomeration, charged processes and state-of the art decomposition processes are the techniques applied for industrial dye removal. However, Kishore et al. [16] reported the demerits of these methods for they require appropriate apparatus, large amount of chemicals and huge electrical energy, and in addition, they release toxic by-products as an outcome of the chemical reactions.

2.2.1 Coagulation–Flocculation

Coagulation–flocculation can be achieved by coagulants like metal salts and polymers facilitate the accumulation of precipitates, thereby resulting in their efficient and quick removal as experimented by Al-Mutairi [17], in the year 2006. Mathuram et al. [18] announced the significance of *Tamarindus indica* and *Azadirachta indica* as innate jelling agents in removing dyes, while Badawi and Zaher [19] conveyed the use magnesium carbonate and hydrated lime that are chemical jelling agents, which had great adsorptive efficiency in separating azo dyes and their fallouts from the industrial effluents.

2.2.2 Advanced Oxidation Processes

Sedentary production of hydroxyl radicals is used to oxidize dyes in waste water. Some of the processes of advanced oxidation methods that remove dyes under rough situations but not leaving any sludge are stimulant using sunlight, using Fenton reagent, charged decomposition reactions. Still these methods are found to be costly, are pH dependent and leave deleterious end products [20]. For photocatalytic reactions, nanoparticles like zinc oxide and titanium peroxide are being employed as light accelerating substances. The organic dyes in effluents are removed by Fenton and photo-Fenton processes [21].

2.2.3 Electrochemical Approaches

Anodic oxidation, electrocoagulation and electro-Fenton are the most sought procedures of the many electrochemical operation systems. Though these techniques make use of limited chemicals, consume much electricity and also are less efficient than the earlier methods but could remove organic dyes from waste water [22].

2.3 Biotic Processes

The physical and chemical methods, when collated with the biological approaches, the biological processes in dyes removal are found to be promising due to a number of valid reasons such as ease of applying, less sludge production, cost-effective, eco-friendly, energy efficient, non-toxic by-products and require less chemical reagents [23]. The potential candidates for biological processes for dye removal are microorganisms like bacteria, fungi, algae and yeast, in addition, the enzymes produced from these microorganisms also have found place in the treatment protocols. Dyes such as Red RBN, C.I. Reactive Green 19A, Navy Blue 2GL, Methyl Orange, C.I. Remazol Red, Reactive Orange M2R, Acid Red 57, Brilliant Green, Red azo dye and methyl red are either biodegraded or bio mineralized effectively by a single microorganism or a consortium of microorganisms.

3 Why Nanoparticles for Bioremediation?

Nanotechnology is a branch of applied science that deals with the pattern, blend, imitation and applications of compounds and tools on a nanoscale, ranging in size from 1 to 100 nm. The varied types of nanomaterials are carbon based, TiO_2 nanoparticles and iron oxides. From mahogany sawdust, Malik [24] produced an efficacious commercial activated carbon-based adsorbent, which is found to remove certain dyes with much efficacy. Similarly, waste materials like skin of orange, coconut leaflet, guava leaves, rice case, jute sticks and spent tea leaves have proved to possess adsorbing ability because of the porous nature, architectural features, and the presence of functional units. Ahmad et al. [25] when they employed activated carbon with leaf of lemon grass to get rid of dyes from industrial effluents, found that within 5 h, an efficient dye removal was achieved for methyl red. However, with a temperature rise, dye concentration and time of exposure, they noticed a further rise in adsorption capacity. In an another study by Ojedokun et al. [26], activated carbon with guava leaves showed a highest removal efficiency for the dyestuff, Congo Red. Doong et al. [27] implanted copper and silver ions on TiO_2 to speed up the removal of methylene blue by photo catalysis. Similarly, Tabassum [28] combined poly fluorine-co-thiopene (PFT) with TiO_2 to promote acid orange reduction with short and long wavelength of light. Luo et al. [29] developed a novel metal semi-conductor by a simple pulse electro deposition technique, where compressed gold nanotubes assemblage was created with minuscule gold substances in the holes of TiO_2 nanomaterial. When an electric potential difference was established at the junction of Au/ TiO_2, AO7 was reduced considerably by photocatalysis. Iron oxide nanoparticles (NPsFeO) were produced from the leaflets of Cucurbita moschata, and the shoots of Beta vulgaris, by green synthesis, are being as adsorbents for the removal of yellow and red dyes [30]. Similarly, experiment done by Saima Noreen et al. [31] in the year 2020 revealed that iron oxide nanoparticles prepared from biological samples such as palm varieties are found to be efficacious adsorbents for the shifting of anionic dyestuff from its initial concentration. Also thermodynamic analysis exposed the fact that the adsorption via nanoadsorbents was an endoergic activity.

Though a number of adsorbents like activated carbon, non-conventional absorbents (agricultural wastes, industrial wastes, natural and synthetic clay and bio-adsorbents) are being deployed for bioremediation of industrial dyes, nanohybrid materials are now finding wide spread utilization because of their considerable working space on the surface and excessive adsorption to mass quotient.

4 Nanohybrid Materials for Dye Removal

The outcome of Richard Feynman's vision "plenty of room at the bottom" resulted in the fabrication and exploiting the materials measuring in nanoscale to attain specific overall characteristics, thereby bringing in broad applications [32]. Since profit-oriented substances require tremendous execution and has to exhibit multi-functionalities consecutively, studies in material science need to change its focal point from singular nanomaterials (NMs) to hybrid materials [33]. These nanoconjoins are given the nomenclature nanohybrids (NHs). These compounded materials may be blends of metals, metalloids and carbon-only nanostructures [34]. The principle intention of fabricating nanohybrid structures is to improvise the characteristics of single compound thereby deploying them for wide-scale applications across disciplines, to name a few, in designing biomedical products and devices, electro chemical fuel cells, super conductors and in environmental remediation.

4.1 Properties of Nanohybrid Materials

Each compound has distinctive physical property based on its chemical origin and this is true for nanoparticles too. These nanoparticles when blended, their aftereffect properties differ and based on the mode of synthesis, one or more components may become superior. The modifications may be expressed in the characteristics of nanohybrid substances like their dimension, appearance, surface chemistry, improved reactivity, suspension ability, capturing ability, energy gap, defiance to decay, electronic properties, mechanical properties and in addition some novel properties. Different shapes of core–shell bimetallic NHs are found to be spherical, cuboidal, rod-like, plate-like, triangular, bipyramidal, polyhedral and dumbbell, and these structures are as a result of the blending synthesis. Alien 3D structures were also obtained when hybridization was found between 0D nanoparticle, 1D nanotube and 2D graphene structures [35]. Thus, researches have proved that novel changes in size, shape and physicochemical properties of these fabricated nanohybrid materials in turn have tremendous implications on their potential, especially in dye removal from waste water.

4.2 Classification of Nanohybrid Materials

The classification of nanohybrid (NH) materials is decided by the parent materials, used for their synthesis. Broadly, NHs are designated into four groups, namely (i) Carbon–Carbon NHs (CCNH), (ii) Carbon–Metal NHs (CMNH), (iii) Metal–Metal NHs (MMNH) and (iv) Organo-Metal–carbon NHs (OMCNH). When many carbonaceous nanoparticles are hybridized, then it will result in the synthesis of CCNHs, some as carbonaceous nanotubes, fullerenes or graphene, which are with one boundary or numerous boundaries [36]. Similarly, when carbonaceous nanoparticles are hybridized with metals, like gold, silver titania, alumina and iron oxides, these blended components are termed as CMNHs [37, 38] (Fu et al.,). MMNHs are as the result of blending between two or more metal-based NMs by chemical reaction or overcoating [39]. Biological molecules, namely drugs, proteins, dyes and enzymes are tagged to metal nanoparticles, then the resultant materials are of OMCNH category. Thus, these nanohybrid materials play pivotal role in a broad range of fields, namely electronics and energy, biomedical, environmental monitoring and remediation, catalysis, construction industry, heat transfer, antimicrobial coating, paints, temperature sensor, etc. Table 1 depicts the classification of nanohybrid materials.

Table 1 Classification of nanohybrid materials with examples under each category

Nanohybrid materials

Carbon-Carbon NHs	Carbon- Metal NHs	Metal-MetalNHs	Organo-Metal-carbon NHs
Examples			
Carbon nanotube (CNT)	Graphene-ZnO	ZnO-Au	Graphene-organic molecule
Graphene nanosheet	MnO_2/CNT	Ag/TiO_2 nanowire	Poly(3-hexylthiophene)-fullerene
Graphene-CNT	Graphene nanosheet/metal nitride	Au@TiO_2 core-shell	Graphene-Ag nanowire
Graphene oxide-CNT peapods	CNT/RuO_2	Au@ZrO_2 core-shell	Fullerene/CNT with porphyrins
Graphene-fullerene	Graphene-Mn_3O_4	Ag@TiO_2 core-shell	Oligothiophene-graphene,
Carbon nano-onions	Graphene-TiO_2	Pt-Pd	Pluronic F-127/graphene
CNT-Graphene nanoplatelet	SWNT-Au	Pt/TiO_2 nanotube	Hematoporphyrin-ZnO
Fullerene-CNT	CNT-fullerene	Fe_3O_4-AuNR nanonecklace	porphyrin-graphene
Graphene wrapped MWNT	ZnO-graphene quantum dots	Fe_3O_4-Au-Fe_3O_4 nanodumbbell	
Azafullerene-CNT peapods	MWNT-ZnO	Au-Pd core-shell structure	
N-doped CNT-graphene peapods	Quantum dot-Fe_3O_4-CNT	Pd-Cu	
	Pt-graphene		
	Ag-graphene oxide		

5 Types of Nanohybrid Materials

5.1 Hybrid Nanocomposite that Are Carbon Based

Hybrid nanocomposites that are carbon based are easily synthesized, cost-effective, non-toxic and highly porous. Many dyes were observed to be removed with superfine nickel/carbon nanohybrid material by Kim et al. [40] Manippady et al. [41] scrutinized the efficacy of iron–carbon hybrid magnetic nanosheets in adsorbing positively charged dyes. When a hybrid nanosubstance was designed using carbon, silicon and nitrogen, efficient adsorption to draw out dyes was made possible [42]. The feasible procedure was due to excellent binding between the adsorbent and dye. Jin et al. [43] proved by their study that a number of dyes, namely CR, RhB, MB are adsorbed by Ni/porous carbon nanotube nanocomposite.

5.2 Hybrid Nanocomposites that Are Activated Carbon Based

To purify dye polluted water, activated carbon-based hybrid nanocomposites play a vital role. For example, konjac glucomannan was frozen with activated carbon aerogel to adsorb MB, as reported by Wang et al. [44]. But over 99% removal of MB in 60 min was accomplished with ZnO nanohybrid particles synthesized by combining with Parthenium weed activated carbon [45]. Gong et al. [46] manufactured a novel hybrid adsorbent from finger citron residue coupled with activated carbon to get rid of deleterious anionic and cationic dyes from the industrial effluent.

5.3 Hybrid Nanomaterials that Are Carbon Nanotube Based

Hybrid nanomaterials that are carbon nanotube (CNT) based are gaining impressive recognition in recent times due to their hollow nature and thereby having more specific surface area. This enabled their efficient adsorption of organic dyes as experimented by Gong et al. [46], for the highly receptive red dyes. Yao et al. produced hybrid carbon nanotube with numerous confines, by which they demonstrated maximum removal ability of MO dye from polluted water. However, when this hybrid carbon nanotube with numerous confines was blended with calcium alginate, the highest removal ability was noted for MB and MO by Sui et al. Another approach was using nanohybrid comprising HNO_3/NaClO/CNT with numerous confines [47] and magnetic substance or CNT with numerous confines by Ai et al. [48], for the take-off of bromothymol blue and MB dyes individually.

5.4 Nanocomposites that Are Graphene Based

Nanocomposites that are graphene based have proved to take up a number of dyes by adsorption from industrial effluents. Zheng et al. [49] manufactured a three-dimensional graded nanocomposite, which has a stuffed design, with graphene in the inside and layered on both the sides with nickel–iron hydroxide, thus concealing graphene oxide with slender nickel–iron hydroxide nanosheets. This structure served as an efficient adsorbing material for the removal of dyes, due to ionic interactions between the dye and the hybrid adsorbent. Further exploration by Mu et al. [50]

had revealed CR dye adsorption employing graphene/polyaniline (PANI)/Fe_3O_4 as nanohybrid material. Wang et al. [51] used a porous core–shell graphene/SiO_2 nanocomposites as a GO-based nanocomposite to remove cationic neutral red dye. Other nanohybrid materials that are GO based include porous wood filters permeated with graphene, Fe_3O_4/GO nanocomposite and metal-ferrite-based GO191.

5.5 *Nanohybrids Made of Natural and Synthetic Clay*

Ruiz-Hitzky et al. [52] studied the characteristics of nanohybrids made of natural and synthetic clay as low-priced, safe, thermally stable, porous, capable of ion exchange and based on the functionalities can be remodelled, and these characteristics make these materials as ideal adsorbents. Cross-linked chitosan/sepiolite clay showed the greatest removal efficiency for MB and reactive orange, in 30 h exposure period as reported by Marrakchi et al. [53] However, a nanohybrid, which double hydroxide layered with Mg–Al–NO_3 material showed an effective drawing out of green based dyes [54].

5.6 *Fly Ash-Based Nanohybrid*

Fly ash, usually a fall out from the thermal power plants using charcoal, when fused with other materials is an excellent hybrid adsorbent (due to its porous nature), in dye removal from waste water. Fly ash geopolymer monoliths in 30 h exposure period, showed superior adsorption of MB dye and in addition the hybrid material could be recycled five times, whereas fly ash-based geopolymer spheres could adsorb MB dye effectively for the same period but could be reused eight times [55, 56]. It was found by Gao et al. [57] that an altered fly ash nanohybrid with $Ca(OH)_2$/ Na_2FeO_4 can adsorb Orange II dye.

5.7 *Hybrids of Bio-adsorbents*

Any biological materials that are altered by combining with non-biological material are called hybrid bio-adsorbents. Thus, a hybrid of bio-adsorbent made of activated oil palm ash zeolite/chitosan exhibited its potential in adsorbing blue-based dyes [58]. Similarly, bio-adsorbents made of cellulose and polyacrylic performed excellently in drawing out Acid Blue 93 and MB dyes as studied by Liu et al. [59] A bio-adsorbent, which is sulphonated obtained from the waste kernel of hawthorn, served as a potential adsorbent in removing MB dye within 6 h from the dye contaminated waste water [60].

5.8 Nanohybrid Materials that Are Magnetic and Non-magnetic Metal Oxide Based

Nanohybrid materials, which are synthesized from metal oxides, are gaining popularity because of their characteristics like extended facet, size in nanometres, ready reactive nature, capacity to be conjoined and powerful movability in fluid. In addition, magnetic metal oxide nanocomposites display extra applications by the use of magnetic field.

Wheat straw, after treatment with NaOH and blended with Fe_3O_4 nanoparticles, showed the highest efficacy in the take-off of MB from industrial effluent, where the endothermic adsorption process was found to pH and temperature dependant as studied by Ebrahimian Pirbazari et al. [61] When Fe_2O_3–Al_2O_3 nanohybrid was deployed by Mahapatra et al. [62] for remediation of dye polluted water, the highest adsorption capacity for CR dye was observed. When Li et al. [63] used a non-magnetic metal oxide nanocomposite, e.g. Co/Cr-codoped ZnO, the maximum captivation of dyes was reported to be 1057.9 mg g^{-1}, primarily because of the positive charge on its surface and high specific surface area.

5.9 Nanohybrid that Are Derived from Metals and Organic Compounds

Metal–organic frameworks (MOFs) have captured recognition as adsorbents due to their excellent alterable porosity and restricted contour. In addition, the uniqueness of these nanohybrid materials is because of the extended surface areas, multiple use, polar/polarisable bonds and the likelihood in the alteration of the functional groups in the metal and the organic compound, thereby facilitating substrate-product interactivity [64]. The MOF formed of MIL-53(Al)-NH_2 was found to hastily bind with MB and MG, and this fast adsorption was found to be because of the strong adhesion between the dye and the adsorbent. A number of dyes were also said to be adsorbed by MIL-Ti MOFs by strong covalent bonds and weak hydrogen bonds [65]. Metal–organic framework, that is zirconium based (Zr-MOF) when used by Zhao et al. [66] showed the highest adsorption ability of 63.38 (for crystal violet) and 67.73 mg g^{-1} (for RhB), for the shifting of dyes present in polluted industrial effluent.

5.10 Polymers and Their Nanocomposites

Polymer-based adsorbents, because of their characteristics such as extended surface area, specific affinity, ease in designing, fascinating doping/de-doping chemistry,

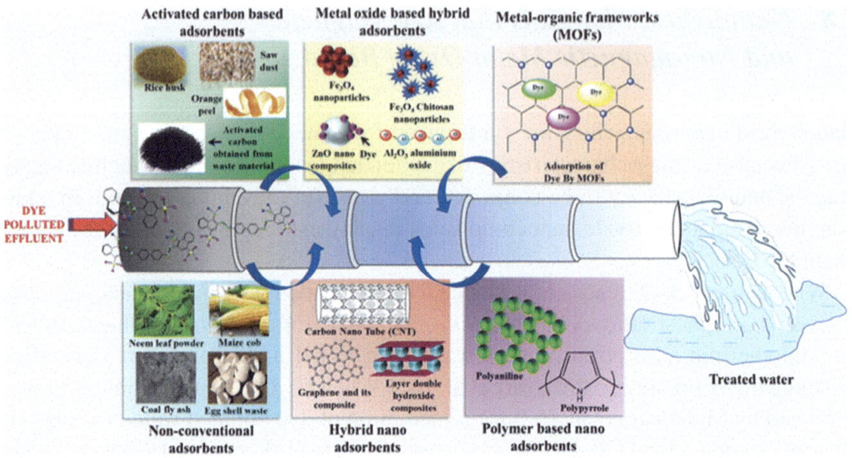

Fig. 1 Various adsorbents used in removal of dyes from industrial effluent. Reproduced with permission from Mater. Ad., 2021, 2, 4522

electric transmissibility and porous nature, are being extensively employed in remediation of dye polluted water. Polypyrrole (PPY) and PANI272 is one such example by which sunset yellow and CR, which are Azo dyes, were attracted by the nanohybrid adsorbent [67]. Similarly, chitosan-based nanoparticles like a fibrous porter with confined chitosan nanoparticles enabled the adsorption of some deleterious dyes by Ion-exchange and hydrophobic binding as studied by Lipatova et al. [68] Polyaniline and polypyrrole-based nanocomposites due to their elevated adsorption power have gained attention in current days. Agarwal et al. [69] made the use of polyvinyl alcohol (PVA) for remediation of polluted effluent. The removal efficiency was observed for bromothymol and MB dyes in 10 min (Fig. 1 and Table 2).

6 Fabrication and Characterization of Nanohybrid Materials

Different conformation of nanohybrid materials are being synthesized, characterized and their multifaceted potential in treating dye polluted industrial effluent are studied. For example, fabrication of Fe_3O_4/Ag nanoparticles was done by covering the exterior of Fe_3O_4 compound with Ag. The elemental composition of Ag was 44.21% and the Fe content was 40.13% in the finally synthesized Fe_3O_4/Ag nanohybrid material [70]. Fang et al. [71] in the year 2017, mentioned the detailed protocol for the synthesis of polyoxometalates (POMs), from the mixture containing $Cu(ClO_4)_2 \cdot 6H_2O$, pyridine-2-carboxamide and $Na_2MoO_4 \cdot 2H_2O$. The blue precipitate that was obtained on whisking the mixture for 0.5 h at normal temperature to yield POMs, following clarification, drying and heating. Then, Fe_3O_4/Ag/POMs

Table 2 Types of nanohybrid materials used for dye removal, their experimental conditions, removal efficiency and the mechanism of removal

Nanohybrid material	Dyes	Experimental conditions	Removal efficiency	Mechanism of removal
Carbon-based hybrid nanocomposite	RhB and MB	Dose: 2 g L^{-1} C0: 5 mg L^{-1} Time: 2 h lmax: 554 and 664 nm, respectively	5.269 and 7.415 mg g^{-1}, respectively	Chemisorption process
Iron–carbon hybrid magnetic nanosheets	CR and MB	Time: 24 min pH: 2 and 8, respectively; Dose: 6 mg/15 mL lmax: 662 and 664 nm, respectively	531.9 and 185.2 mg g^{-1}, respectively	Chemisorption process
Porous silicon–carbon–nitrogen hybrid materials	Methyl blue and acid fuchsin	C0: 300 mg L^{-1} and 200 mg L^{-1}, respectively	1327.7 mg g^{-1} and 1084.5 mg g^{-1}, respectively	Chemisorption process
Ni/porous carbon-CNT	MG, CR, RhB, MB, and MO	C0: 20 mg L^{-1} Time:60 min	898 mg g^{-1}, 818 mg g^{-1}, 395 mg g^{-1}, 312 mg g^{-1}, and 271 mg g^{-1}, respectively	Multilayer adsorption mechanism
Activated carbon-based hybrid Konjac glucomannan activated carbon aerogel	MB	C0: 140 mg L^{-1} Dose: 10 mg/20 mL T: 313 K	416.67 mg g^{-1}	Density gradient force and hydrogen bond interaction
ZnO-NP-loaded Parthenium weed activated carbon (ZnONPs-PWAC)	MB	C0: 100 mg L^{-1} pH: 6 Dose: 50 mg/100 mL Time: 60 min	~ 99%	–
Finger-citron-residue-based activated carbon	MO and MB	C0: 50–500 mg L^{-1} lmax: 464 nm and 664 nm, respectively	934.58 and 581.40 mg g^{-1}, respectively	p–p stacking interaction and electrostatic attraction

(continued)

Table 2 (continued)

Nanohybrid material	Dyes	Experimental conditions	Removal efficiency	Mechanism of removal
Multi-wall carbon nanotubes	MO	C0: 20 mg L^{-1} T: 25 1C pH: B7 Time: 2 h Dose: 15 mg/50 mL lmax: 460 nm	51.74 mg g^{-1}	External diffusion, boundary layer diffusion and intra-particle diffusion
Calcium alginate/ multi-walled carbon nanotubes	MB and MO	pH: 4–12 for MB and < 2 for MO	606.1 and 12.5 mg g^{-1}, respectively	Opposite charge attraction
Magnetite/MWCNTs	Bromothymol blue (BTB)	C0: 10–70 mg L^{-1} Dose: 0.02 g/25 mL Time: 0–11 min pH: 1	55 mg g^{-1}	Electrostatic interactions, p–p dispersion interaction, hydrogen bonding and electron donor–acceptor complex formation
Magnetite-loaded multi-walled carbon nanotubes	MB	C0: 20 mg L-148.06 mg g^{-1}. Dose: 0.02 g/50 mL T: 25 ± 1 1C pH: 7	48.06 mg g^{-1}	Electrostatic attraction and p–p stacking interactions
Graphene oxide, reduced graphene oxide and their nanocomposites. 3D hierarchical GO–NiFe LDH composite	CR and MO	Time: 225 min and 135 min, respectively	489 and 438 mg g^{-1}, respectively	Electrostatic attraction, ions exchange and p–p stacking interaction

(continued)

Table 2 (continued)

Nanohybrid material	Dyes	Experimental conditions	Removal efficiency	Mechanism of removal
Graphene/PANI/Fe_3O_4	CR	C0: 100 mg L^{-1} Time: 2 h Dose: 25 mg/25 mL lmax: 625 nm and 498 nm, respectively	248.76 mg g^{-1}	Electrostatic interaction, hydrogen bond and p–p stacking interaction
Hybrids of natural and synthetic clay, cross-linked chitosan/sepiolite clay composite	MB and reactive orange 16	C0: 100 mg L^{-1} Dose: 2 g L^{-1} Time: 30 h pH: 49 for MB and 3 for reactive orange 16 lmax: 665 nm and 496 nm, respectively	40.986 mg g^{-1} and 190.965 mg g^{-1}, respectively	Physisorption
Mg–Al–NO_3 layered double hydroxides nanohybrid	Amaranth, diamine, green B and brilliant green	T: Room temperature pH: 7–9.5 lmax: 520 nm, 623 nm and 624 nm, respectively	0.8 mmol g^{-1}, 1.089 mmol g^{-1}, 1.418 mmol g^{-1}, respectively	Ion exchange
Hybrids of fly ash Biomass fly ash geopolymer monoliths	MB	C0: 1–50 mg L^{-1} Time: 30 h lmax: 664 nm	15.4 mg g^{-1} Reused up to five cycles	Electrostatic interaction
Porous biomass fly ash-based geopolymer spheres	MB	C0: 10–125 mg L^{-1} Time: 30 h lmax: 664 nm	79.7 mg g^{-1} Reused up to eight cycles	Multilayer adsorption
$Ca(OH)_2$/Na_2FeO_4 modified fly ash	Orange II	C0: 50 mg L^{-1} Dose: 2000 mg L^{-1}	24.8 mg g^{-1}	Electrostatic interaction, hydrogen bonding
Hybrids of bioadsorbents cellulose-g-poly(acrylic acid co-acrylamide)	Acid blue 93 (AB93) and MB	C0: 200 mg L^{-1} Time: 90 min pH: 7	1602 and 1814 mg g^{-1}, respectively	Electrostatic interaction

(continued)

Table 2 (continued)

Nanohybrid material	Dyes	Experimental conditions	Removal efficiency	Mechanism of removal
Sulphonated bio-adsorbent from waste hawthorn kernel	MB	C0: 25–400 mg L^{-1} Time: 0–360 min T: 25–60 1C Dose: 0.1 g in 50 mL lmax: 665 nm	151.5 mg g^{-1}	Ion exchange
Magnetic-based bio-adsorbents Fe_3O_4–wheat straw	MB	Dose: 0.01–0.2 g/100 mL, pH: 7 lmax: 668 nm	1374.6 mg g^{-1}	Complexation formation and ion exchange
Iron oxide–alumina	CR	pH: 7	416.66 mg g^{-1}	Multilayer adsorption
Non-magnetic oxide nanocomposites Co/Cr-codoped ZnO	MO	Dose: 10 mg Time: 120 min lmax: 200 nm	1057.90 mg g^{-1}	Chemisorption and electrostatic interaction
MIL-53(Al)-NH_2	MB and MG	C0: 5 mg L^{-1} Dose: 10 mg/100 mL lmax: 663 and 619 nm, respectively	208.3 mg g^{-1} and 164.9 mg g^{-1}, respectively	Electrostatic attraction or repulsion
MIL-Ti	Basic Red 46, basic blue 41 and MB	Time: 30 min pH: 6.3 lmax: 530, 605, and 665 nm, respectively	1250, 1428, and 833 mg g^{-1}, respectively	p–p interaction, hydrogen bonding, electrostatic interactions and acid–base interactions
Zirconium-based MOFs	CV and RhB	C0: 10 mg L^{-1} pH: 11 and 7, respectively lmax: 585 and 554 nm, respectively	63.38 mg g^{-1} and 67.73 mg g^{-1}, respectively	Chemisorption

(continued)

Table 2 (continued)

Nanohybrid material	Dyes	Experimental conditions	Removal efficiency	Mechanism of removal
PPY and PANI	Azo dyes: sunset yellow and CR	pH: 2 lmax: 470 and 497 nm, respectively	212.1 and 147 mg g^{-1}, respectively	p–p electron donor–acceptor interaction and electrostatic attraction
Chitosan-based adsorbents chitosan nanoparticles immobilized on a fibrous carrier	Direct blue-86, photosens, theraphthal and C.I reactive blue 21	pH: 4–6 C0: 10–130 mg L^{-1} lmax: 665, 679, 675, and 664 nm, respectively	1097 ± 55, 1049 ± 43, 367 ± 22, and 296 ± 18 mg g^{-1}, respectively	Ion-exchange processes and hydrophobic binding
Starch/polyaniline	Reactive black and reactive v4	lmax: 597 and 577 nm, respectively pH: 3	811.30 and 578.39 mg g^{-1}, respectively	Electrostatic interaction
Miscellaneous polymer adsorbents Polyvinyl alcohol	Bromothymol blue (BTB) and MB	pH: 6 Time: 10 min	276.2 mg g^{-1} and 123.3 mg g^{-1}, respectively 98.65% and 61.32%, respectively	Chemisorption

C0: initial dye concentration; T: temperature; and lmax: maximum wavelength. Reproduced (some contents of the table) with permission from Mater. Ad., 2021, 2, 4511–4520

Note the types of nanomaterials used in dyes removal, their experimental conditions, removal efficiency and the mechanism of removal nanocomposites were prepared by conjoining these nanoparticles with POMs. Zhang et al. [72] designed a novel category of nanohybrid particles termed "nanocages" of reactive molecular matrices/networks, where the functional groups will interact with the nanoparticles at specific location of the networks. Silver nanoparticles designed by this process have smaller diameter of 4.34 nm and the size could be altered by managing silver ions concentration and the molecular network functioning groups and are found to be stability. Therefore, silver nanohybrid particles have proved to exhibit extensive demand in the field of biomedicine, catalysis, and optoelectronics.

Nanohybrid materials are characterized by a number of techniques, namely (i) UV–Vis Spectrophotometer, where the spectral values are measured at different wavelengths. (ii) Transmission Electron Microscopy and Scanning Electron Microscopy to get the images of the samples. (iii) Fourier-transform infrared spectroscopy to determine the functional groups. (iv) X-ray photoelectron spectroscopic analysis to study the internal and exterior chemistry of the material. (v) Thermogravimetry to analyse the stability of the substance. (vi) X-ray diffraction to find their purity and spatial orientation. (vii) RAMAN Spectroscopy for qualitative analysis of nanohybrid materials.

7 Mechanism of Dye Removal by Nanohybrid Materials

Nanohybrid materials are employed in the dye removal from the textile industry effluents and the mode of removal by the process of adsorption differs based on the dyes and the nature of the nanohybrid materials. Different mechanisms are said to be operative such as ion-exchange, chemisorption, physisorption, complexation, multilayer adsorption, diffusion, density gradient force, weak forces, covalent bonding and hydrophobic interaction. Acid–base reaction is also one mechanism of dye removal by adsorption. Intermolecular interaction and electrostatic attraction favour the adsorption of dyes like Bromothymol blue, CR, MB and MO. Ion-exchange mechanism is prevalent between liquid dye and the adsorbent in the solid nature. Multilayered adsorption is done by initial adsorption of the dye to nanohybrid material and then eventually resulting in aggregation due to rearranging of the substance on the exterior of the adsorbent. Likewise p–p bonding happens between positively and negatively charged molecules present in dyes and adsorbents. Hydrogen bonding mechanism is promoted by the –OH group present in the adsorbent and the acceptor of the dye molecule. The adsorption mechanism is physisorption, where the dyes are held onto the nanohybrid materials through weak forces and in chemisorption the chemical bonds facilitate the attachment of the dyes on the adsorbents. Acid–base reaction mechanism favours the removal of acidic dyes using alkaline materials as adsorbents [7].

8 Conclusion and Future Perspective

Thus, there exists an expanding anxiety regarding the discharge of dyes from textile industries not only for it spoils the aesthetic value of the aquatic ecosystems, but also for the threat of health ailments, they cause due to their persistence in the water resources. Though conventional methods have proved to be effective in the shifting of the dyes from polluted effluents, due to the many disadvantages they pose, not used extensively. Among the techniques used for dye removal, adsorption was found to exhibit superior efficiency and was used in laboratory study and in large field applications. This is because of their selective removal of dyes, highest adsorption rate, recyclability, easy application, stability and economic feasibility. A number of adsorbents are being used, of which nanoparticles are found to be superior to other materials, for they are porous with large surface area and functional groups. Currently, nanohybrid materials are attaining impulse due to their multifunctional properties. A range of nanohybrid materials, namely Carbon–Carbon NHs (CCNH), Carbon–Metal NHs (CMNH), Metal–Metal NHs (MMNH), Organo-Metal–carbon NHs (OMCNH), etc. are being synthesized from the parent material and so far their ability to adsorb have considerably enhanced than singular nanoparticles. The list of nanohybrid materials will still lengthen by the addition of new materials. Different protocols are being experimented to fabricate new nanohybrid materials, thus novel nanohybrid materials with different size, shape and properties are being produced. However, the toxicity of these materials is still a matter of concern. Hence, the focus is now shifted towards green synthesis of nanoparticles by living organisms and blending them with the chemical compounds. As a step still forward, blending nanoparticles synthesized from microorganisms with them to fabricate purely biological nanohybrid materials are also in the trial. Thus, nanohybrid materials have futuristic role in remediating the polluted water.

References

1. Zollinger H (1987) Colour chemistry-synthesis, properties of organic dyes and pigments. VCH Publishers, New York, pp 92–100
2. Shore J (1996) Advances in direct dyes. Indian J Fib Text Res 21:1–29
3. Srinivasan SV, Murthy DVS (2009) Statistical optimization for decolorization of textile dyes using Trametes versicolor. J Hazard Mater
4. Kale RD, Kane PB (2016) Colour removal using nanoparticles. Text Clothing Sustain 2(4). https://doi.org/10.1186/s40689-016-0015-4
5. Hanger K (2003) Industrial dyes: chemistry, properties and applications, health and safety aspects
6. O'Neill C, Hawkes FR, Hawkes DL, Lourenço ND, Pinheiro HM, Delée W (1999) Colour in textile effluents—sources, measurement, discharge consents and simulation: a review. J Chem Technol Biotechnol. https://doi.org/10.1002/(SICI)1097-4660(199911)74:11%3c1009::AID-JCTB153%3e3.0.CO;2-N

7. Dutta S, Gupta B, Srivastava SK, Gupta AK (2021) Recent advances on the removal of dyes from wastewater using various adsorbents: a critical review. Mater Adv 2:4497. https://doi.org/10.1039/d1ma00354b
8. Al-Tohamya R, Ali SS, Li F, Okasha KM, Mahmoud YA-G, Elsamahy T, Jiao H, Fu Y, Sun J (2022) A critical review on the treatment of dye-containing wastewater: ecotoxicological and health concerns of textile dyes and possible remediation approaches for environmental safety. Ecotoxicol Environ Saf 231.https://doi.org/10.1016/j.ecoenv.2021.113160
9. Samsami S, Mohamadizaniani M, Sarrafzadeh MH, Rene ER, Firoozbahr M (2020) Recent advances in the treatment of dye-containing wastewater from textile industries: overview and perspectives. Process Saf Environ Prot 143:138–163
10. Cao G, Wang R, Ju Y, Jing B, Duan X, Ao Z, Jiang J, Li F, Ho SH (2021) Synchronous removal of emulsions and soluble organic contaminants via a microalgae-based membrane system: performance and mechanisms. Water Res 206:117741
11. Jadhav AC, Jadhav NC (2021) Treatment of textile wastewater using adsorption and adsorbents. In: Sustainable technologies for textile wastewater treatments. Woodhead Publishing, pp 235–273
12. Brião GV, Jahn SL, Foletto EL, Dotto GL (2018) Highly efficient and reusable mesoporous zeolite synthetized from a biopolymer for cationic dyes adsorption. Colloids Surf A Physicochem Eng Asp 556:43–50
13. Madan S, Shaw R, Tiwari S, Tiwari SK (2019) Adsorption dynamics of Congo red dye removal using ZnO functionalized high silica zeolitic particles. Appl Surf Sci 487:907–917
14. Martin NM, Pascu LF, Demba A, Nita-Lazar M, Badea IA, Aboul-Enein HY (2019) Removal of the Acid Orange 10 by ion exchange and microbiological methods. Int J Environ Sci Technol 16(10):6357–6366
15. Bayramoglu G, Kunduzcu G, Arica MY (2020) Preparation and characterization of strong cation exchange terpolymer resin as effective adsorbent for removal of disperse dyes. Polym Eng Sci 60(1):192–201
16. Kishor R, Purchase D, Saratale GD, Saratale RG, Ferreira LFR, Bilal M, Chandra R, Bharagava RN (2021) Ecotoxicological and health concerns of persistent coloring pollutants of textile industry wastewater and treatment approaches for environmental safety. J Environ Chem Eng 9:105012
17. Al-Mutairi NZ (2006) Coagulant toxicity and effectiveness in a slaughterhouse wastewater treatment plant. Ecotoxicol Environ Saf 65(1):74–83
18. Mathuram M, Meera R, Vijayaraghavan G (2018) Application of locally sourced plants as natural coagulants for dye removal from wastewater: a review. J Mater Environ Sci 9(7):2058–2070
19. Badawi AK, Zaher K (2021) Hybrid treatment system for real textile wastewater remediation based on coagulation/flocculation, adsorption and filtration processes: performance and economic evaluation. J Water Process Eng 40:101963
20. Zhao H, Cao P, Lu L, Li F, Pang CH, Wu T (2022) Co-regulation of dispersion, exposure and defect sites on CeO_2 (III) surface for catalytic oxidation of Hg0. J Hazard Mater 424:126566
21. Zhong J, Yang B, Gao FZ, Xiong Q, Feng Y, Li Y, Zhang JN, Ying GG (2021) Performance and mechanism in degradation of typical antibiotics and antibiotic resistance genes by magnetic resin-mediated UV-Fenton process. Ecotoxicol Environ Saf 227:112908
22. Zhang X, Li F, Wang J, Zhao H, Yu XF (2021) Strategy for improving the activity and selectivity of CO_2 electroreduction on flexible carbon materials for carbon neutral. Appl Energy 298:117196
23. Ali SS, Al-Tohamy R, Sun J (2022) Performance of Meyerozyma caribbica as a novel manganese peroxidase-producing yeast inhabiting wood-feeding termite gut symbionts for azo dye decolorization and detoxification. Sci Total Environ 806:150665
24. Malik PK (2004) Dye removal from wastewater using activated carbon developed from sawdust: adsorption equilibrium and kinetics. J Hazard Mater 113:81–88. https://doi.org/10.1016/j.jhazmat.2004.05.022

25. Ahmad A, Mohd-Setapar SH, Chuong CS, Khatoon A, Wani WA, Kumar R, Rafatullah M (2015) Recent advances in new generation dye removal technologies: novel search for approaches to reprocess wastewater. RSC Adv 5(39):30801–30818
26. Ojedokun AT, Bello OS (2017) Appl Water Sci 7:1965–1977
27. Dong R-A, Hsieh T-C, Huang C-P (2010) Photoassisted reduction of metal ions and organic dye by titanium dioxide nanoparticles in aqueous solution under anoxic conditions. Sci Total Environ 408:3334–3341. https://doi.org/10.1016/j.scitotenv.2010.03.032
28. Tabassam N, Mutahir S, Khan MA, Khan IU, Habiba U, Refat (2022) Facile synthesis of cinnamic acid sensitized rice husk biochar for removal of organic dyes from wastewaters: batch experimental and theoretical studies. Mater Chem Phys 288:126327. https://doi.org/10.1016/j.matchemphys.2022.126327
29. Luo X, Ma K, Jiao T, Xing R, Zhang L, Zhou J, Li B (2017) Graphene oxide-polymer composite Langmuir films constructed by interfacial thiol-ene photopolymerization. Nanoscale Res Lett 12:99. https://doi.org/10.1186/s11671-017-1864-8
30. de Lima Barizãoa AC, Silva MF, Andrade M, Brito FC, Gomesc RG, Bergamasco R (2020) J Environ Chem Eng 8(1):103618
31. Noreen S, Mustafa G, Ibrahim SM, Naz S, Iqbal M, Yaseen M, Nisar J (2020) Iron oxide (Fe_2O_3) prepared via green route and adsorption efficiency evaluation for an anionic dye: kinetics, isotherms and thermodynamics studies. J Mater Res Technol 9(3):4206–4217. https://doi.org/10.1016/j.jmrt.2020.02.047
32. Feynman RP (1960) There's plenty of room at the bottom. Eng Sci 23:22–36
33. Muszynski R, Seger B, Kamat PV (2008) Decorating graphene sheets with gold nanoparticles. J Phys Chem C 112:5263–5266
34. Carril M, Fernandez I, Rodriguez J, Garcia I, Penades S (2013) Gold coated iron oxide glyconanoparticles for MRI, CT, and US multimodal imaging. Part Part Syst Charact 31:81–87
35. Saleh NB, Nabiul Afrooz ARM, Bisesi JH, Aich N, Jr. Plazas Tuttle J, Sabo Attwood T (2014) Emergent properties and toxicological considerations of nanomaterials in aquatic systems. Nanomaterials 4:372–407.https://doi.org/10.3390/nano4020372
36. Nasibulin AG, Pikhitsa PV, Jiang H, Brown DP, Krasheninnikov AV, Anisimov AS, Queipo P, Moisala A, Gonzalez D, Lientschnig G et al (2007) Anovel hybrid carbon material. Nat Nanotechnol 2:156–161
37. Rahmam GMA, Guldi DM, Zambon E, Pasquato L, Tagmatarchis N, Prato M (2005) Dispersible carbon nanotube/gold nanohybrids: evidence for strong electronic interactions. Small 1:527–530
38. Tsoufis T, Tomou A, Gournis D, Douvalis AP, Panagiotopoulos I, Kooi B, Georgakilas V, Arfaoui I, Bakas T (2008) Novel nanohybrid derived from the attachment of FePt nanoparticles on carbon nanotubes. J Nanosci Nanotechnol 8:5942–5951
39. Cui Y, Ren B, Yao J-L, Gu R-A, Tian Z-Q (2006) Synthesis of $Ag_{core}Au_{shell}$ bimetallic nanoparticles for immunoassay based on surface enhanced raman spectroscopy. J Phys Chem B 110:4002–4006
40. Kim TS, Song HJ, Dar MA, Lee HJ, Kim DW (2018) Appl Surf Sci 439:364–370
41. Manippady SR, Singh A, Basavaraja BM, Samal AK, Srivastava S, Saxena M (2020) ACS Appl Nano Mater 3:1571–1582
42. Meng L, Zhang X, Tang Y, Su K (2015) J Kong Sci Rep 5:1–16
43. Jin L, Zhao X, Qian X, Dong M (2018) J Colloid Interface Sci 509:245–253
44. Wang Y, Li Y, Li H, Zheng H, Du Q (2019) J Polym Environ 27:1342–1351
45. Kamaraj M, Srinivasan NR, Assefa G, Adugna AT, Kebede M (2020) Environ Technol Innov 17:100540
46. Gong R, Ye J, Dai W, Yan X, Hu J, Hu X, Li S, Huang H (2013) Ind Eng Chem Res 52:14297–14303
47. Ghaedi M, Khajehsharifi H, Yadkuri AH, Roosta M, Asghari A (2012) Toxicol Environ Chem 94:873–883
48. Ai L, Zhang C, Liao F, Wang Y, Li M, Meng L, Jiang J (2011) J Hazard Mater 198:282–290
49. Zheng Y, Cheng B, You W, Yu J, Ho W (2019) J Hazard Mater 369:214–225

50. Mu JT, Zhang L, Wang A (2017) Sci Rep 7:1–12
51. Wang J, Chen T, Xu B, Chen Y (2020) Appl Sci 10:8529
52. Ruiz-Hitzky E, Aranda P, Darder M, Rytwo G (2010) J Mater Chem 20:9306–9315
53. Marrakchi F, Khanday WA, Asif M, Hameed BH (2016) Int J Biol Macromol 93:1231–1239
54. Abdellaoui K, Pavlovic I, Barriga C (2019) ChemEngineering 3:41
55. Novais RM, Ascensão G, Tobaldi DM, Seabra MP, Labrincha JA (2018) J Cleaner Prod 171:783–794
56. Novais RM, Carvalheiras J, Tobaldi DM, Seabra MP, Pullar RC, Labrincha JA (2019) J Cleaner Prod 207:350–362
57. Gao S, Zhang W, Zhou H, Chen D (2018) J Rare Earths 36:986–993
58. Khanday WA, Asif M, Hameed BH (2017) Int J Biol Macromol 95:895–902
59. Liu L, Gao ZY, Su XP, Chen X, Jiang L, Yao JM (2015) ACS Sustain Chem Eng 3:432–442
60. Akko Y, Cos-kun R, Delibas A (2019) J Mol Liq 287:110988
61. Ebrahimian Pirbazari A, Saberikhah E, Habibzadeh Kozani SS (2014) Water Resour Ind 7–8:23–37
62. Mahapatra A, Mishra BG, Hota G (2013) Ceram Int 39:5443–5451
63. Li Z, Sun Y, Xing J, Xing Y, Meng A (2018) J Hazard Mater 352:204–214
64. Xiang W, Zhang Y, Chen Y, Liu CJ, Tu X (2020) J Mater Chem A 8:21526–21546
65. Oveisi M, Asli MA, Mahmoodi NM (2018) J Hazard Mater 347:123–140
66. Zhao X, Liu S, Tang Z, Niu H, Cai Y, Meng W, Wu F, Giesy JP (2015) Sci Rep 5:1–10
67. Aliabadi RS, Mahmoodi NO (2018) J Cleaner Prod 179:235–245
68. Lipatova IM, Makarova LI, Yusova AA (2018) Chemosphere 212:1155–1162
69. Agarwal S, Sadegh H, Monajjemi M, Hamdy AS, Ali GAM, Memar AOH, Shahryari-Ghoshekandi R, Tyagi I, Gupta VK (2016) J Mol Liq 218:191–197
70. Li C, Zhou L, Yang H, Lv R, Tian P, Li X, Zhang Y, Chen Z, Lin F (2017) Self-assembled exopolysaccharide nanoparticles for bioremediation and green synthesis of noble metal nanoparticles. ACS Appl Mater Interfaces 9:22808–22818. https://doi.org/10.1021/acsami.7b02908
71. Fang N, Ji YM, Li CY, Wu YY, Ma CG, Liu HL, Li MX (2017) Synthesis and adsorption properties of $[Cu(L)_2(H_2O)]H_2[Cu(L)_2(P_2Mo_5O_3)]\cdot 4H_2O/Fe_3O_4$ nanocomposites. RSC Adv 7:25325–25333
72. Zhang DS, Xiong J, Li C, Zhang Y, Toh GW, Lin H, Chen Y (2014) Synthesis of size tunable gold nanoparticles polymeric hybrid based on molecular nanocages. Micro Nano Lett 9(4):235–238
73. Janaki V, Oh BT, Shanthi K, Lee KJ, Ramasamy AK, Kamala-Kannan S (2012) Synth Met 162:974–980

Textile Dyes and Their Effect on Human Beings

N. Hemashenpagam and S. Selvajeyanthi

Abstract Water is a crucial resource for both human growth and life on earth. The primary moral obligation to provide humanity with the very minimal requirements for a quality life is to preserve the water. In order to produce the color pigment for textile dyes, a variety of heavy metals and aromatic compounds including mercury, chromium, cadmium, lead and arsenic are used as dyes. Water contamination is further exacerbated by the dumping of untreated textile wastewater into water bodies. A wide variety of organic contaminants are present in the untreated effluents that the textile sector releases and severe ecotoxicological risks that have toxic consequences on living things. Azo dyes are significant aquatic contaminants that have a detrimental effect on humans and ecological predators. Any acute toxicity to textile dyes is induced by oral consumption and inhalation, irritated skin and eyes, and exposure of humans and other species. Any acute toxicity to textile is dangerous. Genotoxic effects of textile dyes and a few mutagenic incidents are crucial to the definition of cancer. Allergic dermatitis may result from the disposal of aromatic amines. By increasing biochemicals and chemical oxygen demand, reducing photosynthesis, preventing plant growth, entering the food chain, providing recalcitrance and bioaccumulation and perhaps promoting toxicity, textile dyes decrease the aesthetic quality of bodies of water. The use of eco-friendly technologies to treat the textile industry is necessary to prevent harm to the environment and human health.

Keywords Textile effluent · Chemicals · Harmful · Mankind · Carcinogenic

N. Hemashenpagam (✉)
Department of Microbiology, Hindusthan College of Arts & Science, Coimbatore, Tamil Nadu 641028, India
e-mail: hemashenpagam.n@hicas.ac.in

S. Selvajeyanthi
Department of Microbiology, Shri Nehru Maha Vidyalaya College of Arts and Science, Coimbatore, Tamil Nadu 641050, India

© The Author(s), under exclusive license to Springer Nature Singapore Pte Ltd. 2023
A. Ahmad et al. (eds.), *Nanohybrid Materials for Treatment of Textiles Dyes*,
Smart Nanomaterials Technology, https://doi.org/10.1007/978-981-99-3901-5_3

1 Introduction

An integral part of economic growth and daily living for humans is textile. Environmental activists have been urging every industry to include sustainability in every aspect of their daily operations. When compared to other industries, the textile industry is thought to be a large contributor to environmental pollution. It is also vulnerable to a number of ecological problems along the entire supply chain, from the production of fiber to the finishing of clothes, including the contamination of water sources, the accumulation of waste and air pollution. Globally, the textile sector generates about $1 trillion annually, accounts for 7% of all exports and supports about 35 million people [1]. Only a handful of the exceedingly dangerous substances utilized by textile industries (TIs) at various stages include sizing, whitening, anti-creasing, sequestering, stabilizers, softening, washing and finishing chemicals. The process of dying TIs involves the application of thousands of synthetic dyes. The industrial effluent produced by the textile industry is regarded as one of the biggest pollutants of our ecosystems. It poses risks to living things that are carcinogenic, mutagenic, genotoxic, cytotoxic and allergic [2]. In ancient time, people make their clothing vibrant and colorful using natural dyes from flower such as tesu flowers and indigo, logwood, madder from plants, tyrian purple, kermes, cochineal from animal sources were included in natural dye preparation. Since these colors were almost biodegradable, they posed no risk. However, the production and usage of these dyes are exceedingly labor-intensive and expensive. Thus, the use chemical source to produce synthetic dyes, which are preferably less expensive and more readily available as dyes started [3]. Synthetic dyes defined as chromogenic substances that are crystal in structure, when applied to a substrate provide color [4, 5]. Dyes are characterized by their application and chemical structure. They are made up of a class of atoms called chromophores that give the pigment its color. Various functional groups, including azo, anthraquinone, methine, nitro, aryl methane, carbonyl, etc., form the basis of these chromophore-containing dyes. Auxochromes are electrons removed or donated from substituent groups to create or enhance the color of the chromophore. Amines, carboxyls, sulfonic acids, and hydroxyls are the four most common auxochromes [6]. The environment is polluted by release of crude and unprocessed wastewater from textile industry into the aquatic ecosystem, industrial effluent contains toxic xenobiotic compounds produced by textile industries. Pollutants include hazardous metals, dyes, dissolved solids and suspended particles which are commonly found in textile wastewater. Total dissolved solids are the primary aspect of textile effluent to be taken into account (TDS). The water that gets polluted with these chemicals are unfit for domestic, commercial and agricultural purposes. Textile dye effluent needs to be treated before discharge, in order to minimize its negative environmental consequences. To remove toxins from wastewater, many effluent treatment techniques, including physical, chemical and biological methods are used. Untreated effluent carries notably poisonous and carcinogenic residences of dyes, which show up over time. Their presence and endurance in effluents discharged into aquatic environments and their relative biodegradability have an unfavorable effect on the surroundings and appreciably on aquatic ecosystems and mankind. Accumulate throughout the aquatic meals net of aquatic wildlife, on the only hand and dysfunctional physiological strategies of aquatic flora (plants, diatoms and algae) on the other, with the aid of sabotaging their photosynthetic processes with the absence of oxygen stream and absorption in aquatic environments and additionally ends in human Illnesses can broaden both without delay with the aid of using inhalation, like breathing issues, allergies, nausea, pores and skin and eye infection and dermatitis or in a roundabout way via the meals chain, which includes tuberculosis, cancer, hemorrhage, genetic abnormalities and coronary heart disease.

2 Classification of Dyes

The strands of the cloth are equally treated with colors during the dyeing process. Printing is the process of imparting color to a specified area of fabric. While bleaching is a finishing procedure that decolorizes (removes dye color from) textile fibers, finishing techniques also include crosslinking, softening and waterproofing [7].

2.1 Based on Origin

According to where they come from, dyes can be divided into two categories. Natural dyes, which have been used since antiquity, are primarily derived from plants, while synthetic dyes are created by the artificial synthesis of chemical molecules.

2.2 Synthetic Dyes

In accordance with the characteristics of the created fiber, synthetic dyes are separated into three groups. These include colors for cellulose fibers, protein fibers and synthetic fibers [8, 9].

2.3 Based on the Textile Fiber's Coating

2.3.1 Cellulose Fiber Dyes

Hemp, lyocell, rayon, rayon blend, linen, cotton and rumie are cellulose-producing plants. With reactive dyes, direct dyes, indigo dyes and sulfur dyes, these fabrics produce flawless results [10]. Reactive dyes, the most of which can also be employed on protein fibers, are commonly used on cellulose fibers. They are renowned for their high pigmentation, permanence, ease of manipulation throughout a wide temperature range and adaptability since they have a variety of reactive groups that can create covalent bonds with various fibers [11]. Direct dyes, although being very cheap, typically fail to adhere to cellulose fibers and instead tend to stay in an aqueous state (they can be used with certain synthetic fibers as well). They are therefore combined with inorganic electrolytes and anionic salts in the form of sodium sulfate (Na_2SO_4) or sodium chloride to enhance their fabric binding capabilities (NaCI). Therefore, it is suggested that you wash them on the cold cycle and with other articles of the same color [12]. Indigo and other deep blue shades fall under the category of vat dyes, which were once insoluble in water but became so following an alkaline reduction. Indigo is first employed in its water-soluble, or leuco form to achieve a proper contact between the dye and the cloth. Then when this form is exposed to oxygen, it oxidizes and returns to its initial, insoluble or keto form. The primary purpose of indigo dyes which account for their massive global production [13, 14] is to color blue denim. Because of their superior dyeing capabilities, simplicity of use and low cost, sulfur dyes are a small but significant class. They feature an intricate structure that includes a disulfide (S–S) bridge. They are classified as vat dyes and sodium sulfide is used to change them from the keto form to the leuco form. Leuco sulfur dissolves in water to generate the appropriate coloring results [15, 16.

2.3.2 Protein Fiber Dyes

Animals provide the protein for textiles like wool, silk, cashmere, angora and mohair. Since they are sensitive to high pH levels, they are colored using an insoluble dye molecule on the fiber using a water-soluble acid dyestuff [17]. After azo dyes, anthraquinone, triarylmethane and phthalocyanine dyes make up the largest group of acid dyes [18, 19]. Azo dyes dominate the synthetic dyes market (60–70%) because of their adaptability, affordability, ease of use, high stability and excellent color intensity [20, 21]. The significant chromophore (N–N-) structure of the dye ensures both the dyes' solubility in water and their attachment to the fiber [22, 23]. The number of azo groups in a dye structure determines which of the three groups it belongs to (mono, di or poly). A carboxyl, sulphonyl or aliphatic group binds an aromatic or heterocyclic chemical to an unsaturated heterocycle on one side and an unsaturated heterocycle on the other.

The textile dying industry, particularly highly depends on the anthraquinone group, has long used the red dyestuff in particular [24]. These dyes are recognized for their water solubility, vivid colors and exceptional fastness features. The anthraquinone structure can also be used to create junctions with azo dyes [8, 25]. The textile industry regularly use the triphenylmethane dyes, which are composed of two groups of sulfonic acid to color wool and silk protein fibers (SO_3H). They can be used as indicators if they only have one sulfonic acid (SO_3H) auxochrome in their chemical structure. These dyestuffs are recognized for both their wide range of bright colors and their water solubility [26]. The substance 1.4-Dicyanobenzene combines with a metallic atom to produce the green and blue colors of the phthalocyanine family of dyes (Nickel, Cobalt, Copper, etc.). High colorfastness to light, resistance to oxidation and solubility in chemical stability and water are inherent properties [27, 28].

2.3.3 Synthetic Fiber Dyes

Spandex, polyester, acrylic polyamide polyacetate, polypropylene, ingeo and acetate are examples of synthetic fibers. They are used in 60% of the world's fiber manufacturing because of their diverse purposes. These fibers are colored using direct dyes, basic dyes and disperse dyes [29, 30]. Disperse dyes contain the smallest molecules of any dye. Although these pigments cannot dissolve in water, they can withstand high temperatures without losing their stability. The high temperature dyeing solution is produced by blending dye powder with the dispersion agent. Basic dyes are also referred to as cationic dyes because they transform into vivid cationic salts that color fabrics made with anionic fibers. These dyes can only be used to color modified polyesters and paper nylon because they are light-sensitive. Their main constituents include cyanine, triarylmethane, anthraquinone, diarylmethane, diaz hemicyanine, oxazine hemicyanine, thiazine and hemicyanine [11].

3 Chemical Classification of Dyes

It is based on the chemical composition of the dyes and more particularly, chromophore grouping allows dyeing plants to predict the chemical reactions between the dye and reducing agents, oxidants and other chemicals. Azo dyes are the largest family of synthetic dyes that produce 70% of the world's annual output. The monoazo, diazo and triazo dyes are those that have one or more non-coupled azo groups coupled to auxochrome groups of the OH or NH_2 type. Anthraquinone dyes are the most important group of colors, followed by azo dyes. Their general formulations show that the chromophore group is an attachable hydroxyl or amino group to a quinone nucleus and that they were developed from anthracene [31]. This group gives the dye molecule excellent light resistance. A precursor of indigo dyes is indigo. Because of this, selenium, sulfur and oxygen homologues with tones ranging from orange to turquoise and indigo blue exhibit visible hypochromic effects. The most well-known substance that gives xanthene colors their brilliant fluorescence is fluorescein. Rarely used for dyeing, these dyes are instead used as tracers in subterranean rivers or markers in maritime accidents as part of the marking process. Dicyanobenzene interacts with one of the metals Cu, Ni, Co or Pt to form the phthalocyanines class of colors. The substances phthalocyanine nucleus gives it good light resistance. The substances phthalocyanine nucleus provides it good light resistance. Due to its remarkable chemical stability, copper phthalocyanine is the most commonly used dye in this group [32].

It is a small and relatively old group of dyes known as nitrorated and nitrosated dyes, which are inexpensive and have a simple chemical structure, which is indicated by the presence of a (Hydroxyl or amino groups) electron donor group in ortho position to a nitro group ($-NO_2$) [33]. Diphenylmethane and triphenylmethane, which are the sources of earliest colors dyes like fuchsin and malachite green are derivatives of auramine, make up the least significant category of synthetic dyes [34]. Triphenylmethane can be sulfonated to create acid dyes. Acid dyes are produced when triphenylmethane is sulfonated. Carboxyl groups proximity produce metallized colours known as OH-auxochromes. Dyes containing diphenylmethane and triphenylmethane. Polymethine dyes are composed of a polymethine chain carried at the terminals of several heterocyclics. They are also referred to as cyanines.

4 Classification of Dyes Based on Solubility

The combination of auxochromes used to define dye classification is of relevance to dyers who want a classification based on areas of use. This division is based on the solubility of the dye in the dye bath, its affinity for different fibers and the kind of fixation [35, 36]. Two groups of dyes—soluble and water-insoluble dyes—are included in this classification [37, 38].

4.1 Water-Soluble Colors

4.1.1 Acid or Anionic Dyes

They are used to color fibers like wool, polyamide, silk and modacrylic that include an amino group (NH_2). These dyes should be placed on the NH_4 cations of these fibers in an acidic medium. They are water soluble due to the presence of either a chromophore group or one or more sulfonates groups [17]. The majority of these dyes are triaryl methanes, anthraquinones or azo dyes [39, 40].

4.1.2 Basic Dyes or Cationic Dyes

Which are organic base salts that can only color fibers with anionic sites after being treated with metal salts. They are made up of azo, anthraquinone, diarylmethane, triarylmethane and/or these structures [41, 42].

4.1.3 Metalliferous Dyes

The goal has been to integrate the metal further into dye itself by generating the metalliferous complex prior rather than precipitating it into the fiber in a subsequent manner in order to simplify the dyer's work by eliminating the etching operation. Therefore, metalliferous dyes are chemically created by adding a metal atom (Cr, Cu, Ni or Co) to the structure of the dye [43]. A metalliferous complex 1/1, in which the metal atom can be linked to a single dye molecule, and a metalliferous complexity 1/2, in which the metal atom can be linked to two dye molecules, can be identified as two complex versions of these dyes.

4.1.4 Reactive Dyes

Reactive dyes frequently contain chromophore groups that come from the azo, anthraquinone and phthalocyanine families [35]. The term refers to the fact that they possess a chemically reactive component, like a triazine or vinyl sulfone, which enables the formation of a strong covalent bond with the fibers. They are more soluble in water, when dyeing cotton, and perhaps when dyeing wool and polyamides as well.

4.1.5 Direct or Substantive Dyes

The positive or negative charges produced by direct or substantive dyes, which are large molecules, are electrostatically drawn to the charges on the fiber. They can be distinguished by their affinity for cellulosic fibers without the need for a mordant, which is related to the planar structure of their molecule [33]. They are used to color cellulosic fibers. Before adding the colors, the salt (sodium chloride or sodium sulfate) and additional ingredients that support the fiber's wettability and dispersion effect are added to the water. Anionic and nonionic surfactant mixtures are used for this. These dyes' main advantages are their wide range of color possibilities, ease of usage and low cost. But their biggest flaw is their low wet strength.

4.2 Insoluble Dyes in Water

4.2.1 Vat Dyes

The vat dyes are insoluble in water when they are fixed to the fibers, but they become soluble and tangible when they are reduced in an extremely alkaline environment (in a vat) using leuco derivatives that are soluble in water [37, 44, 45]. However, they are created inside the fiber and re-solubilized either naturally or with the aid of an oxidizing agent [46]. They are well-known for their superior resistance to deteriorating agents (such washing and sunshine) and their preference for specific fabrics, including as cotton, linen, wool, silk, rayon and other cellulosic fibers like indigo for coloring denim (or denim). Indigo known as vat blue 1 is one of the vat dyes that is utilized, and blue indanthrene RS is called as vat blue 4 and vat green 1.

4.2.2 Sulfur Dyes

Sulfur dyes are applied as relatively similar to vat dyes, but they represent molecules that contain sulfur and have different chemical structures and large molecular weight. They are insoluble in water, but after being reduced by sodium sulfide in an alkaline solution, they can be used as a soluble derivative. Following reoxidation, they become insoluble and return to the fiber. To produce competitively cost dark hues with good wash and light fastness, cotton is typically colored with sulfur [47].

4.2.3 Disperse or Dispersible Dyes

These substances, often referred to as plastosolubles, are utilized as a fine powder that is dispersed throughout the dye bath and are incredibly water insoluble. High temperatures are used to dye them while they are stable, allowing the dye to permeate synthetic fibers before being fixed. Disperse dyes are widely used when a dispersing agent is present, which is always added to the solution, to dye the majority of manufactured fibers, especially polyester and polyamide [33].

4.2.4 Pigments

The pigments are colorful substances that can dissolve in basic and acidic solvents but are insoluble in water. They are also devoid of any groups that would interact with the cloth fiber when a binder is used. In pigment printing techniques, they are frequently used [48, 15]. Natural colors are typically benzoic derivatives, even though the majority of inorganic pigments (minerals) are derivatives of metals including titanium, zinc, barium, lead, iron, molybdenum, antimony, zirconium, calcium, aluminum, magnesium, cadmium and chromium [49]. With the support of dispersants, such as the transparent, flexible layer formed during heat treatment (drying), which offers the pigments good resistance to mechanical tests while maintaining their vibrancy and firmness, they must be very finely separated in order to be kept in suspension.

4.2.5 Fixing Different Dye Classes

A greater or smaller amount of dyes and pigments are lost during the numerous dyeing and printing procedures due to their allure and for surfaces, taking into account the usefulness of solution agents [50, 51]. It is accurate to say that interactions with

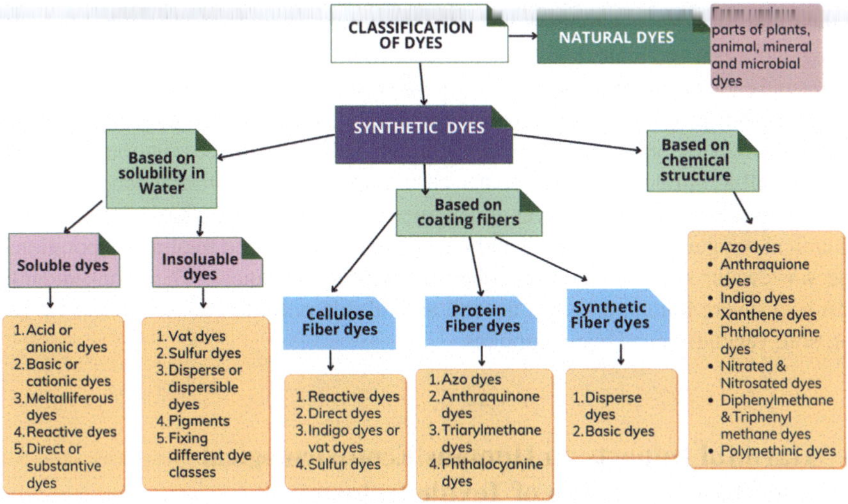

Fig. 1 Classification of textile dyes

hydrogen, hydrophobic and molecules of the Van der Waals type increase in the bonding of dyes to fabric threads. The way a dye interacts with a fiber depends on its kind, chemical content and nature. Further electrostatic interaction enhances the binding between the dye and the fiber inside the complex when they have opposite charges [35, 52]. Based on the composition of the fabric fibers to be dyed, it provides a relative evaluation of the rates of fixation and rejection of various dyestuffs [33, 53]. Furthermore, it appears that a significant amount of the dyes are released together with the effluents, which are frequently discharged into streams without being treated. This negatively impacts the aquatic environment. Because dye has minimal affinity for the cloth surfaces that need to be colored or dyed, a lot of dye is wasted. This necessitates the use of a better regeneration strategy or method (Fig. 1).

5 Characteristics of Dye Containing Textile Industry Wastewater

The majority of synthetic dyes are derived from petrochemical compounds and they are commercially available in various forms such as liquid, powder, granules and paste. They are very potential in fast and uniform coloring of different fabric materials and large range of coloration, ease of manipulation, stability against various external conditions and economically energy consuming. The large amount of chemicals and heavy metals when release into the environment in partially treated or untreated form can cause pollution such as water pollution, air pollution and soil pollution. The majority of synthetic dyes have serious negative impact on environment. Textile industries utilize large amount of water for the processes includes dying, fixing, softening and washing. In the dying process, 15% of the applied color was washed out in water and ended up as effluent [18]. The massive quantity of TIs effluent inappropriate discharge is repeatedly rejected, because a significant amount of industrial effluents are really high in organic and inorganic pollutants like chlorinated compounds, heavy metals, sulfur, nitrates, naphthol, soaps, chromium derivatives, formaldehyde, benzidine, sequestering agents and dyes and pigments. They also have a high biological and chemical oxygen demand (BOD and COD). Even after some treatment operations, the wastewater still contains a number of harmful substances. They consequently result in severe human ailments as well as multiple pollution consequences on the air, soil, vegetation and water supplies.

6 Harmful Impacts on Humans Ecotoxicological and Health Concern of Textile Industry

Industries were discharged untreated industrial effluents directly into water bodies posing serious ecotoxicological threats to organisms. Untreated effluent contains a wide range of toxic heavy metals such as mercury, chromium, cadmium, lead and arsenic. These chemical compounds degrade the natural quality of water by increasing biochemical and chemical oxygen demand in the ecosystem as well as potentially promoting toxicity to humans [54].

6.1 Harmful Impact on Ecosystem

The majority of textile businesses utilize azo dyes, which are released as untreated industrial effluent directly into water bodies. Because azo dyes block light, they have an adverse effect on aquatic life, including fish and predators as well as algae and aquatic plants. The aquatic environment's photic zone is created by these chemical substances, which also have harmful effects on aquatic flora and fauna, cytotoxic effects on plants, etc. This causes a number of health problems for people because of the food chain. When azo dye is present in waste water, fish experiences decreased growth, neurosensory damage, increased metabolic pressure and passing, while plants experience increased growth and productivity. Very high levels of minerals in the water, the presence and physiological accumulation of environmentally detectable pigments, and aquatic plants (algae, mosses, pteridophytes, spermatophytes and vascular plants) inhibit photosynthesis watercourses [55].

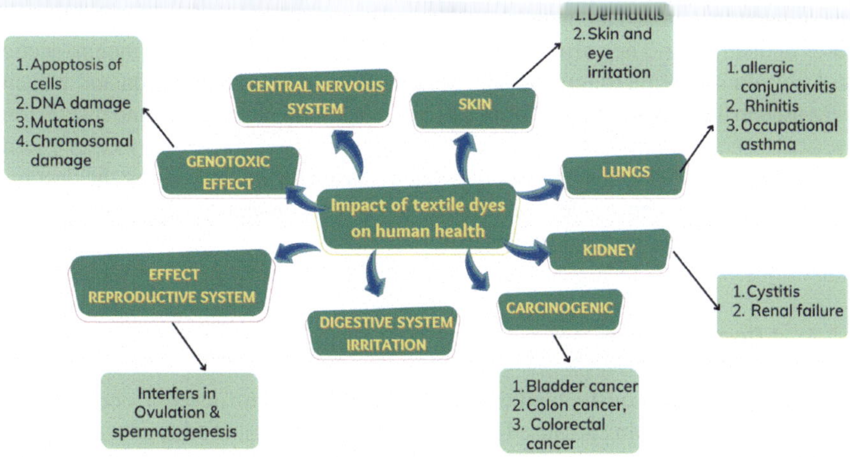

Fig. 2 Harmful impacts on humans health concern of textile dyes

6.2 Harmful Impact on Human Health

All forms of life are negatively influenced by the use of synthetic dyes [56]. Sulfur, naphthol, colors from vats, nitrates, acetic acid, soaps, enzymes and textile effluent are extremely toxic due to the heavy metals compounds containing chromium, copper, arsenic, lead, cadmium, mercury, nickel and cobalt as well as several auxiliary compounds [57]. Figure 2 shows the various health issues caused by untreated textile dye effluents on humans.

6.2.1 Effect on Skin

The skin's primary function is protection and it is the largest organ in the human body. The skin and its byproducts, such as hair, nails, perspiration gland and oil glands, make up the integumentary system. The body is protected by the skin from environmental threats such as diseases, toxins and extreme temperatures. Numerous metals are released throughout various phases of preparation in the textile industry. These metals, such as Cr, As, Cu, and Zn, are present in the effluents and can create many capable of creating a range of health problems, such as dermatitis, skin ulceration and hemorrhage [58]. Textile dyes can irritate the skin and eyes when consumed or inhaled directly or indirectly [59]. 1,4-diamino benzene is an amine used in textile industries are parent azo dyes can aggravate skin and cause contact dermatitis, cause chemosis, lacrimation, exophthalmos, permanently impair vision, cause rhabdomyolysis, cause severe cylindrical corruption supervene, cause acute and chronic gastritis, cause hypertension, cause vertigo, and upon ingestion, cause pulmonary edema of the face, neck, pharynx, tongue and larynx alongside respiratory trouble such as asthma and chronic obstructive pulmonary disease [60].

Dye-related irritations can include stinging eyes, scratchy skin, irritated or blocked noses and sniffling. They also contain formaldehyde-based resins, ammonia, acetic acid, certain shrink resistants, optical whiteners, soda ash, caustic soda and bleach were used in TIs. These are also known to exist in several reactive, vat and dispersion dyes cause negative impact on skin, e.g., triphenylmethane dyes, such as Basic Red 9 is carcinogenic to human because they release aromatic amines during the decomposition of the dye under anaerobic conditions, which can induce allergic dermatitis, skin irritation, mutations and cancer [61].

6.2.2 Effect on Lungs

When workers exposed to textile dyestuffs are known to develop occupational asthma. Painful breathing is caused by inhaling dye particles and other adverse effects include tingling, wet eyes, sniffling and asthmatic symptoms including wheezing and hacking. The dye particles are the most frequent risk associated with reactive dyes. In extreme situations, this can result in a dramatic reaction from the body the next time the person inhales the dye. They can sometimes impair a person's immune system and also cause itching, watery eyes, sneezing and asthmatic symptoms including coughing and wheezing are all signs of respiratory sensitization. Lung function decreases may be observed together with an intermittent cough. In addition to chest pain, textile workers also have cough and bronchitis. These symptoms most likely indicate different types of the bronchial irritation caused by inhaling dust [62]. Rhinitis and allergic conjunctivitis is an inflammatory reactions of the conjunctiva to an allergen of fabric dyes in the air. When a compound of human serum albumin and the reactive dye is generated, reactive dyes appear as an antigen and Immunoglobulin E (IgE). Antibodies have created that bind to histamine and results in inflammation. Employees who paintings with reactive dyes run the threat of contracting hypersensitivity illnesses consisting of touch dermatitis, allergic conjunctivitis, rhinitis and occupational asthma [63].

6.2.3 Impact on Kidney

Water dissolvable azo dyes become risky when processed by liver proteins. In humans, crystal violet or gentian violet is blue monochloride salt, which may result in chemical cystitis, irritation of the skin and digestive system, as well as respiratory and renal failure reported by Mani and Bharagava [64].

6.2.4 Carcinogenic to Human

In 1992, logical studies found that publicity to numerous sweet-smelling amines, especially benzidine, 2-naphthylamine and 4-aminobiphenyl, drastically increased the threat of bladder malignant development. Additionally, splenic sarcomas and hepatocarcinoma have been connected to fragrant amines which can be restrained within the EU (Report with the aid of using LGC [58]). Numerous research have proven a connection among fabric enterprise pollution and lots of cancers, consisting of breast, colorectal, bladder and lung cancers. More than one hundred of the 4000 dyes that have been assessed for toxicity are nonetheless to be had available in the marketplace and feature the ability to supply carcinogenic amines [65]. Many studies affirmed the nearness of most cancer-inflicting aromatic amines in azo dyes. Several organs, consisting of the vagina and bladder, have proven proof of reticular molecular sarcoma in affiliation with this extraordinarily carcinogenic chemical [54]. Disperse Yellow 7 had been destroyed in herbal our bodies of water and produced amines, that are regarded as carcinogenic retailers and Direct Blue 14, had been validated to broaden a carcinogenic amine while uncovered to bacterial species customary on human skin [66]. Basic Red 9 causes cancer. Long-time period publicity to those materials, however, is unfavorable to human health, growing the threat of colorectal most cancers, bladder most cancers and colon most cancers in addition to decreasing human immunity [67]. This certain dyes, Azo dyes, have the cap potential to motivate mutations and it became determined that intestinal microbes catalyze the enzymatic conversion of Sudan I, an azo-lipophilic dye used inside the fabric and meals sectors, into carcinogenic fragrant amines. According to Piątkowska et al. [68], each dye and its metabolites have the ability to commit most cancers. Benzidine-primarily based totally azo dyes and their derivatives have been notably studied for their toxicity, which has been associated with human bladder cancers broadly used in the fabric industries and paper and leather-based industries [67]. Disperse dyes that aren't ionized inside the aqueous nonionic dyes and azo, anthraquinone, simple and reactive dyes are the major cationic dyes. The maximum risky dye is artificial, made of materials that lead to most cancers, consisting of benzidine and different fragrant substituents [69]. The most toxic artificial natural dyes are the ones which can be azo dyes, in particular, diazo and cationic dyes, and those azo businesses have electron-retreating assets and motive electric deficits, those kinds of dyes in large part motive most cancers in human beings and animals.

6.2.5 Genotoxic Effect of Dyes

Reactive Green 19, Disperse Red 1 and Reactive Blue 2 are genotoxic to humans when exposed for an extended period of time. While Reactive Blue 2 and Disperse Red 1 were not proven to be genotoxic, Reactive Green 19 was found to be in a dose-dependent manner [70]. Triarylmethane dyes including Crystal Violet are poison mitosis, causing chromosomal damage and aberrant metaphase accumulation [64]. Methylene Blue(MB) is a cationic dye and Azure B are the predominant metabolite of MB, which could intercalate with the helical shape of DNA and duplex RNA [71]. Disperse Red 1 dye was also found to be mutagenic in human hepatoma (HepG2) cells and human lymphocytes and Disperse Orange 1 dye causes DNA damage by causing base-pair replacement and frameshift mutations that change the reading frame and due to its capacity to enhance the frequency of micronuclei, which is indicative of mutagenesis activity at the chromosome level [72].

6.2.6 Central Nervous System and Cellular Effect

Textile dyes were connected to the entirety from dermatitis to principal fearful gadget issues. Additionally, this dye has the capability to partition to the lipid membrane of animal cells, wherein it'd injure cells via way of means of serving as a powerful reversible inhibitor of the intracellular enzyme monoamine oxidase A, that's essential for human behavior [73]. Azure B's capability to dam glutathione reductase, a critical enzyme for keeping mobile redox balance [74] and Disperse Orange 1 dye has a cytotoxic effect and induces apoptosis in HepG2 cells while it comes into touch with the cells [75].

6.2.7 Impact on Reproductive System and Digestive System Irritation

According to Suryavathi et al. [76], workers of the textile industry are exposed to various potentially harmful compounds that could affect sperm production and ovulation. Amin et al. [77] reported that gut microbiome breakdown azo dyes into their parent amines, which are easily absorbed by intestine were found in both human and animal urine.

7 Effective Control Measures

Because of the desire for cheap labor on the international market, small-scale textile industries all over the world are in threat to cause major pollution in the environment such as air pollution, water pollution and soil pollution. Water pollution by textile industry effluent creates very serious issues in ecosystems and human health concern. The pollutants are removed by various approaches [78] shown in Fig. 3.

Remediation technology provides answers which are intrinsically related to a dedication to sustainable improvement in the face of this complicated issue, which poses extreme dangers to ecosystems and human beings worldwide. In other words, bioremediation technology sells each financial increase. This is environmentally suitable and advanced human well-being. Physical–chemical remedies are used to eliminate dye debris from the economic waste via way of means of adsorption coagulation and flocculation techniques. Biological remedy processes are eco-friendly. Microorganisms are worried about decolonization, degrading, detoxifying and mineralizing the chemical via way of means of biosorption manner [18]. It will lessen Chemical Oxygen Demand (COD), Biological Oxygen Demand (BOD), Total dissolved solid (TDS),Total suspended solids (TSS), Total natural carbon (TOC), turbidity and detoxify the artificial chemical compounds which might be gift withinside the

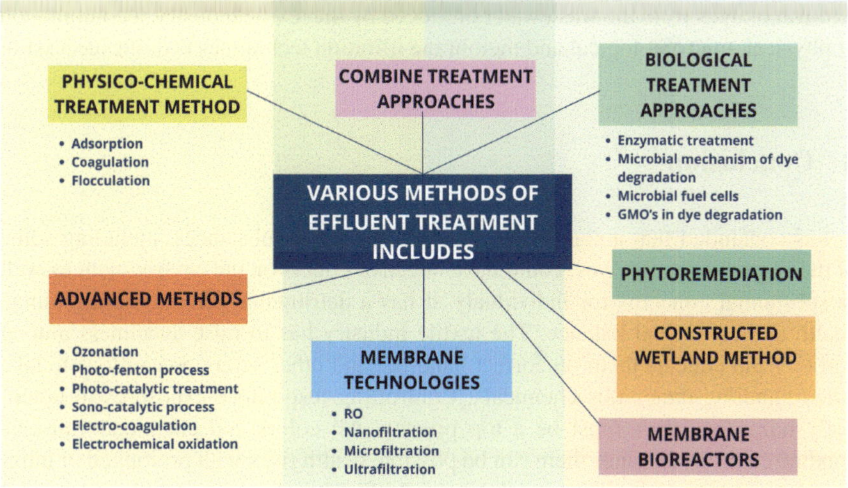

Fig. 3 Shows the various methods of effluent treatment

commercial effluent. Microorganisms along with *Pseudomonas* sp., *Serratia* sp., *Bacillus* sp., Algae, microalgae, fungi worried in organic remediation method. Microbial enzymatic techniques are maximum effective for the degradation and cleansing of artificial dyes via way of means of intracellular and extracellular enzymes along with Azoreductase, laccase, peroxidase and polyphenol oxidase. Halomonas and Bacillus have been capable of mystery azoreductase to reduce azodye in wastewater. Lactase enzymes have been synthesized via *Phomopsis* sp., which are actively degraded the crystal violet, methyl violet and cotton blue dyes. Other enzymes along with lignin peroxidase, veratryl alcohol oxidase, tyrosinase, NADH-DCIP reductase and polyphenol oxidase also are worried in fabric enterprise wastewater remedy. Consortium of diverse micro-organism, fungi, algae decolorize the dyes in wastewater via way of means of exclusive microbial mechanisms. Microbial gasoline cells (MFCs) method is a warm subject matter globally. MFCs are a brand new bioelectrochemical manner that pursuits to supply bicurrent via the usage of electrons derived from biochemical reactions catalyzed via way of means of microorganisms utilizing the pollution of wastewater containing minerals and heavy metals for their energy. Genetically Modified organisms are utilized in the bioremediation of commercial effluents. For instance, powerful remediation of reactive dyes along with reactive black, Congo red, BLack WON, removal of incredible blue R has been dealt with genetically changed *Pichia pastoris*. To effectively remove reactive dyes, the LacTT gene from *Thermus thermophilus* (SG0.5JP17-16) was taken and inserted into Pichia pastoris. The effluent treatment strategy also includes other techniques like phytoremediation, engineered wetland methods and cutting-edge approaches like ozonation, photo-fenton process, photocatalytic approach, sono catalysis, electrocoagulation and electrochemical oxidation. Effective chemical reduction techniques using membrane technologies include RO, nanofiltration and microfiltration. For the removal of dyes from the wastewater produced by the textile industry, a combination of physical, microbiological and membrane filtration techniques is quite successful.

8 Conclusion

It was concluded that a significant number of organic substances, including some of the synthetic textile dyes, could have negative effects on the environment as well as substantial concerns for individuals. It has a detrimental effect on both human health and ecological balance. The textile industry has to raise awareness among workers and educate them on correct handling and other safety precautions to take when handling dangerous chemicals. Controlling respiratory exposures to potentially hazardous dyes must be a top priority. All colors and chemicals must be handled carefully because there can be possible health risks with prolonged or unintentional overexposure. The complexity and difficulty of treating textile wastewater have increased, prompting a continuous research for new, efficient and practical approaches. However, no effective technique that can eliminate both the color and the harmful properties of the dyes released into the environment has been discovered up to this point. A combined method could lower the harmful elements in the wastewater from the textile sector.

References

1. Desore A, Narula SA (2018) An overview on corporate response towards sustainability issues in textile industry. Environ Dev Sustain 20(4):1439–1459
2. Kishor R, Purchase D, Saratale G, Saratale R, Ferreira LF, Bilal M, Chandra R, Bharagava R (2021) Ecotoxicological and health concerns of persistent coloring pollutants of textile industry wastewater and treatment approaches for environmental safety. J Environ Chem Eng 9:105012. https://doi.org/10.1016/j.jece.2020.105012
3. Mathur N, Bhatnagar P, Sharma P (2012) Review of the mutagenicity of textile dye products. Univ J Environ Res Technol 2(2):1–18. www.environmentaljournal.org
4. Bafana A, Devi SS, Chakrabarti T (2011) Azo dyes: past, present and the future. Environ Rev 19:350–370
5. Kirk-Othmer (2004) Encyclopedia of chemical technology, vol 7, 5th edn. Wiley-Interscience
6. Kumar A, Dixit U, Singh K, Gupta SP, Beg MSJ (2021) Structure and properties of dyes and pigments. In: Dyes and pigments—novel applications and waste treatment. IntechOpen. https://doi.org/10.5772/intechopen.97104
7. Bhatia D, Sharma NR, Singh J, Kanwar RS (2017) Biological methods for textile dye removal from wastewater: a review. Crit Rev Environ Sci Technol 47:1836–1876
8. Barathi S, Karthik C, Nadanasabapathi S, Padikasan IA (2019). Biodegradation of textile dye reactive blue 160 by bacillus firmus (Bacillaceae: Bacillales) and non-target toxicity screening of their degraded products. Toxicol Rep 7:16–22. https://doi.org/10.1016/j.toxrep.2019.11.017
9. Gita S, Hussan A, Choudhury TG (2017) Impact of textile dyes waste on aquatic environments and its treatment. Environ Ecol 35:2349–2353

10. Ghaly AE, Ananthashankar R, Alhattab M, Ramakrishnan VV (2014) Production, characterization and treatment of textile effluents: A critical review. J Chem Eng Process Technol 5:1–18
11. Thakur S, Chauhan MS (2018) Treatment of dye wastewater from textile industry by electrocoagulation and fenton oxidation: A review. In Proceedings of the Water Quality Management. In: Singh VP, Yadav S, Yadava RN (eds) Springer, Singapore, pp 117–129
12. Burkinshaw SM, Salihu G (2019) The role of auxiliaries in the immersion dyeing of textile fibres: Part 4 theoretical model to describe the role of liquor ratio in dyeing cellulosic fibres with direct dyes in the absence and presence of inorganic electrolyte. Dyes Pigm 161:565–580. ISSN 0143-7208. https://doi.org/10.1016/j.dyepig.2017.08.052; https://www.sciencedirect.com/science/article/pii/S0143720817316728
13. Paz A, Carballo J, Pérez MJ, Domínguez JM (2017) Biological treatment of model dyes and textile wastewaters. Chemosphere 181:168–177
14. Chowdhury MF, Khandaker S, Sarker F, Islam A, Rahman MT, Awual MR (2020) Current treatment technologies and mechanisms for removal of indigo carmine dyes from wastewater: A review. J Mol Liq 318:114061
15. Nguyen TA, Fu C-C, Juang R-S (2016) Effective removal of sulfur dyes from water by biosorption and subsequent immobilized laccase degradation on crosslinked chitosan beads. Chem Eng J 304:313–324
16. Khattab TA, Abdelrahman MS, Rehan M (2020) Textile dyeing industry: Environmental impacts and remediation. Environ Sci Pollut Res 27:3803–3818
17. Santos Pisoni DD, Marluza PDA, Cesar LP, Severo Rodembusch F, Campo LF (2013) Synthesis, photo physical study and BSA association of water-insoluble squaraine dyes. J Photochem Photob Chem 252:77–83
18. Slama H, Chenari Bouket A, Pourhassan Z, Alenezi F, Silini A, Cherif-Silini H, Oszako T, Luptakova L, Golinska P, Lassaad B (2021) Diversity of synthetic dyes from textile industries, discharge impacts and treatment methods. Appl Sci 11.https://doi.org/10.3390/app11146255
19. Wan Z, Li D, Jiao Y, Ouyang X, Chang L, Wang X (2017) Bifunctional MoS2 coated melamine-formaldehyde sponges for efficient oil–water separation and water-soluble dye removal. Appl Mater Today 9:551–559
20. Li W, Mu B, Yang Y (2019) Feasibility of industrial-scale treatment of dye wastewater via bio-adsorption technology. Bioresour Technol 277:157–170
21. Zhang L, Shao Q, Xu C (2019) Enhanced azo dye removal from wastewater by coupling sulfidated zero-valent iron with a chelator. J Clean Prod 213:753–761
22. Singh RL, Singh PK, Singh RP (2015) Enzymatic decolorization and degradation of azo dyes—A review. Int Biodeterior Biodegrad 104:21–31
23. Louati I, Elloumi-Mseddi J, Cheikhrouhou W, Hadrich B, Nasri M, Aifa S, Woodward S, Mechichi T (2020) Simultaneous cleanup of reactive black 5 and cadmium by a desert soil bacterium. Ecotoxicol Environ Saf 190:110103
24. Shahid M, Wertz J, Degano I, Aceto M, Khan MI, Quye A (2019) Analytical methods for determination of anthraquinone dyes in historical textiles: A review. Anal Chim Acta 1083:58–87
25. Novotný C, Dias N, Kapanen A, Malachov ˇ á K, Vándrovcová M, Itävaara M, Lima N (2006) Comparative use of bacterial, algal and protozoan tests to study toxicity of azo-and anthraquinone dyes. Chemosphere 63:1436–1442
26. Cao DJ, Wang JJ, Zhang Q, Wen YZ, Dong B, Liu RJ, Yang X, Geng G (2019) Biodegradation of triphenylmethane dye crystal violet by cedecea davisae. Spectrochim Acta Part A Mol Biomol Spectrosc 210:9–13
27. Silva MC, Corrêa AD, Amorim MT, Parpot P, Torres JA, Chagas PM (2012) Decolorization of the phthalocyanine dye reactive blue 21 by turnip peroxidase and assessment of its oxidation products. J Mol Catal B Enzym 77:9–14
28. Dinda GB,¸Sahin Z, Cengiz Yatmaz H,˙I¸sci Ü (2019) Cobalt phthalocyanine-TiO2 nanocomposites for photocatalytic remediation of textile dyes under visible light irradiation. J Porphyr Phthalocyanines 23:561–568

29. Cesa FS, Turra A, Baruque-Ramos J (2017) Synthetic fibers as microplastics in the marine environment: A review from textile perspective with a focus on domestic washings. Sci Total Environ 598:1116–1129
30. Almroth BM, Åström L, Roslund S, Petersson H, Johansson M, Persson NK (2018) Quantifying shedding of synthetic fibers from textiles; a source of microplastics released into the environment. Environ Sci Pollut Res 25:1191–1199
31. Berradi M, Hsissou R, Khudhair M, Assouag M, Cherkaoui O, El Bachiri A, El Harfi A (2019) Textile finishing dyes and their impact on aquatic environs. Heliyon 5(11):e02711. https://doi.org/10.1016/j.heliyon.2019.e02711
32. Yamagami A, Kawano K, Futaki S, Kuramochi K, Tsubaki K (2017) Syntheses and properties of second-generation V-shaped xanthene dyes with piperidino groups. Tetrahedron 73:7061–7066
33. Lim SL, Chu WL, Phang SM (2010) Use of Chlorella vulgaris for bioremediation of textile wastewater. Bioresour Technol 1:7314–7322
34. Wang X, Guan W, Pan J (2011) Preparation of magnetic imprinted polymer particles via microwave heating initiated polymerization for selective enrichment of 2-amino-4-nitrophenol from aqueous solution. Chem Eng J 178:85–92
35. Mijin DZ, Avramov ML, Antonije I, Onjia E, Grgur BN (2012) Decolorization of textile dye CI basic yellow 28 with electrochemically generated active chlorine. Chem Eng J 204–206:151–157
36. Webster S, Fu J, Padilha LA, Przhonska OA, Hagan DJ, Van Stryland EW, Bondar MV, Slominsky YL, Kachkovski AD (2008) Comparison of nonlinear absorption in three similar dyes: polymethine, squaraine and tetra one. Chem Phys 348:143–151
37. Chakraborty JN (2010) Dye-fibre interaction. In: Fundamentals and practices in colouration of textiles. pp 20–28
38. Siddiqua UH, Shaukat Z, Munawar I, Tanveer H (2017) Relationship between structures and dyeing properties of reactive dyes for cotton dyeing. J Mol Liq 241:839–844
39. Alkan M, Celikcapa S, Demirbasx OZ, Dogan M (2005) Removal of reactive blue 221 and acid blue 62 anionic dyes from aqueous solutions by sepiolite. Dyes Pigm 65:251–259
40. Wan T, Li D, Jiao Y, Ouyang X, Chang L, Wang X (2017) Bifunctional MoS_2 coated melamine-formaldehyde sponges for efficient oil-water separation and water-soluble dye removal. Appl Mat Tod 9:551–559
41. Umukoro EH, Peleyeju MG, Ngila JC, Arotiba OA (2016) Photocatalytic degradation of acid blue 74 in water using Ag-Ag_2O-ZnO nanostructures anchored on grapheme oxide. Solid State Sci 51:66–73
42. Puttaswamy Vinod KN, Ninge Gowda KN (2008) Oxidation of C.I. acid red 27 by chloramine-T in perchloric acid medium: spectrophotometric, kinetic and mechanistic approaches. Dyes Pigm 78:131–138
43. Iyim TB, Güçlü G (2009) Removal of basic dyes from aqueous solutions using natural clay. Desalination 249:1377–1379
44. Arun Prasad AS, Satyanarayana SV, Bhaskara Rao KV (2013) Biotransformation of direct blue 1 by a moderately halophilicbacterium marinobacter sp. strain HBRA and toxicity assessment of degraded metabolites. J Hazard Mater 262:674–684
45. Hassaan MA, El Nemr A, Madkour FF (2017) Testing the advanced oxidation processes on the degradation of direct blue 86 dye in wastewater, Egypt. J Aqua Res 43(1):11–19
46. Burkinshaw SM, Son YA (2010) The dyeing of supermicrofibre nylon with acid and vat dyes. Dyes Pigm 87:132–138
47. Chaari I, Feki M, Medhioub M, Bouzid J, Fakhfakh E, Jamoussi F (2009) Adsorption of a textile dye "indanthrene blue RS (C.I. vat blue 4)" from aqueous solutions onto smectite-rich clayey rock. J Hazard Mater 172:1623–1628
48. Zhang B, Dong X, Yu D, He J (2012) Stabilization mechanisms of C.I. disperses red 60 dispersions in the presence of its dye-polyether derivatives. Coll Surf Physicochem Eng Asp 405:65–72
49. Es-sahbany H, Berradi M, Nkhili S, Bassir D, Belfaquir M, El Youbi MS (2018) Valorization of Moroccan clay: application to the adsorption of cobalt ions contained in wastewater synthesized. Moroc J Chem 6(1):173–179

50. Hossain L, Kanti Sarker S, Samad Khan M (2018) Evaluation of present and future wastewater impacts of textile dyeing industries in Bangladesh. Envir Develop 26:23–33
51. Khattab TA, Rehan M, Hamouda T (2018) Smart textile framework: photochromic and fluorescent cellulosic fabric printed by strontium aluminate pigment. Carbohydr Polym 195:143–152
52. Rajasimman M, Babu SV, Rajamohan N (2017) Biodegradation of textile dyeing industry wastewater using modified anaerobic sequential batch reactor—start-up, parameter optimization and performance analysis. J Taiw Inst Chem Eng 72:171–181
53. Belpaire C, Reyns T, Geeraerts C, Van Loco J (2015) Toxic textile dyes accumulate in wild European eel Anguilla. Chemosphere 138:784–791
54. Lellis B, Fávaro-Polonio CZ, Pamphile JA, Polonio JC (2019) Effects of textile dyes on health and the environment and bioremediation potential of living organisms. Biotechnol Res Innov 3(2):275–290
55. Hossain MS, Das SC, Islam JM, Al Mamun MA, Khan MA (2018) Reuse of textile mill ETP sludge in environmental friendly bricks–effect of gamma radiation. Radiat Phys Chem 151:77–83
56. Al tohamy R, Ali S, Li F, Okasha K, Mahmoud Y, Elsamahy T, Jiao H, Fu Y, Sun J (2022) A critical review on the treatment of dye-containing wastewater: ecotoxicological and health concerns of textile dyes and possible remediation approaches for environmental safety. Ecotoxicol Environ Saf 231:113160. https://doi.org/10.1016/j.ecoenv.2021.113160
57. Manzoor J, Sharma M (2020) Impact of textile dyes on human health and environment. https://doi.org/10.4018/978-1-7998-0311-9.ch008
58. Report by LGC (1999) The risk of cancer caused by textiles and leather goods coloured with azo dyes. Presented, CSTEE plenary meeting, Brussels, Belgium
59. Clark M (2011) Handbook of textile and industrial dyeing: principles, processes and types of dyes. Elsevier
60. Puvaneswari N, Muthukrishnan J, Gunasekaran P (2006) Toxicity assessment and microbial degradation of azo dyes. Indian J Exp Biol 44:618626
61. Sivarajasekar N, Baskar R (2014) Adsorption of basic red 9 on activated waste Gossypium hirsutum seeds: process modeling, analysis and optimization using statistical design. J Ind Eng Chem 20(5):2699–2709
62. Bouhuys A, Schoenberg JB, Beck GJ, Schilling RSF (1977) Epidemiology of chronic lung disease in a cotton mill community. Lung 154:167–186
63. Hanger K (2003) Industrial dyes: chemistry, properties and applications, health and safety aspects
64. Mani S, Bharagava RN (2016) Exposure to crystal violet, its toxic, genotoxic and carcinogenic effects on environment and its degradation and detoxification for environmental safety. Rev Environ Contam Toxicol 237:71–104
65. Lacasse K, Baumann W (2012) Textile chemicals: environmental data and facts. Springer Science & Business Media
66. Balakrishnan VK, Shirin S, Aman AM, de Solla SR, Mathieu-Denoncourt J, Langlois VS (2016) Genotoxic and carcinogenic products arising from reductive transformations of the azo dye, Disperse Yellow 7. Chemosphere 146:206–215
67. Tounsadi H, Metarfi Y, Taleb M, El Rhazi K, Rais Z (2020) Impact of chemical substances used in textile industry on the employee's health: epidemiological study. Ecotoxicol Environ Saf 197:110594
68. Piątkowska M, Jedziniak P, Olejnik M, Zmudzki J, Posyniak A (2018) Absence of evidence or evidence of absence? A transfer and depletion study of Sudan I in eggs. Food Chem 239:598–602
69. Carmen Z, Daniela S (2012) Textile organic dyes—characteristics, polluting effects and separation/Elimination procedures from industrial effluents—a critical overview. In: Puzyn T, Mostrag-Szlichtyng A (eds) Organic pollutants ten years after the Stockholm convention—environmental and analytical update. IntechOpen. https://doi.org/10.5772/32373

70. Leme DM, Primo FL, Gobo GG, da Costa CRV, Tedesco AC, de Oliveira DP (2015) Genotoxicity assessment of reactive and disperse textile dyes using human dermal equivalent (3D cell culture system). J Toxicol Environ Health A 78(7):466–480
71. Haq I, Raj A (2018) Biodegradation of Azure-B dye by Serratia liquefaciens and its validation by phytotoxicity, genotoxicity and cytotoxicity studies. Chemosphere 196:58–68
72. Chequer FMD, Angeli JPF, Ferraz ERA, Tsuboy MS, Marcarini JC, Mantovani MS, de Oliveira DP (2009) The azo dyes Disperse Red 1 and Disperse Orange 1 increase the micronuclei frequencies in human lymphocytes and in HepG2 cells. Mutat Res Genet Toxicol Environ Mutagen 676(1–2):83–86
73. Li H, Zhang R, Tang L, Zhang J, Mao Z (2014) Evaluation of Bacillus sp. MZS10 for decolorizing Azure B dye and its decolorization mechanism. J Environ Sci 26(5):1125–1134
74. Couto N, Wood J, Barber J (2016) The role of glutathione reductase and related enzymes on cellular redox homoeostasis network. Free Radic Biol Med 95:27–42
75. Ferraz ER, Grando MD, Oliveira DP (2011) The azo dye Disperse Orange 1 induces DNA damage and cytotoxic effects but does not cause ecotoxic effects in Daphnia similis and Vibrio fischeri. J Hazard Mater 192(2):628–633
76. Suryavathi V, Sharma S, Sharma S, Saxena P, Pandey S, Grover R, Kumar S, Sharma KP (2005) Acute toxicity of textile dye wastewaters (untreated and treated) of Sanganer on male reproductive systems of albino rats and mice. Reprod Toxicol 19(4):547–556
77. Amin T, Afrin M, Haque Z, Islam MR (2016) Effects of textile dye waste water on reproductive system of mice and their progeny. Int J Sci Eng Res 7:565–570
78. Sardar M, Manna M, Maharana M, Sen S (2021) Remediation of dyes from industrial wastewater using low-cost adsorbents. https://doi.org/10.1007/978-3-030-474003_15

Synthesis and Characterization of Nanohybrid Materials

Mustapha Omenesa Idris, Mohamad Nasir Mohamad Ibrahim, Akil Ahmad, and Mohammed B. Alshammari

Abstract The combination of multiple nanomaterials has recently become the focus of materials development. This novel material class is known as nanohybrids. Recently, there has been a focus on the conjugation of two or more nanomaterials in order to achieve increased multifunctionality while also enabling for next generation materials with improved performance. The importance of studying novel nanohybrids with their multidimensional and complex behaviour is enormous, and there is a growing interest in their applications. With the increased production and potential use of nanohybrids comes concern about their properties and characteristics, which are responsible for their improved performance in a broader range of applications. Hence, this chapter session provides a comprehensive report on the various types of nanohybrids and their synthesis methods. Their broader applicability was discussed, with a focus on their use in water treatment. Furthermore, this chapter discusses the prospects of nanohybrids in material science.

Keywords Nanomaterial · Carbon–metal · Organic molecule-coated nanohybrids · Water treatment

1 Introduction

Recent advances in nanotechnology have centred on the development and application of multifunctional and superior nanohybrid materials. Nanomaterial development has progressed from single particle synthesis to multicomponent assemblies

M. O. Idris · M. N. M. Ibrahim
School of Chemical Sciences, Universiti Sains Malaysia, 11800 Pulau Pinang, Malaysia

M. O. Idris
Department of Pure and Industrial Chemistry, Kogi State University, P.M.B. 1008, Anyigba, Kogi State, Nigeria

A. Ahmad (✉) · M. B. Alshammari
Department of Chemistry, College of Science and Humanities in Al-Kharj, Prince Sattam Bin Abdulaziz University, Al-Kharj 11942, Saudi Arabia
e-mail: aj.ahmad@psau.edu.sa

© The Author(s), under exclusive license to Springer Nature Singapore Pte Ltd. 2023

A. Ahmad et al. (eds.), *Nanohybrid Materials for Treatment of Textiles Dyes*, Smart Nanomaterials Technology, https://doi.org/10.1007/978-981-99-3901-5_4

or hierarchical structures in which two or more pre-synthesized nanomaterials are conjugated to extract multifunctionality [1]. These groups are known as nanohybrids. Nanohybrid materials are defined as synthetic materials that contain organic or inorganic constituents that are covalently or non-covalently bonded together at the nanometer scale. In more detail, a nanohybrid is formed when multiple nanomaterials of varying dimensionality are conjugated by linking in terms of molecule or macromolecule. Furthermore, nanohybrids can form when one nanomaterial overcoats another with a distinct chemical identity and is engineered to chemically bind to nanomaterial surfaces, enhancing the existing multifunctional usage. However, this definition limits the material class to nanohybrids with unpredictable and unique toxicity levels, environmental fate, and transport [2, 3]. The underlying focus of nanohybrid synthesis is material property modification, which results in changes to inherent properties such as size, composition, structure, and surface chemistry. These changes also result in the emergence of novel emerging properties that are not observed during traditional nanomaterial evaluation [4, 5]. This new approach in nanohybrid synthesis and application thus presents novel challenges and necessitates thorough evaluation. The development of new materials is fraught with uncertainty about their possible environmental and biological consequences. Material release can occur during the manufacture and use of nano-laden products and devices [6]. Nanomaterials are transported and transformed in either occupational or environmental settings after release [7]. The material properties and the mode of release have a significant impact on such processes.

In general, the desire for multifunctionality has led to physical and chemical modifications to nanomaterials. To attain hierarchical and heterostructures, size and shape modulation, as well as physical or chemical functionalization, are used [8, 9]. Such functionalization has changed the inherent surface properties of nanoscale materials and extracted novel electronic configurations, intrinsic hydrophobicity, dissolution properties, and so on. The success of such distortions encouraged researchers to achieve a higher level of functionality by combining multiple nanomaterials, each with unique advantages [10]. Nanogold, nanoscale iron oxide, and graphene nanosheets, for example, each have plasmon resonance, para-magnetism, and superior charge carrying capability. However, as seen in the development of the first sets of bimetallic nanohybrids, careful combination of two or more of these materials has improved their functional performance [11]. Graphene nanosheets were combined with titanium dioxide (TiO_2) and zinc oxide (ZnO) nanoparticles to develop nanohybrid materials with improved properties and applications in fuel cells [12–14]. For example, it is estimated that by 2050, at least 10 million kg of platinum-modified titania multiwalled carbon nanotube (MWNT) nanohybrids will be utilized in fuel cells for automobiles alone, assuming 20% platinum by mass [11, 15]. Because of their underlying hydrophobicity and significant van der Waals interactions, carbonaceous nanomaterials such as fullerenes and carbon nanotubes have a high assemblage tendency, whereas metallic nanomaterials such as zinc oxide have unique dissolution and complex formation properties [16–18]. When combined, the behaviour of metal–carbonaceous conjugates can exhibit either dominant hydrophobicity or dissolution–complexation reactions, depending on the nature of the conjugation.

This chapter provides a comprehensive definition of nanohybrid materials, as well as types and synthesis routes for nanohybrid materials. Review of the literature on nanohybrids was extensively used and discussed. In addition, the synthesis of biochar-based nanohybrid materials was highlighted. The chapter highlighted nanohybridization as a potential method for improving material properties. This chapter described the characterization and application of nanohybrid materials in water treatment. Overall, this chapter will serve as a reference point in the development and advancement of nanohybrids, hopefully emphasising the importance of evaluating these nano-ensembles for improved material properties and applications.

2 Types and Synthesis of Nanohybrid Materials

Nanohybrids are classified into several types based on their parent materials. Carbon–carbon nanohybrids, carbon–metal nanohybrids, metal–metal nanohybrids, and organic molecule-coated nanohybrids are the four main types of nanohybrids. This type of classification is useful for identifying key properties of nanohybrids that are relevant to their safety, but it is not the only basis for classifying these nano-ensembles. However, nanohybrids can also be classified based on their origin in biomass material, such as biochar-based nanohybrids. Table 1 summarizes the various types of nanohybrids and their applications in previous studies.

2.1 Carbon–Carbon Nanohybrids

Carbon-based nanohybrids are made up of a combination of three major carbon nanostructures: zero-dimensional fullerenes, one-dimensional carbon nanotubes, and two-dimensional graphene and carbon nanohorns. Their open-ended hollow structures and fullerene cage-like structures provide unique advantages for producing endohedral and exohedral nanohybrids [32]. When fullerenes are encapsulated within nanohybrid structures via thermal annealing or deposition reactions, they are referred to as nano-peapods [33]. Similar synthesis processes, such as the water-assisted electric arc process, can result in the formation of an exotic multilayered hybrid fullerene structure known as the carbon nano-onion [34]. Exohedral conjugational structures and fullerenes, on the other hand, use long-range electrostatic or short-range specific interactions to drive the ensemble process, where conjugating molecules are covalently functionalized. These functionalization include oxidation of the structures to attach polar carboxyl or hydroxyl surface groups, as well as the attachment of chemically active polymeric assemblies [35, 36]. Fullerenes functionalized with porphyrin derivatives, for example, are refluxed with acidic-treated carbon nanotubes–COOH suspensions to produce fullerene–carbon nanotube nanohybrids via a reaction between the carbon nanotubes' carboxyl functionality and the amine functional group on the porphyrin molecules. Catalytic reaction processes involving

Table 1 Major classification of nanohybrids with their application in past studies

Types of nanohybrids	Specifications	Applications	Reference
Carbon–carbon nanohybrids	Fullerene–CNT peapods	Field-effect transistors	[19]
	Graphene oxide–CNT peapods	Supercapacitors	[20]
	Graphene–CNT hybrid	Transistors	[21]
	Carbon nano-onions	Lithium-ion batteries	[22]
Carbon–metal nanohybrids	Graphene–ZnO and graphene–TiO_2 hybrid	Fuel cells	[23]
	Graphene nanosheet/metal nitride hybrid	Field-effect transistors	[24]
	ZnO–graphene quantum dots	Optical limiting devices	[25]
Metal–metal nanohybrids	ZnO–Au nanohybrid	Lithium-ion batteries/ storage	[26]
	Ag/TiO_2 nanowire	Conductive films	[27]
	Pd–Cu nanohybrids	Fuel cells	[28]
Organic molecule-coated nanohybrids	Graphene–organic molecule hybrid	Semiconductor transistor	[29]
	Fullerene/CNT/ and porphyrins/phthalocyanines	Transparent conductor	[30]
	Oligothiophene–graphene	Fuel cells	[31]

vapour phase reactant molecules are typically used to achieve seamless exohedral bonding between carbon nanotubes and graphene or fullerene via covalent modification. Furthermore, by ionic and non-covalently bonded interactions, other methods of these graphitic nanomaterials can produce layered assemblies of nanohybrid-based thin layers of films [37].

2.2 *Carbon–Metal Nanohybrids*

Carbon–metal nanohybrids are synthesized by combining carbon nanomaterials with various metals or metallic oxide nanomaterials. They include assemblies containing a wide range of metallic nanomaterials, including noble metals such as Pt, Au, Ag, Pd, Rh, and others, as well as lanthanide series metals, metal oxide nanomaterials such as ZnO, TiO_2 nanoparticles, semiconducting quantum dots, and ligand-based metallic compounds [38]. Carbon–metal nanohybrids can be synthesized following four key pathways—they include (1) filling the inner cavities of carbon-based nanomaterials with metal-based nanomaterials using any of vapour deposition, thermal annealing, arc discharge, or wet chemical approach, (2) liking metal nanomaterials onto carbon nanotube surfaces functionalized with pyrene derivatives, (3) sol–gel and hydrothermal decoration approach of carbon nanomaterial with metal nanomaterials, (4) in-situ growth of metal-based nanomaterial on carbon nanomaterial surfaces by electrochemical deposition and redox reactions [39]. Combinations of graphene-based and metallic nanoparticles yield unique and improved general material properties for enhanced application in a variety of fields. The stability property of carbon-based nanoparticles with greater active surface area is particularly promising in the development of high performance and durable Li-ion storage units [40]. Similarly, when ZnO or TiO_2 are conjugated with carbon-based nanomaterials, their antibacterial properties are enhanced, allowing them to be used in detoxification applications [41]. Endohedral metallofullerenes, on the other hand, have great potential with extremely high-water reflexivity when encapsulated inside carbon nanohybrids [42]. Such a diverse set of applications has spurred significant research efforts in nanohybrid development and applications.

2.3 *Metal–Metal Nanohybrids*

Metal–metal nanohybrids are formed when metallic nanomaterials are conjugated to form multimetallic ensembles. The synthesis processes used to create conjugated metallic nanomaterials are determined by the desired structural properties, hybridity, and applications. The most used synthesis techniques are wet chemical processes such as solvothermal, epitaxial growth, ion implantation, and sol–gel methods that involve the thermal decomposition of metal salts [43]. Core–shell nanostructures can be created through sequential reduction, in which a previously formed metal nanomaterial acts as a "seed" for the subsequent growth of another nanomaterial with a different chemical root. To achieve patterned growth, optical lithography is also combined with common methods [44]. Organic or inorganic spacers can be found between core–shell metallic layers. Green synthesis methods for metal–metal nanohybrids have also been developed, with natural extracts such as fruit wastes serving as solvents [45]. Property synergies enable their use in a wide range of applications. Co-axial Ag–TiO_2 core–shell nanowire arrays, for example, with high specific surface area and rapid electron transport, can improve electron collection efficiency for use in dye-sensitized solar cells [46]. Bio-applications based on plasmonic, semiconducting, and magnetic metal–metal nanomaterials have been investigated. Combining the plasmonic properties of Ag and Au results in high efficiency localized surface plasmon resonance and surface enhanced Raman scattering. When semiconducting quantum dots are combined with plasmonic particles, their photoluminescent properties are enhanced and can be used for bioimaging microscopy [47]. Combining TiO_2, ZnO, or Au with other metal nanomaterials has been shown to improve photocatalytic activities and bandgap modulation, making them ideal candidates for organic pollutant degradation [48–50]. Such diverse applications, particularly in biomedicine, increase the material relevance of metal–metal nanohybrids.

2.4 Organic Molecule-Coated Nanohybrids

A substantial volume of literature identifies nanohybrids as carbon-based, meta-based, or polymeric nanomaterials coated with organic molecules. Nanohybrids are another name for hierarchical thin film layers. Organic molecule-coated nanohybrids can be synthesized using a variety of techniques, including physiosorption of organic molecules, electrochemical protein immobilization, polymer grafting to nanomaterial surfaces, ion exchange, and emulsification techniques. Coated nanomaterials are being studied in a variety of applications [51]. Carbon nanomaterials are surface functionalized with phthalocyanine and other molecules to improve charge transfer efficiency in electrochemical applications. Magnetic or plasmonic particles are similarly grafted or coated with organic polymers to improve their solubility and material properties. Metal-based nanomaterials are also linked to organic fluorophores to improve tagging and contrast. Because most of these materials appear to be simply coated nanomaterials, they may not necessitate systematic evaluation for accurate risk assessment. The physisorbed coatings have already been studied in terms of environmental fate and toxicology. Citrate, gum arabic, copolymers, and other additives are commonly adsorbed onto nanomaterials to improve dispersion in a desired solvent [52]. Surface modification approaches for nanomaterials, on the other hand, are carried out with rather complex heterocyclic structures that are covalently bound to the nanomaterial surfaces [53]. Chemically bound coatings of this type will alter nano-EHS behaviour; additionally, conformational differences of organic molecules present on the nanomaterial surface are known to exhibit distinct fate, transformation, and toxicity behaviour [54]. Thus, systematic evaluation of the nano-EHS behaviour of these complex chemically coated nanohybrid is required.

2.5 Biochar-Based Nanohybrid Material

Biochar is a well-known carbon material with a wide range of functions and excellent physicochemical properties, as well as a high wastewater treatment potential. The chemical composition and structure of biochar vary greatly [55]. Various approaches have been developed to improve the key functions of biochar with enhanced surface area, pore sizes, and functional groups of biochar by modifying the material and overcoming some of biochar's limitations. Different biochar-based nanohybrid materials are combined with various modification approaches, such as biomass pre-treatment and biochar chemical treatment. Magnetic biochar nanohybrids, hydroxide biochar nanohybrids, generally functionalized biochar, and nanometallic oxides are well-known and have been engineered to provide desired functionalities [56]. Table 2 summarizes several recent studies on the advancement of biochar-based nanohybrids and the techniques used to develop them.

Table 2 Previous studies on the advancement of biochar-based nanohybrids and the techniques

Biomass	Pyrolysis temperature (°C)	Nanomaterials used	Applied techniques	Reference
Brown marine microalgae	600	Fe_3O_4	One-step electromagnetization	[64]
Baobab fruit epicarp	700	ZnO	Biomass impregnation in $ZnCl_2$ for 2 h, with pyrolysis	[65]
Paper and wheat husk	500	TiO_2	Sol–gel method with $Ti(OBu)_4$ precursor	[66]
Pulp and paper sludge	750	Fe_2O_3	Biomass + $FeCl_3.6H_2O$ (80%, w/v) 1:3, with pyrolysis	[57]
Wheat husk and paper	500	ZrO_2	Sol–gel method with Zr $(OC_3H_7)_4$ precursor	[67]
Rice husk	700	Fe	Incipient saturation method	[68]
Paper waste and wheat husk	500	CeO_2–H	Biochar dilution with CeO_2 in acetone (CeO_2 loading)	[69]
Jute fibres	700	ZnO	Pyrolysis of jute fibres with $Zn(OAc)_2{\cdot}2H_2O$	[70]
Beetle killed pines	600	$CaWO_4$	Co-precipitating $CaCl_2$/sodium tungstate with biomass	[71]
Wakame (*undaria*)	800	$NiCl_2$	Dipping technique with KOH as the activating agent	[72]
Spirogyra alga	400	La/Cu/Zr Tri-metallic	Greener microwave method	[73]
Banana peels	180	Silver nanoparticles	Biochar in Ag nanoparticle and ultrasonication	[74]
Hickory wood chip	600	CuO	Ball-milling technique	[75]

There are two approaches to producing a magnetic biochar-based nanohybrid: post-loading of magnetic matter and biomass pre-treatment. This framework technique is based on recent research. The synthetic co-precipitation of Fe^{3+}/Fe^{2+} with the biomass is an example of biomass pre-treatment [57]. This method for developing magnetic biochar-based nanohybrids uses little energy and results in convenient pyrolysis [58, 59]. The obtained nanohybrids have a highly evolved porous structure with numerous advantageous properties. Magnetic biochar-based nanohybrids have shown to be superior in terms of crystal morphology and size. Furthermore,

Fe-modified biochar nanohybrid can be prepared using a simple co-precipitation method [60]. Biochar has been coupled with nanometallic oxides or hydroxide to enhance the material's functionality. This strategy of the nanometallic oxides/ hydroxide biochar nanohybrid can yield efficient biochar-based material. Similarly, two synthetic routes are used: post-pyrolysis immobilization of nanometallic oxide/ hydroxide and pre-treatment to synthesize the nanometallic oxides nanohybrid with biochar [61]. Electrochemical modification has also resulted in the development of biochar containing MgO nanohybrid (PE-MgO/biochar) with novel porosity [62]. The biochar-based nanohybrid was created by using a marine macroalgal feedstock as the host material and $MgCl_2$ as an electrolyte. A sol–gel method was used to immobilize the nanometal-based hydroxide/oxide onto biochar. The procedure entails producing novel TiO_2 supported on coconut shell biochar at a pyrolysis temperature of 450 °C [63]. The obtained nanohybrid demonstrated excellent decolorization efficiency under strong alkali and acidic conditions. Similarly, the advantages of ZrO_2 and biochar were combined to create the new functionalized biochar nanohybrid. According to the previously mentioned report, the modified sol–gel method was used to synthesize ZrO_2 nanoparticles on biochar.

3 Characterization of Nanohybrid Materials

A systematic assessment of nanomaterial toxicity necessitates extensive physicochemical characterization. Most of the characterization techniques used for single nanomaterials are also relevant to nanohybrids [32, 76]. There are several key physicochemical parameters that are important in evaluating material properties. Several spectroscopic and scanning techniques are used to characterize these parameters, including UV–Visible, atomic absorption spectroscopy (AAS), Raman spectroscopy, X-ray photoelectron spectroscopy (XPS), Fourier transformed infrared (FTIR), energy dispersive X-ray (EDX), scanning electron microscope (SEM), and transmission electron microscope (TEM). Hybridization, on the other hand, is likely to change some of these inherent properties that are not typically manifested by singular nanomaterials. As a result, traditional characterization techniques used for single nanomaterial characterization must be appropriately adjusted to monitor the altered properties of these conjugated nanohybrids. In general, two distinct concept approaches are distinguished in nanohybrid characterization [77]: the first refers to methods for studying the morphology and elemental composition of nanohybrid-based materials, such as SEM coupled with EDX, TEM, XPS, and so on. The second section is concerned with quantitative methods such as AAS, flame emission, and inductively coupled plasma atomic emission spectroscopy (ICP-AES). It should be noted that the analytical tools seek to establish multifaceted characterization of nanohybrids across elemental analysis. Characterizations such as contact angle measurements, FTIR, and Raman spectroscopy are used for nano-objects less frequently. Interestingly, the characterization methods aim to describe nano-based samples and critically visualize them for various application purposes [78].

4 Application of Nanohybrid Material in Water Treatment

Water treatment with nanohybrids has already progressed from concept to model compound survey. Adsorption, catalytic degradation, and membrane separation of toxic pollutants have all been successfully demonstrated using nanohybrids. Adsorption is a simple but effective method for removing toxic pollutants from contaminated water [79, 80]. Toxic contaminants in wastewater can thus be removed by contacting the effluent with an appropriate nanohybrid-based adsorbent. When compared to conventional activated carbon, chitosan, fly ash, biomass, and other adsorbents, nanohybrid-based adsorbents improve adsorption. The synergistic effect of covalent bonding between the parent and impregnated materials results in a connection of numerous functional groups on the surface of nanomaterial constituents of the nanohybrids and exhibits improved surface area, electrostatic repulsion, and hydrophilicity [81]. Several studies have shown that nanohybrid materials deliver a higher adsorption capacity than the individual nanomaterials [82]. For example, it was reported that using nanohybrid beads instead of pure nanomaterial resulted in improved pollutant adsorption in an experiment where the maximum adsorption capacity of 119.5 mg/g was obtained by using 2.5 g/L of nanohybrid adsorbent [83]. These findings suggest that nanohybrids have a higher adsorption capacity for toxic pollutant removal from water.

Furthermore, toxic organic pollutants in water can be efficiently degraded catalytically using nanostructure hybrid materials in an advanced oxidation process. Interestingly, the catalytic degradation has an advantage over adsorption in that it can decompose organic contaminants at very low concentrations without producing secondary pollutants [84]. Fu et al. [85] synthesized GO/Titania nanohybrids for phenolic pollutant degradation and reported that GO/Titania exhibits higher phenolic degradation than pristine Titania and GO. In general, nanohybrids increase the degradation rate, effectively treating the wastewater. Membrane technology is a popular option for wastewater treatment because of its single module and high purity of treated water at a low cost. The technology has some limitations, such as fouling, which reduces water flux. Because of their high surface area, mechanical and thermal stability, advanced materials such as nanohybrids can improve fouling reduction to a certain extent. The incorporation of nanohybrids into polymeric membranes significantly improves their anti-fouling properties by reducing surface roughness, increasing rejection efficiency, and boosting hydrophilicity [86]. Cheng et al. [87] impregnated a polymeric-based nanohybrid with an ultra-filtration membrane and discovered that the developed membrane outperformed a pure nanomaterial membrane in terms of water flux and effluent rejection. The formation of hydrogen bonds between both nanomaterials improves the nanohybrid membrane's permeability and anti-fouling resistance. An ideal nanohybrid membrane, in theory, provides nearly frictionless water flow and excellent retention. Water permeability, solute rejection, and mechanical strength can all be improved with nanohybrid membranes.

5 Conclusion

Materials science has progressed from the synthesis and functionalization of single nanomaterials to stratified configurations of more multifaceted nanohybrids. For many applications, multifunctionality is almost a requirement. As a result, nanohybrids represent the prospect of nanoscale materials with enhanced functions. This chapter explained nanohybrid types and synthesis, as well as the emergent properties that make it a better material class. Conjugation's novel properties are already changing fundamental physicochemical properties. This is plausible that the hybridization of nanomaterials will result in the introduction of novel properties that have not yet been characterized. It is important for the research community to develop open standards for accurate nanohybrid characterization and property assessments. There is a need for systematic studies to assess variations in such properties and to bridge our understanding of these properties as they relate to their applications, particularly in wastewater treatment. Because the production of such nanohybrids with emergent properties is imminent, it is also necessary to assess their biological behaviour. The next group of nanohybrids will have more complex hierarchical structures, with nanomaterials from more than two chemical roots conjugated. Thus, the ability to combine and functionalize nanomaterials in an infinite number of ways creates an almost intractable framework for evaluating their properties. In general, advanced nanohybrid materials improve performance in a variety of applications, such as water treatment when used as an adsorbent or as a catalyst in pollutant degradation.

References

1. Jo YK, Lee JM, Son S, Hwang S-J (2019) 2D inorganic nanosheet-based hybrid photocatalysts: design, applications, and perspectives. J Photochem Photobiol, C 40:150–190
2. Liu L-H, Métivier R, Wang S, Wang H (2012) Advanced nanohybrid materials: surface modification and applications. Hindawi
3. Saleh NB, Afrooz AN, Bisesi JH Jr, Aich N, Plazas-Tuttle J, Sabo-Attwood T (2014) Emergent properties and toxicological considerations for nanohybrid materials in aquatic systems. Nanomaterials 4:372–407
4. Yaqoob AA, Parveen T, Umar K, Mohamad Ibrahim MN (2020) Role of nanomaterials in the treatment of wastewater: a review. Water 12:495
5. Yaqoob AA, Ahmad H, Parveen T, Ahmad A, Oves M, Ismail IM, Qari HA, Umar K, Mohamad Ibrahim MN (2020) Recent advances in metal decorated nanomaterials and their various biological applications: a review. Front Chem 8:341
6. Klaine SJ, Alvarez PJ, Batley GE, Fernandes TF, Handy RD, Lyon DY, Mahendra S, McLaughlin MJ, Lead JR (2008) Nanomaterials in the environment: behavior, fate, bioavailability, and effects. Environ Toxicol Chem: Int J 27:1825–1851
7. Wiesner MR, Bottero J-Y (2011) A risk forecasting process for nanostructured materials, and nanomanufacturing. C R Phys 12:659–668
8. Gottardo S, Mech A, Drbohlavová J, Małyska A, Bøwadt S, Sintes JR, Rauscher H (2021) Towards safe and sustainable innovation in nanotechnology: state-of-play for smart nanomaterials. NanoImpact 21:100297

9. Karousis N, Tagmatarchis N, Tasis D (2010) Current progress on the chemical modification of carbon nanotubes. Chem Rev 110:5366–5397
10. Sumaila A, Ndamitso M, Iyaka Y, Abdulkareem A, Tijani J, Idris M (2020) Extraction and characterization of chitosan from crab shells: kinetic and thermodynamic studies of arsenic and copper adsorption from electroplating wastewater. Iraqi J Sci 2156–2171
11. Aich N, Plazas-Tuttle J, Lead JR, Saleh NB (2014) A critical review of nanohybrids: synthesis, applications and environmental implications. Environ Chem 11:609–623
12. Yaqoob AA, Ibrahim MNM, Yaakop AS, Umar K, Ahmad A (2021) Modified graphene oxide anode: a bioinspired waste material for bioremediation of Pb^{2+} with energy generation through microbial fuel cells. Chem Eng J 417:128052
13. Idris MO, Guerrero–Barajas C, Kim H-C, Yaqoob AA, Ibrahim MNM (2022) Scalability of biomass-derived graphene derivative materials as viable anode electrode for a commercialized microbial fuel cell: a systematic review. Chin J Chem Eng
14. Yaqoob AA, Mohamad Ibrahim MN, Umar K, Bhawani SA, Khan A, Asiri AM, Khan MR, Azam M, AlAmmari AM (2020) Cellulose derived graphene/polyaniline nanocomposite anode for energy generation and bioremediation of toxic metals via benthic microbial fuel cells. Polymers (Basel) 13
15. Rigdon WA, Sightler JJ, Larrabee D, McPherson E, Huang X (2013) Titania and carbon nanotube composite catalyst supports for durable electrocatalyst performance. ECS Trans 50:1681
16. Astefanei A, Núñez O, Galceran MT (2015) Characterisation and determination of fullerenes: a critical review. Anal Chim Acta 882:1–21
17. Smith SC, Rodrigues DF (2015) Carbon-based nanomaterials for removal of chemical and biological contaminants from water: a review of mechanisms and applications. Carbon 91:122–143
18. Ma R, Levard C, Michel FM, Brown GE Jr, Lowry GV (2013) Sulfidation mechanism for zinc oxide nanoparticles and the effect of sulfidation on their solubility. Environ Sci Technol 47:2527–2534
19. Li Y, Kaneko T, Hatakeyama R (2008) Electrical transport properties of fullerene peapods interacting with light. Nanotechnology 19:415201
20. Wu C, Huang X, Wu X, Xie L, Yang K, Jiang P (2013) Graphene oxide-encapsulated carbon nanotube hybrids for high dielectric performance nanocomposites with enhanced energy storage density. Nanoscale 5:3847–3855
21. Bol AA, Chandra B, Kasry A, Maarouf A, Martyna GJ, Tulevski GS (2015) Carbon nanotube-graphene hybrid transparent conductor and field effect transistor. Google Patents
22. Han F-D, Yao B, Bai Y-J (2011) Preparation of carbon nano-onions and their application as anode materials for rechargeable lithium-ion batteries. J Phys Chem C 115:8923–8927
23. Yaqoob AA, Ibrahim MNM, Umar K (2021) Biomass-derived composite anode electrode: synthesis, characterizations, and application in microbial fuel cells (MFCs). J Environ Chem Eng 9:106111
24. Chen J, Mao S, Lu G (2017) Graphene-based field-effect transistor biosensors. Google Patents
25. Son DI, Kwon BW, Park DH, Seo W-S, Yi Y, Angadi B, Lee C-L, Choi WK (2012) Emissive ZnO–graphene quantum dots for white-light-emitting diodes. Nat Nanotechnol 7:465–471
26. Ahmad M, Yingying S, Nisar A, Sun H, Shen W, Wei M, Zhu J (2011) Synthesis of hierarchical flower-like ZnO nanostructures and their functionalization by Au nanoparticles for improved photocatalytic and high performance Li-ion battery anodes. J Mater Chem 21:7723–7729
27. Kholmanov IN, Stoller MD, Edgeworth J, Lee WH, Li H, Lee J, Barnhart C, Potts JR, Piner R, Akinwande D (2012) Nanostructured hybrid transparent conductive films with antibacterial properties. ACS Nano 6:5157–5163
28. Xiong L, Huang Y-X, Liu X-W, Sheng G-P, Li W-W, Yu H-Q (2013) Three-dimensional bimetallic Pd–Cu nanodendrites with superior electrochemical performance for oxygen reduction reaction. Electrochim Acta 89:24–28
29. Ha T-J, Akinwande D, Dodabalapur A (2012) Hybrid graphene/organic semiconductor field-effect transistors. Appl Phys Lett 101:033309

30. Tung VC, Huang J-H, Tevis I, Kim F, Kim J, Chu C-W, Stupp SI, Huang J (2011) Surfactant-free water-processable photoconductive all-carbon composite. J Am Chem Soc 133:4940–4947
31. Liu Y, Zhou J, Zhang X, Liu Z, Wan X, Tian J, Wang T, Chen Y (2009) Synthesis, characterization and optical limiting property of covalently oligothiophene-functionalized graphene material. Carbon 47:3113–3121
32. Vizuete M, Barrejón M, Gómez-Escalonilla MJ, Langa F (2012) Endohedral and exohedral hybrids involving fullerenes and carbon nanotubes. Nanoscale 4:4370–4381
33. Devi S, Singh S, Kanojia S, Tripathi UK, Roy D, Prasad NE (2021) Dynamics of fullerenes confined in nanotube: temperature-modulated Raman scattering and X-ray diffraction studies, The. Eur Phys J D 75:1–8
34. Butsyk O, Olejnik P, Romero E, Plonska-Brzezinska ME (2019) Postsynthetic treatment of carbon nano-onions: surface modification by heteroatoms to enhance their capacitive and electrocatalytic properties. Carbon 147:90–104
35. Yu D, Dai L (2010) Self-assembled graphene/carbon nanotube hybrid films for supercapacitors. J Phys Chem Lett 1:467–470
36. Yaqoob AA, Serrà A, Ibrahim MNM, Yaakop AS (2021) Self-assembled oil palm biomass-derived modified graphene oxide anode: an efficient medium for energy transportation and bioremediating Cd (II) via microbial fuel cells. Arab J Chem 14:103121
37. Wang J, Wang Y, He D, Wu H, Wang H, Zhou P, Fu M (2012) Influence of polymer/fullerene-graphene structure on organic polymer solar devices. Integr Ferroelectr 137:1–9
38. Fu Y, Zeng G, Lai C, Huang D, Qin L, Yi H, Liu X, Zhang M, Li B, Liu S (2020) Hybrid architectures based on noble metals and carbon-based dots nanomaterials: a review of recent progress in synthesis and applications. Chem Eng J 399:125743
39. Wang D, Saleh NB, Sun W, Park CM, Shen C, Aich N, Peijnenburg WJ, Zhang W, Jin Y, Su C (2019) Next-generation multifunctional carbon–metal nanohybrids for energy and environmental applications. Environ Sci Technol 53:7265–7287
40. Tan Q, Kong Z, Chen X, Zhang L, Hu X, Mu M, Sun H, Shao X, Guan X, Gao M (2019) Synthesis of SnO_2/graphene composite anode materials for lithium-ion batteries. Appl Surf Sci 485:314–322
41. Nia ZK, Chen J-Y, Tang B, Yuan B, Wang X-G, Li J-L (2017) Optimizing the free radical content of graphene oxide by controlling its reduction. Carbon 116:703–712
42. Nan Y, Li B, Song X, Sano N (2019) Optimization of pore-opening condition in single-walled carbon nanohorns to achieve high capacity in double layer capacitor at high charge-discharge rate: critical effect of their hierarchical pore structures. Carbon 142:150–155
43. Zaleska-Medynska A, Marchelek M, Diak M, Grabowska E (2016) Noble metal-based bimetallic nanoparticles: the effect of the structure on the optical, catalytic and photocatalytic properties. Adv Coll Interface Sci 229:80–107
44. Zeng J-B, Cao Y-Y, Chen J-J, Wang X-D, Yu J-F, Yu B-B, Yan Z-F, Chen X (2014) Au@ Ag core/shell nanoparticles as colorimetric probes for cyanide sensing. Nanoscale 6:9939–9943
45. Shankar SS, Rai A, Ahmad A, Sastry M (2004) Rapid synthesis of Au, Ag, and bimetallic Au core–Ag shell nanoparticles using Neem (Azadirachta indica) leaf broth. J Colloid Interface Sci 275:496–502
46. Sun M, Fu W, Yang H, Sui Y, Zhao B, Yin G, Li Q, Zhao H, Zou G (2011) One-step synthesis of coaxial Ag/TiO_2 nanowire arrays on transparent conducting substrates: enhanced electron collection in dye-sensitized solar cells. Electrochem Commun 13:1324–1327
47. Khlebtsov NG, Dykman LA (2010) Optical properties and biomedical applications of plasmonic nanoparticles. J Quant Spectrosc Radiat Transfer 111:1–35
48. Dong H, Zeng G, Tang L, Fan C, Zhang C, He X, He Y (2015) An overview on limitations of TiO_2-based particles for photocatalytic degradation of organic pollutants and the corresponding countermeasures. Water Res 79:128–146
49. Idris MO, Kim H-C, Yaqoob AA, Ibrahim MNM (2022) Exploring the effectiveness of microbial fuel cell for the degradation of organic pollutants coupled with bio-energy generation. Sustain Energy Technol Assess 52:102183

50. Yaqoob AA, Ibrahim MNM, Umar K, Parveen T, Ahmad A, Lokhat D, Setapar SHM (2021) A glimpse into the microbial fuel cells for wastewater treatment with energy generation. Desalin Water Treat 214:379–389
51. Das S, Presselt M (2019) Progress and development in structural and optoelectronic tunability of supramolecular nonbonded fullerene assemblies. J Mater Chem C 7:6194–6216
52. Tejamaya M, Römer I, Merrifield RC, Lead JR (2012) Stability of citrate PVP, and PEG coated silver nanoparticles in ecotoxicology media. Environ Sci Technol 46:7011–7017
53. Wróbel D, Graja A (2011) Photoinduced electron transfer processes in fullerene–organic chromophore systems. Coord Chem Rev 255:2555–2577
54. Xie Y, Dong H, Zeng G, Tang L, Jiang Z, Zhang C, Deng J, Zhang L, Zhang Y (2017) The interactions between nanoscale zero-valent iron and microbes in the subsurface environment: a review. J Hazard Mater 321:390–407
55. Tan X, Liu Y, Zeng G, Wang X, Hu X, Gu Y, Yang Z (2015) Application of biochar for the removal of pollutants from aqueous solutions. Chemosphere 125:70–85
56. Zhao Y, Qamar SA, Qamar M, Bilal M, Iqbal HM (2021) Sustainable remediation of hazardous environmental pollutants using biochar-based nanohybrid materials. J Environ Manage 300:113762
57. Chaukura N, Murimba EC, Gwenzi W (2017) Synthesis, characterisation and methyl orange adsorption capacity of ferric oxide–biochar nano-composites derived from pulp and paper sludge. Appl Water Sci 7:2175–2186
58. Nizamuddin S, Baloch HA, Griffin GJ, Mubarak NM, Bhutto AW, Abro R, Mazari SA, Ali BS (2017) An overview of effect of process parameters on hydrothermal carbonization of biomass. Renew Sustain Energy Rev 73:1289–1299
59. Yaqoob AA, Serrà A, Bhawani SA, Ibrahim MNM, Khan A, Alorfi HS, Asiri AM, Hussein MA, Khan I, Umar K (2022) Utilizing biomass-based graphene oxide–polyaniline–Ag electrodes in microbial fuel cells to boost energy generation and heavy metal removal. Polymers 14:845
60. Kayan B, Kalderis D, Kulaksız E, Gözmen B (2017) Adsorption of malachite green on Fe-modified biochar: influencing factors and process optimization. Desalin Water Treat 74:383–394
61. Sun P, Hui C, Azim Khan R, Du J, Zhang Q, Zhao Y-H (2015) Efficient removal of crystal violet using Fe_3O_4-coated biochar: the role of the Fe_3O_4 nanoparticles and modeling study their adsorption behaviour. Sci Rep 5:1–12
62. Jung K-W, Ahn K-H (2016) Fabrication of porosity-enhanced MgO/biochar for removal of phosphate from aqueous solution: application of a novel combined electrochemical modification method. Biores Technol 200:1029–1032
63. Zhang S, Lu X (2018) Treatment of wastewater containing reactive brilliant blue KN-R using TiO_2/BC composite as heterogeneous photocatalyst and adsorbent. Chemosphere 206:777–783
64. Jung K-W, Choi BH, Jeong T-U, Ahn K-H (2016) Facile synthesis of magnetic biochar/Fe_3O_4 nanocomposites using electro-magnetization technique and its application on the removal of acid orange 7 from aqueous media. Biores Technol 220:672–676
65. Cederlund H, Börjesson E, Lundberg D, Stenström J (2016) Adsorption of pesticides with different chemical properties to a wood biochar treated with heat and iron. Water Air Soil Pollut 227:1–12
66. Khataee A, Kayan B, Gholami P, Kalderis D, Akay S (2017) Sonocatalytic degradation of an anthraquinone dye using TiO_2-biochar nanocomposite. Ultrason Sonochem 39:120–128
67. Khataee A, Kayan B, Gholami P, Kalderis D, Akay S, Dinpazhoh L (2017) Sonocatalytic degradation of reactive Yellow 39 using synthesized ZrO_2 nanoparticles on biochar. Ultrason Sonochem 39:540–549
68. Rubeena K, Reddy PHP, Laiju A, Nidheesh P (2018) Iron impregnated biochars as heterogeneous Fenton catalyst for the degradation of acid red 1 dye. J Environ Manage 226:320–328
69. Khataee A, Gholami P, Kalderis D, Pachatouridou E, Konsolakis M (2018) Preparation of novel CeO_2-biochar nanocomposite for sonocatalytic degradation of a textile dye. Ultrason Sonochem 41:503–513

70. Chen M, Bao C, Hu D, Jin X, Huang Q (2019) Facile and low-cost fabrication of ZnO/biochar nanocomposites from jute fibers for efficient and stable photodegradation of methylene blue dye. J Anal Appl Pyrol 139:319–332
71. Zhang L, Loh K-C, Zhang J (2019) Enhanced biogas production from anaerobic digestion of solid organic wastes: current status and prospects. Bioresource Technol Rep 5:280–296
72. Yao X, Ji L, Guo J, Ge S, Lu W, Cai L, Wang Y, Song W, Zhang H (2020) Magnetic activated biochar nanocomposites derived from wakame and its application in methylene blue adsorption. Biores Technol 302:122842
73. Sharma G, Bhogal S, Gupta VK, Agarwal S, Kumar A, Pathania D, Mola GT, Stadler FJ (2019) Algal biochar reinforced trimetallic nanocomposite as adsorptional/photocatalyst for remediation of malachite green from aqueous medium. J Mol Liq 275:499–509
74. Kaushal I, Saharan P, Kumar V, Sharma AK, Umar A (2019) Superb sono-adsorption and energy storage potential of multifunctional Ag-biochar composite. J Alloy Compd 785:240–249
75. Wei X, Wang X, Gao B, Zou W, Dong L (2020) Facile ball-milling synthesis of CuO/biochar nanocomposites for efficient removal of reactive red 120. ACS Omega 5:5748–5755
76. Yaqoob AA, Noor M, Serrà A, Mohamad Ibrahim MN (2020) Advances and challenges in developing efficient graphene oxide-based ZnO photocatalysts for dye photo-oxidation. Nanomaterials 10:932
77. Faucher S, Le Coustumer P, Lespes G (2019) Nanoanalytics: history, concepts, and specificities. Environ Sci Pollut Res 26:5267–5281
78. Silina YE, Gernaey KV, Semenova D, Iatsunskyi I (2020) Application of organic-inorganic hybrids in chemical analysis, bio-and environmental monitoring. Appl Sci 10:1458
79. Idris MO, Yaqoob AA, Ibrahim MNM, Ahmad A, Alshammari MB (2023) Introduction of adsorption techniques for heavy metals remediation. In: Emerging techniques for treatment of toxic metals from wastewater. Elsevier, pp 1–18
80. Kanwal A, Yaqoob AA, Siddique A, Ibrahim MNM, Ahmad A (2023) Polyethersulfone (PES) nanofiltration membrane for treatment of toxic metal contaminated water. In: Emerging techniques for treatment of toxic metals from wastewater. Elsevier, pp 319–341
81. Selvaraj M, Hai A, Banat F, Haija MA (2020) Application and prospects of carbon nanostructured materials in water treatment: a review. J Water Process Eng 33:100996
82. Yaqoob AA, Kanwal A, Ibrahim MNM, Ahmad A (2023) Application and fabrication of nanofiltration membrane for separation of metal ions from wastewater. In: Emerging techniques for treatment of toxic metals from wastewater. Elsevier, pp 365–398
83. Xu L, Wang Z, Ye S, Sui X (2017) Removal of p-chlorophenol from aqueous solutions by carbon nanotube hybrid polymer adsorbents. Chem Eng Res Des 123:76–83
84. Kumar A, Pandey G (2017) A review on the factors affecting the photocatalytic degradation of hazardous materials. Mater Sci Eng Int J 1:1–10
85. Fu C-C, Juang R-S, Huq MM, Hsieh C-T (2016) Enhanced adsorption and photodegradation of phenol in aqueous suspensions of titania/graphene oxide composite catalysts. J Taiwan Inst Chem Eng 67:338–345
86. Ursino C, Castro-Muñoz R, Drioli E, Gzara L, Albeirutty MH, Figoli A (2018) Progress of nanocomposite membranes for water treatment. Membranes 8:18
87. Chang X, Wang Z, Quan S, Xu Y, Jiang Z, Shao L (2014) Exploring the synergetic effects of graphene oxide (GO) and polyvinylpyrrodione (PVP) on poly (vinylylidenefluoride) (PVDF) ultrafiltration membrane performance. Appl Surf Sci 316:537–548

Graphene-Supported Nanohybrid Materials for Removal of Textile Dyes

Mustapha Omenesa Idris, Najwa Najihah Mohamad Daud, Mohamad Nasir Mohamad Ibrahim, and Abdulmumuni Sumaila

Abstract Through the activities of textile industries, there is an increasing release of dye-based contaminants into water bodies, which has eventually become one of the main sources of wastewater pollution in the environment, having a negative impact on human and aquatic life. Textile dye contamination in aqueous solution has been treated with a variety of methods and materials, including flocculation, ion exchange, adsorption techniques, membrane filtration and electrochemical methods, the use of low-cost adsorbents, biochar-based activated carbon, and nanocomposite materials. Graphene-supported nanohybrid materials have been utilized as adsorbents for dye removal from wastewater for over a decade. These materials were characterized using a variety of corresponding characterization tools, some of which are discussed in this chapter. Various nanomaterials and techniques have been reviewed, with a focus on textile dye treatment using GO-based nanocomposite. This chapter provides more insight on the use of graphene-supported nanohybrids for the treatment of textile dyes from dye-contaminated wastewater.

Keywords Nanohybrid · Graphene · Characterization techniques · Textile dyes · Wastewater treatment

1 Introduction

The rising trend in industrialization has significantly exacerbated the water pollution issue. The textile industry is the primary source of this problem because of its extensive use of organic dyes as coloring materials [1]. Because these dyes are resistant to degradation by physio-chemical treatments, the presence of color in wastewater is a major environmental concern. During the dyeing process, a portion of the dyes remain

M. O. Idris · N. N. M. Daud · M. N. M. Ibrahim (✉)
School of Chemical Sciences, Universiti Sains Malaysia, 11800 Gelugor, Penang, Malaysia
e-mail: mnm@usm.my

M. O. Idris · A. Sumaila
Department of Pure and Industrial Chemistry, Kogi State University, P.M.B. 1008, Anyigba, Kogi State, Nigeria

© The Author(s), under exclusive license to Springer Nature Singapore Pte Ltd. 2023
A. Ahmad et al. (eds.), *Nanohybrid Materials for Treatment of Textiles Dyes*,
Smart Nanomaterials Technology, https://doi.org/10.1007/978-981-99-3901-5_5

unattached to the fabrics and are washed out. Unfixed dyes have been found in high concentrations in textile effluents. These effluents contain a high concentration of dyes and chemicals, some of which are non-biodegradable and carcinogenic, posing a serious threat to human health and the environment if not treated appropriately [2, 3]. As a result, in order to mitigate the negative effects on living beings, treatment of these pollutants from contaminating water has become mandatory before discarding or applying for useful purposes.

To treat these effluents, a variety of treatment methods and materials have been used. These includes flocculation, coagulation, sedimentation and gravity separation, advanced oxidation, aerobic activated sludge, photocatalytic degradation, reverse osmosis, electrolysis, and membrane filtration [4]. However, these treatments have not been found to be effective in removing all dyes and chemicals used in the industry. These effluents not only contain high concentrations of dyes, but also the chemicals used in the various stages of processing. They contain trace metals such as chromium, arsenic, copper, and zinc, which can cause a variety of health problems [5]. Other organic and microbial impurities have been discovered in textile effluents. Adsorption is regarded as one of the most important tools for the treatment of organic and inorganic pollutants from textile wastewaters due to its low cost and ease processes [6]. However, adsorption has some drawbacks, such as adsorbents obtained in powder form not being suitable for column operation or thermal and chemical stability [7]. Furthermore, single adsorbents cannot be used to remove and recover all types of pollutants. Miscellaneous adsorbents are used to treat various types of pollutants based on the presence of functional groups. Although the adsorption process has a number of limitations, it has been found to be the most cost-effective method [8].

Adsorption is the most widely used method due to its effectiveness and simplicity. Because of its high porosity and large surface area, carbon-based adsorbents are commonly used in the textile industry for dye removal [9]. However, due to high production costs, they became relatively expensive. Furthermore, regeneration of the carbon-based adsorbents necessitates a high-pressure stream, which adds to the operating costs of this treatment system [10]. This high cost has prompted the search for alternative dye removal adsorbents that are both economical and efficient. The introduction of nanotechnology and its application in various fields, including water treatment, has resulted in drastic changes in water technology over the last two decades [11, 12]. Several nanomaterials-based nanohybrids have been developed for the treatment and recovery of environmental pollutants. The have been extensively studied, and their high adsorption capacities raise the expectation that these materials will be useful in wastewater treatment [13–15]. This chapter deals with the removal of textile dyes using graphene-supported nanohybrids.

2 Nanohybrid Materials

Nanohybrid materials are synthetic materials composed of organic or inorganic constituents that are covalently or non-covalently bonded together at the nanometer scale. A nanohybrid is formed when multiple nanomaterials of varying dimensionality are conjugated via molecule or macromolecule linking [16]. Based on their parent materials, nanohybrids are classified into several types. The four major types of nanohybrids are carbon–carbon nanohybrids, carbon–metal nanohybrids, metal–metal nanohybrids, and organic molecule-coated nanohybrids. This classification is useful for identifying key properties of nanohybrids relevant to their safety, but it is not the only basis for classifying these nano-ensembles. Carbon–carbon nanohybrids are composed of three types of carbon nanostructures: zero-dimensional fullerenes, one-dimensional carbon nanotubes, and two-dimensional graphene and carbon nanohorns. Their open-ended hollow structures and fullerene cage-like structures offer distinct advantages in the production of endohedral and exohedral nanohybrids [17]. Carbon–metal nanohybrids are created by fusing carbon nanomaterials with different metals or metallic oxide nanomaterials. They include assemblies containing a variety of metallic nanomaterials, such as noble metals, lanthanide series metals, metal oxide nanomaterials, semiconducting quantum dots, and ligand-based metallic compounds [18]. When metallic nanomaterials are conjugated to form multi-metallic ensembles, metal–metal nanohybrids are formed. The desired structural properties, hybridity, and applications determine the synthesis processes used to create conjugated metallic nanomaterials. Wet chemical processes such as solvothermal, epitaxial growth, ion-implantation, and sol–gel methods that involve the thermal decomposition of metal salts are the most commonly used synthesis techniques [19]. A considerable volume of the previous research identifies nanohybrids as carbon-based, metal-based, or polymeric nanomaterials coated with organic molecules. Physisorption of organic molecules, electrochemical protein immobilization, polymer grafting to nanomaterial surfaces, ion exchange, and emulsification techniques can all be used to create organic molecule-coated nanohybrids. Coated nanomaterials are being researched for a wide range of applications [20]. Nanohybrid-based adsorbents have a high potential to adsorb pollutants because of their high surface area and porosity [21]. Furthermore, when compared to commercially available composite adsorbents, these nanohybrids have significant properties in terms of catalytic activity and high functionality. The high surface area of nano adsorbents contributes to more active sites, resulting in greater interaction of targeting species [22]. These nanomaterials cannot only treat pollutants, but they also allow engineering processes to use low-cost precursor material without leaching any harmful entities [23]. These nano adsorbents have considerable toxicants binding efficiency, and weary adsorbents can be effectively used again after chemical treatment which is the most endearing quality of nanohybrid adsorbent [24]. Thus, remediation of pollutants using nanohybrid adsorbent is regarded as an alternative strategy for wastewater treatment [25]. The nano adsorbents, which included graphene-supported hybrids and cellulose nanomaterials, performed admirably in terms of water pollution control [26].

3 Graphene-Supported Nanohybrids

Graphene, the most recently developed carbon-based nanomaterial, is organized in the form of two-dimensional sheets. The electromechanical and catalytic properties of graphene-based nanomaterials are outstanding [27]. However, graphene synthesis is expensive, so many efforts have been made to develop cost-effective materials for mass production, such as GO which can further be converted into rGO using various reduction processes [28]. A graphene layer composed of carbon atoms with sp2 hybridization is densely and regularly arranged to form a two-dimensional honeycomb structure [29]. Sigma bonds connect two equivalent sub carbon atom lattices in a hexagonal unit of graphene. Because each carbon atom has the ability to provide pi electrons to a delocalized channel of electrons, graphene is more stable than other nanosheets [30]. It has a large specific surface area and superior mechanical hardness, permeability to gases, a fusion of elevated three-dimensional aspect ratio, versatility, remarkable optical transmittance, exceptional electro-thermal conductivities, and other properties based on a delocalized graphene network. Because of its single-layered structure and two basal planes for pollutant adsorption, graphene has been discovered to be an effective alternative for CNT in the process of treating industrial dyes [31]. However, the inner structure of CNT is not attainable for pollutant adsorption [32]. Graphene-based nanohybrids were used to modify a variety of organic polymers with various functional groups. GO and rGO have also been synthesized by chemical exfoliation of graphite powder in the absence of a catalyst (metal-based). As a result of the lack of a metallic catalyst, no further purification is required. Because GO contains many oxygen-containing active sites (functional groups), no special acid treatment is required to improve its hydrophilicity [33]. The dye molecules have the potency to adsorb pollutants on the surface of GO sheets because of the existence of active sites on their surface, resulting in higher adsorption.

The graphene-supported materials undergo adsorption ability due to intermolecular forces. The potency of interaction of graphene-based materials with pollutants is determined by features such as pore volume and size, as well as specific surface area. Graphene materials performed admirably in removing industrial dyes in both pristine and hybrid forms. Figure 1 demonstrates the possible interactions occurring during adsorption of graphene-supported nanohybrids with textile dyes. GO is produced mainly using Hummer's process and is treated as a single layer of graphite oxide. To explain the structure of graphite oxide, various structural designs have been proposed [34]. According to Szabó et al. [35], the carbon network is made up of two types of regions: trans linked cyclohexane chairs and flat hexagonal ribbon forms with various functional groups. The occurrence of functional substituents accounted for GO's planar acidity. The oxygen substituents on the GO surface were reduced in rGO using advancement of reduction methods. Reduction promotes graphitic nature in GO by reducing defects caused by the addition of numerous functional moieties [36].

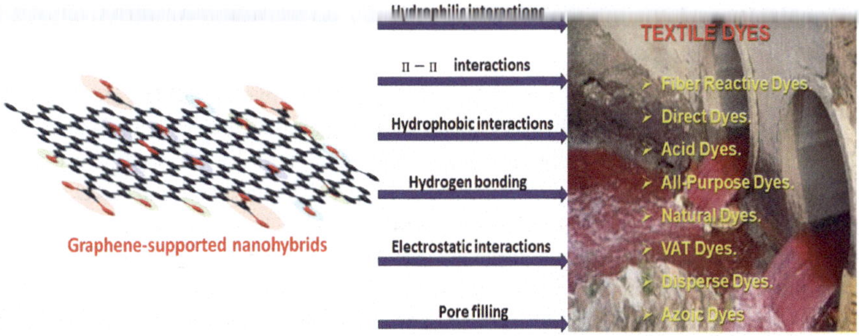

Fig. 1 Adsorption interactions for textile wastewater dye removal using graphene-based nanohybrids as adsorbent

4 Characterization Techniques

Several graphene-based material characterization tools for a wide range of applications have been reported. We focused on some of the spectroscopic and scanning techniques used to characterize these parameters in this chapter, such as UV–Visible, atomic absorption spectroscopy (AAS), Raman spectroscopy, X-ray photoelectron spectroscopy (XPS), Fourier transformed infrared (FTIR), energy dispersive x-ray (EDX), scanning electron microscope (SEM), and transmission electron microscope (TEM). Characterization techniques help to reveal many properties of the synthesized graphene-based materials, including surface morphology, functional groups, optical activities, elemental composition, and overall physical properties.

4.1 SEM and TEM

SEM and TEM are frequently used to visualize the morphology and structure of graphene-based materials, revealing a wealth of information about their surface roughness. Sometimes, SEM is used in conjunction with energy dispersive X-ray spectroscopy (EDX) to examine the surface morphology along with the elemental composition of a material using an electron beam with a maximum voltage of about 5 kV. The material is being heated to 80 °C for 10 min then cooled to 25 °C and allowed to crystallize for 6 h during the experiments [37]. Individual graphene morphologies revealed a smoother perimeter with a flat surface and notable topographical features [38–42].

4.2 XRD and Raman

X-ray diffraction (XRD) is a useful tool for examining the crystalline structure of graphene-based materials, which is an important factor in determining whether a material can be used as an adsorbent. The peak at 2 theta of 26° for pristine graphite and the peak at 2 theta of 11° for graphene oxide are the most common XRD results for graphene-based materials. XRD is commonly used to analyze the crystalline phase of graphene-based nanocomposites and to assess purity and crystalline size. XRD plays an important role in the characterization of graphene-based materials by confirming the reduction of GO to rGO. The use of XRD for phase determination and confirming the reduction of GO to rGO is the most convenient of all X-ray techniques [43]. Furthermore, Raman spectroscopy is a powerful tool for investigating the structure and quality of graphene-based materials. It is a powerful, quick, sensitive, and non-destructive analytical method that produces both qualitative and quantitative data. Raman spectroscopy can be used to investigate the material's bulk and surface structure, such as surface functional groups, chemical modification of the graphene matrix, number of layers, and structural damage caused by heteroatom doping. Graphene has six normal modes and two atoms per unit cell at the Brillouin zone center [44]. In Raman spectroscopy, the E2g phonons are active.

4.3 FTIR, Contact Angle, and UV–Vis

FTIR is one of the most effective and concise methods for determining residual functional groups. Synthesis involves oxidation steps, which are required for graphite oxide exfoliation. As a result, many functional groups may remain in graphene-based materials, affecting the material's properties significantly [45]. O–H stretch (3400 cm^{-1}), C–H stretch (2910 cm^{-1}), C=O stretch (1687–1710 cm^{-1}), C=C stretch (1542–1568 cm^{-1}), C–O stretch (1208 cm^{-1}), C–OH stretch (1113 cm^{-1}), C–O–H bend (1409 cm^{-1}), and C–H system stretch are major peaks in graphene-based material FTIR spectra (2875 cm^{-1}). Because of differences in preparation conditions and graphene supports, the positions of the characteristic peaks from the functional groups may shift slightly. Contact angle testing is a simpler type of characterization than most others. Several tests are carried out with different polar and dispersive components. The polar and dispersive surface energy components used in contact angle testing are water, diiodomethane, ethylene glycol, and glycerol. Several models for converting contact angle data to surface energy have been proposed, and their validity is still being debated. Kozbial et al. [46] calculated surface energy from contact angle data using the Fowkes model, the Owens–Wendt model, and the Neumann model. The fresh graphene-based material surface has significant polar components, according to the Fowkes and Owen-Wendt models, providing additional insights into the nature of the substrate's inherent mildly hydrophilic. Furthermore, UV–vis spectroscopy can also be used to confirm the successful synthesis of graphene-based materials. A graphitic structure has an absorption peak at 202 nm, whereas a monolayer of GO has an absorption peak at 230 nm, which is attributed to C–C bond Π-Π^* transitions. As a result, graphene-based product quality can be easily determined [47].

5 Treatment of Textile Dyes Using Graphene-Based Nanohybrids

Graphene-supported nanohybrids have been used in a variety of applications, including the removal of various textile dyes contained in industrial wastewaters [48]. The low cost of graphene-based nanomaterials is a significant advantage. Recently, there has been a plethora of studies on the fabrication of low-cost materials using waste. However, graphene-supported adsorbents improve physicochemical characteristics (possibility of pollutant diffusion inside graphene pores, porosity, and surface area) [49]. The addition of GO (1 wt%) enhanced the adsorption rate of nanohybrid materials significantly [50]. The presence of a large number of negatively charged oxygen groups is the principal cause for the maximum adsorption of GO. As a result, negatively charged GO is more probable to adsorb cationic dyes than anionic dyes. Adsorption of cationic dyes on GO produces excellent electrostatic attraction, whereas GO is not observed to be favorable toward anionic dyes due to electrostatic repulsion. Bradder et al. [51] used GO to investigate the removal rate of methylene blue (MB) as well as malachite green (MG) in aqueous system. MB and MG adsorption capacities on GO were determined to be 351.0 mg/g and 248.0 mg/g, respectively. Electrostatic attraction was proposed as the adsorption mechanism. Ionic strength and the ability to dissolve organic matter can also improve MB removal efficiency. For example, Liu et al. [52] investigated the removal of MB and methyl violet (MV) using 3D-GO sponge, and the material demonstrated considerable adsorption performance (397.0 mg/g and 467.0 mg/g) for MB and MV, respectively.

The energy levels of adsorption activation via robust Π-Π stacking and anion-cation interaction have been investigated [52]. In basic medium, carboxyl group deprotonation increased adsorption capacity, so GO performed better in terms of adsorption. Cationic dye adsorption on GO was also discovered to be effective at low temperatures and high pH. Ramesha et al. [53], on the other hand, conducted a comparative adsorption study from both cationic and anionic dyes on GO and obtained 95% adsorption efficiency for cationic dyes while anionic dyes had negligible removal efficiency. To conquer the electrostatic repulsion between GO and anionic dyes, GO must be modified or subjected to reduction. The reduction of GO to rGO enhanced anionic dye removal efficiency. Sun et al. [54] utilized an in situ reduction method to prepare GO with sodium hydrosulfite in order to increase the adsorption capacity of the cationic dye acridine orange. GO had a 1.4 g/g adsorption capacity, whereas rGO had a 3.3 g/g adsorption capacity. rGO had also been found to be effective in removing MG [55]. Several characterization tools were used to determine the interaction between MG and rGO. The electrostatic interaction of MG with the Π electron cloud, charged negative oxygen on rGO, and the Π-Π interaction of MG rings increases the large surface area of rGO, causing greater MG adsorption on rGO. The optimized adsorption efficiency of MG on rGO was 476.2 mg/g, with adsorption being endothermic and spontaneous.

A few GO-based nanohybrids were also created to remove dye from wastewater. Multiwalled carbon nanotube (MWCNT)-graphene hybrid aerogel was used to remove rhodamine B (RB), MB, and acid fuchsine [56]. All dye adsorbates with GO/c-MWCNT as adsorbent demonstrated a high adsorption potential. Kharismadewi et al. [57] developed a bio-compatible and thermo-responsive polymer and grafted it onto the GO surface to remove cationic dye (MB). The GO-based polymer forms a hydrogel in water, overcoming the issue with carbonaceous adsorbents. This adsorbent was produced in a green way by scattering polymerization in supercritical carbon dioxide. The impact of adsorbent dosage, dye concentration, contact time, pH, and recyclability on MB adsorption was studied. The Freundlich adsorption isotherm fits the data best, with a maximum adsorption capacity of 39.41 mg/g. Adsorption kinetics as well as thermodynamics were also explored; adsorption behavior is pseudo-second order and endothermic, with a significant adsorption at temperature of 45 °C. The adsorbent based on GO-polymer has the capacity to degrade 99.8% MB from wastewater in 45 min. Similarly, Nguyen-Phan et al. [58] developed a reduced graphene oxide/titanate nanohybrid with an enhanced surface area and pore volume for the removal of MB from wastewater. Other graphene-supported nanohybrids have been reported as dye adsorbents (MB and RB) and have proven excellent dye removal potentials [59].

A chitosan-modified GO nanohybrids were also used as an adsorbent to remove dyes from wastewater. The researchers investigated the efficiency of chitosan modified on GO and rGO for the removal of reactive red dye [60]. Various characterization tools were used to characterize the structural properties of the developed adsorbent. According to the Langmuir model, the maximum adsorption capacity is 32.16 mg/g. The proposed model in the adsorption kinetics was found to be a pseudo-second order model, implying that adsorption could be managed by a chemical rate—limiting step influenced by covalent forces. In a similar study earlier, with an initial RB5 concentration of 1.0 mg/mL, hydrophilic and bio-compatible 3D chitosan-graphene mesostructured were produced for reactive black 5 (RB5) removal and achieved a removal efficiency of 97.5% [61]. Alizarine Red was studied using GO as an adsorbent, and its adsorption efficiency was evaluated by comparing to bare graphite powder. The dye removal achieved a maximum uptake of 70.75 mg/g at very low pH. However, as the pH rises, the dye uptake decreases to 24.62 mg/g. As a result, the optimum pH for dye removal was discovered to be low. Because of the electrostatic interaction between carboxylic groups, lower pH ranges were found to be favorable for higher adsorption capacity. Therefore, a lower pH range was discovered to be favorable for anionic dye adsorption. At higher pH, on the other hand, the layer of GO sheets becomes negatively charged due to the presence of excess oxygenated groups, which promotes the uptake of cationic dye. Several magnetic GO-based nanocomposites with superparamagnetic properties were also synthesized for the removal of dyes in wastewater [62]. The material was used as an adsorbent for MB and neutral red adsorption. Adsorption equilibrium was reached in 0.5 h and 1.5 h for MB and neutral red, with adsorption capacities of 167.2 mg/g and 171.3 mg/g, respectively. Geng et al. [63] generated a responsive hybrid of rGO/ferric oxide nanocomposites and discovered that the hybrid material has a great and flexible adsorption capacity for various dye types. Similarly, rGO-supported with ferrite hybrids was prepared using a solvothermal method. When the hybrids' concentration was 0.6 g/L, it was discovered that 92% of RB and 100% of MB with just a concentration of 5 mg/L can be removed in a very brief time (2 min). Furthermore, hybrid materials demonstrated photocatalytic activity in the degradation of RB and MB. In synthetic aqueous dye, real industrial effluents, and lake water, the graphene-supported Fe_3O_4/rGO nanocomposites removed dye with an efficiency greater than 90% [64].

A team of scientists developed a more complex adsorbent by combining amino silane-powered GO with a high molecular weight copolymer. The nanocomposite was examined for the removal of cationic dyes from aqueous solution, including crystal violet and methyl blue. The maximum adsorption capacity for crystal violet intake was found to be 440.97 mg/g, while methyl blue was observed to be 416.06 mg/g. It was proposed that the adsorption was chemisorption, endothermic, and spontaneous [65]. Fly ash was used to make a GO-based adsorbent known as FCGO, which was then cross-linked with chitosan to remove Acid Red GR dye (ARG) and Cationic Red X-5 GN (CRX). The maximum adsorption capacities for CRX removal were 64.50 mg/g, while ARG removal yielded 38.87 mg/g. As expected, the ratio of removal of ARG with FCGO decreased as the pH increases in acidic medium; however, pH greater than 6 had no effect. As a result, the general hypothesis that lower initial pH improved removal efficiencies is supported [66]. An in situ polymerization method was used to power a GO with polymer for anionic dye removal [67]. The zeta voltage was changed from − 36.5 in the case of unmodified GO to 41.5 in the case of polymeric modified GO, which increased the adsorption capacities of anionic Orange G (OG) dye compared to the unmodified GO. According to the Langmuir model, the maximum adsorption capacity of anionic dye OG is 609.8 mg/g. The adsorption mechanism is a series of intraparticle diffusion and surface adsorption with electrostatic interaction as the primary driving force. The functionalization of GO with magnetic chitosan was studied in order to develop a nanocomposite material for reactive black 5 dye (RB5) adsorption [68]. The mechanism of adsorption was investigated using appropriate characterization tools. According to the FTIR analysis, the amino group was reduced in the loaded adsorbent compares favorably to the virgin ones, which explained the dye's mechanism of interaction. The reduction of amino groups causes the dye molecule to be attracted to the group of adsorbents. Chitosan active sites are functionalized in an acidic pH medium. The introduction of Ferric nanomaterial coupled with GO improved the nanohybrid's adsorption efficiency. Because of the nanoparticles' affinity for dye oxygen atoms, the electroactivity of the surface area of the nanohybrids was improved. Adsorbents made of graphene and Au nanohybrids have been used to treat RhB dye with visible light as catalyst [69]. The dye molecule was excited to an electron transfer system from dye to adsorbent, and the excited electron was transferred to Au elemental nanoparticles that were stuck by oxygen to generate multiple reactive oxidative species, and the dye was finally degraded. In the absence of the catalyst, dye stability was discovered under visible light. The nanohybrid, on the other hand, demonstrated a higher degradation rate as well as a higher rate constant. Similarly, orange II red dye and MB were examined using the same hybrid adsorbent and showed similar trends of high rate constant and removal efficiency.

Furthermore, ZnO/rGO hybrids were produced for the treatment of aqueous dye solution. When compared to unmodified ZnO, the photocatalytic efficiency of ZnO/rGO for the treatment of organic dye was found to be the highest [70]. Under UV irradiation, the photodegradation of MB, Rhodamine dye, and methylene orange was studied using a ZnO/rGO hybrid. In comparison with other studies, the results of the degradation study established total removal of MB. The ZnO/rGO hybrid's four cyclic photodegradation stability results showed no loss in photodegradation capacity, leading to significant photocatalytic stability. Kavitha et al. [71] fabricated graphene/ZnO nanoparticles for photocatalytic MB degradation in ethanol. The results of a degradation study reported a 70% MB removal efficiency within 3 h when exposed to UV light in ethanol. A comparison of photocatalytic degradation of MB using modified hybrids and unmodified ZnO materials was performed. Adsorption intensities findings demonstrate that the hybrids accelerated the degradation of MB over time, with complete degradation occurring in 0.5 h. In reference to pure ZnO nanoparticles, the graphene/ZnO nanocomposite demonstrated three times the photodegradation efficiency of MB. For the removal of MB under UV, visible, and direct sunlight irradiation, a vanadium oxide graphene doped supported nanohybrid material was prepared. When compared to UV emissions, photodegradation revealed that MB degraded quickly under visible light within 150 min. However, rapid degradation was seen when there was of sunlight versus UV and visible light, which could be attributed to the relatively high intensity of sunlight [72]. The effect of hybrid composites on the degradation of MB under visible light was also investigated using pure vanadium oxide. Because of the presence of graphene in the hybrid material, it was discovered that graphene/vanadium oxide hybrids degraded faster than pure metal oxide particles. Furthermore, when compared to hydrothermally synthesized graphene/vanadium oxide, the nanohybrid demonstrated high photodegradation [73]. Table 1 summarizes the various textile dye removal methods using GO-supported nanocomposites.

6 Conclusion and Future Perspective

This chapter has extensively discussed the various graphene-based supported nanohybrids that have been successfully used for the treatment of textile dyes. However, the focus of this research has been on the removal of textile dyes using graphene-based nanohybrids via adsorption techniques as well as photocatalytic degradation. A broad range of graphene-based nanocomposites has been investigated for the removal of pollutants from industrial wastewater. Many pure nanomaterials have demonstrated a significant alternative to traditional remediation technology among

Table 1 Some GO-supported nanocomposites reported for the removal of textile dyes

Graphene-based nanohybrid	Characterization tools	Dye removed	Adsorption capacity (mg/g)	References
MWCNT/graphene hybrid	Raman, FTIR	Rhodamine B, methylene blue	191 152	[56]
GO/PHEMA	–	Methylene blue	39	[57]
rGO/Ti	FESEM, EDX	Methylene blue	83.3	[58]
GO/Fe_3O_4	FTIR, Raman, SEM, TGA, XPS, TEM, EDX	Methylene blue Neutral red	167 171	[74]
GO/Fe_3O_4	–	Crystal violet	416	[65]
GO/APTS	TGA, XPS, TEM, EDX	Methylene blue	441	[75]
FC/GO	FTIR, SEM	Acid red G Cationic red X	39 65	[66]
GO/PDMAEMA	FTIR, SEM	Orange G	610	[76]
GO/Chm	XRD, EDAX, DTA, SEM, DTG	Rhodamine B	391	[68]
Polymeric GO nanocomposites	FTIR, SEM	Acid dye	54.34	[77]
Polyaniline/GO	FTIR, SEM, XRD, TGA	Allura red	98%	[78]
Fe-TiO_2/rGO	XRD, Raman, FTIR, BET, DRS, TEM, FESEM, EDS	Rhodamine B	91%	[79]
GO/methylene bis-acrylamide	SEM, FTIR, XRD	Methylene blue	95.76%	[80]
Montmorillonite/ GO/$CoFe_2O_4$	FTIR, TGA, XRD	Methyl violet dye	97.26	[81]
GO/Cu-MOF	SEM, UV-Vis, BET, XRD	Methylene blue	262	[82]

these GO-supported nanocomposites due to their greater adsorption capacity, notable degradation efficiency, and preferential functionality. Other useful characteristics of these pure nanomaterials, however, are still in the research phase due to technical issues such as fabrication approach, cost implications, system build-up, and effects on the environment. Furthermore, mass production and vast quantities of pure nanomaterials at commercially viable prices could be a significant disadvantage for industrial effluent treatment. Another issue that may give promising importance is the lack of monitoring system to detect the quality and quantity of nanocomposite in the detectable volume of wastewater in practical applications. Nonetheless, because of their higher textile dye treatment efficiency, GO-supported nanohybrids have a high potential for use in real-world wastewater treatment systems.

References

1. Ghaly A, Ananthashankar R, Alhattab M, Ramakrishnan V (2014) Production, characterization and treatment of textile effluents: a critical review. J Chem Eng Process Technol 5:1–19
2. Idris MO, Noh NA, Ibrahim MNM, Yaqoob AA (2022) Sustainable microbial fuel cell functionalized with a bio-waste: a feasible route to formaldehyde bioremediation along with bioelectricity generation. Chem Eng J 5:140781
3. Ahmad A, Setapar SHM, Yaqoob AA, Ibrahim MNM (2021) Synthesis and characterization of GO-Ag nanocomposite for removal of malachite dye from aqueous solution. Mater Today Proc 47:1359–1365
4. Rashid R, Shafiq I, Akhter P, Iqbal MJ, Hussain M (2021) A state-of-the-art review on wastewater treatment techniques: the effectiveness of adsorption method. Environ Sci Pollut Res 28:9050–9066
5. Meese AF, Kim DJ, Wu X, Le L, Napier C, Hernandez MT, Laroco N, Linden KG, Cox J, Kurup P (2021) Opportunities and challenges for industrial water treatment and reuse. ACS ES&T Eng 2:465–488
6. Idris MO, Yaqoob AA, Ibrahim MNM, Ahmad A, Alshammari MB (2023) Introduction of adsorption techniques for heavy metals remediation. Elsevier, Emerging Techniques for Treatment of Toxic Metals from Wastewater, pp 1–18
7. Dotto GL, McKay G (2020) Current scenario and challenges in adsorption for water treatment. J Environ Chem Eng 8:103988
8. Idris MO, Williams EE, Suleiman AI, Okieimen FE (2021) Scalability of palm kernel shell derived activated carbon for the remediation of Pb2+ and 2, 4-dichlorophenol. EAS J Pharm Pharmacol 3:168–175
9. Tan KB, Vakili M, Horri BA, Poh PE, Abdullah AZ, Salamatinia B (2015) Adsorption of dyes by nanomaterials: recent developments and adsorption mechanisms. Sep Purif Technol 150:229–242
10. Marsh H, Rodríguez-Reinoso F (2006) Production and reference material. Activated Carbon 454
11. Yaqoob AA, Parveen T, Umar K, Mohamad Ibrahim MN (2020) Role of nanomaterials in the treatment of wastewater: a review. Water 12:495
12. Kanwal A, Yaqoob AA, Siddique A, Bhawani SA, Ibrahim MNM, Umar K (2021) Hybrid nanocomposites based on graphene and its derivatives: from preparation to applications, Graphene and nanoparticles hybrid nanocomposites. Springer, pp 261–281
13. Xu F, Xie S, Cao R, Ren C, Wang L (2017) Prepare poly-dopamine coated graphene@ silver nanohybrid for improved surface enhanced Raman scattering detection of dyes. Sens Actuators, B Chem 243:609–616
14. Sumaila A, Ndamitso M, Iyaka Y, Abdulkareem A, Tijani J, Idris M (2020) Extraction and characterization of chitosan from crab shells: kinetic and thermodynamic studies of arsenic and copper adsorption from electroplating wastewater. Iraqi J Sci 2156–2171
15. Emmanuel JL, Idris MO, Usman AO, Musa Q, Suleiman AI, Sambo P (2021) Biomass-derived activated carbon: a viable material for remediation of pb2+ and 2, 4-dichlorophenol (2, 4 DCP) through adsorption. J Adv Res Appl Sci Eng Technol 25:37–45
16. Yaqoob AA, Noor M, Serrà A, Mohamad Ibrahim MN (2020) Advances and challenges in developing efficient graphene oxide-based ZnO photocatalysts for dye photo-oxidation. Nanomaterials 10:932
17. Vizuete M, Barrejón M, Gómez-Escalonilla MJ, Langa F (2012) Endohedral and exohedral hybrids involving fullerenes and carbon nanotubes. Nanoscale 4:4370–4381
18. Fu Y, Zeng G, Lai C, Huang D, Qin L, Yi H, Liu X, Zhang M, Li B, Liu S (2020) Hybrid architectures based on noble metals and carbon-based dots nanomaterials: a review of recent progress in synthesis and applications. Chem Eng J 399:125743

19. Zaleska-Medynska A, Marchelek M, Diak M, Grabowska E (2016) Noble metal-based bimetallic nanoparticles: the effect of the structure on the optical, catalytic and photocatalytic properties. Adv Coll Interface Sci 229:80–107
20. Das S, Presselt M (2019) Progress and development in structural and optoelectronic tunability of supramolecular nonbonded fullerene assemblies. J Mater Chem C 7:6194–6216
21. Shahadat M, Khan MZ, Rupani PF, Embrandiri A, Sultana S, Ahammad SZ, Ali SW, Sreekrishnan T (2017) A critical review on the prospect of polyaniline-grafted biodegradable nanocomposite. Adv Coll Interface Sci 249:2–16
22. Osagie C, Othmani A, Ghosh S, Malloum A, Esfahani ZK, Ahmadi S (2021) Dyes adsorption from aqueous media through the nanotechnology: a review. J Market Res 14:2195–2218
23. Pacheco S, Medina M, Valencia F, Tapia J (2006) Removal of inorganic mercury from polluted water using structured nanoparticles. J Environ Eng 132:342–349
24. Yang K, Xing B (2007) Desorption of polycyclic aromatic hydrocarbons from carbon nanomaterials in water. Environ Pollut 145:529–537
25. Lu F, Astruc D (2020) Nanocatalysts and other nanomaterials for water remediation from organic pollutants. Coord Chem Rev 408:213180
26. Cong H-P, Ren X-C, Wang P, Yu S-H (2012) Macroscopic multifunctional graphene-based hydrogels and aerogels by a metal ion induced self-assembly process. ACS Nano 6:2693–2703
27. Klaine SJ, Alvarez PJ, Batley GE, Fernandes TF, Handy RD, Lyon DY, Mahendra S, McLaughlin MJ, Lead JR (2008) Nanomaterials in the environment: behavior, fate, bioavailability, and effects. Environ Toxicol Chem Int J 27:1825–1851
28. Pei S, Cheng H-M (2012) The reduction of graphene oxide. Carbon 50:3210–3228
29. Ivanovskii AL (2012) Graphene-based and graphene-like materials. Russ Chem Rev 81:571
30. Zhu Y, Murali S, Cai W, Li X, Suk JW, Potts JR, Ruoff RS (2010) Graphene and graphene oxide: synthesis, properties, and applications. Adv Mater 22:3906–3924
31. Ali I, Mbianda X, Burakov A, Galunin E, Burakova I, Mkrtchyan E, Tkachev A, Grachev V (2019) Graphene based adsorbents for remediation of noxious pollutants from wastewater. Environ Int 127:160–180
32. Sitko R, Zawisza B, Malicka E (2013) Graphene as a new sorbent in analytical chemistry. TrAC, Trends Anal Chem 51:33–43
33. Zhang T, Wang W, Zhao Y, Bai H, Wen T, Kang S, Song G, Song S, Komarneni S (2021) Removal of heavy metals and dyes by clay-based adsorbents: from natural clays to 1D and 2D nano-composites. Chem Eng J 420:127574
34. Hidalgo-Manrique P, Lei X, Xu R, Zhou M, Kinloch IA, Young RJ (2019) Copper/graphene composites: a review. J Mater Sci 54:12236–12289
35. Szabó T, Berkesi O, Forgó P, Josepovits K, Sanakis Y, Petridis D, Dékány I (2006) Evolution of surface functional groups in a series of progressively oxidized graphite oxides. Chem Mater 18:2740–2749
36. Huang X-M, Liu L-Z, Zhou S, Zhao J-J (2020) Physical properties and device applications of graphene oxide. Front Phys 15:1–70
37. Pielichowska K, Nowak M, Szatkowski P, Macherzyńska B (2016) The influence of chain extender on properties of polyurethane-based phase change materials modified with graphene. Appl Energy 162:1024–1033
38. Idris MO, Guerrero-Barajas C, Kim H-C, Yaqoob AA, Ibrahim MNM (2022) Scalability of biomass-derived graphene derivative materials as viable anode electrode for a commercialized microbial fuel cell: a systematic review. Chin J Chem Eng
39. Yaqoob AA, Ibrahim MNM, Yaakop AS, Umar K, Ahmad A (2021) Modified graphene oxide anode: a bioinspired waste material for bioremediation of Pb2+ with energy generation through microbial fuel cells. Chem Eng J 417:128052
40. Yaqoob AA, Mohamad Ibrahim MN, Umar K, Bhawani SA, Khan A, Asiri AM, Khan MR, Azam M, AlAmmari AM (2020) Cellulose derived graphene/polyaniline nanocomposite anode for energy generation and bioremediation of toxic metals via benthic microbial fuel cells. Polymers (Basel) 13

41. Yaqoob AA, Serrà A, Bhawani SA, Ibrahim MNM, Khan A, Alorfi HS, Asiri AM, Hussein MA, Khan I, Umar K (2022) Utilizing biomass-based graphene oxide–polyaniline–Ag electrodes in microbial fuel cells to boost energy generation and heavy metal removal. Polymers 14:845
42. Yaqoob AA, Serrà A, Ibrahim MNM, Yaakop AS (2021) Self-assembled oil palm biomass-derived modified graphene oxide anode: an efficient medium for energy transportation and bioremediating Cd (II) via microbial fuel cells. Arab J Chem 14:103121
43. Shen C, Oyadiji SO (2020) The processing and analysis of graphene and the strength enhancement effect of graphene-based filler materials: a review. Mater Today Phys 15:100257
44. Nanda SS, Kim MJ, Yeom KS, An SSA, Ju H, Yi DK (2016) Raman spectrum of graphene with its versatile future perspectives. TrAC, Trends Anal Chem 80:125–131
45. Kai M, Zhang L, Liew K (2019) Graphene and graphene oxide in calcium silicate hydrates: chemical reactions, mechanical behavior and interfacial sliding. Carbon 146:181–193
46. Kozbial A, Li Z, Conaway C, McGinley R, Dhingra S, Vahdat V, Zhou F, D'Urso B, Liu H, Li L (2014) Study on the surface energy of graphene by contact angle measurements. Langmuir 30:8598–8606
47. Zhou Y, Bao Q, Tang LAL, Zhong Y, Loh KP (2009) Hydrothermal dehydration for the "green" reduction of exfoliated graphene oxide to graphene and demonstration of tunable optical limiting properties. Chem Mater 21:2950–2956
48. Yaqoob AA, Ahmad H, Parveen T, Ahmad A, Oves M, Ismail IM, Qari HA, Umar K, Mohamad Ibrahim MN (2020) Recent advances in metal decorated nanomaterials and their various biological applications: a review. Front Chem 8:341
49. Perreault F, De Faria AF, Elimelech M (2015) Environmental applications of graphene-based nanomaterials. Chem Soc Rev 44:5861–5896
50. Zhang Y, Tang Z-R, Fu X, Xu Y-J (2010) TiO_2−graphene nanocomposites for gas-phase photocatalytic degradation of volatile aromatic pollutant: is TiO_2−graphene truly different from other TiO_2−carbon composite materials? ACS Nano 4:7303–7314
51. Bradder P, Ling SK, Wang S, Liu S (2011) Dye adsorption on layered graphite oxide. J Chem Eng Data 56:138–141
52. Liu F, Chung S, Oh G, Seo TS (2012) Three-dimensional graphene oxide nanostructure for fast and efficient water-soluble dye removal. ACS Appl Mater Interfaces 4:922–927
53. Ramesha G, Kumara AV, Muralidhara H, Sampath S (2011) Graphene and graphene oxide as effective adsorbents toward anionic and cationic dyes. J Colloid Interface Sci 361:270–277
54. Sun L, Yu H, Fugetsu B (2012) Graphene oxide adsorption enhanced by in situ reduction with sodium hydrosulfite to remove acridine orange from aqueous solution. J Hazard Mater 203:101–110
55. Gupta K, Khatri OP (2017) Reduced graphene oxide as an effective adsorbent for removal of malachite green dye: plausible adsorption pathways. J Colloid Interface Sci 501:11–21
56. Sui Z, Meng Q, Zhang X, Ma R, Cao B (2012) Green synthesis of carbon nanotube–graphene hybrid aerogels and their use as versatile agents for water purification. J Mater Chem 22:8767–8771
57. Kharismadewi D, Haldorai Y, Nguyen VH, Tuma D, Shim J-J (2016) Synthesis of graphene oxide-poly (2-hydroxyethyl methacrylate) composite by dispersion polymerization in supercritical CO_2: adsorption behavior for the removal of organic dye. Compos Interfaces 23:719–739
58. Nguyen-Phan T-D, Pham VH, Kim EJ, Oh E-S, Hur SH, Chung JS, Lee B, Shin EW (2012) Reduced graphene oxide–titanate hybrids: morphologic evolution by alkali-solvothermal treatment and applications in water purification. Appl Surf Sci 258:4551–4557
59. Vo TS, Vo TTBC (2022) Graphene oxide-covered melamine foam utilizing as a hybrid foam toward organic dye removal and recyclability. Prog Nat Sci Mater Int
60. Guo X, Qu L, Tian M, Zhu S, Zhang X, Tang X, Sun K (2016) Chitosan/graphene oxide composite as an effective adsorbent for reactive red dye removal. Water Environ Res 88:579–588
61. Cheng J-S, Du J, Zhu W (2012) Facile synthesis of three-dimensional chitosan–graphene mesostructures for reactive black 5 removal. Carbohyd Polym 88:61–67

62. Carmalin Sophia A, Arfin T, Lima EC (2019) Recent developments in adsorption of dyes using graphene based nanomaterials, A new generation material graphene: applications in water technology 439–471
63. Geng Z, Lin Y, Yu X, Shen Q, Ma L, Li Z, Pan N, Wang X (2012) Highly efficient dye adsorption and removal: a functional hybrid of reduced graphene oxide–Fe_3O_4 nanoparticles as an easily regenerative adsorbent. J Mater Chem 22:3527–3535
64. Kumar M, Dosanjh HS, Singh J, Monir K, Singh H (2020) Review on magnetic nanoferrites and their composites as alternatives in waste water treatment: synthesis, modifications and applications. Environ Sci Water Res Technol 6:491–514
65. Sahraei R, Hemmati K, Ghaemy M (2016) Adsorptive removal of toxic metals and cationic dyes by magnetic adsorbent based on functionalized graphene oxide from water. RSC Adv 6:72487–72499
66. Sheng G, Zhu S, Wang S, Wang Z (2016) Removal of dyes by a novel fly ash–chitosan–graphene oxide composite adsorbent. RSC Adv 6:17987–17994
67. Kyzas GZ, Deliyanni EA, Bikiaris DN, Mitropoulos AC (2018) Graphene composites as dye adsorbents. Chem Eng Res Des 129:75–88
68. Travlou NA, Kyzas GZ, Lazaridis NK, Deliyanni EA (2013) Functionalization of graphite oxide with magnetic chitosan for the preparation of a nanocomposite dye adsorbent. Langmuir 29:1657–1668
69. Xiong Z, Zhang LL, Ma J, Zhao X (2010) Photocatalytic degradation of dyes over graphene–gold nanocomposites under visible light irradiation. Chem Commun 46:6099–6101
70. Fan F, Wang X, Ma Y, Fu K, Yang Y (2015) Enhanced photocatalytic degradation of dye wastewater using ZnO/reduced graphene oxide hybrids. Fullerenes Nanotubes Carbon Nanostruct 23:917–921
71. Kavitha T, Gopalan AI, Lee K-P, Park S-Y (2012) Glucose sensing, photocatalytic and antibacterial properties of graphene–ZnO nanoparticle hybrids. Carbon 50:2994–3000
72. Shanmugam M, Alsalme A, Alghamdi A, Jayavel R (2015) Enhanced photocatalytic performance of the graphene-V_2O_5 nanocomposite in the degradation of methylene blue dye under direct sunlight. ACS Appl Mater Interfaces 7:14905–14911
73. Rakkesh RA, Durgalakshmi D, Balakumar S (2015) Nanostructuring of a GNS-V 2 O 5–TiO_2 core–shell photocatalyst for water remediation applications under sun-light irradiation. RSC Adv 5:18633–18641
74. Xie G, Xi P, Liu H, Chen F, Huang L, Shi Y, Hou F, Zeng Z, Shao C, Wang J (2012) A facile chemical method to produce superparamagnetic graphene oxide–Fe_3O_4 hybrid composite and its application in the removal of dyes from aqueous solution. J Mater Chem 22:1033–1039
75. Li B, Cao H, Yin G (2011) Mg (OH) 2@ reduced graphene oxide composite for removal of dyes from water. J Mater Chem 21:13765–13768
76. Li C, Zhu H, She X, Wang T, She F, Kong L (2016) Selective removal of anionic dyes using poly (N, N-dimethyl amino ethylmethacrylate) functionalized graphene oxide. RSC Adv 6:67242–67251
77. Noreen S, Tahira M, Ghamkhar M, Hafiz I, Bhatti HN, Nadeem R, Murtaza MA, Yaseen M, Sheikh AA, Naseem Z (2021) Treatment of textile wastewater containing acid dye using novel polymeric graphene oxide nanocomposites (GO/PAN, GO/PPy, GO/PSty). J Market Res 14:25–35
78. Nawaz H, Umar M, Ullah A, Razzaq H, Zia KM, Liu X (2021) Polyvinylidene fluoride nanocomposite super hydrophilic membrane integrated with polyaniline-graphene oxide nano fillers for treatment of textile effluents. J Hazard Mater 403:123587
79. Isari AA, Payan A, Fattahi M, Jorfi S, Kakavandi B (2018) Photocatalytic degradation of rhodamine B and real textile wastewater using Fe-doped TiO_2 anchored on reduced graphene oxide (Fe-TiO_2/rGO): characterization and feasibility, mechanism and pathway studies. Appl Surf Sci 462:549–564
80. Fallah S, Mamaghani HR, Yegani R, Hajinajaf N, Pourabbas B (2020) Use of graphene substrates for wastewater treatment of textile industries. Adv Compos Hybrid Mater 3:187–193

81. Foroutan R, Mohammadi R, MousaKhanloo F, Sahebi S, Ramavandi B, Kumar PS, Vardhan KH (2020) Performance of montmorillonite/graphene oxide/$CoFe_2O_4$ as a magnetic and recyclable nanocomposite for cleaning methyl violet dye-laden wastewater. Adv Powder Technol 31:3993–4004
82. Firouzjaei MD, Afkhami FA, Esfahani MR, Turner CH, Nejati S (2020) Experimental and molecular dynamics study on dye removal from water by a graphene oxide-copper-metal organic framework nanocomposite. J Water Process Eng 34:101180

Synthesis and Characterization of Nanohybrid Materials for Anionic Dye Removal

Alain R. Picos-Benítez, María M. Ramírez-Alaniz, Pablo Emilio Escamilla-García, and Blanca L. Martínez-Vargas

Abstract Polluted water resources, particularly those polluted with industrial effluent dyes, are carcinogenic and severely threaten sustainable and long-standing worldwide development. Heterogeneous photocatalysis and adsorption processes can efficiently remove these types of pollutants. The process employs hybrid nanomaterials based on inorganic, organic, polymeric, and ceramic molecules. This chapter presents a bibliographical review that focuses on some synthesis methods of hybrid nanomaterials, mentioning some characterization techniques that should be used to study optical, mechanical, and structural properties. Finally, the applications of specific nanomaterials in removing anionic contaminants are shown. It concludes with a reflection focused on sustainability in using this type of hybrid nanomaterials to treat anionic dyes from the textile industry.

Keywords Nanohybrid materials · Textile dyes · Anionic dye · Water treatment · Sustainability

1 Nanohybrid Material

A nanohybrid material is defined as a multiphase system with at least one nano-dimensional phase, and these materials are also known as a nanocomposite. These materials have enhanced characteristics compared with the commercial contra-part. Also, the characteristics of these materials change radically when their diameter drops below a specific diameter. The material starts showing its enhanced properties when its scale reaches the nanoscale. The materials can be designed by using a wide variety of phases; these are made with the objective of obtaining a new material

A. R. Picos-Benítez · M. M. Ramírez-Alaniz · B. L. Martínez-Vargas (✉)
Centro de Estudios Científicos y Tecnológicos No. 18, Instituto Politécnico Nacional, 98160 Zac., Zacatecas, Mexico
e-mail: bmartinezv@ipn.mx

P. E. Escamilla-García
CECYT 13, Instituto Politécnico Nacional, 1620 Taxqueña Ave., 04250 Coyoacán, México City, Mexico

© The Author(s), under exclusive license to Springer Nature Singapore Pte Ltd. 2023

A. Ahmad et al. (eds.), *Nanohybrid Materials for Treatment of Textiles Dyes*, Smart Nanomaterials Technology, https://doi.org/10.1007/978-981-99-3901-5_6

with a set of desired properties that can be used for specific applications. Among the properties that can be changed in the nanomaterials, we can find mechanical strength [47], photocatalytic [20, 77, 79, 98, 102], electric [81, 106, 113], and magnetic [30, 40] properties among others. These modifications allow the use of the designed nanomaterials in the pharmaceutical [95], medical [3], electronic [25, 81], construction [55, 117], detection of specific compounds [97], and field environmental remediation [14, 36, 70].

2 Types of Nanohybrids

Typically, nanohybrids are composed of two phases, matrix and fillers. According to the required properties, the nanohybrid materials can be synthesized by meticulously selecting the matrix and fillers. The classification of the nanohybrid materials can be found in Fig. 1. The nanohybrid materials can be classified according to the type of matrix or based on the type of fillers, so we can find polymer, metal, and ceramic matrix nanohybrids.

2.1 Polymer Matrix Nanohybrids

As mentioned before, the type of nanohybrid materials will be defined by the chosen matrix. Thus, we can find hybrid materials whose main composition is based on a

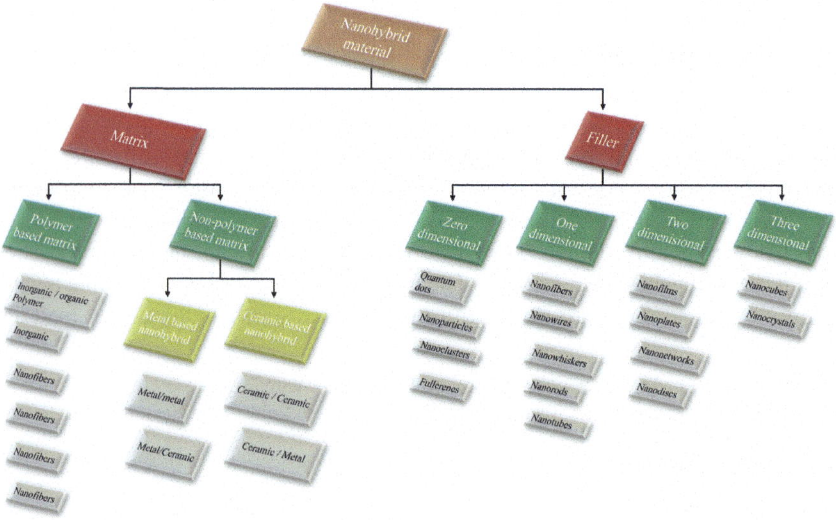

Fig. 1 Classification of nanohybrid materials

polymeric matrix. The resulting materials show enhanced physical and mechanical properties compared with the initial material. The polymeric-based nanomaterials can be complemented with nanofillers to improve optical and electrical properties [16]. These nanomaterials have a wide variety of applications, such as toxic gas detectors [97], drug delivery for cancer treatment [3], and a mechanical enhancer of cement base paste [55]. Also, Haniffa et al. [30] reported several biomedical applications for cellulose base nanohybrids, such as antibacterial material, drug delivery, scaffolding for bone tissue engineering, glucose detection, medical diagnosis, enzyme immobilization, biosensor, among others.

2.2 *Ceramic Matrix Nanohybrids*

These nanohybrids are composed of ceramic materials with a nanophase filler made of silicates, graphene, or a combination of both. The main objective of introducing a filler is to increase the mechanical strength of the nanohybrid material. By incorporating these nanofillers, the growth of cracks is avoided. Other properties are also enhanced, like shock resistance, thermal and electrical conductivity, bioactivity, and wear resistance [16]. The improvement of these properties has attracted attention for their use in fabricating new electronic devices with flexible parts [83]. Also, it is reported that it can be used as an enhancer of the mechanical structure in the fabrication of asphalt mixes [24], also can be used to burst the properties of commercial molten solder and produce a new bulk nanocomposite solder with enhanced characteristics [39], Blasques et al. [17] used a newly synthesized nanomaterial to develop an electrochemical sensor finally, as well as with the polymer matrix nanohybrids, ceramic matrix nanohybrids can be used as photocatalyst or adsorbent for the removal of pollutants from water [49, 121].

2.3 *Metal Matrix Nanohybrids*

These types of nanohybrids are composed mainly of nano-reinforcements that are dispersed in a metal matrix. Some of the used materials include metals or alloys and usually can be combined with metals or ceramics nanocomposites, where titanium, zinc, magnesium, and aluminum are the most used matrix metals. The combinations of the metal matrix and the fillers produce materials with a wide variety of characteristics that can be used in several applications [16]. Among the reported uses of these nanomaterials, we can find supercapacitors [25, 54], biomedical applications like an antimicrobial agent, anti-carcinogenic, antioxidant activity, biosensor and drug delivery [69], and as a non-enzymatic glucose sensor [72]. Recently, this type of nanomaterials has gained the attention of the scientific community for their capabilities to remove several pollutants like heavy metals [40], real wastewater streams from industrial activities such as rubber fabrication [47], textile and pharmaceutical industry [43, 67], and several hardly degradable pollutants like synthetic dyes [43, 67, 106].

Nanohybrid materials are a particular class of advanced new-age materials that have gained wide-scale popularity because of their multi-functionalities. Nanohybrid materials offer biodegradability, flexibility, processibility, and beneficial optical and mechanical properties. Much research has focused on offering cheap, reliable, and sustainable synthesis methods that offer materials of good crystallographic quality and multiple properties. In this section, they present the most used synthesis methods and the techniques required for the characterization of nanohybrid materials [9, 38].

3 Synthetic Methods

There are currently different synthesis methods to obtain nanohybrid materials with different properties (Fig. 2). Some of the important and commonly used synthetic methods of nanohybrids materials have been discussed in this section.

3.1 Colloidal Methods (One-Pot Synthesis)

It is a simple and extensively used technique for synthesizing different nanohybrid materials. Its basic principle relies on mixing the reduction of precursor and colloidal solutions of the fillers dispersed uniformly in an appropriate solvent. This method commonly uses solvents, including water, ethanol, methanol, dimethylformamide, toluene, and chloroform [56, 57]. The process of blending solutions and subsequent

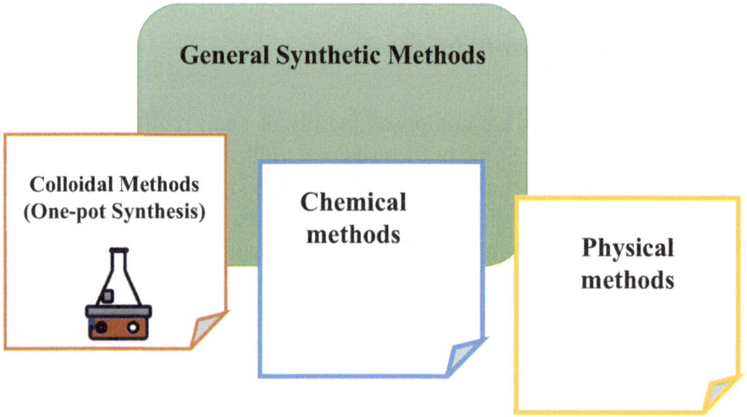

Fig. 2 Synthesis methods to obtain nanohybrid materials

elimination of the solvent results in the formation of nanohybrid materials. The final product can be acquired by a simple method of precipitation followed by filtration. However, selecting an appropriate solvent poses some limitations to this method for fabricating non-hybrid materials [58].

3.2 Chemical Methods

The chemical methods are co-precipitation, hydrothermal, chemical vapor deposition, and sol–gel. Chemical vapor deposition (CVD) is the most employed for nanohybrid materials synthesis, which is used to generate quality, highly efficient solid materials applications in the semiconductors industry. The process involves chemical interactions between halide or organometallic compounds and other gases to generate non-volatile thin films on the surface of substrates. CVD is widely employed for the preparation of nanohybrid materials. The benefits of CVD as compared to other techniques involve: (1) flexibility of selection of a wide range of nanoparticle solutions and chemically suitable precursors; (2) broad applicability in industrial sector like glazing, microelectronics, etc., (3) ease of execution that does not need critical handling steps and aging time [16, 68, 79].

Furthermore, the sol–gel method is extensively used to fabricate inorganic and polymeric nanohybrid materials. The method involves the dissolution of nanomaterial precursor in an appropriate solvent and polymeric matrix. The solution phase (sol) is activated through radiation or heating to form a solid–liquid phase (gel). The gel is cured, and the nanohybrid material is obtained. In this, the polymer initiates the nucleation process and, thereby, the formation of required nanomaterials. The synthetic technique does not involve a steady energy requirement for the distribution of the nanomaterials in the polymeric matrix [17, 37, 85].

3.3 Physical Methods

Of the synthesis methods for obtaining hybrid nanomaterials, mechanical grinding is used, which consists of the mixture of the base materials for the heterojunction, and they are mixed mechanically with the use of a ball mill. The mixtures are generally proportional with respect to the most efficient photocatalytic material and whose properties it is desired to improve [51, 89].

4 Characterization Techniques for Nanohybrid Materials

To ensure that these materials have the desired properties, the characterization of the nanohybrid materials is necessary. For this, specific techniques such as those shown in Fig. 3 are used. In this section, the basic principle of the characterization techniques is described; diffuse reflectance UV-Vis spectroscopy (UV-Vis), X-ray diffraction (XRD), Fourier transform infrared spectroscopy (FT-IR), scanning electron microscopy (SEM), and high-resolution transmission electron microscopy (HR-TEM). The techniques that are used must provide essential parameters that describe the optical, physicochemical, and structural properties of hybrid nanomaterials.

4.1 Ultraviolet-Visible Spectroscopy with Diffuse Reflectance (UV-Vis)

UV-Vis absorption spectroscopy is based on analyzing the amount of electromagnetic radiation in the ultraviolet and visible wavelength range (200–800 nm) that a sample can absorb or transmit depending on the amount of substance present. In the case of solid samples, the technique used is diffuse reflectance. Reflectance measurements under normal conditions contain the two components of reflection: the specular and diffuse components. The specular component contains very little information about the composition, so its contribution to the measurements is minimized with the position of the detector concerning the sample. On the contrary, the diffuse component is the one that provides valuable information about the sample. Diffuse reflectance is explained by the Kubelka–Munk theory [71]. This theory assumes that the radiation that falls on a scattering medium simultaneously undergoes an absorption and scattering process so that the reflected radiation can be defined as a function of the absorption (k) and scattering (s) constants:

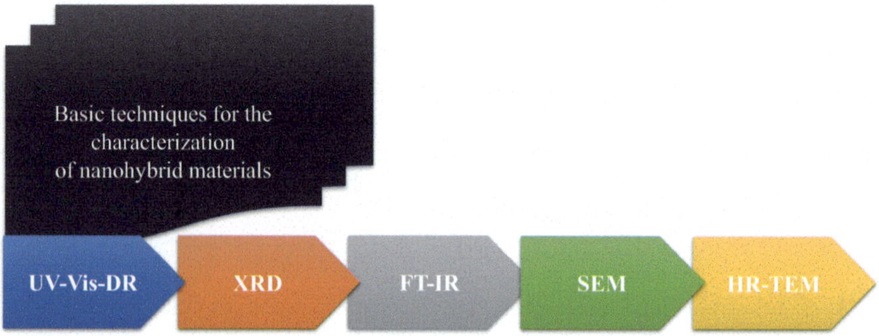

Fig. 3 Basic techniques employed in characterization of nanohybrid materials

$$f(R_\infty) = \frac{(1 - R_\infty)^2}{2R_\infty} = \frac{k}{s} \tag{1}$$

In practice, relative reflectance is used instead of absolute reflectance, which is the relationship between the light intensities reflected by the sample and by a standard [71]. The standard used in this case is barium sulfate. The Kubelka–Munk function written in terms of relative reflectance is as follows:

$$f(R) = \frac{(1 - R)^2}{2R} = \frac{k}{s} \tag{2}$$

Measurement of diffuse reflectance with a UV-visible spectrophotometer is a standard technique in the determination of the absorption properties of materials. In the case of semiconductors for water splitting, the measured properties with diffuse reflectance are the band gap energy (also referred to as the band gap) and the absorption coefficient. Determination of the band gap from the measurement of the diffuse reflectance of a powder sample is a standard technique [71, 85].

4.2 X-Ray Diffraction (XRD)

The physical characteristics of nanohybrid materials, their shape, size, and structural state, govern the properties and applications of that material. In the case of crystals, it is directly related to the expression of specific crystallographic planes that show different surface structures and atomic configurations. The arrangement of the atoms on the surface of the materials, and their intrinsic coordination, determines the molecule's adsorption and reactivity. The photoexcited and reactive molecules can be transferred through the solution from the surface, which is a critical step for removing the pollutants. The difference in structure can influence the properties and, therefore, the photocatalytic behavior.

X-ray diffraction is an analytical technique that allows us to identify crystalline structures from solid samples. X-rays are a short-wavelength form of electromagnetic radiation produced by decelerating high-energy electrons or by electronic transitions involving electrons in the inner orbitals of atoms. When the beam of X-rays is irradiated to a solid sample, it diffuses in all directions; this diffraction is caused by the electrons associated with the atoms or ions it encounters on the way. However, the rest of the beam can give rise to the phenomenon of X-ray diffraction (Fig. 4).

X-ray diffraction occurs if there is an orderly arrangement of atoms and if Bragg's Law is satisfied, which relates the X-ray wavelength and interatomic distance to the angle of incidence of the diffracted beam (Eq. 3).

$$n\lambda = 2d_{hkl}\,\mathrm{Sen}\,\theta \tag{3}$$

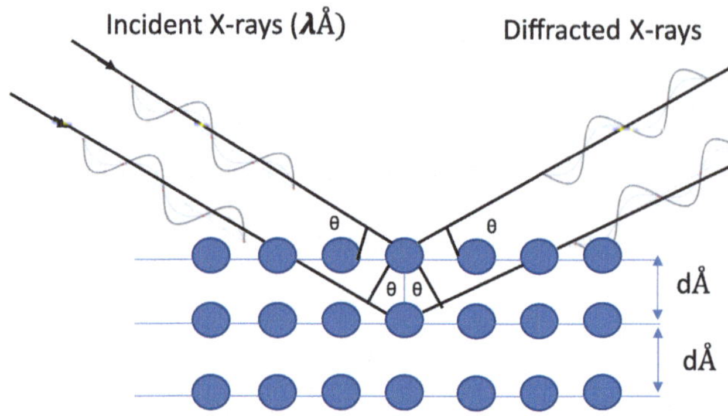

Fig. 4 Representation of the diffraction phenomenon

where d is the distance between the network planes in the direction of incidence; θ is the angle between the incident beam and the crystal surface; n, diffraction order (values of 1, 2, 3 ...), and λ is the wavelength.

An X-ray diffraction pattern is obtained by scanning at an angle of 2θ and measuring the intensity of the radiation using an electronic counter. The diffraction patterns of nanostructured materials are wider than those obtained with bulk materials, so this quality is used to measure their size using the Debye–Scherrer formula (Eq. 4).

The diffraction peak width is negligible for particles larger than 200 nm, so Scherrer's formula only applies to particles smaller than this size.

$$D = \frac{4}{3}\frac{0.9\lambda}{B\mathrm{Cos}\theta} \tag{4}$$

where D is the diameter of the particle, k is the Scherrer constant that takes values from 0.87 to 1.0 (1.0 is commonly used for spherical particles), and B is taken directly from the diffraction pattern on the 2θ axis and represents the peak width. Intensity diffraction is measured at half its height (in radians), θ the angle (in radians) at which the strongest peak is observed, and λ is the wavelength of the incident X-rays [23, 45, 51].

4.3 Fourier Transform Infrared Spectroscopy (FT-IR)

Since most molecules absorb infrared light, infrared spectroscopy converts it into molecular vibration, which provides information about the functional groups present. This technique complements the information obtained with X-ray diffraction when analyzing hybrid nanomaterials [17, 123].

4.4 Scanning Electron Microscopy (SEM)

Scanning electron microscopy (SEM) is a technique that allows obtaining images that are widely used in materials investigations due to its high resolution and ability to analyze morphological, structural, and chemical characteristics of the samples being studied. The detector collects secondary electrons scattered by the atoms on the surface, providing information about the structure and morphology of the surface [33, 34].

4.5 High-Resolution Transmission Electron Microscopy (HR-TEM)

A transmission electron microscope is a potent tool for material science. A high-energy beam of electrons is shone through a very thin sample. The interactions between the electrons and atoms can be used to observe features such as the crystal structure and features in the structure like dislocations and grain boundaries. Chemical analysis can also be performed. TEM can be used to study the growth of layers, their composition, and defects in semiconductors. High resolution can be used to analyze the quality, shape, size, and density of quantum wells, wires, and dots [19, 85, 123].

These are only some techniques that are described; each researcher, depending on the hybrid nanomaterial that he synthesizes, chooses the characterization to check the obtaining, structure, and properties. Once the material is obtained, and its properties have been verified, its application is studied.

5 Applications of Nanohybrid Materials for Pollutant Removal

One of the main interests in nanohybrid materials is their multi-functionality. Their capacity to be synthesized into multiple shapes and the feasibility of mixing with other compounds or elements and improving their properties have allowed the nanohybrid materials to be used in a wide range of applications. Some of the most pronounced applications of the different types of nanomaterials have been discussed in this section. The newly synthesized nanomaterials with enhanced optical properties are mostly used for the photocatalytic degradation of hardly removable pollutants. However, using nanohybrid materials to remove synthetic pollutants is reported with adequate efficiencies by absorption processes.

5.1 *The Use of Heterogenous Photocatalysis for Dye Degradation*

Heterogenous photocatalysis is an advanced oxidation process (AOP) that consists of using an organic or inorganic compound in a liquid or a gaseous solution and putting it in contact with a semiconductor. The mixture is irradiated with UV or solar light or energy that must be equivalent to or higher than the band gap of the semiconductor for its excitation. This process will provide the promotion of an electron from the valence band to the conduction band, creating electron–hole pairs (Reaction 5) and the posterior production of highly oxidizing species, like hydroxyl radicals ($^{\cdot}OH$), with the purpose of the total pollutants removal by the oxidation process [56, 115].

$$\text{Semiconductor} + hv \rightarrow e_{CB}^{-} + h_{VB}^{+} \tag{5}$$

The objective pollutants can be either immediately oxidized in the generated pair-electron holes (Reaction 6) or by highly reactive and non-selective oxygen species (ROS) that are generated in the process mentioned above [96]. Another complicated mechanism consists of the reaction of the h_{VB}^{+} holes of the valence band with hydroxyl ions (OH^{-}) or water molecules that are adsorbed on the surface of the catalyst forming $^{\cdot}OH$ (Reaction 7 and 8). Meanwhile, the reaction of e_{CB}^{-} with dissolved oxygen (O_2) can produce the superoxide radical ($^{\cdot}O_2^{-}$), as shown in Reaction 9. Another way to obtain $^{\cdot}OH$ can be by the combination of the $^{\cdot}O_2^{-}$ radical with water (H_2O) to produce $^{\cdot}OH$ and the hydroperoxy radical ($^{\cdot}HO_2^{-}$), see Reaction 10, which combined with another $^{\cdot}OH$ can produce hydrogen peroxide (H_2O_2), Reaction 11, which can produce more $^{\cdot}OH$ by Reaction 12. Finally, both oxidizing species mineralize the pollutants according to Reaction 13 [108].

$$h_{VB}^{+} + M_s \rightarrow \cdot M_s^{+} \tag{6}$$

$$h_{VB}^{+} + H_2O \rightarrow {}^{\cdot}OH + H^{+} \tag{7}$$

$$h_{VB}^{+} + OH^{-} \rightarrow {}^{\cdot}OH \tag{8}$$

$$e_{CB}^{-} + O_2 \rightarrow {}^{\cdot}O_2^{-} \tag{9}$$

$${}^{\cdot}O_2^{-} + H_2O \rightarrow {}^{\cdot}HO_2^{-} + {}^{\cdot}OH^{-} \tag{10}$$

$${}^{\cdot}HO_2^{-} + {}^{\cdot}OH^{-} \rightarrow H_2O_2 \tag{11}$$

$$H_2O_2 \rightarrow 2\,{}^{\cdot}OH \tag{12}$$

$$^{\cdot}OH + ^{\cdot}O_2^{-} + \text{pollutant} \rightarrow CO_2 + H_2O \quad (13)$$

5.2 The Use of Nanohybrid Materials for the Removal of Harmful Components from Waters

As mentioned above, heterogenous photocatalysis is an AOP that has recently been used widely in the water and wastewater treatment of harmful pollutants. Among the removed pollutants reported in the literature, we can find heavy metals, persistent organic compounds, drugs, pesticides, and synthetic organic dyes. Several authors reported using nanomaterials in the application of heterogeneous photocatalysis for wastewater treatment purposes, especially for those wastewaters with synthetic organic pollutants such as synthetic dyes [78]. In addition to heterogeneous photocatalysis, adsorption has been reported to be an efficient process of removing synthetic dyes. Adsorption implies the presence of an adsorbate, which can be a compound of interest that has to be removed from a gaseous or aqueous phase. The adsorbate will stick to the surface of a solid or, rarely, a liquid called an adsorbent [10]. This process can occur in 2 ways, by physisorption, where Van der Waals interactions can occur. This process is easily reversible with increments of temperature. On the other hand, chemisorption involves a covalent bond between the adsorbate and the adsorbent. It is an activated process and is more stable at higher temperatures [10].

5.3 Nanohybrid Materials in the Photocatalytic Degradation of Anionic Dyes

A new type of pollutant has recently gained the attention of the scientific community, emergent pollutants (EP). These pollutants can be described as those present in the natural environment at very low concentrations. No regulations prohibit their discharged indiscriminately, and because of their low concentration, these pollutants are not considered an ecological hazard for the environment or human populations [11]. A wide variety of these compounds can be found in wastewater treatment facilities. Because of their low concentration and complexity, conventional wastewater treatment methods are inefficient in removing these, causing their discharge and allowing them to reach natural water streams or bodies. Several types of compounds are reported to be part of these pollutants, like pharmaceutical drugs, antibiotics, analgesics, antidepressants, and hormones, among others [12, 42, 74, 82, 110, 111, 118, 120]. Also, some synthetic dyes are reported [1, 19, 46, 65, 66, 75, 93, 100, 101, 103, 119]. Synthetic dyes are one of the most used chemicals worldwide since their use is indispensable for the production of clothes in the textile industry.

The use of synthetic dye has been increasing yearly because of its high availability, low cost of production, and wide variety of colors [86, 87]. Furthermore, some of these compounds are also used in the pharmaceutical and food industry. The problem these synthetic dyes represent is the inefficient degradation of the conventional wastewater treatment methods against these pollutants, the hazard that represents the aquatic and/or terrestrial organisms [61]. Some of these synthetic dyes are considered endocrine disruptors. It also has to be mentioned that some of them are considered non-toxic; however, if these synthetic dyes affect natural water bodies, these compounds can be degraded or metabolized to hazardous compounds (mutagenic, carcinogenic, or genotoxic) by some natural processes, also, can provoke some aesthetic problems, or prevent the pass of solar light affecting the quality of the water body by reducing the photosynthetic activity and due this the dissolved oxygen concentration and, as a result, causing the mortality of some microorganisms such fishes [61, 86, 87].

Synthetic dyes are easily mineralized from water by some advanced oxidation process, such as heterogenous photocatalysis, because of the generation of non-selective and highly oxidative radicals heterogeneous photocatalysis [19, 37, 44, 68, 77, 80, 99–101, 122]. Several studies reported the use of different materials for heterogenous photocatalysis applications. Among all these compounds, we can find titanium dioxide (TiO_2) [7, 27, 45, 89], strontium [26], zinc as zinc oxide [91] or as a combined salt [76], copper as oxide [104] or as a combined salt [32], and some other transition metals like iron [7, 123] and manganese [109] (see Table 1).

5.4 Factors Involved in the Photocatalytic Degradation of Anionic Dyes

Several mechanisms influence the photocatalytic degradation of anionic dyes, the reported mechanism involves the direct electron transfer from the photocatalyst to the pollutant, and this can be enhanced by the presence of some transition metals causing a synergistic effect, e.g., in the presence of Cr VI [107] or Cu [32]. Also, it has been reported that when trying the degradation of more than one dye, a competition of absorption in the surface of the photocatalyst will occur between the different dyes presented in the solution. This is caused by the different chemical and physical properties of synthetic dyes [7]. Additionally, the amount of the anionic dye will negatively affect the removal efficiency of the process since the molecules of the dye can obstruct the irradiated photons from reaching the surface of the photocatalytic material, thus reducing the amount of the produced $^{\cdot}OH$ radicals [59]. However, as long as a saturation of the dye molecule does not occur, the absorption of the dye can continue until the anionic dye's complete coverage of the photocatalyst occurs. The adsorption of the anionic dye on the surface of the catalyst occurs thanks to electrostatic attraction and the interaction of covalent and hydrogen bondings [91]. These interactions between the photocatalyst and the anionic dye will occur until the

Table 1 Nanohybrid materials used in the photocatalytic removal of anionic dyes

Author	Material	Azo dye	Removal
Espinola-Portilla et al. [27]	TiO_2/BDD	Acid blue	96
El-Sayed et al. [26]	NPs de SrO y G-SrO	Anionic dye (carmine) and (methylene blue)	99.5
Rostamzadeh and Sadeghi [91]	Nickel-doped zinc oxide nanoparticles ($Ni_{0.02}Zn_{0.98}O$)	Congo red (CR), methyl orange (MO) (anionic azo dyes) and crystal violet (CV)	$\geq$ 90
Ouni et al. [76]	(ZnS) nanocrystals	Methylene blue (MB) and methyl orange (MO)	99.43
Singh and Bansal [104]	Oxide (CuO) nanoneedles	Victoria blue and direct red 81	> 80
Hoseini and Fath [32]	Cu(0)/CuS nanoneedles	Methylene blue	90
Mathiarasu et al. [59]	Via rare earth element (REE) lanthanum substituted $CaTiO_3$ perovskite catalysts	Reactive black, reactive red, and reactive yellow	> 80
Wan et al. [109]	MnO_2 core-shell nanospheres	Methylene blue (MB), Congo red (CR), and rhodamine B (RhB)	99.5
Rokni et al. [89]	TiO_2 nanoparticles dispersed in chitosan-grafted polyacrylamide matrix	Sirius yellow K-CF	96.81
Mathiarasu et al. [59]	Via rare earth element (REE) lanthanum substituted $CaTiO_3$ perovskite catalysts	Reactive black, reactive red, and reactive yellow	> 80
Wan et al. [109]	MnO_2 core-shell nanospheres	Methylene blue (MB), Congo red (CR), and rhodamine B (RhB)	99.5
Mathiarasu et al. [59]	Via rare earth element (REE) lanthanum substituted $CaTiO_3$ perovskite catalysts	Reactive black, reactive red, and reactive yellow	> 80
Wan et al. [109]	MnO_2 core-shell nanospheres	Methylene blue (MB), Congo red (CR), and rhodamine B (RhB)	99.5
Rokni et al. [89]	TiO_2 nanoparticles dispersed in chitosan-grafted polyacrylamide matrix	Sirius yellow K-CF	96.81
Alzahrani [7]	$Fe_3O_4/SiO_2/TiO_2$ nanospheres	Methyl orange (anionic dye, MO) and methylene blue (cationic dye, MB)	> 30
Thomas and Alexander [107]	Ferrite nanospheres	Methyl orange (MO)	> 80

(continued)

Table 1 (continued)

Author	Material	Azo dye	Removal
Amin et al. [9]	ZIF-8/Chitosan hybrid nanoparticles	Rhodamine B (RB), anionic methyl orange (MO), and cationic methylene blue (MB)	99.9
Khan et al. [45]	TiO_2 nanostructures	Methyl blue, methyl violet, and methyl orange	> 85
Albalwi et al. [6]	Silver nanocomposite alginate	Basic blue 3	70
Mathiarasu et al. [59]	Via rare earth element (REE) lanthanum substituted $CaTiO_3$ perovskite catalysts	Reactive black, reactive red, and reactive yellow	> 80
Wan et al. [109]	MnO_2 core-shell nanospheres	Methylene blue (MB), Congo red (CR), and rhodamine B (RhB)	99.5
Rokni et al. [89]	TiO_2 nanoparticles dispersed in chitosan-grafted polyacrylamide matrix	Sirius yellow K-CF	96.81
Alzahrani [7]	$Fe_3O_4/SiO_2/TiO_2$ nanospheres	Methyl orange (anionic dye, MO) and methylene blue (cationic dye, MB)	> 30
Thomas and Alexander [107]	Ferrite nanospheres	Methyl orange (MO)	> 80
Amin et al. [9]	ZIF-8/Chitosan hybrid nanoparticles	Rhodamine B (RB), anionic methyl orange (MO), and cationic methylene blue (MB)	99.9
Khan et al. [45]	TiO_2 nanostructures	Methyl blue, methyl violet, and methyl orange	> 85

saturation of the surface of the catalyst with the synthetic dye. The saturation of the material can occur at concentrations as low as 10 mg/L for a fixed dose of the catalyst [33]. This effect can reduce the removal efficiency of the anionic dye, caused by the absorption of the irradiated energy to the photocatalyst by the anionic dye, which will act as a blocker [105]. Though, if the amount of the catalyst is increased, the promotion of the oxidation of water will occur, causing an increment in the generation of $^{\cdot}OH$ radicals.

Moreover, a large number of other oxidizing agents, such the superoxide radicals, can be formed by the reaction between the excited electrons and the dissolved oxygen [26, 32, 76, 104, 109]. The generation of these oxidant species also can occur by the formation of $^{\cdot}OH$ radicals on the surface of the catalyst or by the decomposition of H_2O_2 [32]. Also, when a mixture of dyes exists, the solar irradiation of the molecules of one synthetic dye can be excited and form active species that can react and promote the degradation of anionic dyes and, in combination with other reactive species ($^{\cdot}OH$ radicals, $^{\cdot}O_2^-$) can improve their degradation [76].

5.5 Nanohybrid Materials for the Removal of Anionic Dyes by Adsorption Process

Several nanohybrid materials have been reported in the literature for the effective removal of anionic dyes by the adsorption process. These materials are usually mixed with metals or clays to enhance the removal of the anionic dyes by photocatalytic degradation or by improving the materials properties and increasing the actives sites, porosity, or surface area of the material. Among the nanohybrid materials used as adsorbents, we can find some polymeric matrix nanohybrids like graphene oxide, cellulose, chitosan, or some polyacrylic materials [2, 4, 15, 29, 34, 52, 53, 73, 84, 90, 123]. A brief compilation of the used nanohybrid materials is reported in Table 2.

5.6 Factors Involved in the Adsorption of Anionic Dyes

As well as with photocatalytic degradation of anionic dyes, several factors can influence the appropriate removal of anionic dye by the adsorption process. It has been reported that some parameters can affect positively or negatively the removal of anionic dye. Among these parameters, we can find pH, dye and adsorbent dosage, contact time, and temperature [5, 28, 29, 52, 53, 63, 74]. pH can influence the removal of anionic dyes by increasing the protonation or hydroxylation of the surface of the material. Also, pH can reduce the amount of positive or negative charged groups [52]. Some authors reported better adsorption capacities at acidic pH values [5, 15, 28, 34, 52, 53]. However, some authors have reported that the best adsorption efficiencies can be reached at basic pH values [2, 4, 63], but also, it has been reported that nanohybrid materials can remove anionic dyes despite the pH values [90].

The adsorbent dosage has been studied as a controlled parameter in the removal of anionic dyes, according to Li et al. [52], at low adsorbent dosage, the removal of the anionic dye is increased. Increasing the quantity of the adsorbent provokes the rapid taking of the actives sites with low energy, resulting in low availability of the high-energy actives site. Also, the increase in adsorbent doses results in a loss of the surface area by an overlapping of the active sites of the adsorbent [5, 29]. Several authors have reported that an increase in the adsorbent dosage will result in a decrease in adsorption capacity, due to the increasing availability of active sites [15, 84]. On the other hand, it can be thought that increasing the anionic dye concentration will result in a decrease in the adsorption capacity, but no, according to the literature, high initial concentrations of the anionic dye will result in high adsorption efficiencies when a threshold is reached these efficiencies drops drastically, because of the overlapping of the active sites of the adsorbent [29]. It is obvious that the limited concentration of the synthetic dye will be restricted by several factors, such as the characteristics of the material, the pH values, temperature, type of synthetic dye, and contact time [52]. Contact time must be controlled during adsorption experiments since it is a crucial parameter for rapidly increasing adsorption capacity. However, after a specific contact

Table 2 Adsorption studies for the removal of anionic dyes

Author	Material	Azo dye	Removal
Hossain et al. [33]	Nanohybrids kaolinite, TiO_2, and ZnO	Orange G	97.4%
Rajesh et al. [84]	Nanoneedles on graphene oxide sheets (GO-CuONNs)	Brilliant blue, methylene blue, Congo red, and amido black 10B	$\geq$ 50%
Zourou et al. [123]	Graphene oxide (GO)-copper ferrite ($CuFe_2O_4$) nanohybrid	Congo red	> 30%
Ahlawat et al. [4]	Graphene oxide with activated carbon (AC)	Methylene blue and methyl orange	Removal efficiency of 92 and 85% and maximum adsorption capacities of 423.15 and 346.51 mg g^{-1}, respectively
Li et al. [52]	Poly(4-vinylpyridine)–graphene oxide–Fe_3O	Methyl blue	99%
Namvari and Namazi [73]	graphene oxide and Fe_3O_4 nanoparticles	Methylene blue and Congo red	Adsorption capacities of 109.5 and 98.8 mg g^{-1}, respectively
Ferfera-Harrar et al. [29]	Chitosan-grafted-hydrolyzed polyacrylamide as matrix and montmorillonite clay as nanofiller	Basic red 46 textile dye	Adsorption capacities of 1553 mg g^{-1}
Akpomie and Conradie [5]	Solanum tuberosum peel impregnated with silver nanoparticles	Bromophenol blue	Removal of 88.5% of the synthetic dye
Fathy and El-Shafey [28]	Carbon nanotubes/carbon xerogel (CNTs/CX) hybrid loaded with bimetallic catalysts of Fe–Ni	Reactive yellow 160 dye	Adsorption capacities of 167 mg g^{-1}
Ling et al. [53]	Porous magnetic cellulose-based ionic liquid adsorbent	Congo red and methyl blue	Adsorption capacities of 1299.3 and 1068.1 mg g^{-1}, respectively
Hossein Beyki et al. [34]	Poly (O-phenylenediamine)–MgAl@$CaFe_2O_4$ nanohybrid	Congo red	Removal efficiency of 96% and adsorption capacities of 500 mg g^{-1}
Mirzaei et al. [63]	ZIF-8/OND hybrid nanostructures with different percentages of oxidized nanodiamond	Methylene blue	Maximum adsorption capacity of 343 mg g^{-1}

(continued)

Table 2 (continued)

Author	Material	Azo dye	Removal
Rong et al. [90]	Three-dimensional magnetic sulfur/nitrogen co-doped reduced graphene oxide nanomaterials	Methylene blue, Congo red, and neutral red	Maximum adsorption capacities of 171.53, 909.9, and 877.19 mg g^{-1} respectively
Aboelfetoh et al. [2]	Cu_xO nanoparticles, (where, $x = 1, 2$), blended with reduced graphene oxide	Methyl blue and crystal violet	Maximum adsorption capacities of 1200 and 238.6 mg g^{-1}, respectively
Binaeian et al. [15]	TiO_2 nanoparticles dispersed in chitosan-grafted polyacrylamide matrix	Sirius yellow K-CF	Removal efficiency of 96.81% and maximum adsorption capacities of 1000 mg g^{-1}

time, this time can depend on the active sites of the nanohybrid material, temperature, anionic or synthetic dye, etc. The increment in the adsorption of the adsorbate will be a function of the active sites present on the surface of the adsorbent. After these actives sites are gradually occupied, the adsorption capacity of the nanohybrid material will decrease drastically until the equilibrium is reached [5, 15, 29, 33, 84]. Finally, the temperature can increase or decrease the removal efficiency of the anionic dyes. An increase in this parameter can lead to an increment of the collision between the pollutant and the surface of the adsorbent, and this can cause the rapid adsorption of the synthetic dye or reduce their uptake [5, 15, 29, 52, 53].

6 Conclusion. The Sustainable Approach to the Removal of Textile Dyes

Water is the primary medium through which the effects of climate change can be perceived. This element is essential for the planet's health and life means that if its natural cycle is altered, all life is affected. Such alteration involves the natural disturbances caused by global warming and the increasing contamination of water bodies. Water pollution led by anthropogenic actions has drastically increased, given the technification of industrial activities. Water quality is addressed in Sustainable Development Goal 6 (SDG), which focuses on examining different health impacts caused by water-related disease transmission and pollution. The world water needs are drastically increasing, with an estimated growth of up to 50% about current levels for 2050 [112]. This increment implies an urgent necessity to recover water since approximately 80% of industrial and municipal wastewater does not receive proper treatment [13]. The untreated wastewater discharge not only results in environmental affectations such as the destruction of biodiversity and shortage of drinking water but is also directly linked to food chain contamination and disease proliferation [62]. The lack of water supply and sanitation access directly impacts child mortality from infectious diseases in marginalized regions [112]. Even though gastrointestinal diseases are essential concerns, the health effects of polluted water may escalate to graver medical conditions. Parasites, bacteria, and viruses may be present in polluted water, however, chemicals may also be present, increasing the toxicity of the water. Several health issues are associated with the presence of toxic compounds in liquids. Among these, we can name blindness resulting from onchocerciasis [21], neurotrauma and degeneration of brain cells, which affects motor skills [18] and mainly different types of cancer including lymphoma, mesothelioma, and myeloma [22, 35, 114]. The chemical contamination of water occurs when m*etals, sulfates, and different byproducts from industrial activities enter the aquatic ecosystem. Among these industrial activities, the textile industry dramatically impacts the use and contamination of water resources. In terms of utilization,* it is estimated that 93 billion m^3 of water are used by the textile and clothing industry every year. This use implies that a piece of clothing requires, on average, 3.7 m^3 of water [94].

On environmental effects, the textile industry produces 20% of the world's wastewater and 10% of the global CO_2 emissions, with a mean value of 33.4 kg of carbon equivalent per piece [112]. Also, the industry generates 92 million tonnes of waste that are incinerated or landfilled, and around 35% of microplastics are released to the environment [64]. Nevertheless, one of the leading environmental impacts of textiles comes from dyeing processes. The intensive use of synthetic dyes has resulted in the accumulation of sulfur, naphthol, nitrates, acetic acid, enzymes, chromium compounds, and heavy metals such as copper, arsenic, lead, cadmium, mercury, nickel, and cobalt [48]. Other harmful chemicals in dye water can include formaldehyde-based dye fixing agents, chlorinated stain removers, hydrocarbon-based fabric softeners, and non-biodegradable dye chemicals [116]. If dyed water is not purified correctly and treated, the colloidal matter, the coloring substances, and the oily foam increase the turbidity of the water, causing odors [50]. In addition, turbidity prevents the sunlight penetration necessary for the photosynthesis process, affecting the oxygen transfer mechanism between the water and the atmosphere [31]. In this regard, applying natural dyes in artisanal techniques would significantly reduce health and environmental effects. However, this is economically unfeasible since the dyeing per kg of fabric with industrial dyes is estimated to be 350% cheaper than artisan dyeing (Impact and Good).

Therefore, as industrial dyeing continues to dominate in the textile sector, the focus is on wastewater treatment. Although some treatment methods for removing these pollutants from water have been documented, most of the existing methodologies still need to provide a more circular and sustainable approach to avoid side effects on ecosystems and generate economic feasibility to scale toward market applications. As discussed in this chapter, a key element to improve the technical efficiency in dye removal is the type of materials used in the chemical reactions. Since the final goal from the sustainability perspective is not only technical effectiveness but also a reduction in the socio-environmental impacts and financial profitability, wastewater treatment must seek the application of materials with the proper balance of synthetic elements with organic and inorganic components. Due to this need, nanohybrid materials have gained special attention due to their flexibility and capability to be adapted to existing methodologies for removing hazardous elements in the water. Different nanohybrid materials have recently reported sustainable results in dye removal. Hossain et al. [33] used chitosan and a ternary photocatalyst fabricated with kaolinite to remove anionic dyes from water and to obtain a high recyclability rate. Also, efficient dye desalination has been achieved by applying graphene metal nanohybrids and nanocomposite membranes [60].

Moreover, elements such as biochar-based nanohybrid materials have shown good technical features and low-cost results with environmental benefits due to the reduction of energy consumption. Finally, nanohybrid materials have reported economic feasibility in biological oxidation with sequential batch reactors reducing operating costs in the range of 24–39% [88]. In general, high removal efficiencies and long-term reusability with less environmental impact have been achieved by using nanohybrid materials [8, 33, 41, 102].

The environmental pollution resulting from textile products can be reduced by implementing nanohybrid materials to treat water used in dyeing activities. The nanohybrid materials represent a promising advance in sustainable methods for removing hazardous azo dyes from wastewater. It must be noted that since the recycling of textiles represents significant constraints due to the limitations in separating fabrics made from blended fibers, the circular economy approach must be based on the implementation of technologies aimed at reducing the environmental impact of the production process. Also, natural resources must be gradually integrated into dyeing processes. Although the scope remains reduced, there is evidence of the dyeing and printing of knitted fabric with natural dyes with efficient and sustainable results [92]. Therefore, such manufacturing systems can be gradually introduced in sewn fabrics with higher industrialized processes. Based on nanohybrid materials, these remediation techniques would significantly reduce the environmental impacts of the fashion industry.

References

1. Abega AV, Ngomo HM, Nongwe I et al (2019) Easy and convenient synthesis of CNT/TiO_2 nanohybrid by in-surface oxidation of Ti^{3+} ions and application in the photocatalytic degradation of organic contaminants in water. Synth Met 251:1–14. https://doi.org/10.1016/j.synthmet.2019.03.012
2. Aboelfetoh EF, Elhelaly AA, Gemeay AH (2018) Synergistic effect of Cu(II) in the one-pot synthesis of reduced graphene oxide (rGO/CuxO) nanohybrids as adsorbents for cationic and anionic dyes. J Environ Chem Eng 6:623–634. https://doi.org/10.1016/j.jece.2017.12.047
3. Aghazadeh Asl E, Pooresmaeil M, Namazi H (2023) Chitosan coated MOF/GO nanohybrid as a co-anticancer drug delivery vehicle: synthesis, characterization, and drug delivery application. Mater Chem Phys 293:126933. https://doi.org/10.1016/j.matchemphys.2022.126933
4. Ahlawat W, Dilbaghi N, Kumar S (2021) Evaluation of graphene oxide and its composite as potential sorbent for removal of cationic and anionic dyes. Mater Today Proc 45:5500–5505. https://doi.org/10.1016/j.matpr.2021.02.215
5. Akpomie KG, Conradie J (2020) Biogenic and chemically synthesized Solanum tuberosum peel–silver nanoparticle hybrid for the ultrasonic aided adsorption of bromophenol blue dye. Sci Rep 10:1–18. https://doi.org/10.1038/s41598-020-74254-y
6. Albalwi H, Abou El Fadl FI, Ibrahim MM, Abou Taleb MF (2022) Catalytic activity of silver nanocomposite alginate beads for degradation of basic dye: kinetic and isothermal study. Appl Organomet Chem 36(1). https://doi.org/10.1002/aoc.v36.1; https://doi.org/10.1002/aoc.6490
7. Alzahrani E (2017) Photodegradation of binary azo dyes using core-shell $Fe_3O_4SiO_2TiO_2$ nanospheres. Am J Anal Chem 08:95–115. https://doi.org/10.4236/ajac.2017.81008
8. Amin MO, Al-Hetlani E (2021) Development of efficient SALDI substrate based on Au–TiO_2 nanohybrids for environmental and forensic detection of dyes and NSAIDs. Talanta 233:122530. https://doi.org/10.1016/j.talanta.2021.122530
9. Amin P, Shojaei A, Hamzehlouyan T (2022) ZIF-8/Chitosan hybrid nanoparticles with tunable morphologies as superior adsorbents towards both anionic and cationic dyes for a broad range of acidic and basic environments. Microporous Mesoporous Mater 343:112149. https://doi.org/10.1016/j.micromeso.2022.112149
10. Artioli Y (2008) The chemistry of adsorption 60–65

11. Barrios-Estrada C, de Jesús R-A, Muñoz-Gutiérrez BD et al (2018) Emergent contaminants: endocrine disruptors and their laccase-assisted degradation—a review. Sci Total Environ 612:1516–1531. https://doi.org/10.1016/j.scitotenv.2017.09.013
12. Bastami TR, Ahmadpour A (2016) Preparation of magnetic photocatalyst nanohybrid decorated by polyoxometalate for the degradation of a pharmaceutical pollutant under solar light. Environ Sci Pollut Res 23:8849–8860. https://doi.org/10.1007/s11356-015-5985-2
13. Bijekar S, Padariya HD, Yadav VK et al (2022) The state of the art and emerging trends in the wastewater treatment in developing nations. Water (Switzerland) 14:1–19. https://doi.org/10.3390/w14162537
14. Bilal M, Rizwan K, Rahdar A et al (2022) Graphene-based porous nanohybrid architectures for adsorptive and photocatalytic abatement of volatile organic compounds. Environ Pollut 309:119805. https://doi.org/10.1016/j.envpol.2022.119805
15. Binaeian E, Babaee Zadvarzi S, Yuan D (2020) Anionic dye uptake via composite using chitosan-polyacrylamide hydrogel as matrix containing TiO_2 nanoparticles: comprehensive adsorption studies. Int J Biol Macromol 162:150–162. https://doi.org/10.1016/j.ijbiomac.2020.06.158
16. Biswal A, Swain SK (2022) Nanohybrid materials. In: Swain S (ed) Nanohybrid materials for water purification. Springer Nature, Singapore, pp 23–46
17. Blasques RV, Pereira MAA, Mendes AMRV et al (2020) Synthesis and characterization of a new ceramic nanomaterial SiO_2/NPsSm$_2$O$_3$/C-graphite for the development of electrochemical sensors. Mater Chem Phys 243.https://doi.org/10.1016/j.matchemphys.2019.122255
18. Bondy SC, Campbell A (2018) Water quality and brain function. Int J Environ Res Public Health 15.https://doi.org/10.3390/ijerph15010002
19. Chen L, Chuang Y, Nguyen TB et al (2020) Novel molybdenum disulfide heterostructure nanohybrids with enhanced visible-light-induced photocatalytic activity towards organic dyes. J Alloys Compd 848:156448. https://doi.org/10.1016/j.jallcom.2020.156448
20. Chennimalai M, Vijayalakshmi V, Senthil TS, Sivakumar N (2021) One-step green synthesis of ZnO nanoparticles using Opuntia humifus fruit extract and their antibacterial activities. Mater Today Proc 47:1842–1846. https://doi.org/10.1016/j.matpr.2021.03.409
21. Colebunders R, Kaiser C, Basáñez MG et al (2022) Reducing onchocerciasis-associated morbidity in onchocerciasis-endemic foci with high ongoing transmission: a focus on the children. Int J Infect Dis 116:302–305. https://doi.org/10.1016/j.ijid.2022.01.042
22. Cui Y, Liang L, Zhong Q et al (2017) The association of cancer risks with pentachlorophenol exposure: focusing on community population in the areas along certain section of Yangtze River in China. Environ Pollut 224:729–738. https://doi.org/10.1016/j.envpol.2016.12.011
23. Cullity BD (1956) Elements of X-ray diffraction Addison-Wesley metallurgy series. J Chem Inf Model 53:1689–1699
24. Debbarma K, Debnath B, Sarkar PP (2022) A comprehensive review on the usage of nanomaterials in asphalt mixes. Constr Build Mater 361:129634. https://doi.org/10.1016/j.conbuildmat.2022.129634
25. Dhandapani P, Balan B, Dinadayalane T, Angaiah S (2022) In-situ grown of $FeCo_2O_4$ @ 2D-Carbyne coated nickel foam—a newer nanohybrid electrode for high performance asymmetric supercapacitors. J Energy Storage 56:105943. https://doi.org/10.1016/j.est.2022.105943
26. El-Sayed F, Hussien MSA, Alabdulaal TH et al (2022) Study of catalytic activity of G-SrO nanoparticles for degradation of cationic and anionic dye and comparative study photocatalytic and electro & photo-electrocatalytic of anionic dye degradation. J Mater Res Technol 20:959–975. https://doi.org/10.1016/j.jmrt.2022.07.108
27. Espinola-Portilla F, Navarro-Mendoza R, Gutiérrez-Granados S et al (2017) A simple process for the deposition of TiO_2 onto BDD by electrophoresis and its application to the photoelectrocatalysis of Acid Blue 80 dye. J Electroanal Chem 802:57–63. https://doi.org/10.1016/j.jelechem.2017.08.041
28. Fathy NA, El-Shafey S (2022) Carbon-based nanohybrid fabricated in-situ and boosted the adsorption of anionic reactive yellow dye. Int J Environ Sci Technol.https://doi.org/10.1007/s13762-022-04061-7

29. Ferfera-Harrar H, Benhalima T, Sadi A (2022) Development of functional chitosan-based superabsorbent hydrogel nanocomposites for adsorptive removal of basic red 46 textile dye. Springer, Berlin
30. Haniffa MACM, Munawar K, Chee CY et al (2021) Cellulose supported magnetic nanohybrids: synthesis, physicomagnetic properties and biomedical applications-A review. Carbohydr Polym 267:118136. https://doi.org/10.1016/j.carbpol.2021.118136
31. Hassaan MA, Nemr A El (2017) To cite this article: Mohamed A. Hassaan, Ahmed El Nemr. Health and environmental impacts of dyes: mini review. Am J Environ Sci Eng 1:64–67. https://doi.org/10.11648/j.ajese.20170103.11
32. Hoseini SJ, Fath RH (2016) Formation of nanoneedle Cu(0)/CuS nanohybrid thin film by the disproportionation of a copper(i) complex at an oil-water interface and its application for dye degradation. RSC Adv 6:76964–76971. https://doi.org/10.1039/c6ra14995b
33. Hossain MR, Rashid TU, Lata NP et al (2022) Fabrication of novel nanohybrid material for the removal of Azo Dyes from wastewater. J Compos Sci 6.https://doi.org/10.3390/jcs6100304
34. Hossein Beyki M, Alijani H, Fazli Y (2016) Poly o-phenylenediamine-MgAl@$CaFe_2O_4$ nanohybrid for effective removing of lead(II), chromium(III) and anionic azo dye. Process Saf Environ Prot 102:687–699. https://doi.org/10.1016/j.psep.2016.04.027
35. Hu J, Liu J, Li J et al (2021) Metal contamination, bioaccumulation, ROS generation, and epigenotoxicity influences on zebrafish exposed to river water polluted by mining activities. J Hazard Mater 405:124150. https://doi.org/10.1016/j.jhazmat.2020.124150
36. Intisar A, Ramzan A, Sawaira T et al (2022) Occurrence, toxic effects, and mitigation of pesticides as emerging environmental pollutants using robust nanomaterials—a review. Chemosphere 293:133538. https://doi.org/10.1016/j.chemosphere.2022.133538
37. Jafari H, Afshar S, Zabihi O, Naebe M (2016) Enhanced photocatalytic activities of TiO_2-SiO_2 nanohybrids immobilized on cement-based materials for dye degradation. Res Chem Intermed 42:2963–2978. https://doi.org/10.1007/s11164-015-2190-3
38. Jones K, Boxall C, McCabe R et al (2007) Nanocomposites for water treatment. Springer Nature, Singapore
39. Jung DH, Sharma A, Jung JP (2018) Influence of dual ceramic nanomaterials on the solderability and interfacial reactions between lead-free Sn-Ag-Cu and a Cu conductor. J Alloys Compd 743:300–313. https://doi.org/10.1016/j.jallcom.2018.02.017
40. Kalimuthu P, Kim Y, Subbaiah MP et al (2022) Novel magnetic Fe@NSC nanohybrid material for arsenic removal from aqueous media. Chemosphere 308:136450. https://doi.org/10.1016/j.chemosphere.2022.136450
41. Karami K, Beram SM, Bayat P et al (2022) A novel nanohybrid based on metal–organic framework MIL101−Cr/PANI/Ag for the adsorption of cationic methylene blue dye from aqueous solution. J Mol Struct 1247:131352. https://doi.org/10.1016/j.molstruc.2021.131352
42. Katheresan V, Kansedo J, Lau SY (2018) Efficiency of various recent wastewater dye removal methods: a review. J Environ Chem Eng 6:4676–4697. https://doi.org/10.1016/j.jece.2018.06.060
43. Khaki MRD, Shafeeyan MS, Raman AAA, Daud WMAW (2017) Application of doped photocatalysts for organic pollutant degradation—a review. J Environ Manage 198:78–94. https://doi.org/10.1016/j.jenvman.2017.04.099
44. Khan FSA, Mubarak NM, Tan YH et al (2020) Magnetic nanoparticles incorporation into different substrates for dyes and heavy metals removal—a review. Environ Sci Pollut Res 27:43526–43541. https://doi.org/10.1007/s11356-020-10482-z
45. Khan SB, Hou M, Shuang S, Zhang Z (2017) Morphological influence of TiO_2 nanostructures (nanozigzag, nanohelics and nanorod) on photocatalytic degradation of organic dyes. Appl Surf Sci 400:184–193. https://doi.org/10.1016/j.apsusc.2016.12.172
46. Krishnan S, Murugesan S, Vasanthakumar V et al (2021) Facile green synthesis of $ZnFe_2O_4$/rGO nanohybrids and evaluation of its photocatalytic degradation of organic pollutant, photo antibacterial and cytotoxicity activities. Colloids Surfaces A Physicochem Eng Asp 611:125835. https://doi.org/10.1016/j.colsurfa.2020.125835

47. Kusworo TD, Kumoro AC, Utomo DP et al (2021) Performance of the crosslinked PVA coated PES-TiO_2 nano hybrid membrane for the treatment of pretreated natural rubber wastewater involving sequential adsorption—ozonation processes. J Environ Chem Eng 9:104855. https://doi.org/10.1016/j.jece.2020.104855
48. Lara L, Cabral I, Cunha J (2022) Ecological approaches to textile dyeing: a review. Sustain 14.https://doi.org/10.3390/su14148353
49. Lee Y, Cha M, So Y et al (2022) Functionalized boron nitride ceramic nanofiltration membranes for semiconductor wastewater treatment. Sep Purif Technol 300:121945. https://doi.org/10.1016/j.seppur.2022.121945
50. Lellis B, Fávaro-Polonio CZ, Pamphile JA, Polonio JC (2019) Effects of textile dyes on health and the environment and bioremediation potential of living organisms. Biotechnol Res Innov 3:275–290. https://doi.org/10.1016/j.biori.2019.09.001
51. Li L, Liu L, Li Z et al (2019) The synthesis of CB[8]/ZnO composites materials with enhanced photocatalytic activities. Heliyon 5:e01714. https://doi.org/10.1016/j.heliyon.2019.e01714
52. Li Y, Lu H, Wang Y et al (2019) Efficient removal of methyl blue from aqueous solution by using poly(4-vinylpyridine)–graphene oxide–Fe_3O_4 magnetic nanocomposites. J Mater Sci 54:7603–7616. https://doi.org/10.1007/s10853-019-03441-8
53. Ling C, Yimin D, Qi L et al (2022) Novel High-efficiency adsorbent consisting of magnetic Cellulose-based ionic liquid for removal of anionic dyes. J Mol Liq 353.https://doi.org/10.1016/j.molliq.2022.118723
54. Liu BL, Ma YX, Wang JW et al (2022) Fabrication of nickel cobalt bimetallic sulfide doped graphite carbon nanohybrids as electrode materials for supercapacitors. Diam Relat Mater 124:108955. https://doi.org/10.1016/j.diamond.2022.108955
55. Ma J, Shu X, Zheng S et al (2022) Effects of polyurethane–silica nanohybrids as additives on the mechanical performance enhancement of ordinary Portland cement paste. Constr Build Mater 338.https://doi.org/10.1016/j.conbuildmat.2022.127666
56. Martínez-Huitle CA, Scialdone O, Rodrigo MA (2018) Electrochemical water and wastewater treatment. Elsevier Inc.
57. Martínez-Vargas BL, Cruz-Ramírez M, Díaz-Real JA et al (2019) Synthesis and characterization of n-ZnO/p-MnO nanocomposites for the photocatalytic degradation of anthracene. J Photochem Photobiol A Chem 369:85–96. https://doi.org/10.1016/j.jphotochem.2018.10.010
58. Martínez-Vargas BL, Durón-Torres SM, Bahena D et al (2019) One-pot synthesis of ZnO–Ag and ZnO–Co nanohybrid materials for photocatalytic applications. J Phys Chem Solids 135:109120. https://doi.org/10.1016/j.jpcs.2019.109120
59. Mathiarasu RR, Manikandan A, Panneerselvam K et al (2021) Photocatalytic degradation of reactive anionic dyes RB5, RR198 and RY145 via rare earth element (REE) lanthanum substituted CaTiO3 perovskite catalysts. J Mater Res Technol 15:5936–5947. https://doi.org/10.1016/j.jmrt.2021.11.047
60. Mehrabi N, Aich N (2022) Using deep Eutectic solvents for the in-situ synthesis of graphene-metal nanohybrids and nanocomposite membranes for dye desalination. SSRN Electron J 11:109101. https://doi.org/10.2139/ssrn.4111004
61. Meireles G, Daam MA, Sanches ALM et al (2018) Red disperse dyes (DR 60, DR 73 and DR 78) at environmentally realistic concentrations impact biochemical profile of early life stages of zebrafish (*Danio rerio*). Chem Biol Interact 292:94–100. https://doi.org/10.1016/j.cbi.2018.07.007
62. Mendoza-Cano O, Sánchez-Piña RA, Barrón-Quintana J et al (2017) Riesgos potenciales de salud por consumo de agua con arsénico en Colima, México. Salud Publica Mex 59:34. https://doi.org/10.21149/8413
63. Mirzaei K, Mohammadi A, Jafarpour E et al (2022) Improved adsorption performance of ZIF-8 towards methylene blue dye by hybridization with nanodiamond. J Water Process Eng 50.https://doi.org/10.1016/j.jwpe.2022.103254
64. Mishra PK, Izrayeel AMD, Mahur BK et al (2022) A comprehensive review on textile waste valorization techniques and their applications. Environ Sci Pollut Res 29:65962–65977. https://doi.org/10.1007/s11356-022-22222-6

65. Modi A, Bellare J (2019) Efficient removal of dyes from water by high flux and superior antifouling polyethersulfone hollow fiber membranes modified with ZnO/cGO nanohybrid. J Water Process Eng 29:100783. https://doi.org/10.1016/j.jwpe.2019.100783
66. Mohammed MKA (2020) Carbon nanotubes loaded ZnO/Ag ternary nanohybrid with improved visible light photocatalytic activity and stability. Optik (Stuttg) 217:164867. https://doi.org/10.1016/j.ijleo.2020.164867
67. Moreira FC, Boaventura RAR, Brillas E, Vilar VJP (2017) Electrochemical advanced oxidation processes: a review on their application to synthetic and real wastewaters. Appl Catal B Environ 202:217–261. https://doi.org/10.1016/j.apcatb.2016.08.037
68. Mudhoo A, Paliya S, Goswami P et al (2020) Fabrication, functionalization and performance of doped photocatalysts for dye degradation and mineralization: a review. Springer International Publishing
69. Mujahid MH, Upadhyay TK, Khan F et al (2022) Metallic and metal oxide-derived nanohybrid as a tool for biomedical applications. Biomed Pharmacother 155:113791. https://doi.org/10.1016/j.biopha.2022.113791
70. Munir S, Baig MM, Zulfiqar S et al (2022) Synthesis of 2D material based Bi_2O_3/MXene nanohybrids and their applications for the removal of industrial effluents. Ceram Int 48:21676–21689. https://doi.org/10.1016/j.ceramint.2022.04.148
71. Murphy AB (2007) Band-gap determination from diffuse reflectance measurements of semiconductor films, and application to photoelectrochemical water-splitting. Sol Energy Mater Sol Cells 91:1326–1337. https://doi.org/10.1016/j.solmat.2007.05.005
72. Mustafa A, Alsafari IA, Somaily HH et al (2023) Fabrication, characterization of NiO–Co_3O_4/rGO based nanohybrid and application in the development of non-enzymatic glucose sensor. Phys B Condens Matter 648:414404. https://doi.org/10.1016/j.physb.2022.414404
73. Namvari M, Namazi H (2014) Clicking graphene oxide and Fe_3O_4 nanoparticles together: an efficient adsorbent to remove dyes from aqueous solutions. Int J Environ Sci Technol 11:1527–1536. https://doi.org/10.1007/s13762-014-0595-y
74. Nidheesh PV (2017) Graphene-based materials supported advanced oxidation processes for water and wastewater treatment: a review. Environ Sci Pollut Res 24:27047–27069. https://doi.org/10.1007/s11356-017-0481-5
75. Nidheesh PV, Zhou M, Oturan MA (2018) An overview on the removal of synthetic dyes from water by electrochemical advanced oxidation processes. Chemosphere.https://doi.org/10.1016/j.chemosphere.2017.12.195
76. Ouni S, Bel Haj Mohamed N, Bouzidi M et al (2021) High impact of thiol capped ZnS nanocrystals on the degradation of single and binary aqueous solutions of industrial azo dyes under sunlight. J Environ Chem Eng 9:105915. https://doi.org/10.1016/j.jece.2021.105915
77. Oyewo OA, Elemike EE, Onwudiwe DC, Onyango MS (2020) Metal oxide-cellulose nanocomposites for the removal of toxic metals and dyes from wastewater. Int J Biol Macromol 164:2477–2496. https://doi.org/10.1016/j.ijbiomac.2020.08.074
78. Pang Y, Yu J, Tang L et al (2018) Magnetic nanohybrid materials for water-pollutant removal. Elsevier Inc.
79. Patole S, Islam M, Aiyer RC, Mahamuni S (2006) Optical studies of ZnO/Ag nanojunctions. J Mater Sci 41:5602–5607. https://doi.org/10.1007/s10853-006-0296-0
80. Rafaie HA, Nor RM, Azmina MS et al (2017) Decoration of ZnO microstructures with Ag nanoparticles enhanced the catalytic photodegradation of methylene blue dye. J Environ Chem Eng 5:3963–3972. https://doi.org/10.1016/j.jece.2017.07.070
81. Rafiq S, Aadil M, Warsi MF et al (2022) NiO nanoparticles and their nanohybrid with flat rGO sheets: as an ideal electroactive material for hybrid capacitor applications. Ceram Int 48:14596–14605. https://doi.org/10.1016/j.ceramint.2022.01.353
82. Raha S, Ahmaruzzaman M (2020) Enhanced performance of a novel superparamagnetic g-C_3N_4/NiO/ZnO/Fe_3O_4 nanohybrid photocatalyst for removal of esomeprazole: effects of reaction parameters, co-existing substances and water matrices. Chem Eng J 395:124969. https://doi.org/10.1016/j.cej.2020.124969

83. Rajbhar MK, Das P, Satpati B et al (2019) Joining of two different ceramic nanomaterials for bottom-up fabrication of heterojunction devices. Appl Surf Sci 478:651–660. https://doi.org/10.1016/j.apsusc.2019.02.002
84. Rajesh R, Iyer SS, Ezhilan J et al (2016) Graphene oxide supported copper oxide nanoneedles: an efficient hybrid material for removal of toxic azo dyes. Spectrochim Acta—Part A Mol Biomol Spectrosc 166:49–55. https://doi.org/10.1016/j.saa.2016.05.002
85. Ramírez-Ortega D, Meléndez AM, Acevedo-Peña P et al (2014) Semiconducting properties of ZnO/TiO_2 composites by electrochemical measurements and their relationship with photocatalytic activity. Electrochim Acta 140:541–549. https://doi.org/10.1016/j.electacta.2014.06.060
86. Rawat D, Mishra V, Sharma RS (2016) Detoxification of azo dyes in the context of environmental processes. Chemosphere 155:591–605. https://doi.org/10.1016/j.chemosphere.2016.04.068
87. Rawat D, Sharma RS, Karmakar S et al (2018) Ecotoxic potential of a presumably non-toxic azo dye. Ecotoxicol Environ Saf 148:528–537. https://doi.org/10.1016/j.ecoenv.2017.10.049
88. Rodrigues CSD, Madeira LM, Boaventura RAR (2014) Synthetic textile dyeing wastewater treatment by integration of advanced oxidation and biological processes - Performance analysis with costs reduction. J Environ Chem Eng 2:1027–1039. https://doi.org/10.1016/j.jece.2014.03.019
89. Rokni SE, Haji Seyed Mohammad Shirazi R, Miralinaghi M, Moniri E (2020) Efficient adsorption of anionic dyes onto magnetic graphene oxide coated with polyethylenimine: kinetic, isotherm, and thermodynamic studies. Res Chem Intermed 46:2247–2274.https://doi.org/10.1007/s11164-020-04090-2
90. Rong Y, Huang Y, Jin P et al (2020) Highly efficient removal of cationic, anionic and neutral dyes by hierarchically porous structured three-dimensional magnetic sulfur/nitrogen co-doped reduced graphene oxide nanohybrid. J Water Process Eng 37.https://doi.org/10.1016/j.jwpe.2020.101345
91. Rostamzadeh D, Sadeghi S (2022) Ni doped zinc oxide nanoparticles supported bentonite clay for photocatalytic degradation of anionic and cationic synthetic dyes in water treatment. J Photochem Photobiol A Chem 431:113947. https://doi.org/10.1016/j.jphotochem.2022.113947
92. Roy Maulik S, Debnath C, Pandit P (2022) Sustainable dyeing and printing of knitted fabric with natural dyes. In: Maity S, Rana S, Pandit P, Singha K (eds) Advanced knitting technology. Woodhead Publishing, pp 537–565
93. Saini PK, Kumar N, Chandra R et al (2019) Facile synthesis of novel SWCNT/HgS nanohybrid: an effective photocatalyst for degradation of methylene blue. Mater Lett 250:5–8. https://doi.org/10.1016/j.matlet.2019.04.090
94. Šajn N (2019) Environmental impact of the textile and clothing industry. What consumers need to know. Eur Parliam Res Serv
95. Sangtam AR, Richa K, Saikia P et al (2022) Synthesis and characterization of Co(II)–Co(III) LDH and Ac@Co(II)–Co(III) LDH nanohybrid and study of its application as bactericidal agents. Results Chem 4:100671. https://doi.org/10.1016/j.rechem.2022.100671
96. Schneider J, Bahnemann D (2016) Photocatalysis fundamentals and perspectives
97. Shakeel A, Rizwan K, Farooq U et al (2022) Advanced polymeric/inorganic nanohybrids: an integrated platform for gas sensing applications. Chemosphere 294:133772. https://doi.org/10.1016/j.chemosphere.2022.133772
98. Shanker U, Rani M, Jassal V (2017) Degradation of hazardous organic dyes in water by nanomaterials. Environ Chem Lett 15:623–642. https://doi.org/10.1007/s10311-017-0650-2
99. Sharma M, Poddar M, Gupta Y et al (2020) Solar light assisted degradation of dyes and adsorption of heavy metal ions from water by CuO–ZnO tetrapodal hybrid nanocomposite. Mater Today Chem 17:100336. https://doi.org/10.1016/j.mtchem.2020.100336
100. Sharma M, Sondhi H, Krishna R et al (2020) Assessment of GO/ZnO nanocomposite for solar-assisted photocatalytic degradation of industrial dye and textile effluent. Environ Sci Pollut Res 27:32076–32087. https://doi.org/10.1007/s11356-020-08849-3

101. Siddiqui SI, Chaudhry SA (2019) Nanohybrid composite Fe_2O_3-ZrO_2/BC for inhibiting the growth of bacteria and adsorptive removal of arsenic and dyes from water. J Clean Prod 223:849–868. https://doi.org/10.1016/j.jclepro.2019.03.161
102. Singh J, Kumar S, Rishikesh, et al (2020) Fabrication of ZnO–TiO_2 nanohybrids for rapid sunlight driven photodegradation of textile dyes and antibiotic residue molecules. Opt Mater (Amst) 107:110138. https://doi.org/10.1016/j.optmat.2020.110138
103. Singh J, Sahu K, Satpati B et al (2019) Facile synthesis, structural and optical properties of Au-TiO_2 plasmonic nanohybrids for photocatalytic applications. J Phys Chem Solids 135:109100. https://doi.org/10.1016/j.jpcs.2019.109100
104. Singh V, Bansal P (2020) Fabrication and characterization of needle shaped CuO nanoparticles and their application as photocatalyst for degradation of organic pollutants. Mater Lett 261:126929. https://doi.org/10.1016/j.matlet.2019.126929
105. Sultana T, Dey SC, Molla MAI et al (2021) Facile synthesis of TiO_2/Chitosan nanohybrid for adsorption-assisted rapid photodegradation of an azo dye in water. React Kinet Mech Catal 133:1121–1139. https://doi.org/10.1007/s11144-021-02009-5
106. Thangavel S, Pazhamalai P, Krishnamoorthy K et al (2022) Ferroelectric-semiconductor $BaTiO_3$–Ag_2O nanohybrid as an efficient piezo-photocatalytic material. Chemosphere 292.https://doi.org/10.1016/j.chemosphere.2021.133398
107. Thomas B, Alexander LK (2018) Enhanced synergetic effect of Cr(VI) ion removal and anionic dye degradation with superparamagnetic cobalt ferrite meso–macroporous nanospheres. Appl Nanosci 8:125–135. https://doi.org/10.1007/s13204-018-0655-6
108. Uddin T, Hoque E, Bhoumick MC (2020) Facile one-pot synthesis of heterostructure SnO_2/ZnO photocatalyst for enhanced photocatalytic degradation of organic dye. RSC Adv 10:23554–23565. https://doi.org/10.1039/d0ra03233f
109. Wan H, Ge H, Zhang L, Duan T (2019) CS@MnO_2 core-shell nanospheres with enhanced visible light photocatalytic degradation. Mater Lett 237:290–293. https://doi.org/10.1016/j.matlet.2018.11.088
110. Wang J, Wang S (2016) Removal of pharmaceuticals and personal care products (PPCPs) from wastewater: a review. J Environ Manage 182:620–640. https://doi.org/10.1016/j.jenvman.2016.07.049
111. Wang N, Zheng T, Zhang G, Wang P (2016) A review on Fenton-like processes for organic wastewater treatment. J Environ Chem Eng 4:762–787. https://doi.org/10.1016/j.jece.2015.12.016
112. World Bank (2019) Working together for a water-secure world, 1–20
113. Wu W, Zhao Z, Li M et al (2022) Electrified nanohybrid filter for enhanced phosphorus removal from water. Chemosphere 303:135226. https://doi.org/10.1016/j.chemosphere.2022.135226
114. Yasasve M, Manjusha M, Manojj D et al (2022) Unravelling the emerging carcinogenic contaminants from industrial waste water for prospective remediation by electrocoagulation—a review. Chemosphere 307:136017. https://doi.org/10.1016/j.chemosphere.2022.136017
115. Ye S, Chen Y, Yao X, Zhang J (2020) Simultaneous removal of organic pollutants and heavy metals in wastewater by photoelectrocatalysis: a review. Chemosphere 273:128503. https://doi.org/10.1016/j.chemosphere.2020.128503
116. Zhao Y, Qamar SA, Qamar M et al (2021) Sustainable remediation of hazardous environmental pollutants using biochar-based nanohybrid materials. J Environ Manage 300:113762. https://doi.org/10.1016/j.jenvman.2021.113762
117. Zheng S, Qiao W, Ma J et al (2022) Synthesis of a novel poly-TA-Si nanohybrid with dynamic covalent S-S bonds and its effect on the mechanical strength of cement paste. Constr Build Mater 356:129293. https://doi.org/10.1016/j.conbuildmat.2022.129293
118. Zia J, Aazam ES, Riaz U (2020) Synthesis of nanohybrids of polycarbazole with α-MnO_2 derived from *Brassica oleracea*: a comparison of photocatalytic degradation of an antibiotic drug under microwave and UV irradiation. Environ Sci Pollut Res 27:24173–24189. https://doi.org/10.1007/s11356-020-08149-w

119. Zia J, Aazam ES, Riaz U (2020b) Highly efficient visible light driven photocatalytic activity of MnO_2 and polythiophene/MnO_2 nanohybrids against mixed organic pollutants. J Mol Struct 1207.https://doi.org/10.1016/j.molstruc.2020.127790
120. Zia J, Farhat SM, Aazam ES, Riaz U (2021) Highly efficient degradation of metronidazole drug using $CaFe_2O_4$/PNA nanohybrids as metal-organic catalysts under microwave irradiation. Environ Sci Pollut Res 28:4125–4135. https://doi.org/10.1007/s11356-020-10694-3
121. Zinatloo-Ajabshir S, Baladi M, Salavati-Niasari M (2021) Sono-synthesis of $MnWO_4$ ceramic nanomaterials as highly efficient photocatalysts for the decomposition of toxic pollutants. Ceram Int 47:30178–30187. https://doi.org/10.1016/j.ceramint.2021.07.197
122. Zinatloo-Ajabshir S, Salavati-Niasari M (2017) Photo-catalytic degradation of erythrosine and eriochrome black T dyes using $Nd_2Zr_2O_7$ nanostructures prepared by a modified Pechini approach. Sep Purif Technol 179:77–85. https://doi.org/10.1016/j.seppur.2017.01.037
123. Zourou A, Ntziouni A, Adamopoulos N et al (2022) Graphene oxide-$CuFe_2O_4$ nanohybrid material as an adsorbent of Congo red dye. Carbon Trends 7:100147. https://doi.org/10.1016/j.cartre.2022.100147

Decolourization of Textile Dyes Using CNT-Based Hybrid Materials

Rania Edrees Adam Mohammad, Abdullahi Haruna Birniwa, Shehu Sa'ad Abdullahi, and Ahmad Hussaini Jagaba

Abstract The treatment of wastewater is very significant for sustainable development, industrialization and population growth which leads to discarding of a lot of effluents which are harmful to the environment and water bodies. Dyes and pigments are organic-based substances that are carcinogenic in nature, and their presence as industrial effluents cause serious effect s to the aquatic and terrestrial life. Therefore, an eco-friendly and less expensive techniques need to be developed to overcome this menace. Carbon nanotube (CNT) membrane-based absorbent materials offer a lot of promise due to their high specific surface area, mesoporous structure, customizable surface characteristics, and excellent chemical stability. They can tolerate challenging of wastewater conditions including acidic, basic, and salty environments at high concentrations or at high temperatures because of these qualities. Even though CNT nanoparticles have been extensively studied for their ability to remove organic dyes from organic wastewater systems, engineering challenges still stand in the way of their widespread use. One route to the removal of these pollutants is the utilization of CNT composites which includes magnetic carbon nanotubes as adsorbents, carbon nanotube-based membranes, carbon nanotube-based nanocellulose membranes, carbon nanotube-based buck paper, and it also talked about mechanism of pollutant adsorption in carbon-based nanocellulose membranes. This chapter discusses techniques used for wastewater treatment. Second, CNT adsorbents of dyes

R. E. A. Mohammad
School of Chemical Sciences, Universiti Sains Malaysia, USM, 11800 Gelugor, Penang, Malaysia

A. H. Birniwa (✉)
Department of Chemistry, Sule Lamido University Kafin Hausa, Kafin Hausa 048, Jigawa, Nigeria
e-mail: birniwa01@gmail.com

S. S. Abdullahi (✉)
Department of Polymer Technology, Hussaini Adamu Federal Polytechnic Kazaure, Kazaure 5004, Jigawa, Nigeria
e-mail: usman2434@gmail.com

R. E. A. Mohammad
Faculty of Education, Department of Sciences, Open University of Sudan, Khartoum, Sudan

A. H. Jagaba
Department of Civil Engineering, Abubakar Tafawa Balewa University, Bauchi, Nigeria

© The Author(s), under exclusive license to Springer Nature Singapore Pte Ltd. 2023
A. Ahmad et al. (eds.), *Nanohybrid Materials for Treatment of Textiles Dyes*,
Smart Nanomaterials Technology, https://doi.org/10.1007/978-981-99-3901-5_7

for different wastewater systems are introduced, still in this chapter, it also contains different applications of CNT in wastewater treatment, and also the real wastewater applications. After considering the materials' scalability and practical usability, cost factors are finally determined.

Keywords CNT · Decolourization · Textile dyes · Wastewater treatment

1 Introduction

Water is the vital valuable natural resource for sustaining life, yet in the modern day, it is suffering from serious pollution problems due to unchecked human development and unsustainable usage [1, 2]. The origins and features of the pollutants influence how much and to what extent water contamination occurs, and various industrial and municipal discharges are frequent sources [3]. The development of a water quality management system to guarantee proper distribution of scarce water resources has advanced quickly in response to the exponentially increasing necessitate for water for household, industrial, agricultural and other applications [4]. To guarantee that other sectors' needs are met, the industrial effluent was given top priority while aiming for sustainable water usage [5–7]. This viewpoint suggests that industrial wastewater needs more control and effective effluent management [8]. Among other industrial pollutants, the textile industry's effluent attracted special attention because of several carbocyclic and heterocyclic by-product emitted to the water bodies through their drainage systems, hence, they are carcinogenic in nature [6, 7]. The use of numerous colours in the textile industry has increased quickly due to ongoing industrialization and modernization, resulting in a variety of dyes as well as other pollutants being present in their effluents [5]. Due to their immediate and long-term effects that negatively affect both the environment and human wellbeing, these rogue effluents that are discharged into aquatic bodies constitute critical challenges [9, 10]. Even at extremely modest concentrations, the presence of these dyes gives water bodies a strong hue that makes them unsightly, and their breakdown by-products frequently have a tendency to be possible carcinogens and mutagens [11, 12]. Dyes obtain a more significant part in the dye wastewater released because of its powerful colour and poisonous nature [13]. The dyes are varied aromatic (unsaturated) organic compounds with conjugated aromatic structures, auxochromes, and chromophores [14, 15]. These organic materials have the capacity to absorb light while reflecting colour to the visible spectrum (from 380 to 750 nm) [16]. An estimate is that the global textile industry produces 7×10^7 tonnes of synthetic dyes per year, and that after the dying process, 10% of the dyestuffs are released into the environment [17]. As a result, these textile companies discharge massive amounts of very contaminated and hazardous wastewater, contributing 17% to 20% of all industrial water pollution [18, 19]. The dyes that are immediately discharged into the aquatic environment can degrade into hazardous by-products that are thought to be potentially cancer- and mutagenic-causing to living things [20]. This might be explained by the presence of aromatic chemicals like benzidine and naphthalene [21, 22]. The (lethal dosage) LD50 values for the aromatic azo class of dyes are predicted to vary from 100 to 2000 mg/kg body weight, and thus pose a significant danger to health [23]. Additionally, according to the toxicological investigations, 98% of the dyes had (lethal concentration) LC50 values higher than 1 mg/L for fish and 59% had LC50 values greater than 100 mg/L [24]. As a result, the release of such effluents to the aquatic environment is rigorously governed by several severe rules [3]. The industries are now more attentive about regulating effluent discharge limitations because of the strong legislative legislation enacted to address environmental concerns and the sanctions imposed for their violation [25].

Nanomaterials made of carbon are among the most widely used adsorptive materials; carbon nanotubes (CNTs), an allotropic form of carbon with a tubular structure and uses in the biomedical sector, catalysis, agriculture, and food industries, are one such substance [26, 27]. CNT fundamental composition may be described as rolled-up graphene sheets. Initial observations of concentric graphene tubes led to the designation of them as multi-walled carbon nanotubes (MWCNTs) [28]. Following this finding, single-walled carbon nanotubes (SWCNTs) with only one layer of carbon atoms were purposefully synthesized [29]. The roots of nanotechnology are in environmental restoration, and numerous studies have recently examined the possible uses of nanoadsorbents in order to remove dyes from facilities that treat wastewater. In general, nanoadsorbents are defined as having at least of 1-D between 1 and 100 nm [30]. Numerous nanoadsorbents have been prepared by the scientific community; among them, magnetic carbon nanotube (CNT)-based nanoadsorbents are now widely recognized and used adsorbents for the removal of dyes [18]. The magnetic-CNTs nanoadsorbent has the advantages of easy production, low substance loss and excellent adsorption [31]. Furthermore, the separation of harmful contaminants from an aqueous phase necessitates the use of an external magnetic field [1]. The extraordinary qualities of buckypaper membrane have also attracted a lot of attention [32]. Membranes made of buckypaper (BP) are very strong, flexible, and light. Additionally, they have strong chemical, mechanical, and thermal stabilities [33]. As a result, they have a lot of promise for use as reinforcement, actuators, and filter membranes, in composite materials, among other things. Numerous methods have been devised for their production because of their prominence in the scientific community, including drop casting, shear pressing, CNT drawing, and electrophoretic technique [34]. Due to the remarkable characteristics of BP membranes, such as their great range of water flow and fouling resistance, several researches have been reported on their potential and viability for water purification applications [35]. The study investigations showed that BP membranes are capable of eliminating a wide range of organic substances. However, several of the materials are employed in the physical processes that are now available struggle with fouling, low recovery, somewhat lengthy extraction times, and high levels of harmful solvent consumption [35]. It is suggested that carbon nanotubes (CNTs) be added to current physical processes without changing their structural designs to get over these restrictions and produce water that is cleaner [36]. CNTs are taken into consideration because their hydrophobic hollow pores can absorb the majority of aromatic contaminants while still enabling water to flow through them without encountering any resistance [37]. The functionalization of CNTs with different sensors and groups also enables the selective trapping of important contaminants [38]. CNTs are generating a lot of attention as a matrix for immobilization due to their higher specific surface area, superior adsorption capacity, mass transfer resistance, and stronger enzyme-loading capacity when compared to their traditional equivalents like silica and other polymeric materials [39]. They are made of rolled graphite flakes, which are made of carbon (sp^2), to produce a cylindrical structure [40]. They have particular physical and chemical unique features that make them more essential in a variety of sectors, including strong mechanical strength, high conductivity, a greater surface-to-volume ratio, excellent dispersibility, inertness, compatibility, and good biocompatibility [41, 42]. CNTs have a number of benefits over traditional enzyme carriers, including improved electronic characteristics, quick electron transfer kinetics, and a high ratio of surface to volume. Additionally, in contrast to other carbon-based support materials, the design of sensors enhanced by CNTs has shown to contain unique qualities such as increased sensitivity, low-level limit detection, and quick electrode kinetics [43]. Like other nanostructures, CNTs can also undergo extra functionalization, primarily at their surface, by solubilizing chains or concentrating on specific target molecules, enhancing their performance in practical settings such as biofuel cells or the medicinal sector. Other nanostructures lack and lose this capability [44].

2 Water Pollution

Global problems with water continue to exist; numerous causes, such as urbanization, industrialization and population increase, have repeatedly put a strain on hydrological resources. In addition, the growing need for and use of dyes, pigments and chemicals has accelerated the rivers eutrophication and the expansion of dead regions in a variety of ecosystems [3]. In addition, improper wastewater management and a lack of public solutions exacerbated the problem [5, 45]. The condition of natural bodies of pure water and streams throughout the world has been declining for a number of decades, both wild and human life depend on access to clean water [46]. A healthy lifestyle depends on having access to clean drinking water [20]. Numerous studies have claimed that the effects of climate change would exacerbate these water issues and could lead to more severe droughts, flooding, sea level rise, glacier melt, unequal groundwater distribution, etc. [47, 48]. Water sources that are contaminated can affect people because they may expose them to diseases, dangerous chemicals when plants are watered with them, toxic substances consumed by aquatic life, or contaminated surface water utilized for swimming or other recreational activities [49]. As a result, drinking dirty water directly has a significant negative impact on one's health in developing nations [25]. The regular life cycles of people and aquatic creatures are disrupted by pollution with both natural and anthropogenic components, the naturally existing resources contribute significantly to water pollution and pollutants [50]. The disruption of plant debris and terrestrial animals, which is then carried into streams, results in allochthonous pollution, and each canal has its own internal source of allochthonous contaminants, which arise naturally from microbial activity and the decomposition of aquatic plants [5, 51]. Additionally, certain microorganisms like bacteria and viruses as well as individual chemical components that make up natural organic matter are classed as natural water pollutants, and few microorganisms may have significant and pervasive negative impacts on human health, but most chemical substances are considered to be rather innocuous to human health [52]. Worldwide, rivers have recently had a considerable rise in the quantity and variety of manmade chemical species, and a vast variety of synthetic organic and inorganic contaminants have been added to water sources due to growth in agriculture, industry and disposal practices in sophisticated societies [53]. Among the contaminants produced by industry which are heavy metals, dyes, pigments, and plasticizers, it is quite challenging to effectively remove pollutants from water due to the wide variety of industrially introduced contaminants, including organic and inorganic kinds [6, 7].

3 Dyes

Dyes are colourful substances utilized for the purpose of giving clothing and other fashion products long-lasting hues [5]. There are two basic sorts of dyes: natural and artificial. have been used since the thirteenth century, and their origins involve minerals, plants, and animals, While organic chemistry developed in the late nineteenth century was strongly related to the rise of synthetic dyes [5, 54], natural dyes are developed and used for a variety of purposes, including printing inks, food colouring, textiles, and other commodities [55]. Conversely, synthetic dyes are heavily utilized in the fields of photography, cloth dyeing, coal conversion, and oil refining [56]. Additionally, synthetic dyes are classed according to their chemical composition and method of use [13]. Synthetic dye production and use are significant, and however, various manufacturing processes result in undesired waste of poorly biodegradable effluents [1]. Synthetic dyes are widely used, and inefficiencies in dying techniques result in the decrease of a significant quantity of dye waste, which displays a serious threat to the environment [57]. Complex, synthetic, and difficult-to-degrade molecular compounds are present in the dye and textile industries' wastewater [58]. The pH, colour, oxygen demand, suspended particles and salt content of dye effluents are all high [59]. Additionally, dye waste contains compounds that may be poisonous, mutagenic and carcinogenic to many bacteria, people, and aquatic animals [60]. Due to the massive outflow of wastewater containing colours in high quantities, textile dyeing industries are one of the primary contributors of environmental contamination [5]. The presence of these colourants in water is not only unsightly, but harmful to ecological and biological creatures as well as to human bodies [61]. Ecosystems are hampered in their ability to carry out tasks that are necessary for a healthy and sustainable environment because of contaminations brought on by dye wastewater. Restoration or clean-up work might be quite expensive, and due to ever-stricter environmental regulations, the textile manufacturers are compelled to eradicate colourants from their waste effluents [62]. China, for example, caps the COD allowed at 80 mgL^{-1} and the chrominance at 60 [63]. In a similar vein, the USEPA has established a cap on COD of 163 kg $tonne^{-1}$ of fabric per day depending on the requirements for the discharge of gross dyeing and printing effluent [64]. The COD of dye wastewater might be fifteen times more than what is necessary. Consequently, reducing the emission of dye effluent is a difficult challenge. According to estimates, about 0.7 million tonnes of dyes are produced annually for textile fibres, which expand by more than 80 million tonnes globally each year [65]. Even though these dyes are only discharged in little amounts, they nonetheless have a significant negative impact on the environment.

3.1 Types of Dyes

Dyes substances are aromatic compounds that are attracted to electromagnetic energy or light with a wavelength between 350 and 700 nm. Chromophores and auxochromes' delocalization electron configurations with double bonds that are conjugated and that donate electrons substituents are the components of dyes. Common auxochromes include –COOH, –OH, $-SO_3H$ and $-NH_2$, whereas chromophores often include quinoid rings and –C=O, –N=N–, –C=C– and –C=N [66]. The colour index number was created by the Society of Dyers and Colourists and is used to specify the kind of dyes. After a dye's chemical composition has been determined, an index number with five digits for colours is assigned to it. The dye type is represented by the first number, and the dye shade by the second number, as in CI acid yellow 36 [67]. In addition, dyes are grouped according to their composition or application techniques, which are briefly discussed in this section and presented in Figs. 1 and 2, respectively.

Dye sources in the aquatic environment are very harmful to aquatic living organism, because synthetic dyes are widely utilized and have several uses, and they infiltrate many different areas of the ecosystem. The origins of colours in the water bodies are numerous and both direct and indirect. The tanning, paper and textile industries are considered direct sources since they demand a lot of water and contribute to water pollution by discharging a lot of effluent into aquatic ecosystems. For instance, the Ganga river receives daily releases of around 9×10^3 m^3 of crude waste from tanneries in Jajmau, India [26, 27, 69]. During several phases of the dyeing process, wastewater containing dye is produced, and the type of cloth and dye class have a significant impact on dye loss. For instance, reactive dyes and sulphur may be present in textile effluents after dyeing cellulose fibres in amounts of 10–40% and 10–50%, respectively [19]. A 2×10^{11} L of coloured effluents are produced by the textile sector annually, and a significant amount of industrial wastes that include up to 50% dyes are dumped into the aquatic environment [70, 71]. It was estimated that 2.8×10^5 tonnes of dyes are wasted in the textile industry each year and end up in aquatic habitats. According to estimates, 20% of industrial water

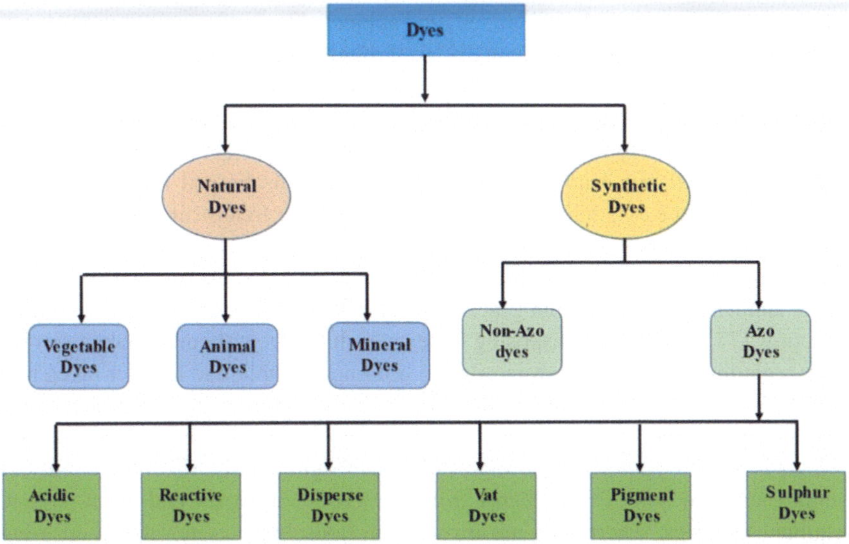

Fig. 1 Various textile's colours used industries [68]. Copy right Elsevier 2022

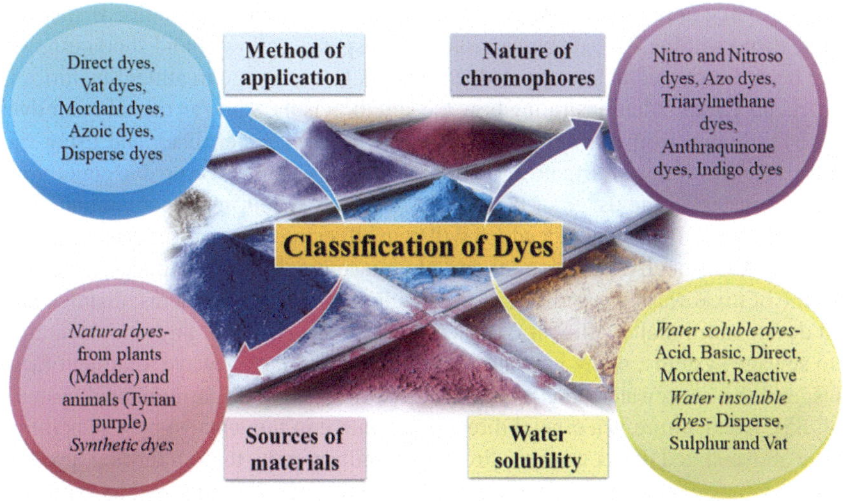

Fig. 2 Classification of dyes. Copyright Elsevier 2021 [37]

pollution originates from the textile industry [72]. Another way that synthetic dyes are introduced into water bodies is through household wastewater, which is created by the release of unused or expired dye, drugs, cosmetics containing dyes, hair dyeing or household chemicals that are released to sewers that lead to the water ecosystem or sewage treatment plants [73, 74]. At sewage treatment facilities, both home and industrial effluents are gathered, mixed, and subjected to first treatment procedures. Industrial wastewater may occasionally go through a pre-treatment process before being released into the municipal sewage system. Primary, secondary and tertiary treatment procedures based on diverse physical, biological and chemical methods are used to treat wastewater in sewage treatment facilities [75]. Sludge, which is primarily dumped into landfills, and wastes that are discharged into water channels directly are the two main by-products of wastewater treatment. The majority of these by-products include various contaminants including dyes, heavy metals and antibiotics [76]. Sewage treatment facilities generate a sizable amount of sewage sludge annually. In 2013, China generated 6.25 108 tonnes of dry sludge, according to records [77]. The proper disposing of sewage sludge is without a doubt a serious issue, regarding this matter, sewage sludge is dumped, applied to land, or taken into consideration for the production of building materials. More than 80% of the sludge generated by sewage treatment plants is found to have been discharged improperly. As a result, dye-polluted sewage sludge can significantly affect the quality of groundwater and soil. The effluents from numerous industries end up at sewage treatment facilities, where they are mixed with household wastewater after going through various procedures. In most cases, synthetic colourants cannot be degraded by traditional sewage treatment facilities. This is why sewage treatment facilities, which encourage compound cycling in the aquatic ecosystem, can be viewed as an indirect source of pollution [78, 79]. Synthetic dyes including rhodamine B, disperse red 1, disperse blue 373, disperse violet 93, disperse yellow 3 and rhodamine 6G have been discovered in sewage sludge/effluents from sewage treatment facilities in Brazil and Taiwan, demonstrating how the situation is the same as that of other dyes [80, 81]. Sludge from sewage treatment plants is used as fertiliser, and using raw industrial wastewater to irrigate agriculture can lead to soil and crop dye contamination. For instance, the usage of tannery effluent may result in the contamination of more than 5.5×10^4 ha of land, which will then affect the efficiency of drinking water [82, 83]. Additionally, the hospital and pharmaceutical sectors produce a sizeable volume of toxic wastewater that is loaded with medications, pigments, disinfectants and dyes. Additionally, effluent is typically discharged without any prior treatment into sewage systems, many hospitals across the world lack sewage treatment facilities, and the wastewater they produce is deposited straight via municipal systems [84, 85]. Aquaculture is a crucial direct dyes source in the water supply in addition to the above-stated direct sources. From 34.6 million units to 66.5 million, production grew between 2001 and 2012, according to records. It is impossible to prevent the expanding usage of medications in the fish farming industry given the scale of output. Few synthetic organic dyes, like malachite green or crystal violet, are easy to produce, inexpensive and very effective, which is why it is illegal to use them to prevent and cure bacterial, fungal and parasite diseases in fish [86]. Malachite green is found in 43% of the fish marketed in Malaysia, whereas studies in Iran found that it is present in almost 60% of fish specimens. Numerous studies demonstrate that despite their known toxicity, synthetic colours are nonetheless appealing to the fish farming industry [87]. Other synthetic colours than malachite green, including

Azure A, Azure B and Victoria pure blue BO have also been discovered in farmed fish [88, 89].

3.2 *Toxic Effects of Dyes*

According to WHO and UNICEF, various water pollutants are responsible for 70–80% of all diseases in underdeveloped countries, mostly impacting women and children. The ability of sunlight to penetrate water is decreased by coloured organic compounds, which may have an impact on aquatic plants' ability to photosynthesis [90]. The toxicity of dyes is mostly caused by several ingredients including aromatics and amines [5, 91]. The majority of dyes are carcinogenic, mutagenic or teratogenic to all living things and have detrimental consequences on human health, including kidney, reproductive, liver and central nervous system dysfunction [92]. According to a United Nations assessment, three out of ten people lack access to safe drinking water [93]. Numerous fatalities due to diarrhoea are caused each year by tainted drinking water. Only 3% of groundwater is clean enough to be used for home and drinking purposes [94]. Half of the world's population by 2025 will reside in areas without access to clean drinking water due to the quantity of numerous novel pollutants that are constantly expanding in hydrological environments and the decreasing supply of clean water. In order to minimize these challenges' unfavourable implications, water remediation will become more difficult [93, 95].

4 Techniques for Wastewater Treatment

There are numerous methods for treating wastewater, and they may be categorized into three categories (physical, biological and chemical methods), along with the benefits and drawbacks of each. Researchers have examined a wide range of microorganisms (fungi, algae, bacteria, etc.) and enzymes for the biodegradation and bioaccumulation of dyes [96]. Algal ingest colour particles during the algae dilapidation process to proliferate. In the enzyme degradation process, colour molecules are broken down by extracted enzyme [96]. *Pseudomonas* sp., *Bacillus* sp. and *Enterococcus* sp. certain aerobic bacteria are employed in the breakdown of dyes. According to reports, the strain *Kurthia* sp. successfully decolourises a range of triphenylmethane dyes, including pararosaniline, magenta, bright green crystal violet, malachite green and ethyl violet (92–100%) [97]. *Escherichia coli* and *Clostridium* sp. are examples of anaerobic bacteria used for colour separation [98]. The removal of different pollutants often involves the employment of chemical procedures such as ozonation, the accelerated oxidation process, the Fenton reaction, electrochemical destruction and photochemical irradiation. UV light is utilized in the ultraviolet irradiation method to disintegrate the contaminants in contaminated water. The wastewater is treated with a variety of oxidizing chemicals, totally converting the dye particles into water and carbon dioxide [91]. The oxidation rate can also be accelerated by employing catalysts. Fenton's reagent eliminates soluble and insoluble pollutants from effluents, but it is not possible to separate dispersion and vat dyes using this procedure due to the lengthy processing time and heavy iron sludge produced. Non-soluble anodes are used in the electrochemical technique of destruction to treat dye effluent [91]. Commercially, chemical contaminants separation techniques are less effective since they are pricy, energy intensive and produce additional problems with dumping noxious secondary pollutants [91]. Adsorption, membrane filtration, ion exchange, reverse osmosis, coagulation or flocculation, radiation, nanofiltration, or ultra-filtration, among other conventional physical pollutant removal techniques, are only a few examples [99]. Compared to approaches for removing biological or chemical pollutants, each of these techniques uses less chemicals [99]. An extremely thin membrane that separates the water from the impurities on one side and the contaminants on the other is used in reverse osmosis, a pressure-dependent process [99]. At the laboratory level, the radiation-based irradiation technique is effective at removing pollutants from polluted water. High-quality treated water was created using the ion exchange method, which is only effective for certain impurities [100]. Adsorption is a remediation technology that is successful, eco-friendly, simple to use and affordable [1]. It can eliminate organic, inorganic and biological contaminants from wastewater in addition to soluble and insoluble pollutants. A schematic of the suggested individual and combination treatments with hybrid materials is shown in Fig. 3 for the treatment of dye wastewater [68].

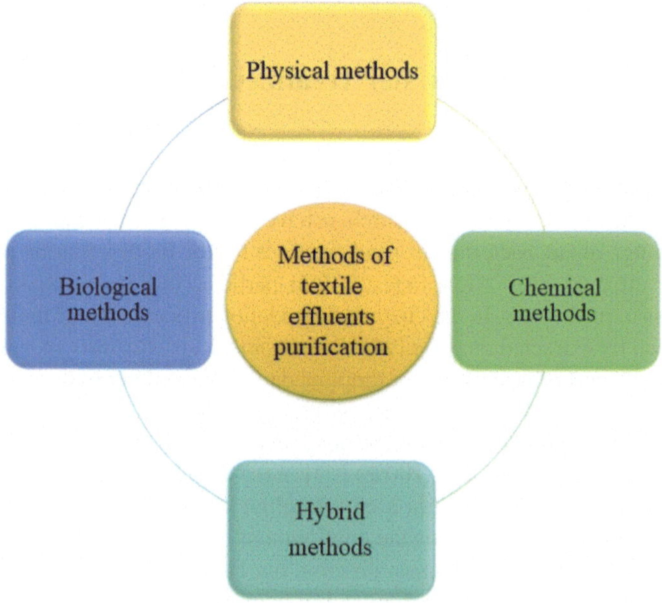

Fig. 3 Available techniques for the treatment of dyes [68]. Copyright Elsevier 2022

4.1 CNT Composites

Composites made of CNTs and graphene over the past decade report that CNTs have been used to remove both organic and inorganic contaminants due to their exceptional qualities, including their quite low weight, strong mechanical properties, stable thermochemistry, and large specific surface area [101]. They have cylindrical tubes made of graphene that have a unique sidewall curvature and a highly hydrophobic surface due to their conjugated structure. They can be categorized as single-walled and multi-walled carbon nanotubes (MWCNT) [101]. The adsorption of the organic and inorganic pollutants is caused by morphology defects and active sites in carbon nanotubes. Chemically modifying CNT to include different functional groups, such as carboxyl, hydroxyl and amine, improves their capacity as active binding sites for removing contaminants [20]. For instance, the oxidation of MWCNTs produces oxygen functional groups that contain on the surface of CNTs. To make them more hydrophilic and to raise the possibility for complexation with pollutants, these chemical functional groups are increased. The adsorption capacity of biowaste was enhanced by the addition of CNTs to plant biomass. Typically, hydrophobic interaction, bonding, hydrogen bonding, covalent and electrostatic interactions are capable for the adsorption of pollutants on CNT composites [102]. According to Hamza et al., a composite made of sugarcane bagasse and MWCNTs works well as an adsorbent to eliminate lead ions. The composite had a monolayer adsorption capacity of 56.6 mg/g, which was twice as much as sugarcane bagasse (23.8 mg/g). The adsorbent's mechanical strength and metal-binding abilities were both boosted by the MWCNTs [103]. The outward appearance changes SEM analysis was used to examine CNTs-A before and after the dyes methyl blue and methylene orange were adsorbed. As shown, the adsorption of dyes on CNTs-A surface results in the production of adsorbent white layer that indicates the formation of dye molecules on the surface as shown in Fig. 4a, b. The diameter of CNTs-A-methylene blue is also larger than that of CNTs-A as shown in Fig. 4c [104].

Comparing untreated hickory biochar, bagasse biochar and CNT-hickory biochar composites for MB, bagasse biochar and CNT composite showed outstanding adsorption performance. The addition of CNTs to bagasse biochar increased pore volume, surface area and thermal stability, resulting in the high adsorption capacity [106]. Due to its distinctive qualities, such as their enormous surface area, durability, exceptional flexibility, electrical conductivity and mechanical resistance, graphene-based composites have become widely employed for wastewater treatment in recent years [107]. However, their limited hydrophilicity restricts their ability to bond with contaminants found in wastewater. As a result, materials having amino, hydroxyl and carboxyl groups are used to treat them to enhance the adsorption application and maybe boost their hydrophilicity. Graphene or its derivatives may be combined with plant-based adsorbents, which are abundant in amino, carboxyl and hydroxyl groups, to create composites that effectively remove pollutants from wastewater. Reactive red 120 was removed using a combination of graphitic carbon nitride (g-C_3N_4) and biochar by photocatalytic degradation and adsorption. The composite was made by

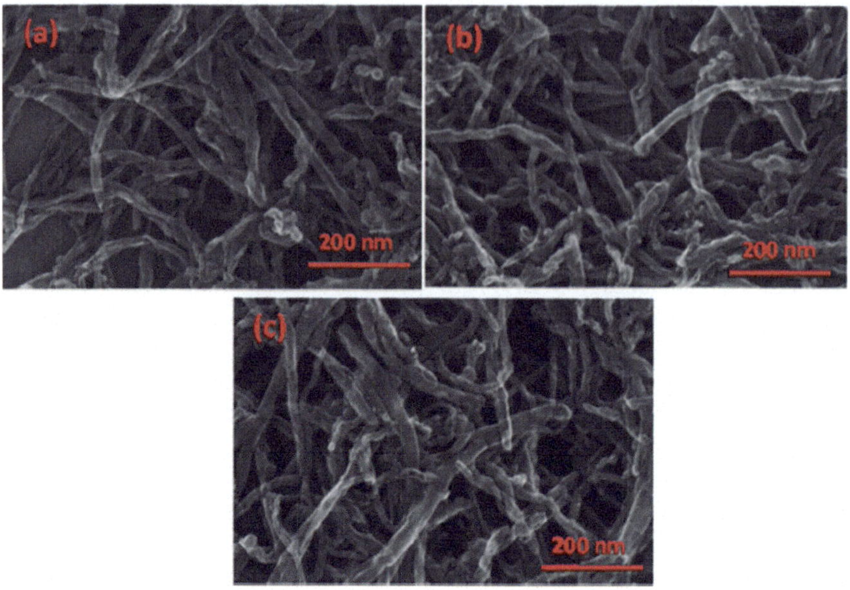

Fig. 4 SEM of **a** CNTs-A, **b** MO-adsorbed CNTs-A, and **c** MB-adsorbed CNTs-A (MO1/4methyl orange, MB1/4methylene blue) [104, 105]. Copyright 2012, American Chemical Society

pyrolyzing urea and hickory chips in a single pot at 520 °C in various ratios. The composite's characteristics showed that the g-C_3N_4 structure, N-containing surface functional groups, decreased surface area and improved thermal stability of the biochar were all introduced during the composite's production. The adsorption of dye occurred due to significant electrostatic affinity that exists between anionic dye and g-C_3N_4 functional groups on the surface of biochar, and modified biochar demonstrated better adsorption capacity than unmodified biochar. The relevance of g-C_3N_4 and N-containing functional groups on the biochar surface to the adsorption process is demonstrated by the fact that the level of dye adsorption on the composites upped as the urea modification ratio was raised [108]. A thorough overview of magnetic-CNTs nanoadsorbent and magnetic-CNT-based BP membranes is given. An overview of the key sources and presence of dyes in the aquatic environment is given first. Then, several methods for removing these contaminants from diverse water sources are presented along with their advantages and disadvantages. The various manufacturing methods for magnetic-CNT-based BP membranes and magnetic-CNT nanoadsorbents have been thoroughly reviewed. The removal of dyes using magnetic-CNT nanoadsorbent and magnetic-CNT-based BP membranes is covered in-depth in this review article based on the literature. Finally, using magnetic-CNT nanoadsorbents and magnetic CNT-based BP membranes as a potential solution for different water contaminants is presented.

4.1.1 Magnetic Carbon Nanotubes for Dyes Removal

Multi-walled carbon nanotubes (MWCNTs) are frequently employed for numerous applications than single-walled carbon nanotubes. The most common method for creating magnetic nanoparticles embedded with MWCNTs is to coat MWCNTs with covalently modified Fe_2O_3 molecules; however, there are a number of other methods for producing MWCNTs [109, 110]. Utilizing an external magnet, a synthesized magnetic-modified MWCNTs may be readily eliminated from water and appropriately dispersed [35]. The use of magnetically modified CNTs for various pollutants found in water sources is now under consideration. Limited research has been done on the relationship between magnetically modified CNTs and pollutants found in wastewater [111, 112].

Magnetic carbon nanotubes (mCNTs) have gained the greatest research attention for a variety of reasons, including its excellent conductivity, high chemical stability and anisotropy and stability. Numerous forms of carbon have been studied by researchers, including buckminsterfullerene, graphene, carbon nanofibre, activated carbon and carbon nanotubes [113]. These materials often exhibit distinctive magnetic properties when coupled with magnetic nanoparticles, and a very capable materials with the potential to be utilized in the progress of several operational domains are magnetic carbon-based materials [114]. Researchers combined carbonaceous materials with magnetic qualities to create porous, stable components that have magnetic properties. Additionally, the porous nature of magnetic carbon-based materials makes them more suitable for use in fields including catalysis, energy generation and storage, and environmental remediation [115, 116]. Notably, magnetic nanoparticles have drawn a lot of concern for use in a variety of applications, including dynamic sealing, magnetic fluid separation, cancer treatment and electrochemical biosensing [117]. As a result, it may be assumed that magnetic materials based on carbon may be capable of being novel magnetic materials. Combining CNTs with nanomagnetic materials may further improve the characteristics [118].

4.1.2 Carbon Nanotube-Based Membranes

According to their method of creation, carbon nanotube-based membranes are commonly divided into two major classes: nanocomposite CNT membranes and free-standing CNT membranes [119]. In applications involving water treatment, free-standing CNT membrane is more far divided into vertically aligned CNT and buckypaper membranes. While CNTs in buckypaper membranes are randomly arranged, they are coupled with cylindrical holes in vertically aligned CNT membranes. The fact that buckypaper has a large specific surface area is one of its advantages [120]. Since CNTs provide low-cost energy options for water treatment, they are appealing in enhanced membrane technologies for water purification. Water travels through CNT-based membranes with almost little friction while retaining a variety of aquatic pollutants. The smooth hydrophobic walls and high aspect ratio enable incredibly efficient water molecule movement [121]. These membranes can improve or take the place of membranes used in the water purification processes of forward and reverse osmosis, nanofiltration and microfiltration. Because of this, reverse osmosis and ultra-filtration membranes may be traded using CNT-based membranes with little to no energy usage [122].

4.1.3 Carbon Nanotube-Based Nanocellulose Membranes

Contrarily, carbon nanotubes are one of the highly studied carbon nanomaterials and have been used in several industries, including biology, water treatment and most recently electricity. CNTs have outstanding mechanical, electrical, thermal and optical properties [119]. CNT-nanocellulose membranes have a lot of potential for use in wastewater treatment when combined with the features of nanocellulose. It is possible to create these composite nanomembranes with a variety of excellent qualities, including high mechanical strength, flexibility, adjustable thermal and electrical conductivity, nanoporous nature and vast adsorption capacity [123]. CNT-based nanocellulose membranes have a lot of potential for use in wastewater treatment when combined with the features of nanocellulose. It is possible to create these composite nanomembranes with a variety of excellent qualities, including high mechanical strength, flexibility, nanoporous nature, high adsorption capacity, adjustable thermal and electrical conductivity. It is significant to highlight that, at the moment, there is little study carried out on the potential use in membranes for removal of contaminants in wastewater; instead, the use of pure nanocellulose is now the centre of attention. At the moment, sensors and energy devices are the main applications for CNT-based nanocellulose membranes [123].

Carbon-Based Nanocellulose Membranes' Pollutant Adsorption Mechanism

Electrostatic interactions between the pollutants and the membrane surface are the fundamental mechanism for the adsorption of pollutants on carbon nanomaterial-based nanocellulose membranes [124]. The integration of nanoparticles in the construction of nanocellulose membranes has been shown to increase the adsorption capacity of the membranes because of the nanoparticles' large surface area and their surface functionalities, which may be customized to target certain contaminants. Functional groups on the surface of membranes, functional nanomaterials and environmental factors like pH all have a significant impact on how pollutants are adsorbed. CNTs and nanomaterials based on graphene are the most often employed carbon-based adsorbents in water treatment. The most often utilized type of graphene is graphene oxide. Depending on their production process and surface functionalization, the nanomaterials includes a lot of oxygen bearing active groups, such as carboxyl, carbonyl, hydroxyl and epoxy [1]. In order to facilitate electrostatic attraction or repulsion between the carbon nanomaterial (GO/CNC) and ionic pollutant molecules, these oxygen active sites provide a negative charge on the nanomaterial's surface. In addition to electrostatic attraction, the bonding of organic pollutants with

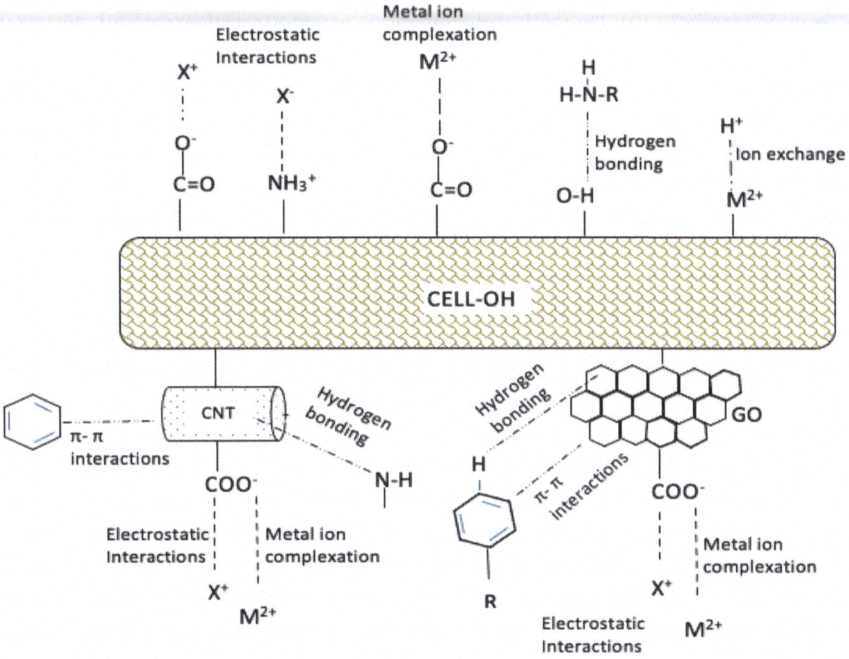

Fig. 5 Adsorption mechanisms in carbon-based nanocellulose membranes [129]. Copyright Elsevier 2022

aromatic molecules in their structural backbone with nanomaterials has also been connected to the adsorption of contaminants, hydrogen bonding and hydrophobic interactions [125–127]. On the other hand, the surface of nanocellulose is likewise heavily embellished with hydroxyl groups as well as other functional groups such as carboxyl and amino groups, by electrostatic interactions, complexation and ion exchange with the desired contaminants; these functional groups are in charge of the adsorption of pollutants [128]. Figure 5 demonstrates some of the likely processes for the adsorption of contaminants from wastewater in carbon-based nanocellulose membranes, because functional groups responsible for binding pollutants are present, the use of CNTs in nanocellulose-based membranes has the benefit of functioning as an adsorbent furthermore to enhancing the mechanical characteristics of the membrane. Since the majority of pollutant adsorption occurs on the membrane surface, it is crucial to consider the closeness of the nanomaterials to the pollutants while creating the membranes [129].

4.1.4 Carbon Nanotube-Based Buckypaper

The simplest type of CNT membrane design is known as "buckypaper," which is a flexible self-supported film made of entangled CNTs and is chemically and physically extremely steady [130]. Due to its advantageous mechanical, thermal and electrical qualities, CNT-based BP is applied in a variety of fields, such as water purification and cold field emission cathodes. The traditional method for creating CNT-based BP includes filtering CNTs via surfactant or solvent solutions, retaining a large surface area and intense self-interactions as a result. It is challenging to construct BP because CNTs agglomerate when subjected to a polymer matrix or solvent [125, 126, 131]. Only a few surfactants, like sodium dodecyl sulphate and triton X-100, have demonstrated effective CNTs dispersion in aqueous phases. Another method for achieving uniform dispersion is chemical modification, commonly known as surface functionalization. The alteration of CNTs increases their hydrophilicity, which in turn causes the CNT suspensions to be stable. Additionally, adding a fluorine or amine group to CNTs has uniform CNT dispersion and cross-links the CNTs inside of a BP [132, 133]. As a result, thorough elimination of surfactants that have been adsorbed at CNTs is a difficult operation since any leftovers might alter the characteristics of BP. Additionally, methods utilized for chemically altering CNTs include many steps, such as sonication and refluxing, and are time consuming [132, 134]. CNTs, particularly single-walled and multi-walled carbon, must be considered since they vary fundamentally from one another and, in the case of SWCNTs, may not be chemically changed without endangering their morphological and electrical characteristics. However, chemical functionalization of MWCNTs is possible without breaking their inner cylinder [125, 126, 135, 136]. Therefore, it could be feasible to incorporate modified MWCNTs into materials while retaining the outstanding properties of CNTs, such as tensile strength and conductivity [137, 138]. Numerous MWCNTs-based membranes have been produced and adapted for a variety of uses, including diodes. Although SWCNTs have shown the potential to be used in light-harvesting applications, the functionalization may not be in the inherent features' favour.

Carbon Nanotube-Based Buckypaper for Dyes Removal

Filtration is avoided due to a number of problems with the currently available membranes, including as their low solute selectivity, brief lifespan and fouling [139, 140]. The formation of novel materials for water purification, desalination, gas separation and many other applications is now receiving a lot of attention [132]. CNTs have gained a plethora of attention for their use as membranes in the filtration of water from various contaminants. CNT-based buckypaper is astonishingly well permeable to gases and liquids, as shown by molecular dynamic simulations; although BP has been known to the scientific community for a long time, its applications in filtration have only recently come to light [141, 142]. As a result, BP may be a suitable option for industrial water treatment [132, 143]. Additionally, several studies have noted that the selectivity of nanoparticles and their ability to filter solute rely on their size. According to published research, the energy consumption of a CNT-based BP membrane, especially for desalination, might be significantly less than that of reverse osmosis because water molecules entered nanotubes at a rate that is 2–5 times greater than predicted by an equation developed by Hagen–Poiseuille [144, 145]. According to a research by Dumée et al., that examined the effectiveness of direct contact membrane distillation using BP membranes based on CNT, the membranes are very porous, thermally conductive and hydrophobic in nature [146]. The produced membranes are useful for desalination since they have a clearance rate of up to 99% for salt from synthetic water [147]. Few studies on the removal of dyes have been identified in the literature since the majority of research has been done on desalination utilizing CNT-based BP membrane [144]. In a different work, Lau et al. [148] synthesized a modified CNT-based BP and added it to polyvinyl alcohol for the methylene blue removal from the aqueous phase. The research found that the anti-fouling CNT-based BP designed for removing humic acid from water removed 99.7% of the methylene blue [149]. The adsorption technique was used to remove humic acid by 65% in this study [150]. In a different work, Lau et al. [148] created enhanced CNT-based BP and integrated it into polyvinyl alcohol for the removal of methylene blue from the aqueous media [148].

5 Applications

Dyes are removed via CNT-based adsorbents, and the elimination of the colours from organic contaminants has received extensive research interest. An overview of the most recent uses of dyes such as organic pollutants will be provided in this section. Several factors make the dyes removal from wastewater crucial: (i) The fundamental reason that the existence of dyes in the water reduces sunlight diffusion because of their colour. This prevents aquatic plants from photosynthesis, which might cause a general disruption in the ecosystems [151]. (ii) It is crucial to remove the dyes in order to have clean drinking water because of their detrimental impact on both humans and wildlife's health [152]. Acid blue 92 (AB 92) was eliminated from water sample using SWCNTs as shown in Fig. 6b. At pH 3 and 0.12 g/L, SWCNTs eliminated 99.4% of the dye in 75 min. Hydrogen bonding, dispersion interactions, dipole–dipole bonds and hydrophobic interactions were all implicated in the adsorption [153]. In a different investigation, reactive yellow 15 (RY 15) and reactive yellow 42 (RY 42) dyes were removed using SWCNTs [154]. Anionic functionalized CNTs were used to remove the cationic dye methylene blue (MB). A combination of chemical and physical mechanisms led to the adsorption [155]. The surface of CNTs is functionalized to add charged groups that can interact with dye molecules that have opposite charges. To eliminate MB, a hydrogel bionanocomposite (HBNC) made of carboxyl-functionalized CNT and tragacanth gum (TG) was designed. In the best circumstances, it demonstrated an extraction efficiency of 80%. The primary benefit of this work is the environmentally friendly nature of TG, a secure and biocompatible polysaccharide derived from plants. When such materials are combined with CNTs, the secondary pollution brought on by the adsorbents is reduced [156]. Methylene violet (MV) and Nile blue (NB) dyes were removed using CNT/MgO/$CuFe_2O_4$ with a specific surface area of 127.52 m^2/g. COD, biochemical oxygen demand (BOD),

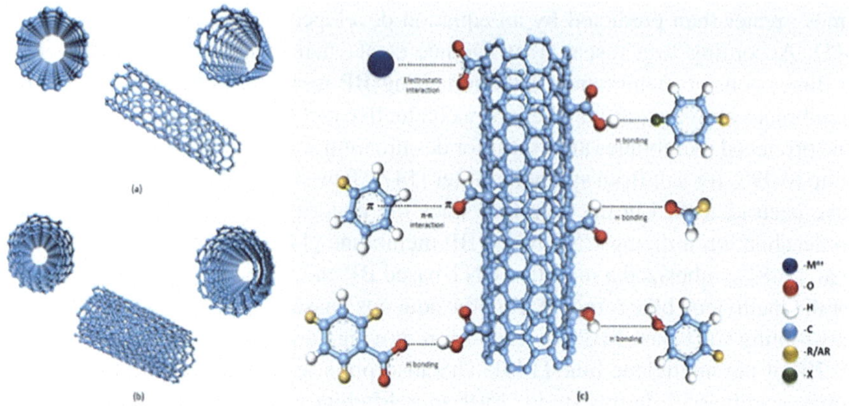

Fig. 6 **a** Single-walled carbon nanotube (SWCNT), **b** multi-walled carbon nanotube (MWCNT) and **c** mechanism of interaction between the chemical functional groups presenting the high adsorption capacity of CNTs for the removal of organic dyes. Reused with permission from Jain et al. [158]. Copyright Elsevier 2021

total organic carbon (TOC) and electrical conductivity levels (ECL) were all much lower after the treatment [36]. Ismate violet 2R (IV 2R) was removed from polluted water using functionalized MWCNTs having a surface area of 182.99 m^2/g. Within 10 min, 85.9% of the dye (10 mg/L) was eliminated by the adsorbent (0.005 g). Since the adsorbent becomes positively charged at low pH values 4, it was better to utilise pH = 4 for removing anionic dye [157].

At pH 12, sulphonic-acid-functionalized CNTs had the excellent MB adsorption, and in this case, the dye is cationic in nature, whereas the adsorbent includes anionic groups. The adsorbent's surface will be positively charged at lower pH levels, and positively charged surface and dye molecules will repel one another electrostatically. Moreover, as pH rises, the surface form negatively charge and may interact with dye molecules more effectively [155]. In another instance, cationic dyes were removed using an adsorbent made of CNT/MgO/$CuFe_2O_4$. When the pH was raised from 2 to 8, the adsorption rate increased, and electrostatic repulsion among the surface and the dye molecules resulted from an abundance of H^+ ions in the solution and positively charged surface of the adsorbent at pH 4, respectively [36]. In addition, cationic dye molecules and H^+ ions are in competition [36]. However, when the pH rises, the amount of H^+ ions in the solution declines, and the surface becomes negatively charged, making it ideal for the dye molecules adsorption. The best adsorption was achieved at pH 8, where the negatively charged adsorbent surface produced electrostatic attractions for CNT/MgO/$CuFe_2O_4$, the point of zero charge equal to 6.8 [36].

For the adsorption of dyes from aqueous media, polymeric materials are a great option due to they provide a large range of pore sizes, chemical stability in diverse environmental conditions, and affinity for aim dyes based on the polymer and dyes functional groups [159]. Methyl orange (MO) dye was removed using a composite made of polyaniline (PANI) and MWCNTs as shown in Fig. 6a. PANI's amine and imine groups can react with the dye molecules' functions. The composite outperformed any individual component in terms of adsorption performance [159]. A flow injection technique rely on mini-column packed 4-aminosalicylic acid functionalized MWCNTs was used to remove the dye crystal violet (CV) from an aqueous solution [159]. The adsorption procedure removed 99% of the dye in 1 min, and after 250 cycles of adsorption and desorption, the sorbent's adsorption behaviour remained unchanged. To enhance the number of active sites for the adsorption of contaminants on CNTs, amine groups were functionalized on the materials and coupled with prophyrine derivatives [160]. When the adsorbent dose was raised from 0.001 to 0.06 g, it demonstrated 100% removal efficiency [160]. To remove MO dye from aqueous environments, poly(acrylonitrile-co-styrene)/MWCNTs were utilized, taking use of the adsorptive qualities of both the polymeric substance and the carbon fibre as shown in Fig. 6c. Polyvinyl alcohol (PVA) and CNTs are mixed because of the hydrogel-like characteristics, biocompatibility and functions that may interact with dyes [161]. To eliminate MB from water, PVA and MWCNTs were combined with the plant material Elaeis guineensis, and after the adsorption procedure, pure CNTs typically have separation issues and need to be either filtered or centrifuged [161]. This influences the process's price and duration, because they are simple to separate with the aid of an external magnet, magnetic MWCNTs are preferred sorbents. Using a magnetic tungsten disulphide/CNTs nanocomposite (WS_2/ Fe_3O_4/CNTs-NC) that was produced, amaranth and brilliant blue FCF dyes were removed [162]. Two-dimensional layered material was used in magnetic composite due of the high specific surface area and several reaction sites [162]. Sonication helped the adsorption process, under ideal circumstances, and this adsorbent was capable of removing 100% [162]. Through a single stage of annealing, a magnetic nanomaterial made of surface oxidized nanocobalt wrapped in nitrogen-doped CNTs (Co@CoO/NC) was synthesized [163]. It had a 679.56 mg/g adsorption ability to eliminate rhodamine B [163]. Fe_3O_4 NPs were added to MWCNTs of various diameters to yield magnetic MWCNT composites, which were then used to remove MB. As CNTs' diameter increased, the adsorption capacity dropped [164]. Chemical vapour deposition was used to develop CNTs from petrochemical waste oil in order to remove cationic and anionic pigments from wastewater [165]. The use of petrochemical waste oil for the adsorbent preparation used for environmental remediation is a beneficial component of this work [165]. Malachite green (MG) and MB were removed using asparagine-functionalized MWCNTs in single and binary systems, and due to its hydroxyl and carboxylic acid groups, the amino acid asparagine (Asp) may facilitate the facile dispersion of CNTs and offer further selectivity towards the target contaminants [166]. Under ideal circumstances, MG and MB both have extremely high adsorption capabilities of 637 and 500 mg/g, respectively, which were established. The adsorption kinetics were incredibly quick, and for MG and MB, equilibrium was reached in 3 and 5 min, respectively [166]. To combine cutting-edge carbon materials with biocompatible and less toxic components, hydrogel composite beads made of alginate/carboxyl functionalized CNT carbon dot (CD) and magnesium fluoro hydroxyapatite (MFA) were obtained [167]. The surfaces of multi-walled CNTs with carboxyl functionalization were altered using MFA and CD, and then, using ultrasonic waves, this was absorbed into the alginate as shown in Fig. 7. Hydrogen bonds, electrostatic interactions and other possible interaction mechanisms were implemented to remove MB [167].

Banerjee et al. described amorphous CNT by a solid-state process and utilized it for the first time to remove the anionic dye methyl orange and the cationic dye rhodamine B from an aqueous solution. Acidic pH was discovered to be favourable for the adsorption of both dyes due to intermolecular H-bonding between COOH groups of rhodamine B and OH groups of CNTs and electrostatic interaction between the positively charged $N(CH_3)_2$ groups of methyl orange and the negatively charged OH group of CNTs as shown in Fig. 8a [168]. Additionally, because of the rise in adsorbent SSA and availability of additional adsorption sites, the quantity of adsorbed dye increases with an increase in adsorbent dose as shown in Fig. 8b. Additionally, lower starting concentrations result in a greater amount of rhodamine B and methyl orange dye being adsorbed on CNTs as shown in Fig. 8c [168].

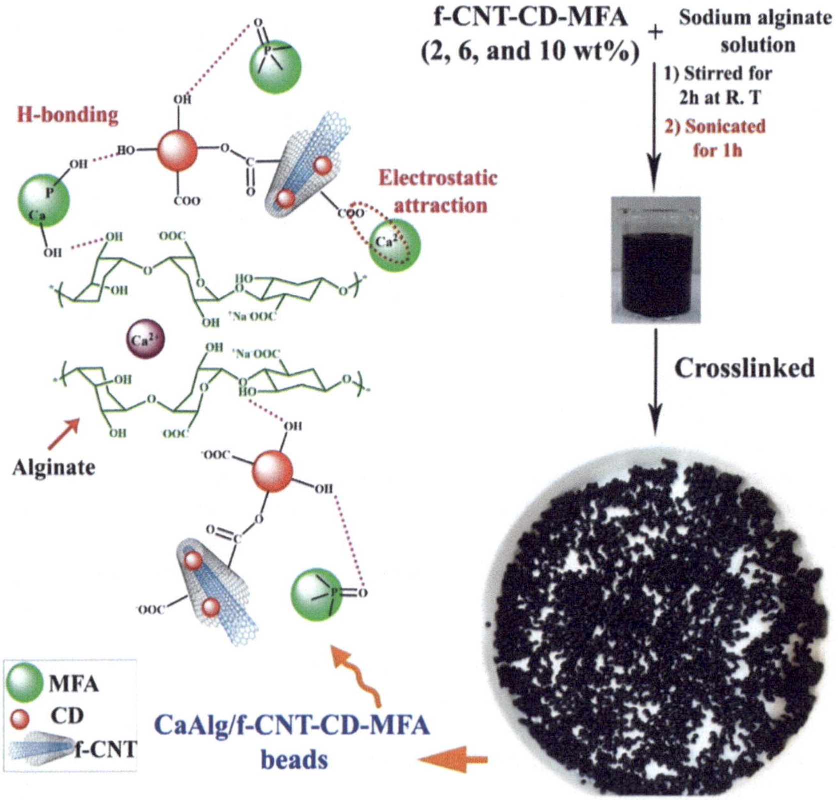

Fig. 7 Preparation procedure of CaAlg/f-CNT-CD-MFA composite beads. Reprinted from Mallakpour et al. [167]

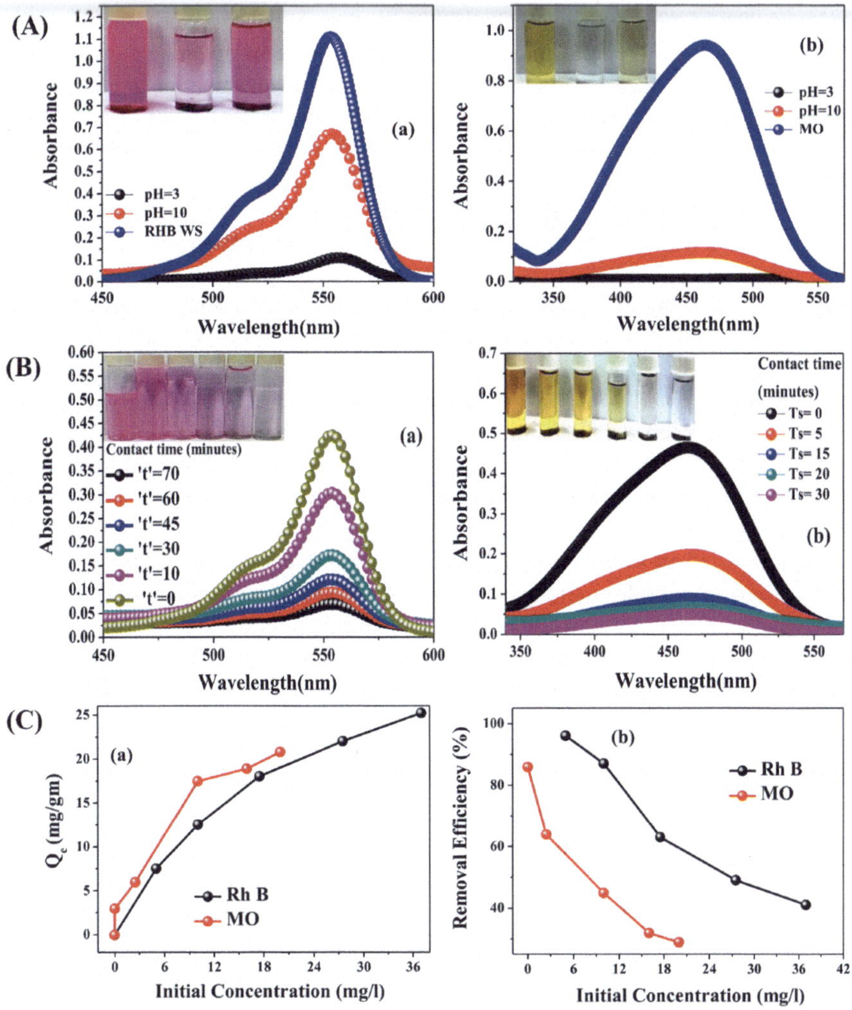

Fig. 8 **A** Effect of solution pH on adsorption of (**a**) RhB and (**b**) MO by a-CNTs, **B** (**a**) effect of contact time on adsorption of (**a**) RhB and (**b**) MO by a-CNTs with regards of colour change of the solution shown inset, and **C** variation of (**a**) RhB and (**b**) MO by a-CNTs dye removal efficiency with initial dye concentration. Copy right Elsevier 2017 [168]

5.1 Activity of CNT-Based Nanofluids on Anti-bacterial and Anti-fungal

In recent years, nanofluids have become widely used in the field of wastewater treatment, showing great promise for antimicrobial treatment and disinfection [1, 169]. The last stage of any general water treatment cycle, disinfection, according to the freshwater rules, is where undesired organic compounds and microbiological contaminants, or more specifically pathogens that might cause waterborne illnesses, are removed. In the general disinfection process, straightforward physical treatment techniques including ozonation, UV irradiation and chlorination are frequently employed [3]. These disinfection techniques each have a unique set of issues and restrictions. Due to serious problems including high total costs and the emergence of hazardous by-products, most infections have become resistant to these conventional treatments (CBPs). The research states that using these outdated standard disinfection techniques results in the formation of more than 600 TBPs. Most bacteria, both gram-positive and gram-negative, are resistant to extreme pressure, temperatures and antibiotics [3]. Therefore, a long-term answer to this unsolved issue may be found in nanoparticles with a greater surface area, smaller particle size and a distinctive crystalline structure. Nanoparticles including CNT, MgO, silver, ZnO, CuO and self-cleaning TiO_2 showed improved behaviour in suppressing microbial activity at low concentrations. Nanoparticles' unfavourable characteristics are overcome by doping them to produce composite or hybrid materials, which function better against a variety of microorganisms. Eco-friendly nanoparticles were created from natural materials to attain biodiversity.

Numerous recent studies contend that the cylindrical form and high volume-to-surface ratio of peptide or lipid nanotubes enable them to penetrate cell barriers and trigger apoptosis. Functionalized SWCNTs might pass-through cell membranes due to their geometrical similarity. In this regard, Kang et al. [170] used extremely well-maintained SWCNTs with a small diameter dispersion to show that contact with SWCNTs can result in significant cellular damage and cell death. Fakhroueian et al. examined the antibacterial effects of a hybrid nanofluid made of ZnO and CNT on gram-positive and gram-negative bacteria [171–174]. A complex matrix was formed because of the synergistic action of nanoparticles, which boosted bacterial cell membrane penetration and, ultimately, reduced bacterial activity. CNT-doped ZnO hybrid nanofluid had superior anti-bacterial activity against both gram-positive and gram-negative bacteria because of its altered surface and enhanced dispersion stability [175]. The natural microbial populations of river water and wastewater effluent, SWCNTs shows antibacterial activity against both gram-positive and gram-negative bacteria. In this study, it was shown that bacterial cytotoxicity is time-dependent, with longer incubation duration, the more toxic bacteria are to cell membranes [129, 175].

5.2 *Real Wastewater Applications*

Real wastewater has a far more complicated matrix than artificial pollution solutions, along with bacteria, wastewater also includes inorganic and organic species. Due to competitive adsorption, these species have a significant influence on the selectivity and adsorption potential of CNTs. The pH of the actual wastewater, which is anticipated to be closer to neutral, will often be used for treatment. It might be difficult to alter the pH of significant amounts of wastewater to basic or acidic conditions. Few research has documented the use of CNTs in actual wastewater, even though many research on contaminants removal by CNTs described in the literature are based on batch experiments on a laboratory scale. The following list includes a few illustrative examples: In different research, MWCNTs and magnetic MWCNTs were used to investigate the removal of dimethyl phthalate and sulphamethazine from municipal effluent. Deionized and tap water had a much higher removal efficiency as a result than the effluent and influent of the secondary treatment. The intricacy of the secondary treatment was cited as the primary cause of the poor removal from the influent and effluent [176]. Under the same circumstances, wastewater removed IV 2R dye at a lesser rate (88%) than deionized water (98.4%), most likely because genuine wastewater contains interfering contaminants. Another research examined the removal of organic contaminants from industrial wastewater using MWCNTs that had been functionalized with cyclodextrin. Before and after treatment, the COD levels in the wastewater were 826 mg/L and 47 mg/L, respectively. After treatment, the wastewater, which had a pH of 8, formed a dark orange tint. A 100 mL sample only needed 0.1 g of the adsorbent, and the contact time was just 10 min [177]. According to the research, CNTs have shown to be used for the remove toxic contaminants from actual wastewater. To evaluate the performance of these fantastic materials in practical applications, additional experimental research and tests on a pilot size are necessary. Table 1 provides a summary of several studies that used CNT to remove dyes wastewater.

6 Future Challenges and Perspectives

Due to their distinctive structural characteristics, CNTs have demonstrated tremendous potential in the adsorption of several contaminants. Additionally, selective functionalization and changes can improve their selectivity towards certain kinds of contaminants. Despite these benefits, it is difficult to produce CNTs on a wide scale from inexpensive sources. In addition to these issues, safety precautions should be explicitly outlined and concerns about the toxicity of CNTs should be addressed on both people and the environment. It is generally accepted that functionalized CNTs are less poisonous than raw CNTs, possibly because of the presence of metallic catalysts. There have been serious concerns raised about human health after exposure to CNTs. To work with such materials, it is necessary to employ suitable safety gear and a laboratory setting that has been properly prepared. In addition to CNTs' toxicity, treating and functionalizing CNTs involves using significant amounts of dangerous acids and other chemicals. Not only should these amounts be decreased, but also greener CNT modification methods should be identified. Additionally, innovative and efficient ways for modifying CNTs should be developed in order to provide them the required qualities suitable for certain applications. The majority of research is restricted to batch-scale tests utilizing pollutant solutions made with deionized water. Due to the fact that these solutions are inaccurate representations of the complexity of actual wastewater, the performance of the adsorbent in selectively

Table 1 Application of CNT on the adsorption of dyes from wastewater

Adsorbent material	Dye type	Experimental condition			Adsorption capacity (mg g^{-1})	Isotherm model	Kinetic model	Ref.
		pH	Dosage	Time, min				
Magnetic carbon nanotubes CNTs/k-Fe_2O_3	Azo dye	3	–	60	66.7	Intra-particle diffusion	Pseudo-second-order	[113]
Multi-walled carbon nanotubes	Azo dyes	–	0.08 g	240	35.4	Temkin	Pseudo-second-order	[178]
Single-walled carbon nanotubes (SWCNTs)	Azo dyes	3	0.2 g	–	179.90	Freundlich and Langmuir	Pseudo-second-order	[154]
CNT/MgO/$CuFe_2O_4$	Methyl violet dye (MVD) and Nile blue dye (NBD)	8	0.001 g	50 and 60	36.46	Intra-particle diffusion	Pseudo-second-order	[36]
Magnetic tungsten disulphide/carbon nanotubes nanocomposite (WS_2/Fe_3O_4/CNTs-NC) for	Amaranth (AM) and brilliant blue FCF (BB FCF)	3	0.015 g	–	174.8–166.7	Langmuir	Pseudo-second-order	[162]
Asparagine multi-walled carbon nanotube (Asp-CNT) was	Cationic dyes	4, 6	0.01 g	24	637	Langmuir	Pseudo-second-order	[166]
Chitosan wrapping magnetic nanosized c-Fe_2O_3 and multi-walled carbon nanotubes (m-CS/c-Fe_2O_3/MWCNTs)	Methyl orange	3.14–6.5	–	120	66.1	Langmuir	Pseudo-second-order	[117]

(continued)

Table 1 (continued)

Adsorbent material	Dye type	Experimental condition			Adsorption capacity (mg g^{-1})	Isotherm model	Kinetic model	Ref.
		pH	Dosage	Time, min				
Single-walled carbon nanotubes	Acid blue 92	3.2	0.12 g/L	75	86.91	Freundlich and Langmuir isotherms	Pseudo-second-order	[153]
Elaeis guineensis/ polyvinyl alcohol/ carbon nanotube composites	Dye	11	3 g/L	300	17.4	Langmuir	Pseudo-second-order	[161]
SWCNTs	Reactive blue 29	2	30 mg	60	496	Liu	Pseudo-second-order	[179]
SWCNTs	Acid red 18	7	0.04 g	180	166.7	Langmuir	Pseudo-second-order	[180]
SWCNTs	Reactive red 120	5	0.01, 0.04 g/L	180	426.49	Langmuir and Freundlich	–	[181]
MWCNTs	Reactive red M-2BE	2	30 mg	60	335.7	Liu	Pseudo-second-order	[179]
MWCNTs	MB	10	40 mg/L	60	95.3	Langmuir	Pseudo-second-order	[182]
Oxidized MWCNTs	Bromothymol blue (BTB)	1	0.02 g	9	55	Langmuir	Pseudo-second-order	[183]

absorbing the target contaminants from complicated wastewater cannot be attained. It should be proven how CNT-based adsorbents may be used to remove contaminants from actual wastewater in pilot-scale investigations employing continuous columns. It will give a general notion of how co-existing pollutants affect the elimination of the intended contaminants. Additionally, these investigations must concentrate on the simultaneous elimination of many contaminants. Another difficulty with using CNTs on a broad scale is their high cost. MWCNTs are said to cost between 0.6 and 25 dollars/g, whereas SWCNTs cost between 25 and 300 dollars/g. In addition, the cost of the CNT-based water treatment process will be determined by the kind of CNT, the design of the water treatment, the simplicity or complexity of CNT separation, the reusability of CNT and the effectiveness of the process. Such adsorbents will be needed in vast numbers to treat the volume of wastewater generated globally. Thus, to overcome the issue of the high cost, inexpensive mass production will be necessary. Additionally, pollutant-oriented CNTs could be created to investigate significant function that CNTs play in adsorption. Reusability of CNT-based adsorbents is significant from both an economic and environmental point of view. Therefore, it is important to investigate desorption conditions for CNT reuse in various situations. Researchers should concentrate on creating simple and effective strategies, so that used CNTs may be recycled and used in numerous cycles. Additionally, CNT-based adsorbents ought to be accessible in different forms, including as filters and membranes, to allow researchers to examine their function in the elimination of diverse types of contaminants. The creation of an automated system should get equal attention. It is important to carry out mathematical modelling and simulation studies to comprehend the adsorption mechanism under various circumstances. Furthermore, the difficulty of creating and constructing biomimetic CNTs is a significant barrier to the environment. The bulk of CNT toxic effects study has discovered that CNTs are harmful and soluble in water. CNTs may occasionally be functionalized with hydroxyl, carboxyl and amine groups to increase their solubility, rate of dissolution and biocompatibility. Hybrid nanofluids based on CNTs are intended for use in wastewater treatment facilities.

7 Conclusions

The potential of CNTs, CNT-based composites and polymer/CNT NCs for the removal of dyes from wastewater has been discussed in this chapter. As a result of numerous chemical functional groups of dyes-CNT interactions, involving covalent and electrostatic interactions, hydrophobic effects and hydrogen bonds, CNTs may effectively adsorb a variety of dyes. Although the primary driving force behind the interaction between CNTs and dyes is molecular structure, significantly influenced by charge, dyes loading and CNTs bonds, pure CNTs may be used to remove dye because of their high adsorption effectiveness, but their expensive cost prevents them from being widely used in industry. Additionally, because of their tiny size and strong aggregation characteristic, CNTs are extremely difficult to separate from aqueous solutions. Researchers have concentrated their efforts on creating NCs of CNTs with polymers, metal oxide, carbon, etc., which serve as a stable matrix for CNTs to address these issues. This work has reported on the influence of several aspects, like physical characteristics of CNTs, the nature of the adsorbate, the CNTs chemical treatment and process variables (such as CNT dose, temperature, solution pH, contact duration and starting dye concentration). These materials have a high capacity for adsorbing different classes of dyes, which implies that they might be used as possible adsorbents to remove colourants from wastewater. The synthesis process, environmental conditions, subsequent treatment and purification of CNTs all affect their adsorption properties, and because more binding sites are created when CNTs are functionalized with hydroxyl, carboxylic acid and other groups following acid treatment, they are better able to adsorb dyes and pigments. Nevertheless, regardless of their size, CNTs are not selective towards dyes and pigments. SWCNTs have a greater adsorption capability than MWCNTs, this is because MWCNTs' active sites are compromised as a result of purifying difficulties. The multi-layered structure of MWCNTs reduces the possibility for diffusion of dyes and pigments. The common processes present in dyes and pigments adsorption on CNTs include surface complexation, physical adsorption, electrostatic interactions, precipitation and ion exchange.

Acknowledgements The authors wish to acknowledge Tertiary Education Trust Fund (TETFUND) Nigeria, and Open University of Sudan for the financial supports.

References

1. Mohammad REA, Elbashir AA, Karim J, Yahaya N, Rahim NY, Miskam M (2021) Adsorptive performances of magnetic graphene oxide adsorbent for the removal of fluoroquinolones in the Langat River Basin, Malaysia. Int J Environ Anal Chem 00(00):1–20. https://doi.org/10.1080/03067319.2021.1957464
2. Parveen T, Umar K, Mohamad Ibrahim MNM (2020) Role of nanomaterials in the treatment of wastewater: a review. Water 12;12(2):495
3. Mohammad REA, Elbashir AA, Karim J, Yahaya N, Rahim NY, Miskam M (2022) Deep eutectic solvent functionalized graphene oxide based ferrofluid for the liquid phase microextraction of fluoroquinolones from water samples. Microchem J 920 KEM(July), 114–121. https://doi.org/10.4028/p-fx1521
4. Mamman S, Mohammad REA, Abdullahi SS, Chadi AS, Birniwa AH (2020) Synthesis, characterization and optimization of magnetite molecularly imprinted polymer for application in the removal of non-steroidal anti inflammatory drugs (NSAIDS). ChemSearch J 11(1):1–8
5. Abdullahi SS, Musa H, Habibu S, Birniwa AH, Mohammad REA (2021) Facile synthesis and dyeing performance of some disperse monomeric and polymeric dyes on nylon and polyester fabrics. Bull Chem Soc Ethiop 35(3):485–497. https://doi.org/10.4314/bcse.v35i3.2
6. Jagaba AH, Kutty SRM, Noor A, Isah AS, Lawal IM, Birniwa AH, Usman AK, Abubakar S (2022) Kinetics of pulp and paper wastewater treatment by high sludge retention time activated sludge process. J Hunan Univ Nat Sci 49(2):242–251. https://doi.org/10.55463/issn.1674-2974.49.2.24

7. Jagaba AH, Rahman S, Kutty M, Abubakar S, Birniwa AH, Al-zaqri N, Al-maswari BM, Nasir M, Ibrahim M, Hussain F (2022) Synthesis, characterization, and performance evaluation of hybrid waste sludge biochar for COD and color removal from agro-industrial effluent, 1–22
8. Jagaba AH, Kutty SRM, Noor A, Birniwa AH, Affam AC, Lawal IM, Kankia MU, Kilaco AU (2021) A systematic literature review of biocarriers: central elements for biofilm formation, organic and nutrients removal in sequencing batch biofilm reactor. J Water Process Eng 42(March):102178. https://doi.org/10.1016/j.jwpe.2021.102178
9. Pereira L, Alves M (2012) Dyes-environmental impact and remediation. Environ Prot Strat Sustain Dev 111–162.https://doi.org/10.1007/978-94-007-1591-2_4
10. Saratale RG, Saratale GD, Chang JS, Govindwar SP (2011) Bacterial decolorization and degradation of azo dyes: a review. J Taiwan Inst Chem Eng 42(1):138–157. https://doi.org/10.1016/j.jtice.2010.06.006
11. Ahmad A, Setapar SH (2021) Synthesis and characterization of GO-Ag nanocomposite for removal of malachite dye from aqueous solution. Mater Today Proc 1(47):1359–1365
12. Ito T, Adachi Y, Yamanashi Y, Shimada Y (2016) Long–term natural remediation process in textile dye–polluted river sediment driven by bacterial community changes. Water Res 100:458–465. https://doi.org/10.1016/j.watres.2016.05.050
13. Balarabe S, Habibu S, Gumel SM, Ladan M, Birniwa AH, Koki IB, Usman SN (2017) Antibacterial finishing of a dyed cotton fabrics using diospyros mespiliformis leaves extracts. Malays J Fundam Appl Sci 13(3). https://doi.org/10.11113/mjfas.v13n3.601
14. Rehman A, Usman M, Bokhari TH, Haq A ul, Saeed M, Rahman HMAU, Siddiq M, Rasheed A, Nisa MU (2020) The application of cationic-nonionic mixed micellar media for enhanced solubilization of direct brown 2 dye. J Mol Liq 301.https://doi.org/10.1016/j.molliq.2019.112408
15. Gumel SM, Abdulmalik A, Habibu S, Ladan M (2015) Influence of auxiliaries in dyeing of wool with acid dyes. Int J Text Sci 4(4):78–83. https://doi.org/10.5923/j.textile.20150404.02
16. Benkhaya S, Mrabet S, Elharfi A (2020) A review on classifications, recent synthesis and applications of textile dyes. Inorg Chem Commun 115:107891. https://doi.org/10.1016/j.inoche.2020.107891
17. Chandanshive VV, Kadam SK, Khandare RV, Kurade MB, Jeon BH, Jadhav JP, Govindwar SP (2018) In situ phytoremediation of dyes from textile wastewater using garden ornamental plants, effect on soil quality and plant growth. Chemosphere 210:968–976. https://doi.org/10.1016/j.chemosphere.2018.07.064
18. Haruna Birniwa A (2014) The evaluation of antimicrobial activity of cotton fabric dyed with eco-friendly natural colourant extract (*Trianthema pentendra*) against bacteria. Discov Innov 25:48–53
19. Kant R (2012) Textile dyeing industry an environmental hazard. Nat Sci 04(01):22–26. https://doi.org/10.4236/ns.2012.41004
20. Haruna Birniwa A, Salisu A, Mahmud HNM, Mohamed Kutty SR, Jagaba AH, Abdullahi S, Zango Z (2022) Application of agricultural wastes for cationic dyes removal from wastewater, pp 239–274. https://doi.org/10.1007/978-981-19-2832-1_9
21. Hao OJ, Kim H, Chiang PC (2000) Decolorization of wastewater. Crit Rev Environ Sci Technol 30(4):449–505. https://doi.org/10.1080/10643380091184237
22. Zaharia C, Suteu D, Muresan A, Muresan R, Popescu A (2009) Textile wastewater treatment by homogeneous oxidation with hydrogen peroxide. Environ Eng Manag J 8(6):1359–1369. https://doi.org/10.30638/eemj.2009.199
23. Reemtsma T (1989) Subject index. Neuroscience 33:iii–vii. https://doi.org/10.1016/0306-4522(89)90252-2
24. Shalla AH, Bhat MA, Yaseen Z (2018) Hydrogels for removal of recalcitrant organic dyes: a conceptual overview. J Environ Chem Eng 6(5):5938–5949. https://doi.org/10.1016/j.jece.2018.08.063
25. Mohamad Yusoff NAN, Rahim NY, Mohammad REA, Yahaya N, Miskam M (2021) Deep eutectic solvent-based emulsification liquid-liquid microextraction for the analysis of phenoxy acid herbicides in paddy field water samples. Royal Soc Open Sci 8(3). https://doi.org/10.1098/rsos.202061

26. Birniwa AH, Abdullahi SS, Yakasai MY, Ismaila A (2021) Studies on physico-mechanical behaviour of kenaf/glass fiber reinforced epoxy hybrid composites. Bull Chem Soc Ethiop 35(1):171–184. https://doi.org/10.4314/BCSE.V35I1.15
27. Birniwa AH, Abubakar AS, Huq AKO, Mahmud HNME (2021) Polypyrrole-polyethyleneimine (PPy-PEI) nanocomposite: an effective adsorbent for nickel ion adsorption from aqueous solution. J Macromol Sci Part A Pure Appl Chem 58(3):206–217. https://doi.org/10.1080/10601325.2020.1840921
28. Abdullahi SS, Birniwa AH, Mohammad REA, Mamman S, Chadi AS (2020) Caliphate Journal of Science & Technology (CaJoST) impact of fibre reinforced polyester (*Adansonia digitata*) stem. Caliphate J Sci Technol (CaJoST) 2(January):94–100
29. Abbasi N, Khan TA (2021) Chapter 34—Progress in carbon nanotubes for water treatment. In: Ahamad A, Siddiqui SI, PBT-C of W Singh (eds). Academic Press, pp 509–518. https://doi.org/10.1016/B978-0-12-824058-8.00038-4
30. Basheer AA (2018) New generation nano-adsorbents for the removal of emerging contaminants in water. J Mol Liq 261:583–593. https://doi.org/10.1016/j.molliq.2018.04.021
31. Palkar VR (1999) Sol-gel derived nanostructured γ-alumina porous spheres as an adsorbent in liquid chromatography. Nanostruct Mater 11(3):369–374. https://doi.org/10.1016/S0965-9773(99)00053-7
32. Khan DM, Kausar A, Saeed A (2017) Research advancement towards polymer/nanodiamond composite in buckypaper: a review. Polym Plast Technol Eng 56(9):946–965. https://doi.org/10.1080/03602559.2016.1247283
33. Cha JE, Kim SY, Lee SH (2016) Effect of continuous multi-walled carbon nanotubes on thermal and mechanical properties of flexible composite film. Nanomaterials 6(10). https://doi.org/10.3390/nano6100182
34. Sadegh H, Ali GAM, Gupta VK, Makhlouf ASH, Shahryari-ghoshekandi R, Nadagouda MN, Sillanpää M, Megiel E (2017) The role of nanomaterials as effective adsorbents and their applications in wastewater treatment. J Nanostruct Chem 7(1):1–14. https://doi.org/10.1007/s40097-017-0219-4
35. Han JH, Zhang H, Chen MJ, Wang GR, Zhang Z (2014) CNT buckypaper/thermoplastic polyurethane composites with enhanced stiffness, strength and toughness. Compos Sci Technol 103:63–71. https://doi.org/10.1016/j.compscitech.2014.08.015
36. Foroutan R, Peighambardoust SJ, Esvandi Z, Khatooni H, Ramavandi B (2021) Evaluation of two cationic dyes removal from aqueous environments using CNT/MgO/$CuFe_2O_4$ magnetic composite powder: a comparative study. J Environ Chem Eng 9(2):104752. https://doi.org/10.1016/j.jece.2020.104752
37. Yadav S, Yadav A, Bagotia N, Sharma AK, Kumar S (2021) Adsorptive potential of modified plant-based adsorbents for sequestration of dyes and heavy metals from wastewater—a review. J Water Process Eng 42(May):102148. https://doi.org/10.1016/j.jwpe.2021.102148
38. Kumari A, Rajeev R, Benny L, Sudhakar YN, Varghese A, Hegde G (2021) Recent advances in carbon nanotubes-based biocatalysts and their applications. Adv Coll Interface Sci 297(October):102542. https://doi.org/10.1016/j.cis.2021.102542
39. Feng W, Ji P (2011) Enzymes immobilized on carbon nanotubes. Biotechnol Adv 29(6):889–895. https://doi.org/10.1016/j.biotechadv.2011.07.007
40. Khan AK, Mubarak NM, Abdullah EC, Khalid M, Nizamuddin S, Baloch HA, Siddiqui MTH (2019) Immobilization of lipase enzyme carbon nanotubes via adsorption. IOP Conf Series Mater Sci Eng 495(1). https://doi.org/10.1088/1757-899X/495/1/012055
41. Mundra RV, Wu X, Sauer J, Dordick JS, Kane RS (2014) Nanotubes in biological applications. Curr Opinion Biotechnol 28(Figure 1):25–32. https://doi.org/10.1016/j.copbio.2013.10.012
42. Ong CB, Annuar MSM (2018) Immobilization of cross-linked tannase enzyme on multiwalled carbon nanotubes and its catalytic behavior. Prep Biochem Biotechnol 48(2):181–187. https://doi.org/10.1080/10826068.2018.1425707
43. Trojanowicz M (2006) Analytical applications of carbon nanotubes: a review. TrAC Trends Anal Chem 25(5):480–489. https://doi.org/10.1016/j.trac.2005.11.008

44. Cang-Rong JT, Pastorin G (2009) The influence of carbon nanotubes on enzyme activity and structure: investigation of different immobilization procedures through enzyme kinetics and circular dichroism studies. Nanotechnology 20(25):255102. https://doi.org/10.1088/0957-4484/20/25/255102
45. Olvera RC, Silva SL, Robles-Belmont E, Lau EZ (2017) Review of nanotechnology value chain for water treatment applications in Mexico. Resource-Efficient Technol 3(1):1–11. https://doi.org/10.1016/j.reffit.2017.01.008
46. Sweetman M, May S, Mebberson N, Pendleton P, Vasilev K, Plush S, Hayball J (2017) Activated carbon, carbon nanotubes and graphene: materials and composites for advanced water purification. C 3(4):18. https://doi.org/10.3390/c3020018
47. Al-Mahbashi N, Kutty SRM, Jagaba AH, Al-Nini A, Ali M, Saeed AAH, Ghaleb AAS, Rathnayake U (2022) Column study for adsorption of copper and cadmium using activated carbon derived from sewage sludge. Adv Civil Eng 2022(March). https://doi.org/10.1155/2022/3590462
48. Noyes PD, McElwee MK, Miller HD, Clark BW, Van Tiem LA, Walcott KC, Erwin KN, Levin ED (2009) The toxicology of climate change: environmental contaminants in a warming world. Environ Int 35(6):971–986. https://doi.org/10.1016/j.envint.2009.02.006
49. Schwarzenbach RP, Egli T, Hofstetter TB, Von Gunten U, Wehrli B (2010) Global water pollution and human health. Annu Rev Environ Resour 35:109–136. https://doi.org/10.1146/annurev-environ-100809-125342
50. Her N, Amy G, McKnight D, Sohn J, Yoon Y (2003) Characterization of DOM as a function of MW by fluorescence EEM and HPLC-SEC using UVA, DOC, and fluorescence detection. Water Res 37(17):4295–4303. https://doi.org/10.1016/S0043-1354(03)00317-8
51. Lozovik PA, Morozov AK, Zobkov MB, Dukhovicheva TA, Osipova LA (2007) Allochthonous and autochthonous organic matter in surface waters in Karelia. Water Resour 34(2):204–216. https://doi.org/10.1134/S009780780702011X
52. King DN, Donohue MJ, Vesper SJ, Villegas EN, Ware MW, Vogel ME, Furlong EF, Kolpin DW, Glassmeyer ST, Pfaller S (2016) Microbial pathogens in source and treated waters from drinking water treatment plants in the United States and implications for human health. Sci Total Environ 562:987–995. https://doi.org/10.1016/j.scitotenv.2016.03.214
53. Kong L, Kadokami K, Wang S, Duong HT, Chau HTC (2015) Monitoring of 1300 organic micro-pollutants in surface waters from Tianjin, North China. Chemosphere 122:125–130. https://doi.org/10.1016/j.chemosphere.2014.11.025
54. Itawi RK, Al-Jobori SM, Jalil M, Farhan SS, Mheemeed A, Saaeed K, Ahmed E (1991) Determination of major, minor and trace elements in textile dyes by INAA. J Radioanal Nucl Chem 149(2):339–343. https://doi.org/10.1007/BF02062062
55. Habibu S (2013) Synthesis, characterization and application of some polymeric dyes derived from monoazo disperse dyes. IOSR J Appl Chemistr 5(3):42–47. https://doi.org/10.9790/5736-0534247
56. Puvaneswari N, Jayarama M, Gunasekaran P (2006) Toxicity assessment and microbial degradation of azo dyes. Indian J Exp Biol 44:618–626
57. Keharia H, Madamwar D (2003) Bioremediation concepts for treatment of dye containing wastewater: a review. Indian J Exp Biol 41(9):1068–1075
58. Al-Aseeri M, Bu-Ali Q, Haji S, Al-Bastaki N (2007) Removal of acid red and sodium chloride mixtures from aqueous solutions using nanofiltration. Desalination 206(1):407–413. https://doi.org/10.1016/j.desal.2006.03.575
59. Yaseen DA, Scholz M (2016) Shallow pond systems planted with Lemna minor treating azo dyes. Ecol Eng 94:295–305. https://doi.org/10.1016/j.ecoleng.2016.05.081
60. Nguyen TA, Juang RS (2013) Treatment of waters and wastewaters containing sulfur dyes: a review. Chem Eng J 219:109–117. https://doi.org/10.1016/j.cej.2012.12.102
61. Wang W, Yue Q, Li R, Song W, Gao B, Shen X (2017) Investigating coagulation behavior of chitosan with different Al species dual-coagulants in dye wastewater treatment. J Taiwan Inst Chem Eng 78:423–430. https://doi.org/10.1016/j.jtice.2017.06.052

62. Vakili M, Rafatullah M, Salamatinia B, Abdullah AZ, Ibrahim MH, Tan KB, Gholami Z, Amouzgar P (2014) Application of chitosan and its derivatives as adsorbents for dye removal from water and wastewater: a review. Carbohyd Polym 113:115–130. https://doi.org/10.1016/j.carbpol.2014.07.007
63. Li W, Mu B, Yang Y (2019) Feasibility of industrial-scale treatment of dye wastewater via bio-adsorption technology. Bioresour Technol 277(December 2018):157–170. https://doi.org/10.1016/j.biortech.2019.01.002
64. Hessel C, Allegre C, Maisseu M, Charbit F, Moulin P (2007) Guidelines and legislation for dye house effluents. J Environ Manage 83(2):171–180. https://doi.org/10.1016/j.jenvman.2006.02.012
65. Al-Qodah Z, Lafi WK, Al-Anber Z, Al-Shannag M, Harahsheh A (2007) Adsorption of methylene blue by acid and heat treated diatomaceous silica. Desalination 217(1–3):212–224. https://doi.org/10.1016/j.desal.2007.03.003
66. van der Zee FP (2002) Anaerobic azo dye reduction. Wageningen University
67. Husain Q, Husain M (2012) In: Malik A, Grohmann E (eds) Peroxidases as a potential tool for the decolorization and removal of synthetic dyes from polluted water BT—environmental protection strategies for sustainable development. Springer, Netherlands, pp 453–498. https://doi.org/10.1007/978-94-007-1591-2_15
68. Shabir M, Yasin M, Hussain M, Shafiq I, Akhter P, Nizami AS, Jeon BH, Park YK (2022) A review on recent advances in the treatment of dye-polluted wastewater. J Ind Eng Chem 112:1–19. https://doi.org/10.1016/j.jiec.2022.05.013
69. Wang C, Yediler A, Lienert D, Wang Z, Kettrup A (2002) Toxicity evaluation of reactive dyestuffs, auxiliaries and selected effluents in textile finishing industry to luminescent bacteria Vibrio fischeri. Chemosphere 46(2):339–344. https://doi.org/10.1016/S0045-6535(01)00086-8
70. Das GS, Sarkar S, Aggarwal R, Sonkar SK, Park JW, Tripathi KM, Kim TY (2019) Fluorescent microspheres of zinc 1,2-dicarbomethoxy-1,2-dithiolate complex decorated with carbon nanotubes. Carbon Lett 29(6):595–603. https://doi.org/10.1007/s42823-019-00058-4
71. Xu W, Wang X, Cai Z (2013) Analytical chemistry of the persistent organic pollutants identified in the Stockholm convention: a review. Anal Chim Acta 790:1–13. https://doi.org/10.1016/j.aca.2013.04.026
72. Maas R, Chaudhari S (2005) Adsorption and biological decolourization of azo dye reactive red 2 in semicontinuous anaerobic reactors. Process Biochem 40(2):699–705. https://doi.org/10.1016/j.procbio.2004.01.038
73. Eriksson E, Auffarth K, Eilersen AM, Henze M, Ledin A (2003) Household chemicals and personal care products as sources for xenobiotic organic compounds in grey wastewater. Water SA 29(2):135–146. https://doi.org/10.4314/wsa.v29i2.4848
74. Zhou Q, Wang M (2010) Adsorption–desorption characteristics and pollution behavior of reactive X-3B red dye in four Chinese typical soils. J Soils Sediments 10(7):1324–1334. https://doi.org/10.1007/s11368-010-0247-x
75. Menezes E (2003) Developments in waste water treatment methods. Chem Wkly. Bombay, 48:187–196. https://www.scopus.com/inward/record.uri?eid=2-s2.0-85101640462&partnerID=40&md5=89998799b982083db72f308537f7e3a2
76. Carneiro PA, Umbuzeiro GA, Oliveira DP, Zanoni MVB (2010) Assessment of water contamination caused by a mutagenic textile effluent/dyehouse effluent bearing disperse dyes. J Hazard Mater 174(1–3):694–699. https://doi.org/10.1016/j.jhazmat.2009.09.106
77. Yang G, Zhang G, Wang H (2015) Current state of sludge production, management, treatment and disposal in China. Water Res 78:60–73. https://doi.org/10.1016/j.watres.2015.04.002
78. Brown AK, Wong CS (2018) Distribution and fate of pharmaceuticals and their metabolite conjugates in a municipal wastewater treatment plant. Water Res 144:774–783. https://doi.org/10.1016/j.watres.2018.08.034
79. Huerta B, Rodriguez-Mozaz S, Nannou C, Nakis L, Ruhí A, Acuña V, Sabater S, Barcelo D (2016) Determination of a broad spectrum of pharmaceuticals and endocrine disruptors in biofilm from a waste water treatment plant-impacted river. Sci Total Environ 540:241–249. https://doi.org/10.1016/j.scitotenv.2015.05.049

80. Göbel A, McArdell CS, Joss A, Siegrist H, Giger W (2007) Fate of sulfonamides, macrolides, and trimethoprim in different wastewater treatment technologies. Sci Total Environ 372(2–3):361–371. https://doi.org/10.1016/j.scitotenv.2006.07.039
81. Zocolo GJ uliã, Pilon dos Santos G, Vendemiatti J, Vacchi FI nforçat, de Umbuzeiro GA, Zanoni MV alnic B (2015) Using SPE-LC-ESI-MS/MS analysis to assess disperse dyes in environmental water samples. J Chromatogr Sci 53(8):1257–1264.https://doi.org/10.1093/chromsci/bmu221
82. Dixit S, Yadav A, Dwivedi PD, Das M (2015) Toxic hazards of leather industry and technologies to combat threat: a review. J Cleaner Prod 87(C):39–49. https://doi.org/10.1016/j.jclepro.2014.10.017
83. Tigini V, Giansanti P, Mangiavillano A, Pannocchia A, Varese GC (2011) Evaluation of toxicity, genotoxicity and environmental risk of simulated textile and tannery wastewaters with a battery of biotests. Ecotoxicol Environ Saf 74(4):866–873. https://doi.org/10.1016/j.ecoenv.2010.12.001
84. dos Santos AB, Cervantes FJ, van Lier JB (2007) Review paper on current technologies for decolourisation of textile wastewaters: perspectives for anaerobic biotechnology. Biores Technol 98(12):2369–2385. https://doi.org/10.1016/j.biortech.2006.11.013
85. Gupta P, Mathur N, Bhatnagar P, Nagar P, Srivastava S (2009) Genotoxicity evaluation of hospital wastewaters. Ecotoxicol Environ Saf 72(7):1925–1932. https://doi.org/10.1016/j.ecoenv.2009.05.012
86. Authority and Prevention (2017) The European Union summary report on trends and sources of zoonoses, zoonotic agents and food-borne outbreaks in 2010. Euro Surveill Bull Euro Sur Les Mal Trans Euro Commun Dis Bull 17(10). https://doi.org/10.2903/j.efsa.2012.2597
87. Andersen WC, Turnipseed SB, Roybal JE (2006) Quantitative and confirmatory analyses of malachite green and leucomalachite green residues in fish and shrimp. J Agric Food Chem 54(13):4517–4523. https://doi.org/10.1021/jf0532258
88. Wang Y, Liao K, Huang X, Yuan D (2015) Simultaneous determination of malachite green, crystal violet and their leuco-metabolites in aquaculture water samples using monolithic fiber-based solid-phase microextraction coupled with high performance liquid chromatography. Anal Methods 7(19):8138–8145. https://doi.org/10.1039/c5ay01611h
89. Xu YJ, Tian XH, Zhang XZ, Gong XH, Liu HH, Zhang HJ, Huang H, Zhang LM (2012) Simultaneous determination of malachite green, crystal violet, methylene blue and the metabolite residues in aquatic products by ultra-performance liquid chromatography with electrospray ionization tandem mass spectrometry. J Chromatogr Sci 50(7):591–597. https://doi.org/10.1093/chromsci/bms054
90. Pearce CI, Lloyd JR, Guthrie JT (2003) The removal of colour from textile wastewater using whole bacterial cells: a review. Dyes Pigm 58(3):179–196. https://doi.org/10.1016/S0143-7208(03)00064-0
91. Zhou Y, Lu J, Zhou Y, Liu Y (2019) Recent advances for dyes removal using novel adsorbents: a review. Environ Pollut 252:352–365. https://doi.org/10.1016/j.envpol.2019.05.072
92. Dai Y, Sun Q, Wang W, Lu L, Liu M, Li J, Yang S, Sun Y, Zhang K, Xu J, Zheng W, Hu Z, Yang Y, Gao Y, Chen Y, Zhang X, Gao F, Zhang Y (2018) Utilizations of agricultural waste as adsorbent for the removal of contaminants: a review. Chemosphere 211:235–253. https://doi.org/10.1016/j.chemosphere.2018.06.179
93. Goodman SM, Bura R, Dichiara AB (2018) Facile impregnation of graphene into porous wood filters for the dynamic removal and recovery of dyes from aqueous solutions. ACS Appl Nano Mater 1(10):5682–5690. https://doi.org/10.1021/acsanm.8b01275
94. Senthil Kumar P, Janet Joshiba G, Femina CC, Varshini P, Priyadharshini S, Arun Karthick MS, Jothirani R (2019) A critical review on recent developments in the low-cost adsorption of dyes from wastewater. Desalin Water Treat 172:395–416. https://doi.org/10.5004/dwt.2019.24613
95. Birniwa AH, Abdullahi SS, Mahmud HNME (2019) Study on physico-mechanical behaviour of *Acacia nilotica* (gum tree) and glass fiber blend reinforced epoxy resin composite study on physico-mechanical behaviour of *Acacia nilotica* (gum tree) and glass fiber blend reinforced

epoxy resin composite. ChemSearch J 10(January 2020):46–53. http://www.ajol.info/index.php/csj%0A
96. Bhatia D, Sharma NR, Singh J, Kanwar RS (2017) Biological methods for textile dye removal from wastewater: a review. Crit Rev Environ Sci Technol 47(19):1836–1876. https://doi.org/10.1080/10643389.2017.1393263
97. Sani RK, Banerjee UC (1999) Decolorization of triphenylmethane dyes and textile and dye-stuff effluent by Kurthia sp. Enzyme Microb Technol 24(7):433–437. https://doi.org/10.1016/S0141-0229(98)00159-8
98. Sarayu K, Sandhya S (2012) Current technologies for biological treatment of textile wastewater-a review. Appl Biochem Biotechnol 167(3):645–661. https://doi.org/10.1007/s12010-012-9716-6
99. Khan NA, Bhadra BN, Jhung SH (2018) Heteropoly acid-loaded ionic liquid@metal-organic frameworks: effective and reusable adsorbents for the desulfurization of a liquid model fuel. Chem Eng J 334(November 2017):2215–2221. https://doi.org/10.1016/j.cej.2017.11.159
100. Katheresan V, Kansedo J, Lau SY (2018) Efficiency of various recent wastewater dye removal methods: a review. J Environ Chem Eng 6(4):4676–4697. https://doi.org/10.1016/j.jece.2018.06.060
101. Zhang J, Huang Z-H, Lv R, Yang Q-H, Kang F (2009) Effect of growing CNTs onto Bamboo charcoals on adsorption of copper ions in aqueous solution. Langmuir 25(1):269–274. https://doi.org/10.1021/la802365w
102. Gupta VK, Kumar R, Nayak A, Saleh TA, Barakat MA (2013) Adsorptive removal of dyes from aqueous solution onto carbon nanotubes: a review. Adv Coll Interface Sci 193–194:24–34. https://doi.org/10.1016/j.cis.2013.03.003
103. Hamza IAA, Martincigh BS, Ngila JC, Nyamori VO (2013) Adsorption studies of aqueous Pb(II) onto a sugarcane bagasse/multi-walled carbon nanotube composite. Phys Chem Earth 66:157–166. https://doi.org/10.1016/j.pce.2013.08.006
104. Ma J, Yu F, Zhou L, Jin L, Yang M, Luan J, Tang Y, Fan H, Yuan Z, Chen J (2012) Enhanced adsorptive removal of methyl orange and methylene blue from aqueous solution by alkali-activated multiwalled carbon nanotubes. ACS Appl Mater Interfaces 4(11):5749–5760. https://doi.org/10.1021/am301053m
105. Mallakpour S, Rashidimoghadam S (2018) Carbon nanotubes for dyes removal. In: Composite nanoadsorbents. Elsevier Inc.https://doi.org/10.1016/B978-0-12-814132-8.00010-1
106. Inyang M, Gao B, Zimmerman A, Zhang M, Chen H (2014) Synthesis, characterization, and dye sorption ability of carbon nanotube-biochar nanocomposites. Chem Eng J 236:39–46. https://doi.org/10.1016/j.cej.2013.09.074
107. Mohammad F, Arfin T, Al-lohedan HA (2019) Enhanced biosorption and electrochemical performance of sugarcane bagasse derived a polylactic acid-graphene oxide-CeO_2 composite. Mater Chem Phys 229(February):117–123. https://doi.org/10.1016/j.matchemphys.2019.02.085
108. Zheng Y, Yang Y, Zhang Y, Zou W, Luo Y, Dong L, Gao B (2019) Facile one-step synthesis of graphitic carbon nitride-modified biochar for the removal of reactive red 120 through adsorption and photocatalytic degradation. Biochar 1(1):89–96. https://doi.org/10.1007/s42773-019-00007-4
109. Huang Z nan, Wang X ling, Yang D suo (2015) Adsorption of Cr(VI) in wastewater using magnetic multi-wall carbon nanotubes. Water Sci Eng 8(3):226–232.https://doi.org/10.1016/j.wse.2015.01.009
110. Qu S, Huang F, Yu S, Chen G, Kong J (2008) Magnetic removal of dyes from aqueous solution using multi-walled carbon nanotubes filled with Fe_2O_3 particles. J Hazard Mater 160(2–3):643–647. https://doi.org/10.1016/j.jhazmat.2008.03.037
111. Sonkar SK, Tripathi KM, Sarkar SY (2014) Ferromagnetic behaviour of anthropogenic multi-walled carbon nanotubes trapped in spider web indoor. J Nanosci Nanotechnol 14(3):2532–2538. https://doi.org/10.1166/jnn.2014.8524
112. Tripathi KM, Begum A, Sonkar SK, Sarkar S (2013) Nanospheres of copper(iii) 1,2-dicarbomethoxy-1,2-dithiolate and its composite with water soluble carbon nanotubes. New J Chem 37(9):2708–2715. https://doi.org/10.1039/c3nj00368j

113. Yu F, Chen J, Chen L, Huai J, Gong W, Yuan Z, Wang J, Ma J (2012) Magnetic carbon nanotubes synthesis by Fenton's reagent method and their potential application for removal of azo dye from aqueous solution. J Colloid Interface Sci 378(1):175–183. https://doi.org/10.1016/j.jcis.2012.04.024
114. Fraczek-Szczypta A, Menaszek E, Blazewicz S (2011) Some observations on carbon nanotubes susceptibility to cell phagocytosis. J Nanomater 2011:11–16. https://doi.org/10.1155/2011/473516
115. Dubey P, Sonkar SK, Majumder S, Tripathi KM, Sarkar S (2013) Isolation of water soluble carbon nanotubes with network structure possessing multipodal junctions and its magnetic property. RSC Adv 3(20):7306–7312. https://doi.org/10.1039/c3ra22933e
116. Yee KF, Ong YT, Mohamed AR, Tan SH (2014) Novel MWCNT-buckypaper/polyvinyl alcohol asymmetric membrane for dehydration of etherification reaction mixture: Fabrication, characterisation and application. J Membr Sci 453:546–555. https://doi.org/10.1016/j.memsci.2013.11.040
117. Zhu HY, Jiang R, Xiao L, Zeng GM (2010) Preparation, characterization, adsorption kinetics and thermodynamics of novel magnetic chitosan enwrapping nanosized γ-Fe_2O_3 and multi-walled carbon nanotubes with enhanced adsorption properties for methyl orange. Biores Technol 101(14):5063–5069. https://doi.org/10.1016/j.biortech.2010.01.107
118. Yu F, Sun S, Han S, Zheng J, Ma J (2016) Adsorption removal of ciprofloxacin by multi-walled carbon nanotubes with different oxygen contents from aqueous solutions. Chem Eng J 285:588–595. https://doi.org/10.1016/j.cej.2015.10.039
119. Das R, Ali ME, Hamid SBA, Ramakrishna S, Chowdhury ZZ (2014) Carbon nanotube membranes for water purification: a bright future in water desalination. Desalination 336(1):97–109. https://doi.org/10.1016/j.desal.2013.12.026
120. Li H, Zou L (2011) Ion-exchange membrane capacitive deionization: a new strategy for brackish water desalination. Desalination 275(1–3):62–66. https://doi.org/10.1016/j.desal.2011.02.027
121. Clara M, Strenn B, Saracevic E, Kreuzinger N (2004) Adsorption of bisphenol-A, 17β-estradiole and 17α-ethinylestradiole to sewage sludge. Chemosphere 56(9):843–851. https://doi.org/10.1016/j.chemosphere.2004.04.048
122. Ahn CH, Baek Y, Lee C, Kim SO, Kim S, Lee S, Kim SH, Bae SS, Park J, Yoon J (2012) Carbon nanotube-based membranes: fabrication and application to desalination. J Ind Eng Chem 18(5):1551–1559. https://doi.org/10.1016/j.jiec.2012.04.005
123. Bacakova L, Pajorova J, Tomkova M, Matejka R, Broz A, Stepanovska J, Prazak S, Skogberg A, Siljander S, Kallio P (2020) Applications of nanocellulose/nanocarbon composites: focus on biotechnology and medicine. Nanomaterials 10(2):1–32. https://doi.org/10.3390/nano10020196
124. Mashkoor F, Nasar A, Inamuddin (2020) Carbon nanotube-based adsorbents for the removal of dyes from waters: a review. Environ Chem Lett 18(3):605–629.https://doi.org/10.1007/s10311-020-00970-6
125. Wang M, Li N, Wang GD, Lu SW, Zhao Q, Di, Liu XL (2021) High-sensitive flexural sensors for health monitoring of composite materials using embedded carbon nanotube (CNT) buckypaper. Compos Struct 261(March 2020), 113280. https://doi.org/10.1016/j.compstruct.2020.113280
126. Wang Y, Yu Z, Dufresne A, Ye Z, Lin N, Zhou J (2021) Quantitative analysis of compatibility and dispersibility in nanocellulose-reinforced composites: Hansen solubility and Raman mapping. ACS Nano 15(12):20148–20163. https://doi.org/10.1021/acsnano.1c08100
127. Zhu C, Liu P, Mathew AP (2017) Self-assembled TEMPO cellulose nanofibers: graphene oxide-based biohybrids for water purification. ACS Appl Mater Interfaces 9(24):21048–21058. https://doi.org/10.1021/acsami.7b06358
128. Salama A, Abouzeid R, Leong WS, Jeevanandam J, Samyn P, Dufresne A, Bechelany M, Barhoum A (2021) Nanocellulose-based materials for water treatment: Adsorption, photocatalytic degradation, disinfection, antifouling, and nanofiltration. Nanomaterials 11(11). https://doi.org/10.3390/nano11113008

129. Moyo S, Gumbi NN, De Kock LA, Nxumalo EN (2022) A mini-review of nanocellulose-based nanofiber membranes incorporating carbon nanomaterials for dye wastewater treatment. Environ Nanotechnol Monit Manag 18(January):100714. https://doi.org/10.1016/j.enmm.2022.100714
130. Špitalský Z, Aggelopoulos C, Tsoukleri G, Tsakiroglou C, Parthenios J, Georga S, Krontiras C, Tasis D, Papagelis K, Galiotis C (2009) The effect of oxidation treatment on the properties of multi-walled carbon nanotube thin films. Mater Sci Eng B Solid-State Mater Adv Technol 165(3):135–138. https://doi.org/10.1016/j.mseb.2009.09.019
131. Yang X, Lee J, Yuan L, Chae SR, Peterson VK, Minett AI, Yin Y, Harris AT (2013) Removal of natural organic matter in water using functionalised carbon nanotube buckypaper. Carbon 59:160–166. https://doi.org/10.1016/j.carbon.2013.03.005
132. Alvarenga DF, Junior MG, Santos MCG, Pinto PS, da Cunha THR, Dias MC, Lavall RL, Ortega PFR (2022) Tuning carbon nanotube-based buckypaper properties by incorporating different cellulose nanofibrils for redox supercapacitor electrodes. J Energy Storage 52(February). https://doi.org/10.1016/j.est.2022.104848
133. Sun J, Gao L (2003) Development of a dispersion process for carbon nanotubes in ceramic matrix by heterocoagulation. Carbon 41(5):1063–1068. https://doi.org/10.1016/S0008-6223(02)00441-4
134. Geng HZ, Lee DS, Kim KK, Han GH, Park HK, Lee YH (2008) Absorption spectroscopy of surfactant-dispersed carbon nanotube film: modulation of electronic structures. Chem Phys Lett 455(4–6):275–278. https://doi.org/10.1016/j.cplett.2008.02.102
135. Endo M, Muramatsu H, Hayashi T, Kim YA, Terrones M, Dresselhaus MS (2005) 'Buckypaper' from coaxial nanotubes. Nature 433(7025):476. https://doi.org/10.1038/433476a
136. Ventura DN, Stone RA, Chen KS, Hariri HH, Riddle KA, Fellers TJ, Yun CS, Strouse GF, Kroto HW, Acquah SFA (2010) Assembly of cross-linked multi-walled carbon nanotube mats. Carbon 48(4):987–994. https://doi.org/10.1016/j.carbon.2009.11.016
137. Khan FSA, Mubarak NM, Tan YH, Khalid M, Karri RR, Walvekar R, Abdullah EC, Nizamuddin S, Mazari SA (2021) A comprehensive review on magnetic carbon nanotubes and carbon nanotube-based buckypaper for removal of heavy metals and dyes. J Hazard Mater 413(November 2020):125375. https://doi.org/10.1016/j.jhazmat.2021.125375
138. Ni B, Andrews R, Jacques D, Qian D, Wijesundara MBJ, Choi Y, Hanley L, Sinnott SB (2001) A combined computational and experimental study of ion-beam modification of carbon nanotube bundles. J Phys Chem B 105(51):12719–12725. https://doi.org/10.1021/jp0123233
139. Lee KP, Arnot TC, Mattia D (2011) A review of reverse osmosis membrane materials for desalination-development to date and future potential. J Membr Sci 370(1–2):1–22. https://doi.org/10.1016/j.memsci.2010.12.036
140. Rai S, De A, Guin M, Singh NB (2022) In: Muthu SS, Khadir A (eds) Carbon materials for dye removal from wastewater BT—textile wastewater treatment: sustainable bio-nano materials and macromolecules, volume 1. Springer Nature, Singapore, pp 141–183. https://doi.org/10.1007/978-981-19-2832-1_7
141. Majumder M, Chopra N, Andrews R, Hinds BJ (2005) Enhanced flow in carbon nanotubes. Nature 438(7064):44. https://doi.org/10.1038/438044a
142. Yin Z, Duoni, Chen H, Wang J, Qian W, Han M, Wei F (2018) Resilient, mesoporous carbon nanotube-based strips as adsorbents of dilute organics in water. Carbon 132:329–334.https://doi.org/10.1016/j.carbon.2018.02.074
143. López-Lorente AI, Simonet BM, Valcárcel M (2010) The potential of carbon nanotube membranes for analytical separations. Anal Chem 82(13):5399–5407. https://doi.org/10.1021/ac902629n
144. Ali S, Rehman SAU, Shah IA, Farid MU, An AK, Huang H (2019) Efficient removal of zinc from water and wastewater effluents by hydroxylated and carboxylated carbon nanotube membranes: behaviors and mechanisms of dynamic filtration. J Hazard Mater 365(19):64–73. https://doi.org/10.1016/j.jhazmat.2018.10.089

145. Holt JK, Park HG, Wang Y, Stadermann M, Artyukhin AB, Grigoropoulos CP, Noy A, Bakajin O (2006) Fast mass transport through sub-2-nanometer carbon nanotubes. Science 312(5776):1034–1037. https://doi.org/10.1126/science.1126298
146. Dumée L, Germain V, Sears K, Schütz J, Finn N, Duke M, Cerneaux S, Cornu D, Gray S (2011) Enhanced durability and hydrophobicity of carbon nanotube bucky paper membranes in membrane distillation. J Membr Sci 376(1–2):241–246. https://doi.org/10.1016/j.memsci.2011.04.024
147. Jun LY, Mubarak NM, Yon LS, Bing CH, Khalid M, Jagadish P, Abdullah EC (2019) Immobilization of peroxidase on functionalized MWCNTs-buckypaper/polyvinyl alcohol nanocomposite membrane. Sci Rep 9(1):1–15. https://doi.org/10.1038/s41598-019-39621-4
148. Lau YJ, Karri RR, Mubarak NM, Lau SY, Chua HB, Khalid M, Jagadish P, Abdullah EC (2020) Removal of dye using peroxidase-immobilized Buckypaper/polyvinyl alcohol membrane in a multi-stage filtration column via RSM and ANFIS. Environ Sci Pollut Res 27(32):40121–40134. https://doi.org/10.1007/s11356-020-10045-2
149. Jafari A, Nasseri S, Nabizadeh R, Mousavi SA, Rezaee R, Mahvi AH (2017) Humic acid removal from water using a novel fabricated antifouling carbon nanotube bucky—paper membrane and effect of operating parameters. Glob Nest J 19(2):217–224. https://doi.org/10.30955/gnj.001694
150. Ihsanullah PF, Khraisheh M, Atieh MA, Laoui T (2017) Novel aluminum oxide-impregnated carbon nanotube membrane for the removal of cadmium from aqueous solution. Materials 10(10). https://doi.org/10.3390/ma10101144
151. Ihsanullah I, Jamal A, Ilyas M, Zubair M, Khan G, Atieh MA (2020) Bioremediation of dyes: current status and prospects. J Water Process Eng 38(June):101680. https://doi.org/10.1016/j.jwpe.2020.101680
152. Hassan AA, Sajid M, Tanimu A, Abdulazeez I, Alhooshani K (2021) Removal of methylene blue and rose bengal dyes from aqueous solutions using 1-naphthylammonium tetrachloroferrate (III). J Mol Liq 322:114966. https://doi.org/10.1016/j.molliq.2020.114966
153. Balarak D, Zafariyan M, Igwegbe CA, Onyechi KK, Ighalo JO (2021) Adsorption of acid blue 92 dye from aqueous solutions by single-walled carbon nanotubes: isothermal, kinetic, and thermodynamic studies. Environ Process 8(2):869–888. https://doi.org/10.1007/s40710-021-00505-3
154. Naghizadeh A, Karimi A, Derakhshani E, Esform A (2022) Single-walled carbon nanotubes (SWCNTs) as an efficient adsorbent for removal of reactive dyes from water solution: equilibrium, kinetic, and thermodynamic. Environ Qual Manage 31(4):133–140. https://doi.org/10.1002/tqem.21753
155. Lei Y, Huang Q, Dou J, Huang H, Yang G, Deng F, Liu M, Li X, Zhang X, Wei Y (2021) Fast adsorptive removal of cationic organic dye by anionic group functionalized carbon nanotubes with high efficiency. Colloids Interface Sci Commun 40(October 2020):100328. https://doi.org/10.1016/j.colcom.2020.100328
156. Mallakpour S, Tabesh F (2021) Green and plant-based adsorbent from tragacanth gum and carboxyl-functionalized carbon nanotube hydrogel bionanocomposite for the super removal of methylene blue dye. Int J Biol Macromol 166:722–729. https://doi.org/10.1016/j.ijbiomac.2020.10.229
157. Abualnaja KM, Alprol AE, Abu-Saied MA, Mansour AT, Ashour M (2021) Studying the adsorptive behavior of poly(Acrylonitrile-co-styrene) and carbon nanotubes (nanocomposites) impregnated with adsorbent materials towards methyl orange dye. Nanomaterials 11(5). https://doi.org/10.3390/nano11051144
158. Jain M, Khan SA, Pandey A, Pant KK, Ziora ZM, Blaskovich MAT (2021) Instructive analysis of engineered carbon materials for potential application in water and wastewater treatment. Sci Total Environ 793:148583. https://doi.org/10.1016/j.scitotenv.2021.148583
159. Pete S, Kattil RA, Thomas L (2021) Polyaniline-multiwalled carbon nanotubes (PANI-MWCNTs) composite revisited: an efficient and reusable material for methyl orange dye removal. Diam Relat Mater 117(February):108455. https://doi.org/10.1016/j.diamond.2021.108455

160. Saleh TA, Elsharif AM, Bin-Dahman OA (2021) Synthesis of amine functionalization carbon nanotube-low symmetry porphyrin derivatives conjugates toward dye and metal ions removal. J Mol Liq 340:117024. https://doi.org/10.1016/j.molliq.2021.117024
161. Zulfiqar M, Rabat NE, Bahadar A, Lashari N, Mahnashi MH, Alqarni AO (2022) Development of *Elaeis guineensis*/polyvinyl alcohol/carbon nanotube composites for efficient adsorption of dye: experimental and theoretical approach. Int J Environ Sci Technol 19(7):6499–6520. https://doi.org/10.1007/s13762-021-03520-x
162. Arabkhani P, Javadian H, Asfaram A, Sadeghfar F, Sadegh F (2021) Synthesis of magnetic tungsten disulfide/carbon nanotubes nanocomposite (WS2/Fe_3O_4/CNTs-NC) for highly efficient ultrasound-assisted rapid removal of amaranth and brilliant blue FCF hazardous dyes. J Hazard Mater 420(July). https://doi.org/10.1016/j.jhazmat.2021.126644
163. Yang G, Li Y, Yang S, Liao J, Cai X, Gao Q, Fang Y, Peng F, Zhang S (2021) Surface oxidized nano-cobalt wrapped by nitrogen-doped carbon nanotubes for efficient purification of organic wastewater. Sep Purif Technol 259(September 2020):118098. https://doi.org/10.1016/j.seppur.2020.118098
164. Song G, Li A, Shi Y, Li W, Wang H, Wang C, Li R, Ding G (2021) Sorptive removal of methylene blue from water by magnetic multi-walled carbon nanotube composites. Environ Sci Pollut Res 28(30):41268–41282. https://doi.org/10.1007/s11356-021-13543-z
165. Amon REC, Lawagon CP (2021) Efficient removal of cationic and anionic dyes from wastewater using carbon nanotubes from petrochemical waste oil. Chem Eng Trans 86(February):349–354. https://doi.org/10.3303/CET2186059
166. Saxena M, Lochab A, Saxena R (2021) Asparagine functionalized MWCNTs for adsorptive removal of hazardous cationic dyes: exploring kinetics, isotherm and mechanism. Surf Interfaces 25(May):101187. https://doi.org/10.1016/j.surfin.2021.101187
167. Mallakpour S, Behranvand V, Mallakpour F (2021) Adsorptive performance of alginate/carbon nanotube-carbon dot-magnesium fluorohydroxyapatite hydrogel for methylene blue-contaminated water. J Environ Chem Eng 9(2):105170. https://doi.org/10.1016/j.jece.2021.105170
168. Banerjee D, Bhowmick P, Pahari D, Santra S, Sarkar S, Das B, Chattopadhyay KK (2017) Pseudo first ordered adsorption of noxious textile dyes by low-temperature synthesized amorphous carbon nanotubes. Physica E: Low-Dimension Syst Nanostruct 87(November 2016):68–76. https://doi.org/10.1016/j.physe.2016.11.024
169. Malika M, Sonawane SS (2019) Review on application of nanofluid/nano particle as water disinfectant. J Indian Association Environ Manag (JIAEM) 39(1–4):21–24. http://op.niscair.res.in/index.php/JIAEM/article/view/30487
170. Kang S, Pinault M, Pfefferle LD, Elimelech M (2007) Single-walled carbon nanotubes exhibit strong antimicrobial activity. Langmuir ACS J Surf Colloids 23(17):8670–8673. https://doi.org/10.1021/la701067r
171. Fakhroueian Z, Katouzian F, Esmaeilzadeh P, Moradi Bidhendi S, Esmaeilzadeh P (2019) Enhanced engineered ZnO nanostructures and their antibacterial activity against urinary, gastrointestinal, respiratory and dermal genital infections. Appl Nanosci 9(8):1759–1773. https://doi.org/10.1007/s13204-019-00996-5
172. Abualnaja KM, Alprol AE, Ashour M, Mansour AT (2021) Influencing multi-walled carbon nanotubes for the removal of ismate violet 2r dye from wastewater: isotherm, kinetics, and thermodynamic studies. Appl Sci (Switzerland) 11(11). https://doi.org/10.3390/app11114786
173. Birniwa AH, Abubakar AS, Mahmud HNME, Kutty SRM, Jagaba AH, Abdullahi SS, Zango ZU (2022) In: Muthu SS, Khadir A (eds) Application of agricultural wastes for cationic dyes removal from Wastewater BT—textile wastewater treatment: sustainable bio-nano materials and macromolecules, volume 1. Springer Nature, Singapore, pp 239–274. https://doi.org/10.1007/978-981-19-2832-1_9
174. Abdullahi SL, Birniwa AH, Audu AA (2020). Modification of Ashaka and tango bentonites with tetraphenyl phosphonium and hexadecyl pyridinium bromides for possible use in polymer-clay nanocomposite preparation. Nano Trends-A J Nano Technol Appl. https://doi.org/10.37591/nanotrends.v22i2.989

175. Kang S, Mauter MS, Elimelech M (2009) Microbial cytotoxicity of carbon-based nanomaterials: implications for river water and wastewater effluent. Environ Sci Technol 43(7):2648–2653. https://doi.org/10.1021/es8031506
176. Zhuang S, Zhu X, Wang J (2020) Adsorptive removal of plasticizer (dimethyl phthalate) and antibiotic (sulfamethazine) from municipal wastewater by magnetic carbon nanotubes. J Mol Liq 319:114267. https://doi.org/10.1016/j.molliq.2020.114267
177. Mohammadi A, Veisi P (2018) High adsorption performance of β-cyclodextrin-functionalized multi-walled carbon nanotubes for the removal of organic dyes from water and industrial wastewater. J Environ Chem Eng 6(4):4634–4643. https://doi.org/10.1016/j.jece.2018.07.002
178. Vuono D, Catizzone E, Aloise A, Policicchio A, Agostino RG, Migliori M, Giordano G (2017) Modelling of adsorption of textile dyes over multi-walled carbon nanotubes: equilibrium and kinetic. Chin J Chem Eng 25(4):523–532. https://doi.org/10.1016/j.cjche.2016.10.021
179. Machado FM, Bergmann CP, Fernandes THM, Lima EC, Royer B, Calvete T, Fagan SB (2011) Adsorption of reactive red M-2BE dye from water solutions by multi-walled carbon nanotubes and activated carbon. J Hazard Mater 192(3):1122–1131. https://doi.org/10.1016/j.jhazmat.2011.06.020
180. Shirmardi M, Mahvi AH, Mesdaghinia A, Nasseri S, Nabizadeh R (2013) Adsorption of acid red18 dye from aqueous solution using single-wall carbon nanotubes: kinetic and equilibrium. Desalin Water Treat 51(34–36):6507–6516. https://doi.org/10.1080/19443994.2013.793915
181. Bazrafshan E, Mostafapour FK, Hosseini AR, Raksh Khorshid A, Mahvi AH (2013) Decolorisation of reactive red 120 dye by using single-walled carbon nanotubes in aqueous solutions. J Chem.https://doi.org/10.1155/2013/938374
182. Selen V, Güler Ö, Özer D, Evin E (2016) Synthesized multi-walled carbon nanotubes as a potential adsorbent for the removal of methylene blue dye: kinetics, isotherms, and thermodynamics. Desalin Water Treat 57(19):8826–8838. https://doi.org/10.1080/19443994.2015.1025851
183. Ghaedi M, Khajehsharifi H, Yadkuri AH, Roosta M, Asghari A (2012) Oxidized multi-walled carbon nanotubes as efficient adsorbent for bromothymol blue. Toxicol Environ Chem 94(5):873–883. https://doi.org/10.1080/02772248.2012.678999

Synthesis and Catalytic Applications of PANI-Based Hybrid Materials for the Catalytic Removal of Organic Dyes from Wastewaters

Sandeep Kaushal and Karina Bano

Abstract Most of the chemical industries which use organic dyes when releasing their residual effluent in flowing water directly affect marine life and human life. This kind of water pollution causes severe diseases to mankind, and there is an urgent need to overcome this situation. For water treatment methods, adsorption is one of the best ways in this regard. Polymer-based adsorbents are most popular as they are effective in their functionality. Here, we studied polyaniline (PANI)-based hybrid materials used for the adsorption of organic dyes. Till now, different modifications were imparted in the basic structure of PANI to enhance its catalytic efficiency and reusability. Effect of doping of different catalytic materials in PANI and their effect on adsorption efficiency has been determined in this study. It is clear from the literature study that PANI-based adsorbent materials are highly emerging and advanced as they showed high specificity, improved catalytic efficiency, and extraordinary stability.

Keywords Polyaniline · Adsorption · Organic dyes · Water pollution

1 Introduction

Environment-related issues including harmful pollutants, global warming, ozone depletion, and waste disposal greatly influence the quality of air, water, and soil. Out of all serious problems, contamination of water is the biggest subject that needs to be resolved at priority. Textile industries dispose of their waste in water bodies which cause bioaccumulation and limit the path of sunlight, damaging aquatic life [1]. Textile industries use both natural and synthetic dyes which include several types of harmful chemicals and upon completion of work, and discharge their waste effluent in rivers, canals, and lakes which contaminate quality of water and damage ecology [2]. These chemicals are highly persistent in nature and travel long distances along

S. Kaushal (✉) · K. Bano
Department of Chemistry, Sri Guru Granth Sahib World University, Fatehgarh Sahib, Punjab, India
e-mail: kaushalsandeep33@gmail.com

© The Author(s), under exclusive license to Springer Nature Singapore Pte Ltd. 2023
A. Ahmad et al. (eds.), *Nanohybrid Materials for Treatment of Textiles Dyes*, Smart Nanomaterials Technology, https://doi.org/10.1007/978-981-99-3901-5_8

with water. Fertility of soil also gets affected when these chemicals stay for a long time period [3]. Thus, there is a critical need for a harmless way to the ecosystem treatment approaches for treating pollutant containing wastewater preceding before its last removal into the environment. Till date, various catalytic techniques such as microbial degradation, membrane filtration technique, chemical oxidation–reduction, and plasma ozonization were reported to eliminate water pollutants [4–6]. Based on the efficiency and cost of different methods, some limitations were found as complexity in synthesis, low efficiency, time-consuming synthesis methods, waste disposal problems, and prohibitive practices. Till date, removal of toxic pollutants via adsorption reaction and photocatalysis has gained wide attention. A lot of research work has been done to report new efficient and eco-friendly catalysts. Particularly, those materials which comprise high specific area, low band gap value, visible light active, highly stable, and environmentally safe are considered as most promising catalysts. Polymer-based adsorbents and photocatalysts are highly efficient for removal of harmful organic dyes because of their easy availability, simple synthesis method, cost-effective, and capable of reusability [7].

On this basis, polyaniline and its corresponding hybrid materials were employed as an efficient adsorbent and photocatalyst for removal of these dyes [8], and in this review, detailed literature study was conducted to gain better knowledge about advanced treatment approaches utilized for wastewater treatment. Polyaniline (usually referred as PANI or Pani) was developed by Ferdinand Runge in 1834. It was also known as aniline black as it was firstly derived by oxidizing aniline. As it contains –NH groups (Fig. 1), polyaniline is among the most conductive polymers and, thus, has gotten a lot of interest from scientists [9]. Because of its ease of synthesis, doping feasibility, unique physicochemical properties, flexibility, high stability, and simple availability of its monomer, this polymer is one of the most investigated polymers. It is widely employed in solar cells, sensors, electromagnetic shielding equipment, batteries, electrical gadgets, etc. In addition, polyaniline has been extensively researched for the production of adsorbents by modification in chemical structure, doping, and composite fabrication. The presence of active groups, namely amine and imine, which interact with molecules of different pollutants found in contaminated water, makes polyaniline a possible adsorbent for wastewater treatment [10]. It is made up of a series of aromatic rings of the benzene diamine and/or quinone diimine kinds, each of which is connected by a nitrogen heteroatom. Polyaniline may be found in three redox states: leucoemeraldine, emeraldine, and pernigraniline, which are completely reduced, partially oxidized, and entirely oxidized, respectively. Out of these forms, emeraldine and pernigraniline are the ones that can be found as salts or bases [11]. Emeraldine is perhaps the most stable and usable form of polyaniline because of excellent stability at room temperature, low cost, eco-friendly, facile synthesis, and simple doping chemistry. It is made up of the same number of oxidized and reduced units. When the imine group of emeraldine base is doped or protonated and exposed to acid, polycation forms. As the degree of protonation grew from 0 to 20%, the conductivity of the emeraldine form rose tremendously. Leucoemeraldine is a colorless, non-conducting type of polyaniline that contains the amine nitrogen

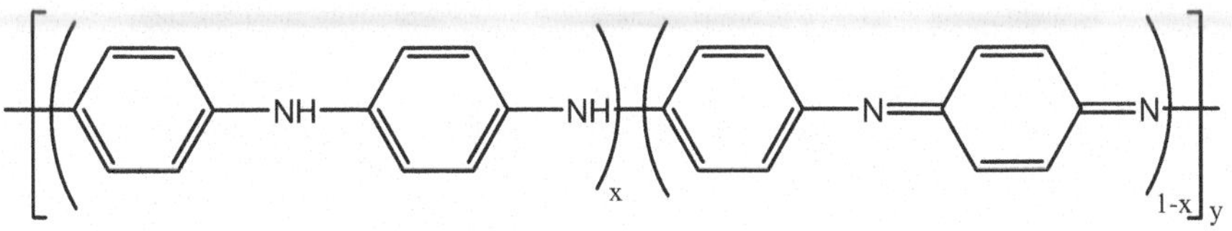

Fig. 1 Basic chemical structure of polyaniline

atom. It oxidizes to emeraldine when exposed to an acidic environment. In the pernigraniline form of polyaniline, imine linkage, aminobenzene, and quinone diimine are all present in varying amounts. Because the quinone diimine fragment is unstable in the presence of a nucleophile, particularly water, pernigraniline and its salts disintegrate in the air [12, 13]. Polyaniline has recently been employed as an excellent precursor, chemical composition modifying agent, or a composite component in the production of a range of dye adsorbent [14, 15].

2 Mechanism of Catalytic Adsorption by PANI-Based Adsorbents

Catalytic adsorption of dyes involves removal of organic dyes from contaminated wastewater by using polyaniline (PANI)-based nanocomposite materials. Polyaniline acted as a good adsorbent as it provided a wide surface to pollutants and resulted in removal of toxic chemicals. When molecules of dye get adsorbed onto the surface of catalyst, various physical interaction occurred between them [16] such as acid–base reactions, hydrogen bonding, van der Waals attraction, ion-exchange reaction, and π–π interactions (shown in Fig. 2). In ion-exchange reaction, interchange of ions takes place between adsorbent and adsorbate. In a study by Zheng et al. [17], adsorption of CR and MO dye was studied on adsorbent GO-NiFe-LDH and possible mechanism showed that high adsorption capacity was gained due to ion-exchange reaction and strong π–π interactions between dye and adsorbent molecules. Shen and Gondal [18] employed coffee powder as an adsorbent for rhodamine B dye and found a high value of adsorption capacity due to intermolecular interactions between the molecules of adsorbent and adsorbate.

Pirbazari et al. [19] reported that during removal of MB dye by wheat straw, two important reactions were involved such as exchange of ions between dye molecules and adsorbent which further leads to a complex molecule. During formation of complex ions on the surface, firstly ions get bonded to different functional groups present on the surface of adsorbent followed by electrostatic interaction between adsorbent and adsorbate. Formation of hydrogen bond between dye molecules and catalyst during adsorption process was explained by Cojocaru et al. [20] by chitosan-based adsorbents. In addition to this, other covalent interactions which are similar to electrostatic forces also occurred during adsorption reactions. At the same time,

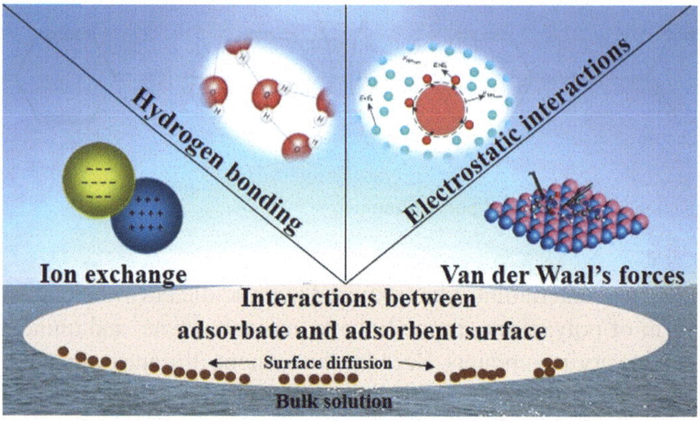

Fig. 2 Different possible interactions during adsorption of dye on adsorbent surfaces

adsorption mechanisms can follow more than one route. This statement was proved by Thamer et al. [21], by performing adsorption reactions of Coomassie Brilliant Blue R 250 dye. This catalytic reaction mechanism involved π–π interactions, H-bonding between molecules, and electrostatic interlinkage.

3 Polyaniline-Based Adsorbents for Dye Adsorption

3.1 Polyaniline/Bipolymer Composites

In a study by Janaki et al. [22], aniline was processed under chemical oxidative method to synthesize starch/polyaniline nanocomposite which was further utilized for degradation of various organic dyes. Experiment results revealed that the synthesized nanocomposite is an efficiency catalyst toward water treatment reactions as it completely removed reactive black 5, reactive violet 4 dyes. In addition to this, prepared catalyst was also employed for decolorization of dye bath effluent and successfully removed 87% color. Kinetic studies declared that dye degradation reactions followed pseudo-second-order, whereas dye bath effluent followed a pseudo-first-order reaction mechanism. An efficient adsorption of dyes from textile effluents with the help of sarch/polyaniline nanocomposite was a successful effort [22]. One another bipolymer catalyst was presented by Janaki et al. for adsorption of organic dyes. This eco-friendly catalyst was synthesized by using chitosan polymer attached to polyaniline. The high catalytic efficiency of this catalyst was also expected due to its rough surface and similar reflection peaks found from SEM and X-ray analysis, respectively. Removal efficiency for Congo red, Coomassie Brilliant Blue and Remazol Brilliant Blue R dye was found to be 95%, 98%, and 99.8%, respectively [23]. After this work, Olad et al. [24] introduced a modification in starch/polyaniline composite so as to compare the efficiency of both samples toward adsorption of organic dyes. In this research work, chemical oxidation method was adopted to polymerize aniline, and to induce modification, starch-montmorillonite dispersion was intermixed, and the nanocomposite was named as St-MMT/PANI. For application purpose, adsorption of reactive blue (RB 194) dye was selected, and value for maximum adsorption capacity was 92 mg g^{-1} and reaction was proceeded via pseudo-second-order reaction kinetics [24].

In a study, Pandiselvi et al. synthesized ZnO-based bipolymer composite by polymerization of aniline hydrochloride and intermixing with $ZnCl_2$ and chitosan. The incorporation of ZnO nanomaterial into chitosan–polyaniline composite successfully increases the porous surface of the catalyst and favors excellent reactive orange 16 dye adsorption. The adsorption experiment found value of adsorption capacity 476.2 mg g^{-1} and also best fit of equilibrium values to the Langmuir isotherm equation, and adsorption reaction followed pseudo-second-order model. From these results, it was concluded that the doping of $ZnCl_2$ into chitosan–polyaniline hybrid enhanced the efficiency of adsorption of organic dyes [25]. In a research work by Sahnoun and his co-workers [26], catalytic adsorption of tartrazine dye was targeted. For this experiment, a chitosan-based bipolymer catalyst was synthesized (Cht-PANI). Maximum adsorption capacity achieved for tartrazine dye was 580 mg g^{-1} with best fit of equilibrium data in the Freundlich isotherm and pseudo-second-order reaction kinetics. Reusability and stability of the catalyst were evaluated for continuous four cycles which proved this catalyst to be efficient as promising nanocomposite toward adsorption mechanisms [26].

In a study by Abasian et al. [27], eco-friendly and cost-effective bipolymers were reported fabricated via chemical oxidation polymerization method. In this series, chitosan-doped polyaniline-based hybrid molecules such as chitosan-*g*-poly (*N*-methylaniline) (CS-*g*-PNMANI), chitosan-*g*-polyaniline (CS-*g*-PANI), and chitosan-*g*-poly (*N*-ethylaniline) (CS-*g*-PNEANI) were fabricated. The maximum adsorption capacities calculated for adsorption of AR4 was 98 mg g^{-1} and for DR23 its value was 112 mg g^{-1}. Kinetic studies revealed that both of these adsorption reactions followed the pseudo-second-order kinetics [27]. Ayazi et al. [28], reported a nanocomposite named Alg–MMT–PANI fabricated by polymerization reaction of aniline and diffusion in Alg–MMT material. This polymeric hybrid composite was utilized for the adsorptive elimination of reactive orange 13 (RO13) dye. With this unique composite catalyst, the maximum value obtained for adsorption efficiency was 111.12 mg g^{-1}. With reference to these experimental results, we found that adsorption reactions proceeded via pseudo-second-order model and data was best fitted into the Langmuir isotherm model [28].

Synthesis of bipolymer nanocomposite has gained huge attention due to their effective catalytic response. Bhaumik et al. [29], reported a bipolymer composite named polypyrrole–polyaniline (PPy–PANI–NFs) fabricated via intermixing PPy^+ and $PANI^+$ without using template free radicals. FESEM analysis confirmed formation of nanostructured fibers firmly connected to each other. Adsorption capacity of synthesized nanocomposite was evaluated for removal of Congo red dye, and we found different values of adsorption capacity for different temperatures such as

222.3 mg g^{-1} at temperature 25 °C and 270.27 mg g^{-1} at 35 °C, followed by pseudo-second-order reaction kinetics model and the equilibrium data for these reactions fitted perfectly to Langmuir model [29]. Shen et al. [30] in a study reported wide surface area nanocomposite of value 126 m^2 g^{-1} by mixing of polyaniline (PANI) adsorbent with poly (2-acrylamido-2-methyl-1-propanesulfonic acid) and named this as PANI-PAMPSA. Adsorption capacity was determined by eliminating MB and RB dye. Large specific surface area of catalyst favors high adsorption capability values of 466.5 mg g^{-1} for MB dye and 440.0 mg g^{-1} for RB dye and proceeds via pseudo-second-order model and the Langmuir adsorption model. Possible mechanisms of adsorption may involve strong π–π interactions and electrostatic interlinkage [30].

3.2 *Polyaniline/Magnetic Composites*

Saha et al. [31] reported PANI@Fe–Mn–Zr hybrid nanocomposite fabricated via mixing of Fe, Mn, and Zr metals in the first step followed by diffusion of these hybrid metals onto the surface of PANI. Catalytic application of this unique hybrid material was evaluated for adsorption of methyl red dye. In this experiment, catalyst dose was taken 0.4 g L^{-1}, concentration of dye solution was 15 mg L^{-1}, and solution was sonicated for 15 min. obtained results showed 98.19% removal of MR dye with adsorption capacity of 434.78 mg g^{-1}. This adsorption capacity of the catalyst was attributed to large electrostatic interactions and strong hydrogen bonding [31]. Out of many different kinds of composites, Teng et al. developed PANI/Fe_3O_4 nanocomposite, which was quite helpful in control of water pollution due to its extraordinarily wide surface area. In this work, authors tried to alter different PANI concentration on Fe_3O_4 nanoparticle surface to enhance catalytic removal efficiency for Congo red (CR) dye. Out of different PANI/Fe_3O_4 composite ratios prepared, PANI/Fe_3O_4 composite at 70/30 ratio showed highest removal of Congo red up to 89.62% as compared to the other ratios of the composite [32]. A MIL-101 Fe/PANI/Pd nanomaterial was designed via a hydrothermal method as a photocatalyst for methylene blue removal. For obtaining optimum results from the composite, the conventional variables like MIL-101 Fe/PANI/Pd mass, pH, concentration of initial dyes, and H_2O_2 concentration were scrutinized and maximum 92% dye degradation rate was achieved with newly synthesized hybrid nanocomposite [33]. Another study by Imgrahan et al. on a polyaniline@Fe-ZSM-5 composite fabricated via an in situ interfacial polymerization method revealed that this composite was effectively adsorbing the orange G (OG) stuff from water for the first time. This composite was showing maximum adsorption capacity of 217 mg g^{-1}, and regeneration investigation of the composite revealed that it could be used effectively up to five cycles [34].

Rajappa et al. demonstrated a simple, easy, and effective solvent-free fast mechanochemical approach to synthesize Fe_2O_3-based hybrid TAZnPc-PANI@Fe_2O_3 nanocomposite, and the catalytic activity of hybrid adsorbent was evaluated toward removal of main pollutants like MB and EY dyes. This dispersion of Fe_2O_3 on tetra amino zinc phthalocyanine embedded PANI (TAZnPc-PANI) was found to be significantly increased toward the photodegradation of methylene blue and Eosin Y dyes, and this photodegradation occurs up to 95% in 80 min under optimized conditions [35]. Alves and his co-workers [36] worked on preparation of hybrids comprising citric acid, polyaniline, and magnetic iron oxide (MOM) and diffused them together via in situ chemical polymerization route, and catalytic activity of nanohybrid was determined for degradation of methylene blue dye (MB) from polluted water. The adsorption catalytic activity was combined with photodegradation, and complete removal of MB was achieved due to synergistic effect between the MOM-polymer phases [36]. In a study by Xiong et al. [37], a novel polyaniline/FeOOH composite was reported, various chemical and physical analysis were performed such as SEM, HRTEM, XRD, FTIR, and BET, to determined morphology, surface area, reflection planes, etc. Catalytic adsorption of acid orange II dye was evaluated with polyaniline/FeOOH hybrid composite. The maximum adsorption efficiency for AOII was 155.8 mg g^{-1}, and reaction proceeds via pseudo-second-order kinetic model and Langmuir isotherm model. On the basis of these results, we found that PANI/FeOOH nanocomposite had better applicability for removal of different types of anionic dyes [37]. Zhang et al. [38] reported a new polymer hybrid by doping Fe_3O_4 in graphene oxide as a template, and obtained sample was PANI-GO-Fe_3O_4 nanosheets. The obtained PANI-GO-Fe_3O_4 hybrid adsorbent was utilized for adsorption of organic dyes and resulted in high value of adsorption capacity such as 252.7 mg g^{-1} for Congo red dye and 181.0 mg g^{-1} for methyl orange dye, followed by quasi-secondary kinetic model, and the experimental data was well fitted in Langmuir model. Stability of nanocomposite was examined for successive five cycles of catalytic adsorption and desorption reaction, and the obtained results showed that the removal rates of CR and MO dyes by PANI-GO-Fe_3O_4 was 88 and 76%. This work gives another course to advance the divisible superior execution adsorbents for eliminating the ionic colors pollution from squander water [38].

3.3 Polyaniline/Metal Based Composites

In a study by Katowah et al. [39], a novel PANI-based composite named PANI@Cs-GO-OXS/CuO was synthesized by following copolymerization method. This cross-linked hybrid composite with high thermal stability was further utilized for removal of rhodamine B dye (RhB). Dye solution was prepared with concentration 2.0×10^{-5}, slight acidic pH, and amount of catalyst taken was 10 mg. With these reaction parameters, 94% of RhB dye was removed within 130 min [39]. The nanocomposite ended up being an effective adsorbent for adsorption of poisonous color from wastewater and furthermore shows extraordinary reusability. In a study by Karami et al. [40], a novel hybrid composite was reported and named as MIL101-Cr/PANI/Ag prepared by using hydrothermal method. The synthesized nanocomposite comprised particle size of 153 nm and BET surface area 2861 $m^2\ g^{-1}$ and exhibited high catalytic efficiency toward adsorption of methylene blue dye (MB). During the catalytic experiment, pH of solution 12, MB dye concentration 25 mg L^{-1}, and catalyst dose 0.03 g were taken which leads to 96% removal of MB dye [40].

In another study, α-MnO_2 nanoparticles were synthesized via biogenic method and coated on polyaniline (PANI) matrix which showed conducting nature, and this mechanism was performed via in situ polymerization method. After this, α-MnO_2/PANI nanocomposite modified pencil graphite electrode (PGE) was specified as microbial fuel cell (MFC) which further helped in removal of methyl red (MR) dye which further generated electricity. Catalytic results revealed that the decolorization efficiency (DE) of α-MnO_2/PANI modified PGE showed 12.47 times higher value as compared to unmodified PGE. This nanocomposite modified anode is responsible for maximum decolorization of methyl red in MFC [41]. Fatima et al. developed WS_2/PANI nanocomposite by "free two birds with one key" strategy for the detection and degradation of toxic pollutants. This nanocomposite successfully removes methyl orange dye with catalytic efficiency of 97% and showed high stability for next least four catalytic cycles. Properties like high photo-induced charge separation of 5.76 ns and band gap energy value of 2.6 eV are responsible for a great photocatalytic activity of WS_2/PANI nanocomposite [42]. Another study by Merangmenla et al. synthesized a CP1/PANI composite. In the first step, 1D copper (II) complex was linked to a ligand (pyridine-2-carboxylic acid), and in the next step, the obtained complex was diffused on PANI via polymerization of aniline. Further, a good photocatalytic activity was observed for the composite for the degradation of methylene blue dye. Coming to optimization of the photocatalyst composition, 10 wt% CPI/PANI composite material exhibits the highest photocatalytic performance with 88% efficiency within 210 min. This great photocatalytic activity of CPI/PANI composite can be attributed to the synergistic effect between the two materials, i.e., CPI and PANI [43].

Park et al. examined the conducting polymers' potential with transition metal oxides as a photocatalyst from an environmental friendly perspective and synthesized a visible-light-responsive PANI-rGO-MnO_2 composite for the photocatalytic degradation of cationic dye, i.e., methylene blue; in an aqueous solution. The successful incorporation and growth of MnO_2 nanoparticles into the PANI-rGO composite were observed, and this PANI-rGO-MnO_2 ternary composite shows an enhanced photocatalytic and catalytic activity under visible light irradiation for up to 90% degradation of methylene blue within two hours. This work provided a deep insight on the study of synthesis and photocatalytic activity of the PANI, PANI-rGO, and the PANI-rGO-MnO_2 ternary composite. Such composites have potential for environmental remediation [44].

Further, Aamir et al. synthesized Co–Zr doped Ni ferrite substituted polyaniline (PANI) nanocomposite via polymerization method. $NiFeCo_{0.5}Zr_{0.5}O_4$ with different concentrations of PANI (12.5%, 25%, 37.5%, and 50% named A1, A2, A3, and A4, respectively) were utilized to get better photocatalytic activity. This photocatalytic efficiency was enhanced with Co–Zr doped Ni ferrite contents in the synthesized nanocomposites. In FTIR spectrum of the composite, the shifting of peaks toward higher wavenumber indicates the interactions between the nanoparticles and polymer.

The efficiency of degradation was enhanced with increase of Co–Zr ferrite contents, which indicated the significant influence of the nanoparticles in such nanocomposite. It was assumed that the synthesized PANI/$NiFeCo_{0.5}Zr_{0.5}O_4$ nanocomposite was efficient photocatalysts for the degradation of MO (methyl orange) dye in water [45]. Rajaji et al. synthesized tin oxide/polyaniline (SnO_2/PANI) nanocomposite via co-precipitation method for the photocatalytic degradation of direct blue 15 (DB15), and the synthesized SnO_2/PANI nanocomposite was characterized by UV–visible analysis. The incorporation of PANI in SnO_2 decreases the crystalline size and enhances surface area in comparison to bare SnO_2 nanoparticles. The band gap energy (3.1–2.7 eV) and the reflectance of the SnO_2 was reduced after substitution with PANI. Also, the SnO_2/PANI nanocomposite shows enhanced photocatalytic dye removal efficiency as compared to that of bare SnO_2. This data showed that SnO_2/PANI nanocomposite give rise to only non-toxic by-products, and these are less toxic as compared to aquatic food web components [46].

At room temperature, the use of one-pot solvothermal method and in situ polymerization of aniline at room temperature resulted in formation of a novel Cu_2O/ZnO-PANI ternary nanocomposite. In this study by Mohammed et al., the novel composite showed excellent adsorption characteristics, rapid photocatalytic activities, and a highly enhanced stability. This modification with PANI resulted in surface area enhancement from 8.82 $m^2\ g^{-1}$ (Cu_2O) and 35.66 $m^2\ g^{-1}$ (Cu_2O/ZnO) to 45.32 $m^2\ g^{-1}$. The Congo red dye was degraded using pure Cu_2O, Cu_2O/ZnO, and Cu_2O/ZnO-PANI composite photocatalysts, and it was observed that it was a model photocatalytic reaction using the composite. Also, the composite was showing degradation efficiency up to fifth cycle, and it was highest (100%) as compared to that of Cu_2O and Cu_2O/ZnO showing 81% and 94% degradation efficiency, respectively. Such an advanced photocatalytic degradation efficiency was due to a large surface area of the photocatalyst and photoluminescence quenching phenomenon resulting in a significant separation of electron/holes and a subsequent decrease in recombination rates of charged carriers. Such work provides an efficient method to develop photocatalysts with improved photocatalytic degradation efficiency using conducting polymers for ternary nanocomposite processing [47].

A PANI-TiO_2/rGO photocatalyst with an efficient charge separation was developed by functionalizing rGO surface with TiO_2 and combining it with polyaniline. This PANI covered with rGO exhibits an increased efficiency of charge separation. It was observed that the TiO_2 nanoparticles, which are dispersed in rGO layers, did not impose any changes on structure of rGO. The rate of hydrogen production and efficiency of photocatalytic degradation of rhodamine B (RhB) by the catalyst was observed to be 0.806 mmol $h^{-1}\ g^{-1}$ and 90.5%, respectively. These values were 1.8 and 10.1 times greater as compared to the values noticed in case of TiO_2/rGO and TiO_2 photocatalysts. The synergistic interactions among rGO, TiO_2, and PANI can be the reason for an improved photocatalytic efficiency. Such synergy results in increased separation of charge carriers as well as improved spectral results. This study shows that generation of such novel photocatalytic systems can be effective against environmental contamination [48]. Sharma et al. synthesized a 3D hierarchical hybrid composite named PANI/CdS. This nanocomposite was synthesized by a two-step process involving hydrothermal method and chemical oxidation. The resulting nanocomposite was showing an eight-fold increment in the catalytic performance of rhodamine B (RhB) degradation as compared to that of CdS nanoparticles. Interestingly, This PANI/CdS photoanode exhibits five-fold higher photocurrent density as compared to that of pristine CdS under visible light irradiation. There is overall enhancement in charge separation, light absorption, surface oxidation, and reduction reactions for effective degradation of rhodamine B. Furthermore, PANI/CdS photoanode shows four times increase in IPCE than that of CdS. The unique morphology of CdS and generation of type-II heterojunction was responsible for the enhanced photocatalytic and PEC performance [49]. As we all know that many industrial sewage contain Congo red dye in their effluent and this is a serious problem for the environment, it is very important to design low-cost photocatalytic systems with high photocatalytic activity to resolve such problems. In previous decades, polyaniline-based composites were designed to eliminate such harmful effluents from the environment. Das et al. performed synthesis of graphite/cobalt sulfide/PANI-based ternary composite for elimination of harmful Congo red from water. Analysis data proved that synergistic interactions were responsible for high catalytic activity of catalyst. The degradation of Congo red dye was achieved 95.5% efficiency in 120 min. The phenomenon of adsorption on the surface of catalyst was responsible for the degradation of dye which showed chemisorption nature [50].

Shankar et al. utilized the reduced graphene oxide (rGO) and polyaniline-based carbon-doped porous ZnO heterojunction nanocomposite for enhancement in photocatalytic performance of commercial pharmaceutical antibiotic drug amoxicillin and methylene blue dye. The efficient utilization of the light-induced electron–hole pair and lifetime of excitation is the main feature that must be present in a light-driven photocatalyst. Formation of such a composite resulted in reduction of band gap energy to 2.8 eV as compared to 3.34 eV for ZnO. This RPZ heterojunction successfully degraded the methylene blue dye and ACP in 100 min under visible light irradiation with photocatalytic efficiency of 95% and 47%, respectively [51]. In another study, the peculiar features of $LaNiSbWO_4$-G-PANI (LNSWGP) polymer composite was described for their utilization as an efficient photocatalyst, and this composite was synthesized using an easy hydrothermal process. The evaporation induced hydrothermal method was employed to cover $LaNiSbWO_4$ (LNSW) having highly conducting polyaniline with thin graphene sheets. This composite nicely degraded anionic methylene blue dye and a phenolic acid, i.e., trihydroxybenzoic acid, and this enhanced photocatalytic performance was attributed to appropriate interfaces of LNSWGP resulting in an enhancement in charge-separation efficacy. This photocatalytic activity was actually quite higher for LNSWGP in the presence of polymer as compared to that of any different type of composite in water as a medium. The easy preparation and highly enhanced photocatalytic performance of this LNSWGP makes it an competitive, reliable, and a safe photocatalyst under visible light irradiation, and such composite is very much operative in field of solar cells and can be employed for environmental remediation [52]. A $Sb_2O_3/CuBi_2O_4$ (Sb_2O_3/CBO) composite was utilized to prepare a photocatalytic membrane for removal of methylene blue and acid blue under natural light. Both methylene blue and acid blue dyes are very harmful for the environment. It was revealed from characterization data that the pH and use of catalyst had a positive impact on degradation of dyes. It was observed that the degradation of methylene blue was greater in basic medium than that of the acidic medium and reverse was true for acid blue dye. The maximum photocatalytic performance for dye removal up to 94.6% was observed for 10% Sb_2O_3/CBO photocatalyst along with use of membrane under optimum conditions [53].

Another composite named Hu/PANI@Ni_2O_3 was synthesized by using natural zeolite heulandite and polyaniline (Hu/PANI) as a catalyst support. Its morphological, chemical, structural, and optical features were analyzed utilizing different techniques. The authors analyzed the degradation of safranin-T dye using Hu/PANI@Ni_2O_3 composite under visible light irradiation and observed excellent photocatalytic performance of this composite. The safranin-T dye was removed within minutes after use of higher concentrations of the composite. At lower concentrations of 0.025, 0.03, and 0.035 g, the composite was resulting into 80, 98, and ~100% removal of 5 mg L^{-1} dye under natural light within one minute. The best results were obtained in alkaline medium. The recyclability of the composite was also remarkably high as it remains up to 84.5% even after five cycles. The Hu/PANI@Ni_2O_3 displayed great results for degradation of other dyes as well as mixed dye solutions with same speed of degradation. Hence, nickel oxide loading onto the catalyst support of hybrid Hu/PANI composite resulted in excellent photocatalytic degradation efficiency [54]. Palliyalil et al., synthesized TiO_2@CS–PANI nanocomposite via heterogeneous chemical polymerization technique resulted in reduction of band gap energy value. Photocatalytic activity of synthesized catalyst was evaluated for removal of methylene blue (MB) and methyl orange (MO) dye, and the obtained results showed that 92.3% MB and 89.5% MO dyes were degraded successfully. The TiO_2@CS–PANI composite showed high stability as it retained its catalytic efficiency for five cycles and proved to be an effective photocatalyst for wastewater treatment [55]. After this, removal of MB dye was also evaluated with a new catalyst named SW–ZnO–PANI synthesized by intermixing seaweed–zinc oxide and polyaniline bonded together via surface electrostatic attractions. The catalytic adsorption reaction was proceeded via pseudo-second-order kinetic model. When experimental data was fitted in the Langmuir and Freundlich adsorption model, it was found that the best fit was obtained using the Freundlich isotherm model [56]. In a study by Hiragond et al. [57], different conducting polymer materials were used to synthesize CdS/polythiophene (CdS/PTh), CdS/polypyrrole (CdS/PPy), and CdS/polyaniline (CdS/PANI) hybrid catalysts, and their catalytic efficiency was evaluated for removal of MB dye. Synergistic effect between CdS and PTh, PPy, or PANI polymers was observed which resulted in high dye removal efficiency [57]. In a study by Ahmad et al. [58], HCl/PANI and PANI/BN nanocomposite were reported synthesized by doping of HCl and boron nitride via polymerization method. The degradation of methylene blue (MB) dye achieved was 57%, whereas for rhodamine B (RhB) dye 71.6% degradation efficiency was obtained [58].

3.4 PANI/Carbon Based Hybrid Nanocomposite

Maruthapandi et al. [59] reported a PANI-based nanocomposite PANI-N@CDs. The first step involved hydrothermal processing of bovine serum albumin to synthesize nitrogen-doped carbon nanodots (N@CDs); then in the next step, PANI-N@CDs nanocomposite was fabricated by ultrasonication route. This multicomponent catalyst was utilized to eliminate Congo red (CR), methylene blue (MB), rhodamine B (RhB), and crystal violet (CV) dyes. From scavenger experiment, results revealed that active charge species such as holes (h+), hydroxyl ($OH^•$), and $O_2^{•-}$ were responsible for complete removal of the CR and CV dyes within 20 min, whereas MB and RhB dyes took longer times for complete removal [59]. In a study by Mitra et al. [60], catalytic removal of malachite green, rhodamine B, and Congo red dyes was evaluated with the help of nanocomposite made up of polyaniline/reduced graphene oxide and synthesized via oxidative polymerization technique. For these three organic dyes, rate of removal was much higher with PANI/RGO composite as compared to PANI. Obtained results showed that degradation of 99.7, 99.4, and 98.7% was achieved for MG, RhB, and CR dye, and time taken for these reactions was 15, 30, and 40 min, respectively. In this work, rGO template was used to hinder the agglomeration of PANI nanoparticles and improve the surface area of the catalyst. Scavenger experiment helped in designing possible pathway of dye degradation as shown in figure. From these findings, it was concluded that PANI-based rGO worked as an excellent template for wastewater treatment [60].

Pete et al. [61] reported CNT-based nanocomposite synthesized by in situ polymerization. PANI-MWCNTs nanocomposite was applied to remove methyl orange dye by catalytic adsorption. Incorporation of MWCNTs in PANI increased specific surface area of catalyst, and from BET analysis, it was found to be 43.892 $m^2\ g^{-1}$. This mesoporous catalyst exhibited high adsorption capacity of 149.25 mg g^{-1}, this reaction proceeds via pseudo-second-order kinetic model, and value of rate constant was found to be 5.265×10^{-4} g mg^{-1} min^{-1}. The absorption mechanism acquired less than 1 value for both Langmuir constant (R_L) and Freundlich constant ($1/n$) which signifies the spontaneous nature and favorable pathway of this reaction [61]. After this, Amina et al. [62] further examined the catalytic adsorption efficiency of methyl orange dye by preparing a highly stable nanocomposite by joining active carbon with PANI through oxidation method. This catalyst exhibited a wide surface area of 332 $m^2\ g^{-1}$ which improved adsorption of dye and also fit best into the Langmuir model. Degradation of methyl orange dye was conducted at pH 6 and 298 K temperature, and obtained results showed 192.52 mg g^{-1} adsorption efficiency for mesoporous PANI@AC. The rate of reaction was found to follow a pseudo-second-order kinetic model, and also reusability and regeneration efficiency of the catalyst were retained for up to five cycles [62].

In a study, Chatterjee et al. [63] polymerized aniline with SWCNT and reported formation of PANI-SWCNT composite. For analysis of physical and chemical properties of sample powder XRD, FESEM analysis, UV–visible spectroscopy, FTIR, XPS analysis, photoluminescence, and BET analysis were conducted. The catalytic applications of the hybrid composites were evaluated for the removal of organic dyes such as rose Bengal (RB) and methyl orange (MO). Maximum degradation efficiencies of 95.91% and 90.34% were achieved for RB dye (10 min) and MO dye (30 min), respectively [63]. In a study, Hasan et al. [64] synthesized a nanocomposite polyaniline/activated carbon via in situ oxidative polymerization method. Adsorptive capacity of POAC composite was evaluated for removal of organic dye methyl orange (MO) from contaminated water. Maximum adsorption capacity of 285 mg g^{-1} was obtained with 70% of removal efficiency within 2 h. This reaction followed pseudo-second-order kinetic model, and experimental data values were best fit in the Langmuir isotherm model. These results indicate that POAC is an excellent and effective adsorbent for removal of dyes [64].

One-pot synthesis of G@PANI-Fe_3O_4 nanocomposite resulted in availability of a large number of porous sites which were active for adsorption of dye molecules. For removal of direct red 23 dye, 95.44% removal rate was observed by using 1 g L^{-1} G@PANI-Fe_3O_4 at pH 4 and could be reused for at least 10 cycles. The proficient dye removal limit of as-prepared nanomaterials could give an inventive stage to various applications including the control of water contamination [65]. Moonis et al. [66] studied the effect of aggregation of graphene oxide particles on their catalytic applications in various fields and found that agglomeration greatly reduced graphene oxide (GO) catalytic performance. To overcome this issue, an oxidation polymerization method was adopted to synthesize polyaniline (PANI) terminals bound on the outline stacked GO sheets. The catalytic performance of this newly synthesized GO-PANI was evaluated for elimination of brilliant green (BG) dye. The equilibrium data was best fit to the Langmuir isotherm and pseudo-second-order kinetic models. Maximum adsorption capacity (q_m) calculated for BG dye was found to be 142.8 mg g^{-1}, and reaction was favored under endothermic conditions. During the adsorption mechanism, the possibility of π–π stacking interaction and electrostatic interaction is most and they played a vital role during BG adsorption on GO-PANI composite [66].

Adsorption of methylene blue dye with the help of two efficient catalysts PANI/GO and PANI/rGO was conducted by Kamel et al. [67]. For the adsorption experiment, concentration of dye solution was taken 50 mg L^{-1} and total time taken for the reaction was 270 min. With all the optimized reaction parameters, maximum adsorption capacity obtained with PANI/GO and PANI/rGO was 14.2 and 19.2 mg g^{-1} with pseudo-first-order kinetics and adsorption data was best fitted into Langmuir adsorption isotherm model. From these findings, we came to the conclusion that PANI/RGO is much more efficient as compared to PANI/GO for the elimination of toxic organic dyes from polluted water [67]. For further improvement in the morphology, structure, and stability of the nanocomposite, some modifications were required. In a study by Liu et al. [68], a 3D nanocomposite was reported prepared by dispersion of polyaniline/dicarboxyl acid cellulose hybrid on the surface of graphene oxide. Physical and chemical analysis confirmed enhancement in stability and catalytic performance of newly formed nanocomposite as compared to PANI alone. Catalytic properties of nanocomposite were examined for degradation of brilliant red dye. In the first phase, adsorption occurred followed by photocatalysis, whereas in the second phase, adsorption and photocatalysis were performed simultaneously. It took total 25 min to attain adsorption equilibrium and calculated adsorption capacity was found to be 447.0 mg g^{-1}. When photocatalysis was performed in the presence of light energy, it resulted into catalytic adsorption of 729.0 mg g^{-1} within 180 min. When degradation was performed involving both adsorption-photocatalytic reaction, the adsorption capacity of 558.1 mg g^{-1} was achieved in 25 min and the total degradation was found to be 733.3 mg g^{-1}. Hence, this modified PANI-DCC@GO nanocomposite showed great reusability and high potential toward treatment of polluted water [68].

Wang et al. [69] reported PANI/GO nanocomposite for the catalytic removal of MB and MO orange dyes. Maximum value obtained for catalytic adsorption was 962 mg g^{-1} and 885 mg g^{-1} for MB and MO dyes, respectively. The stacked structure of graphene oxide is responsible for π–π interaction and high adsorption capability. The stability and reusability of nanocomposite was evaluated for five catalytic cycles, and outstanding results were achieved as no significant loss was observed in adsorption capacity for MB (87.8%) and MO (75.0%) dyes, respectively [69].

3.5 PANI/Biodegradable Waste

In a study, Mane et al. [70] reported a biodegradable adsorbent for wastewater treatment. Rice husk ash was obtained from risk husk chambers where it is deposited in exhaust pipe lines. In this research paper, catalytic removal of brilliant green (BG) dye was performed with the help of biodegradable absorbent rice husk ash (RHA). For BG dye removal, pH = 3, catalyst dose = 6 g L^{-1} was taken and equilibrium was established in 5 h. Catalytic adsorption reaction followed pseudo-second-order kinetics, and data was best fitted into Langmuir and R–P isotherms. All thermodynamic parameters, such as negative Gibbs free energy, positive value for change in entropy (ΔS^0), and heat of adsorption (ΔH^0), favored spontaneous reaction [70]. Gopal et al. [71], made an attempt to synthesize a biogenic nanocomposite based on activated carbon (produced from *Prosopis juliflora* seeds) and polyaniline. The catalytic adsorption efficiency was examined toward degradation of direct red 23 dye. Experimental results confirmed that adsorption reaction follows pseudo-second-order rate kinetics along with the high value of Langmuir monolayer adsorption capacity (Q_0) such as 109.89 mg g^{-1}. Dispersion of carbon onto polyaniline produced high surface area of 1028 m^2 g^{-1} [71].

After this work, Tayebi et al. [72] further studied various possible ways to modify rice-husk-based adsorbent and reported a new hybrid named polyaniline/HCl-MRH. Catalytic efficiency of this new polyaniline/HCl-modified rice husk composite was evaluated for removal of acid red 18 dye. Data of adsorption equilibrium fitted into Langmuir adsorption isotherm and reaction kinetics resulted that adsorption proceeded via pseudo-first-order model. Thermodynamic parameters also favored this reaction as we found a negative value of ΔG° and positive value of ΔH° for adsorption of acid red 18 on polyaniline/HCl-MRH [72]. Kanwal et al. [73]

synthesized PANI/AL nanocomposite where AL stands for *Alstonia scholaris* by following oxidative polymerization method. Catalytic adsorption of diamond green (DG) dye was evaluated for this newly synthesized nanocomposite. To optimize reaction parameters such as contact time, pH of solution, adsorbent amount, and temperature of reaction, batch experiment was conducted. In this experiment, both monolayer and multilayer adsorption was observed, and result found for monolayer adsorption was 8.130 mg g^{-1} and for multilayer adsorption value of adsorption capacity was 0.947 mg g^{-1} which favored both Langmuir and Freundlich adsorption isotherms models [73].

In a study by Lyu et al. [74], sawdust was used to prepare highly effective polyaniline-based nanocomposite material. In this reaction, $FeCl_3$ was used as an oxidant to prepare polyaniline via polymerization technique. These nanofibers were further coated on sawdust biomass, and the resulting catalyst was named as PANI-NFs/SD. It shows high efficiency toward separation of nanoscaled absorbents from the aqueous solution during catalytic activity. The catalytic efficiency of as synthesized nanocomposite was examined for removal of anionic azo dye acid red G (ARG), and the experimental results revealed that reaction was feasible and spontaneous. In addition to this, catalytic adsorption reactions followed pseudo-second-order kinetics and experimental data were best fitted in the Freundlich isotherm model. During adsorption reaction, value of maximum adsorption capacity was found to be 213 mg g^{-1} at 308 K. Moreover, PANI-NFs/SD nanocomposite was highly stable as its catalytic efficiency was maintained after 29 sequential adsorption–desorption cycles [74].

4 Future Perspective

From this detailed literature survey, we found that the adsorption method is the most promising technique due to its easy to handle mechanism, great efficiency, reusability, and cost-effective approach. A catalytic adsorption reaction has been implied from nanoscale to industrial scale level for effective removal of toxic chemicals. Different types of catalysts display independent value for adsorption capacity depending upon their surface area, extent of intermolecular interactions, stability of molecules, and parameters of reaction conditions. Prior requirement for any catalyst for good adsorption is availability of wide surface area, porous structure, and active sites. Nowadays, highly efficient adsorbent materials have been introduced which worked on industrial level for adsorption of toxic dyes, antibiotics, pesticides, and drugs and produced clean water. Among all, semiconductor-based catalyst materials based on metal oxides and their hybrid materials explored a wide area of water mineralization due to their effective band gap value, active site on the surface of catalysts and light activated reactions. After this, metal organic frameworks are also one of the best porous materials which provide large surface area for catalysis. But commercial synthesis of MOFs is not cost effective, and they need more time and energy for synthesis. Moreover, polymer-based catalysts are excellent materials toward treatment of contaminated water due to their complex monomer bindings which expand the catalytic surface area. Our studies mainly focused on polyaniline polymer (PANI) and PANI-based various hybrid materials such as magnetic nanocomposite, metal-based composite materials, and biogenic hybrid materials. Hydrothermal polymerization technique is the most common way to synthesize polymer hybrid materials which was further analyzed by various analytical methods. Catalytic adsorption capability of these materials was also explored for removal of various water contaminants.

During treatment of contaminated water, adsorption is the prior method involved in the primary treatment step. To introduce a new catalyst which is eco-friendly, highly capable, and shows extraordinary stability, numerous challenges are there to meet this requirement. An alternative method of using biogenic materials which do not cause any harm to humans can be implied for fabrication of polymer-based materials. In literature, biogenic synthesis was adopted in a few reports as most of the work has been done on the basis of chemical reactions. For betterment of the future, our catalyst material should have high surface area, low cost of production, and high stability. After catalytic reaction, separation of adsorbent from dye molecules should be an easy process, so that the catalyst can be used for further successive adsorption cycles. In addition to this, safe dumping of spent adsorbent is required or it can be reused by treating with acidic or basic solution. Moreover, high scale use of catalysts in industries should be maintained at a good level by considering all possible improvements in their structure and catalytic activity.

5 Summary

Water pollution due to harmful dyes, pesticides, drugs, and aromatic compounds has been hiked to a large extent. Direct disposal of industrial waste in water makes it poisonous for both aquatic and human beings. An urgent need of water treatment techniques has pushed the researchers toward fabrications of more effective and highly stable catalysts materials which can bring good results. This study compiled various polymer-based adsorbent materials and examined their catalytic adsorption capacity toward removal of organic dyes. Most dyes are used by textile, paper, leather, and cosmetic industries which contribute a large value toward water pollution. Different methods are available for water treatment, but we studied adsorption mechanisms due to its high effectiveness. In this review, synthesis methods of polymer-based adsorbents were discussed along with their characterization methods. During catalytic adsorption, various physical and chemical interactions occurred which are discussed in the mechanism of adsorption. Along with this, kinetic studies of adsorption reaction and fitting of experimental data in different adsorption models were also studied. PANI-based hybrid materials including carbon, metal–metal oxides, graphene oxides, and biogenic materials are included in this chapter. Thus, this chapter summarized PANI-based hybrid nanomaterials, their high catalytic efficiency, stability or reusability, and kinetics studies.

References

1. Bano K, Mittal SK, Singh PP, Kaushal S (2021) Sunlight driven photocatalytic degradation of organic pollutants using a $MnV_2O_6/BiVO_4$ heterojunction: mechanistic perception and degradation pathways. Nanoscale Adv 3:6446–6458. https://doi.org/10.1039/D1NA00499A
2. Parmar S, Daki S, Bhattacharya S, Shrivastav A (2022) Microorganism: an eco-friendly tool for waste management and environmental safety. Dev Wastewater Treat Res Processes 175–193. https://doi.org/10.1016/B978-0-323-85657-7.00001-8
3. Dutta S, Bhattacharjee J (2022) A comparative study between physicochemical and biological methods for effective removal of textile dye from wastewater. Dev Wastewater Treatment Res Processes 1–21. https://doi.org/10.1016/B978-0-323-85657-7.00003-1
4. Rafaqat S, Ali N, Torres C, Rittmann B (2022) Recent progress in treatment of dyes wastewater using microbial-electro-Fenton technology. RSC Adv 12:17104–17137. https://doi.org/10.1039/D2RA01831D
5. Widakdo SJ, Mani S, Austria M, Hung WS, Nagar JK, Hu CC, Lai JY (2022) An eco-friendly and reusable syringe filter membrane for the efficient removal of dyes from water via low pressure filtration assisted self-assembling of graphene oxide and SBA-15/PDA. J Clean Prod 349:131425. https://doi.org/10.1016/j.jclepro.2022.131425
6. Merouani S, Dehane A, Belghit A, Hamdaoui O, Boussalem NH, Daif H (2022) Removal of persistent textile dyes from wastewater by Fe (II)/H_2O_2/H_3NOH^+ integrated system: process performance and limitations. Environ Sci Adv 1:192–207. https://doi.org/10.1039/D2VA00011C
7. Zum G, Ozmen BA, Akgul ET, Yavuz E (2022) Emulsion-templated porous polymers for efficient dye removal. ACS Omega 7:16127−16140. https://doi.org/10.1021/acsomega.2c01472
8. Stejskal J (2020) Interaction of conducting polymers, polyaniline and polypyrrole, with organic dyes: polymer morphology control, dye adsorption and photocatalytic decomposition. Chem Pap 74:1–54. https://doi.org/10.1007/s11696-019-00982-9
9. Saravanan R, Sacari E, Gracia F, Khan MM, Mosquera E, Gupta VK (2016) Conducting PANI stimulated ZnO system for visible light photocatalytic degradation of coloured dyes. J Mol Liq 221:1029–1033. https://doi.org/10.1016/j.molliq.2016.06.074
10. Li J, Huang Y, Shao D (2015) Conjugated polymer-based composites for water purification, Fundamentals of conjugated polymer blends, copolymers and composites: synthesis, properties, and applications, pp 581–620. https://doi.org/10.1002/9781119137160.ch11
11. Nasar A, Mashkoor F (2019) Application of polyaniline-based adsorbents for dye removal from water and wastewater—a review. Environ Sci Pollut Res 26:5333–5356. https://doi.org/10.1007/s11356-018-3990-y
12. Song E, Choi JW (2013) Conducting polyaniline nanowire and its applications in chemiresistive sensing. Nanomaterials 3:498–523. https://doi.org/10.3390/nano3030498
13. Shinde SS, Kher JA (2014) A review on polyaniline and its noble metal composites. Int J Innov Res Sci Eng Technol 03:16570–16576. https://doi.org/10.15680/IJIRSET.2014.0310023
14. Boeva ZA, Sergeyev VG (2014) Polyaniline: synthesis, properties, and application. Polym Sci Ser C 56:144–153. https://doi.org/10.1134/S1811238214010032
15. Tanzifi M, Hosseini SH, Kiadehi AD, Olazar M, Karimipour K, Rezaiemehr R, Ali I (2017) Artificial neural network optimization for methyl orange adsorption onto polyaniline nano-adsorbent: kinetic, isotherm and thermodynamic studies. J Mol Liq 244:189–200. https://doi.org/10.1016/j.molliq.2017.08.122
16. Nikita G, Das S, Biswas G, Haldar PK (2022) Review on some metal oxide nanoparticles as effective adsorbent in wastewater treatment. Water Sci Technol 85:3370–3395. https://doi.org/10.2166/wst.2022.153
17. Zheng Y, Chenga B, Youa W, Yua J, Ho W (2019) 3D hierarchical graphene oxide-NiFe LDH composite with enhanced adsorption affinity to Congo red, methyl orange and Cr(VI) ions. J Hazard Mater 369:214–225. https://doi.org/10.1016/j.jhazmat.2019.02.013

18. Shen K, Gondal MA (2017) Removal of hazardous Rhodamine dye from water by adsorption onto exhausted coffee ground. J Saudi Chem Soc 120–127. https://doi.org/10.1016/j.jscs.2013.11.005
19. Pirbazari AE, Saberikhah E, Kozani SSH (2014) Fe_3O_4–wheat straw: preparation, characterization and its application for methylene blue adsorption. Water Resour Ind 7:23–37. https://doi.org/10.1016/j.wri.2014.09.001
20. Cojocaru C, Samoila P, Pascariu P (2019) Chitosan-based magnetic adsorbent for removal of water-soluble anionic dye: artificial neural network modeling and molecular docking insights. Int J Biol Macromol 123:587–599. https://doi.org/10.1016/j.ijbiomac.2018.11.080
21. Thamer BM, Aldalbahi A, Moydeen M, Hamshary H, Enizi AM, Newehy MH (2019) Effective adsorption of Coomassie brilliant blue dye using poly(phenylene diamine)grafted electrospun carbon nanofibers as a novel adsorbent. Mater Chem Phys 234:133–145. https://doi.org/10.1016/j.matchemphys.2019.05.087
22. Janaki V, Vijayaraghavan K, Oh BT, Lee KJ, Muthuchelian K, Ramasamy AK, Kannan SK (2012) Starch/polyaniline nanocomposite for enhanced removal of reactive dyes from synthetic effluent. Carbohydr Polym 90:1437–1444. https://doi.org/10.1016/j.carbpol.2012.07.012
23. Janaki V, Oh BT, Shanthi K, Lee KJ, Ramasamy AK, Kannan SK (2012) Polyaniline/chitosan composite: an eco-friendly polymer for enhanced removal of dyes from aqueous solution. Synth Met 162:974–980. https://doi.org/10.1016/j.synthmet.2012.04.015
24. Olad A, Azhar FF, Shargh M, Jharfi S (2014) Application of response surface methodology for modeling of reactive dye removal from solution using starch-montmorillonite/polyaniline nanocomposite. Polym Eng Sci 54:1595–1607. https://doi.org/10.1002/pen.23697
25. Kannusamy P, Sivalingam T (2013) Synthesis of porous chitosan–polyaniline/ZnO hybrid composite and application for removal of reactive orange 16 dye. Colloids Surf B 108:229–238. https://doi.org/10.1016/j.colsurfb.2013.03.015
26. Sahnoun S, Boutahala M (2018) Adsorption removal of tartrazine by chitosan/polyaniline composite: kinetics and equilibrium studies. Int J Biol Macromol 114:1345–1353. https://doi.org/10.1016/j.ijbiomac.2018.02.146
27. Abbasian M, Jaymand M, Niroomand P, Habibi AF, Abad SGK (2017) Grafting of aniline derivatives onto chitosan and their applications for removal of reactive dyes from industrial effluents. Int J Biol Macromol 95:393–403. https://doi.org/10.1016/j.ijbiomac.2016.11.075
28. Ayazi Z, Khoshhesab ZM, Azhar FF, Mohajeri Z (2017) Modeling and optimization of adsorption removal of reactive orange 13 on the alginate–montmorillonite–polyaniline nanocomposite via response surface methodology. J Chin Chem Soc 64:627–639. https://doi.org/10.1002/jccs.201600876
29. Bhaumik M, Crindle R, Maity A (2013) Efficient removal of Congo red from aqueous solutions by adsorption onto interconnected polypyrrole–polyaniline nanofibers. J Chem Eng 228:506–515. https://doi.org/10.1016/j.cej.2013.05.026
30. Shen J, Shahid S, Amura I, Sarihan A, Tian M, Emanuelsson E (2018) Enhanced adsorption of cationic and anionic dyes from aqueous solutions by polyacid doped polyaniline. Synth Met 245:151–159. https://doi.org/10.1016/j.synthmet.2018.08.015
31. Saha B, Debnath A, Saha B (2022) Fabrication of PANI@Fe–Mn–Zr hybrid material and assessments in sono-assisted adsorption of methyl red dye: uptake performance and response surface optimization. J Indian Chem Soc 99:100635. https://doi.org/10.1016/j.jics.2022.100635
32. Teng XP, Bryan MYK, Chai PV, Law JY (2021) Preparation of polyaniline iron oxide composite (PANI/Fe_3O_4) for enhanced Congo red removal performance. Mater Today Proc 46:1875–1881. https://doi.org/10.1016/j.matpr.2021.01.766
33. Karami K, Beram SM, Siadatnasab F, Bayat P, Ramezanpour A (2021) An investigation on MIL-101 Fe/PANI/Pd nanohybrid as a novel photocatalyst based on MIL-101(Fe) metal–organic frameworks removing methylene blue dye. J Mol Struct 1231:130007. https://doi.org/10.1016/j.molstruc.2021.130007
34. Imgharn A, Anchoum L, Hsini A, Naciri Y, Laabd M, Mobarak M, Aarab N, Bouziani A, Szunerits S, Boukherroub R, Lakhmiri R, Albourine A (2022) Effectiveness of a novel polyaniline@Fe-ZSM-5 hybrid composite for Orange G dye removal from aqueous media:

experimental study and advanced statistical physics insights. Chemosphere 295:133786. https://doi.org/10.1016/j.chemosphere.2022.133786
35. Rajappa S, Shivarathri PG, Kumari MLA, Swamygowda DK, Devendrachari MC, Kotresh HMN (2022) Mechanochemical route for tetra amino zinc phthalocyanine embedded PANI sensitized Fe_2O_3 heteroarchitecture for photodegradation of dyes under the influence of low power LED light source. Surf Interfaces 29:101720. https://doi.org/10.1016/j.surfin.2022.101720
36. Alves FHO, Araujo OAA, Oliveira AC, Garg VK (2021) Preparation and characterization of PAni(CA)/Magnetic iron oxide hybrids and evaluation in adsorption/photodegradation of blue methylene dye. Surf Interfaces 23:100954. https://doi.org/10.1016/j.surfin.2021.100954
37. Xiong H, Zhang B, Cui C, Xu Y (2022) Polyaniline/FeOOH composite for removal of Acid Orange II from aqueous solutions. Mater Chem Phys 278:125701. https://doi.org/10.1016/j.matchemphys.2022.125701
38. Lv C, Zhang J, Li G, Xi H, Ge M, Goto T (2020) Facile fabrication of self-assembled lamellar PANI-GO-Fe_3O_4 hybrid nanocomposites with enhanced adsorption capacities and easy recyclicity towards ionic dyes. Colloids Surf 585:124147. https://doi.org/10.1016/j.colsurfa.2019.124147
39. Katowah DF, Saleh SM, Mohammed GI, Alkayal NS, Ali R, Hussein MA (2021) Ultra-efficient hybrid material-based cross-linked PANI@Cs-GO-OXS/CuO for the photocatalytic degradation of Rhodamine-B. J Phys Chem Solids 157:110208. https://doi.org/10.1016/j.jpcs.2021.110208
40. Karami K, Beram SM, Bayat P, Siadatnasab F, Ramezanpour A (2022) A novel nanohybrid based on metal–organic framework MIL101−Cr/PANI/Ag for the adsorption of cationic methylene blue dye from aqueous solution. J Mol Struct 1247:131352. https://doi.org/10.1016/j.molstruc.2021.131352
41. Dessie Y, Tadesse S, Adimasu Y (2022) Improving the performance of graphite anode in a microbial fuel cell via PANI encapsulated α-MnO_2 composite modification for efficient power generation and methyl red removal. Adv Chem Eng 10:100283. https://doi.org/10.1016/j.ceja.2022.100283
42. Fatima T, Husain S, Khanuja M (2022) Superior photocatalytic and electrochemical activity of novel WS_2/PANI nanocomposite for the degradation and detection of pollutants: antibiotic, heavy metal ions, and dyes. Adv Chem Eng 12:100373. https://doi.org/10.1016/j.ceja.2022.100373
43. Merangmenla, Nayak B, Baruah S, Puzari A (2022) 1D copper (II) based coordination polymer/PANI composite fabrication for enhanced photocatalytic activity. J Photochem 427:113803. https://doi.org/10.1016/j.jphotochem.2022.113803
44. Park Y, Numan A, Ponomarev N, Iqbal J, Khalid M (2021) Enhanced photocatalytic performance of PANI-rGO-MnO_2 ternary composite for degradation of organic contaminants under visible light. J Environ Chem Eng 9:106006. https://doi.org/10.1016/j.jece.2021.106006
45. Aamir M, Aleem W, Akhtar MN, Din AA, Yasmeen G, Ashiq MN (2022) Synthesis and characterizations of Co–Zr doped Ni ferrite/PANI nanocomposite for photocatalytic methyl orange dye degradation. Physica B Condens Matter 624:413392. https://doi.org/10.1016/j.physb.2021.413392
46. Rajaji U, Rani SEGD, Chen SM, Rajakumar K, Govindasamy M, Alzahrani FM, Alsaiari NS, Ouladsmane M, Lydia IS (2021) Synergistic photocatalytic activity of SnO_2/PANI nanocomposite for the removal of direct blue 15 under UV light irradiation. Ceram Int 47:29225–29231. https://doi.org/10.1016/j.ceramint.2021.07.087
47. Mohammed AM, Mohtar SS, Aziz F, Aziz M, Hamid AU (2021) Cu_2O/ZnO-PANI ternary nanocomposite as an efficient photocatalyst for the photodegradation of Congo Red dye. J Environ Chem Eng 9:105065. https://doi.org/10.1016/j.jece.2021.105065
48. Ma J, Dai J, Duan Y, Zhang J, Qiang L, Xue J (2020) Fabrication of PANI-TiO_2/rGO hybrid composites for enhanced photocatalysis of pollutant removal and hydrogen production. Renew Energy 156:1008–1018. https://doi.org/10.1016/j.renene.2020.04.104

49. Sharma S, Kumar D, Khare N (2021) Hierarchical PANI/CdS nano architecture system for visible light induced photocatalytic dye degradation and photo electrochemical water splitting. Polymer 231:124117. https://doi.org/10.1016/j.polymer.2021.124117
50. Das M, Ray PG, Dhara S, Roy S (2022) Symbiotically Augmented removal of Congo red by polyaniline/cobalt sulfide/graphite composites. Mater Chem Phys 278:125487. https://doi.org/10.1016/j.matchemphys.2021.125487
51. Shankar EG, Aishwarya M, Khan A, Kumar ABVK, Yu JS (2021) Efficient solar light photocatalytic degradation of commercial pharmaceutical drug and dye using rGO-PANI assisted c-ZnO heterojunction nanocomposite. Ceram Int 47:23770–23780. https://doi.org/10.1016/j.ceramint.2021.03.206
52. Biswas RD, Cho KY, Jung CH, Oh WC (2019) Novel synthesis of $LaNiSbWO_4$-G-PANI designed as quaternary type composite for high photocatalytic performance of anionic dye and trihydroxybenzoic acid under visible-light. Process Saf Environ Prot 126:348–355. https://doi.org/10.1016/j.psep.2019.04.022
53. Azimifar M, Ghorbani M, Peyravi M (2022) Fabrication and evaluation of a photocatalytic membrane based on Sb_2O_3/CBO composite for improvement of dye removal efficiency. J Mol Struct 1270:133957. https://doi.org/10.1016/j.molstruc.2022.133957
54. Abukhadra MR, Shaban M, Samad MAA (2018) Enhanced photocatalytic removal of Safranin-T dye under sunlight within minute time intervals using heulandite/polyaniline@ nickel oxide composite as a novel photocatalyst. Ecotoxicol Environ Saf 162:261–271. https://doi.org/10.1016/j.ecoenv.2018.06.081
55. Palliyalil S, Resha Kasim VC, Vigneshwaran S, Poovathumkuzhi NC, Chelaveettil BM, Meenakshi S (2022) Ternary system of TiO_2 confined chitosan–polyaniline heterostructure photocatalyst for the degradation of anionic and cationic dyes. Environ Technol Innov 28:102586. https://doi.org/10.1016/j.eti.2022.102586
56. Thambidurai RPS (2016) Synthesis of seaweed-ZnO-PANI hybrid composite for adsorption of methylene blue dye. J Environ Chem Eng 4:1332–1347. https://doi.org/10.1016/j.jece.2016.01.030
57. Hiragond CB, Khanna PK, More PV (2018) Probing the real-time photocatalytic activity of CdS QDs sensitized conducting polymers: featured PTh, PPy and PANI. Vacuum 155:159–168. https://doi.org/10.1016/j.vacuum.2018.06.009
58. Ahmad S, Sultan A, Raza W, Muneer M, Mohammad F (2016) Boron nitride based polyaniline nanocomposite: preparation, property, and application. J Appl Polym Sci 133. https://doi.org/10.1002/app.43989
59. Maruthapandi M, Saravanan A, Manohar P, Luong JHT, Gedanken A (2021) Photocatalytic degradation of organic dyes and antimicrobial activities by polyaniline–nitrogen-doped carbon dot nanocomposite. Nanomaterials 11:1128. https://doi.org/10.3390/nano11051128
60. Mitra M, Ahamed ST, Ghosh A, Mondal A, Kargupta K, Ganguly S, Banerjee D (2019) Polyaniline/reduced graphene oxide composite-enhanced visible light-driven photocatalytic activity for the degradation of organic dyes. ACS Omega 4:1623−1635. https://doi.org/10.1021/acsomega.8b02941
61. Pete S, Kattil RA, Thomas L (2021) Polyaniline-multiwalled carbon nanotubes (PANI-MWCNTs) composite revisited: an efficient and reusable material for methyl orange dye removal. Diam Relat Mater 117:108455. https://doi.org/10.1016/j.diamond.2021.108455
62. Bekhoukh A, Moulefera I, Zeggai FZ (2022) Anionic methyl orange removal from aqueous solutions by activated carbon reinforced conducting polyaniline as adsorbent: synthesis, characterization, adsorption behavior. Regeneration kinetics study. J Polym Environ 30:886–895. https://doi.org/10.1007/s10924-021-02248-6
63. Chatterjee MJ, Ghosh A, Mondal A, Banerjee D (2017) Polyaniline–single walled carbon nanotube composite—a photocatalyst to degrade rose Bengal and methyl orange dyes under visible-light illumination. RSC Adv 7:36403–36415. https://doi.org/10.1039/C7RA03855K
64. Hasan M, Rashid MM, Hossain MM, Mesfer MK, Arshad M, Danish M, Lee M, Jery A, Kumar N (2019) Fabrication of polyaniline/activated carbon composite and its testing for methyl orange removal: optimization, equilibrium, isotherm and kinetic study. Polym. Test. 77:105909. https://doi.org/10.1016/j.polymertesting.2019.105909

65. Pethsangave DA, Khose RV, Wadekar PH, Kulal DK, Surajit (2020) One-pot synthetic approach for magnetically separable graphene nanocomposite for dye degradation. Chem Select 5:1516–1525. https://doi.org/10.1002/slct.201903966
66. Khan MK, Govindasamy R, Ahmad A, Siddiqui MR, Alshareef SA, Hakami AAH, Rafatullah M (2021) Carbon based polymeric nanocomposites for dye adsorption: synthesis, characterization, and application. Polymers 13:419. https://doi.org/10.3390/polym13030419
67. Sharkaway EA, Kamel RM, Sherbiny IM, Gharib SS (2019) Removal of methylene blue from aqueous solutions using polyaniline/graphene oxide or polyaniline/reduced graphene oxide composites. Environ Technol 41:2854–2862. https://doi.org/10.1080/09593330.2019.1585481
68. Liu T, Wang Z, Wang X, Yang G, Liu Y (2021) Adsorption-photocatalysis performance of polyaniline/dicarboxyl acid cellulose@graphene oxide for dye removal. Int J Biol Macromol 182:492–501. https://doi.org/10.1016/j.ijbiomac.2021.04.038
69. Wang H, Duan M, Guo Y, Wang C, Shi Z, Liu J, Lv J (2018) Graphene oxide edge grafting of polyaniline nanocomposite: an efficient adsorbent for methylene blue and methyl orange. Water Sci Technol 77:2751–2760. https://doi.org/10.2166/wst.2018.250
70. Mane VS, Mall ID, Srivastava VC (2007) Kinetic and equilibrium isotherm studies for the adsorptive removal of Brilliant Green dye from aqueous solution by rice husk ash. J Environ 84:390–400. https://doi.org/10.1016/j.jenvman.2006.06.024
71. Gopal N, Asaithambi M, Sivakumar P, Sivakumar V (2014) Adsorption studies of a direct dye using polyaniline coated activated carbon prepared from *Prosopis juliflora*. J Water Process Eng 2:87–95. https://doi.org/10.1016/j.jwpe.2014.05.008
72. Shabandokht M, Binaeian E, Tayebi HA (2016) Adsorption of food dye Acid red 18 onto polyaniline-modified rice husk composite: isotherm and kinetic analysis, *Desalin.* Water Treat 57:27638–27650. https://doi.org/10.1080/19443994.2016.1172982
73. Kanwal F, Rehman R, Bakhsh IQ (2018) Batch wise sorptive amputation of diamond green dye from aqueous medium by novel Polyaniline-*Alstonia scholaris* leaves composite in ecofriendly way. J Clean Prod 196:350–357. https://doi.org/10.1016/j.jclepro.2018.06.056
74. Lyu W, Yu M, Feng J, Yan W (2018) Highly crystalline polyaniline nanofibers coating with low-cost biomass for easy separation and high efficient removal of anionic dye ARG from aqueous solution. Appl Surf Sci 458:413–424. https://doi.org/10.1016/j.apsusc.2018.07.074

Recovery and Removal of Textile Dyes Through Adsorption Process

Growther Lali, V. Mahalakshmi, M. Seenuvasan, and G. Sarojini

Abstract Textile dyes released in the effluents of textile industries are a major source of environmental pollution. These dyes are non-biodegradable due to the aromatic structure and are recalcitrant in nature. Thus, the metabolic pathways of complete biodegradation are not clearly understood. The increasing concentrations of these dyes are a threat to aquatic life forms and human beings. The bioaccumulation and biomagnifications of these chemicals in the fauna and flora are a major cause of concern. More than 40% of these chemicals are well known carcinogens, and thus, the removal of these dyes from the water bodies is crucial. Biodegradation of these chemicals is possible; however, several factors influence the process. In spite of the use of microorganisms like bacteria, algae and fungi, commercial application of the process is still not promising. An alternative method is biosorption that uses microorganisms and agricultural waste residues to remove textile dyes by adsorption process. The process of adsorption helps removal of these dyes from industrial effluents. Though the mechanism of adsorption is not fully understood, the use of various other adsorbants like clay, coal-based adsorbants, activated carbon, zeolites, alumina, magnetic biosorbants, polyelectrolytes, etc., have been studied during the last three decades. The use of nanomaterials has been reported as an efficient, cost effective and eco-friendly method for textile dye removal. Therefore, the present review summarises the adsorption process, mechanism of adsorption, the types of adsorbants and their potential in large-scale applications.

Keywords Adsorption · Biosorption · Nanomaterials · Textile dyes · Environmental pollution

---

G. Lali (✉)
Department of Microbiology, Hindusthan College of Arts & Science, Coimbatore, India
e-mail: laligrowther@hicas.ac.in

V. Mahalakshmi
Department of Microbiology, Madras Christian College, Chennai, India

M. Seenuvasan · G. Sarojini
Department of Chemical Engineering, Hindusthan College of Engineering & Technology, Coimbatore, India

© The Author(s), under exclusive license to Springer Nature Singapore Pte Ltd. 2023
A. Ahmad et al. (eds.), *Nanohybrid Materials for Treatment of Textiles Dyes*,
Smart Nanomaterials Technology, https://doi.org/10.1007/978-981-99-3901-5_9

1 Introduction

In the textile industry, the leading environmental unease is coloured water caused by the dyeing practice. Presently, there are more than 1,00,000 various commercially available dyes on marketplace, and their yearly outcome has been assessed to be over 7×10^5 tones [1]. Because textile dyes end up in wastewater treatment plants, the textile industry is one of the most significant contributors to environmental degradation. As a direct outcome of the utilisation of these pigments, the waterways have turned out to be polluted. Because of the aromatic nature of these pigments, they are impervious to any and all kinds of chemical transformations and bacteria are unable to break them down. This directly contributes to the fact that we have limited knowledge of the metabolic mechanisms that are required for complete biodegradation. A significant proportion of the countless and varied species of aquatic life are under jeopardy as a direct result of the expanding human population as well as the rising consumption of these hues. A substantial body of study lends credence to the concern that these chemicals may be bioaccumulating and biomagnifying in the local flora and wildlife, which is a worry that is confirmed by the findings of the research [2].

2 Textile Industries

The textile industry is both one of the oldest and one of the most important industries today. India is home to more than 3,400 large textile mills, in addition to an uncountable number of smaller textile businesses and shops. The state of Maharashtra is home to the headquarters of the vast majority of the world's textile companies, including those that manufacture dyes. As textile effluents seize a very elevated concentration of colour, COD, suspended solids and other pollutants, it is persisted as one of the most complicated wastewater to treat. Being a multifarious kind needs the most advanced expensive processing technique which is not affordable to textile industries in developing countries [3]. The textile industries in neighbouring states such as Rajasthan, Ahmedabad and Gujarat have experienced success comparable to that of Maharashtra. Not only in India, but also in every other country in the world, a sizeable number of people make their living in these industries on a daily basis. Additionally, a significant portion of their revenue comes from the export of various goods. In spite of this, there is still a great deal of concern regarding the long-term viability of the livelihoods of those who work in these industries, in addition to the extensive list of adverse effects on the health of both humans and the environment. These companies have a high water footprint because they use a lot of water, which they then pollute and dump into bodies of water without first putting it through the appropriate treatment [4].

3 Dyes Used in Textile Industries

It is possible to differentiate between contaminants in effluent water that are able to dissolve and those that remain suspended in the water. Both types of contaminants can be found in the water. These organic liquid compounds are referred to as dyes, and they are subdivided into a number of subgroups based on the chemical makeup of the chromophoric groups that they contain. Dye pigments can be used to create a wide variety of colours. Direct and basic dyes (methylene blue (MB), basic red 1 or rhodamine 6G), reactive dyes (C.I. reactive red 120, C.I. reactive red 147 and C.I. reactive blue 19), sulphur dyes (sulphur brilliant green, sulphur blue and sulphur black 1) and sulphur dyes (sulphur brilliant green, sulphur blue, and sulphur black 1) these dyes pose a threat to [5]. Textile dyeing effluent contaminated sites desire instantaneous reclamation [6].

4 Ecotoxicological Effects of Textile Dyes

The untreated effluent from the textile industries are the main source of pollution of water bodies [7], which normally constitute 80% of the total emissions produced by this industry [8]. The waste from these textile units contain not only heavy metals but also biochemically resistant, synthetic and soluble dyes that cannot be broken down by natural methods. Though there are larger wastewater treatment facilities in India, only 25% of the waste waters are treated and the remaining waste waters from the small household textile units are discharged directly into the water bodies without any treatment.

The chemical and biochemical oxygen demand (BOD) values found in the wastewater produced by the textile industry are relatively high. More consideration needs to be given to the prevalence of non-biodegradable chemical compounds, particularly those used in the production of textile dyes. These substances have a negative impact on the environment and have the potential to detract from the aesthetic value of bodies of water while also reducing the amount of light that is able to penetrate water [5]. As a direct consequence of this, photosynthesis slows down, which leads to a drop in dissolved oxygen levels, which in turn has an effect on the entire aquatic biota [5].

5 Biomagnification

The textile dyes also act as toxic, mutagenic and carcinogenic agents [9, 10]. They persist as environmental pollutants contaminating the water bodies. They seep into the ground water and enter the next level in the food chain. Higher trophic levels of the dyes or the toxic compounds have been observed in higher level organisms, when compared to their prey, providing a strong evidence for biomagnification [11].

6 Adsorption

The removal or recovery of dyes from textile dye effluents is going to have to play a significant role in order for sustainable development to be achieved. One of the different kinds of physical processes that can be used to get rid of or recover dye is called adsorption. This is praised for being an effective method that is also cost-effective. The process of adsorption has benefits that stand on their own. The process of adsorption is not only flexible but also quick and does not cost very much. It does not require any expensive instrumentation equipment. During this process, there is not a single instance in which an intermediate or by-product that is harmful is generated. This method can be utilised by effluent treatment units of any scale, from the smallest to the largest. It is possible to remove any organic or inorganic contaminants with a high degree of efficacy. It is possible to regenerate the adsorbent, which in turn increases the likelihood of successfully recovering the adsorbate. For the sake of simplicity, absorption can be broken up into two stages. The first thing that needs to happen is for the dye to be moved from the effluent to the adsorbent. During the second stage of the process, the dye makes its way into the pores on the surface of the adsorbent material [12].

Both physisorption and chemisorption are viable explanations for the interaction between the dye and the adsorbent material. The process known as physisorption takes place when the surface of the adsorbent does not undergo any changes. The term "chemisorption" describes the formation of an ionic or covalent chemical link between dyes and the compounds that they adsorb, as well as the modification of the surface properties of the dyes. The two adsorption processes do not behave in the same way as one another. The efficiency of the process can be affected by a number of factors, including the kinetics of the process, the type of bonding, the concentration and the specificity of the dyes, as well as the saturation level of the adsorbent and the specificity of the binding [12].

6.1 Adsorption Isotherms

An adsorption isotherm explains the relationship between interactions of dyes from effluents to the adsorbents at a fixed temperature [12]. The relationship between the substrate concentration adsorbed per unit mass or the pressure in the effluent at constant temperature is explained by Langmuir, Freundlich, Dubinin and Radushkerich (D-R) isotherms [13].

6.1.1 Langmuir Isotherm

According to this hypothesis, the single adhesion layer that connects the dyes in the bulk effluent will get thinner when the dyes are moved away from the adsorbate surface that is present on the homogenous surface of the adsorbent. This is because the adsorbate surface is present on the homogenous surface of the adsorbent. The fact that this method functions most effectively in settings with low pressure or concentrations is, unfortunately, one of its drawbacks [14].

6.1.2 Freundlich Isotherm

The deposition of several layers onto a variety of different sorts of surfaces may be simulated with the help of this model. By characterising the surface coverage in layers, these two simple models are able to characterise the surface in terms of its physical and chemical characteristics, as well as its adsorption capabilities [15].

6.1.3 Dubinin and Radushkerich (D-R) Isotherm

This model provides an explanation for the adsorption isotherms of systems with a single substrate. It postulates that the adsorbent fills the pores of the surface as opposed to covering the surface in layers as traditional methods do [12].

6.2 Mechanism of Adsorption

Adsorption mechanisms have been studied by different authors using different dyes. Electrostatic attraction (adsorption of rhodamine dye; [16]), hydrogen bonding, acid–base reactions, hydrophobic interaction, ion exchange (Congo red, methyl orange; [17]), π–π interactions (methylene blue, Coomassie brilliant blue R 250 dye; [18]) and Van der Waals forces (acid orange 7; [19]) are few examples that have been elucidated by many researchers. Positively charged particles interact with negatively charged particles as adsorbent–adsorbate interaction in π–π interactions [20, 21].

6.3 Adsorbents

There are tried-and-true methods that make use of a wide variety of adsorbents in a variety of different combinations. Every sorbent has its own unique set of advantages and disadvantages, which makes it impossible for any other chemical to compete. Since they are less expensive, waste materials like agricultural waste and other inexpensive waste items are becoming an increasingly essential resource.

Activated carbon's rising popularity over the last few years has led to a rise in the amount of the substance that is being put to use. Adsorption is the process of removing a material from one phase and allowing it to accumulate or concentrate on the surface of another phase. This is accomplished by removing the substance from the first phase. During the adsorption process, the term "adsorbate" refers to the substance that becomes concentrated or adsorbed at the surface of the adsorbing phase [22].

Adsorption may be thought of as a chemical reaction. It is possible to generate the term "adsorption" by combining the words "concentration" and "adsorption." These adsorbents, when seen via a microscope, cling to other substances and behave in a manner that is similar to that of sponges with very small holes. These substances are able to exist in the three different states of matter: solid, liquid and gas. When it comes to sorbents, there is a wide selection of both organic and inorganic forms available to choose from. The removal of certain chemical components from textile effluents is made possible by the use of adsorbents. Adsorbents may be made from a wide variety of materials, the most common of which being zeolites, charcoal, clays and ores; other types of waste may also be used. Coconut shells, rice husks, waste from the petroleum industry, tannin-rich materials, sawdust, waste from the fertiliser industry, fly ash, waste from the sugar industry, blast furnace slag, chitosan, waste from the processing of seafood, seaweed and algae, peat moss, scrap tyres, and fruit wastes are all examples of adsorbents that can be made from waste materials. Other potential sources include waste from the sugar industry and blast furnace. Waste products from the sugar sector and emissions from blast furnaces are two more possible sources [23].

6.4 Classification of Adsorbents

Adsorbents are classified as carbon-based adsorbents, oxygen-containing adsorbents and polymer-based adsorbents [24].

6.4.1 Carbon-Based Adsorbents

These are non-polar and hydrophobic in nature. Examples include activated carbon (AC) and graphite. Activated carbon is a porous form of carbon and manufactured from various carbonaceous raw materials like wood, coal, lignite, peat and coconut shell [24].

6.4.2 Oxygen-Containing Adsorbents

Polar and hydrophilic compounds like silica gel and zeolites. Clay materials are hydrous aluminium silicates sometimes with minor amounts of Fe, Mg and other cations like bentonite, kaolinite, dickite, halloysite (polymorphs of $AlSi2O_5(OH)_4$), illite group clay-micas, sepiolite, attapulgite, China clay, oyster shell etc. [25].

6.4.3 Polymer-Based Adsorbents

Polar or non-polar functional groups in a polymeric matrix, e.g. polymers and resins, are used as adsorbents.

6.4.4 Biosorbents

Polysaccharides, proteins and lipids act as biosorbents. Viable, dead, pre-treated and immobilised biological cells of bacteria, fungi, yeast and Algae are good biosorbents [24].

6.4.5 Polymers

Chitin, chitosan, cellulose, amberlite XAD_4, etc., also had been proved as adsorbents.

6.5 Classification Based on Pore Sizes of Adsorbents [24].

Based on the pore size, adsorbents could be classified as the following types.

Microporous adsorbents (Pore size: 2–20 A).
Mesoporous adsorbents (Pore size: 20–500 A).
Macroporous adsorbents (Pore size >500 A).

6.6 Low-Cost Materials as Potential Adsorbents

6.6.1 Industrial Wastes

Fertilisers can be made from a variety of by-products of industrial processes, such as bagasse from sugarcane, fly ash, bottom ash, waste from the steel and fertiliser industries, sludge from blast furnaces, calcium and aluminium silicates in slag, and red mud. Some of these by-products come from sugarcane, while others come from fly ash, bottom ash, fly ash and bottom ash.

6.6.2 Animal Wastes

Waste materials from animals that could be used as potential adsorbants include hairs, bones, nails, egg shell membrane, chemically treated chicken feathers and chemically treated human hairs.

6.6.3 Agricultural Wastes

Peanut hull, tea leave dust, paddy straw, coir pith, Parthenium plant (flowering plant of Aster family), plumkernels, de-oiled-soya, rice bran, mustard seed, lotus seed, Mytilus Edulis shells, chestnut peel, walnut wood, pomegranate pulp and watermelon rind had been used in the literature as adsorbants.

7 Nanomaterials

7.1 Hybrid Nanomaterials

Because of the increased surface area and high adsorption-to-mass ratio that nanomaterials possess, they can be utilised in the process of water purification. At a pH of 2.5 and a dye concentration of 10 mg L^{-1}, the nanohybrid (TP0.75CS0.25), which was composed of 75% ternary photocatalyst (w/w) and 25% chitosan (w/w), was able to remove 97.4% of the orange G dye in 110 min. This was achieved at a concentration of the dye of 10 mg L^{-1} [66]. The other few nanohybrids proved as efficient adsorbents include Cd (II) semicarbazone and $PM_{12}O_{40}{}^{3-}$ [67], biochar-based nanohybrid materials [68], carrageenan-based nanohybrid constructs which consist of titania/silica nanohybrids [69], Mg–Al LDH intercalated with NO_3-decorated graphene oxide nanohybrid [70], carbon nanotubes, nanorods, nanoclay, nanomembranes, metal organic frameworks, graphene oxide, carrageenan biopolymers, and so on (LGZF) [71], and oligomeric silsesquioxane (POSS) nanohybrid [72].

8 Factors Affecting Adsorption

8.1 pH of Solution

pH of solution greatly affects the performance of hybrid nanomaterials as pH has the tendency to alter the chemistry and characteristics of hybrid nanomaterials [73]. The surface charge of the nanohybrid materials may be altered by adjusting the pH of the solution. This is due to the fact that pH has an effect on the surface charge, which in turn has an effect on the ionic chemistry. The vast majority of studies have shown that

Table 1 Textile dyes adsorbed using different adsorbents

Adsorbent	Material adsorbed	Dye removal efficiency (%)	References
Organic adsorbents			
Unburnt carbon (fly ash with unburnt carbon)	Methylene blue and crystal violet	92–95	Abbaszadeh et al. [26], Wang and Chen [8]
Activated carbon (Kernel tea and palm shell wastes)	Crystal violet	82	
Areca husk carbon	Crystal violet	70	Binoj et al. [27]
Saw dust	Methylene blue, basic dye	60	Careddu et al. [28]
Orange peel	Direct red 23 (DR23) and direct red 80 (DR80)	75	Wu et al. [23], Arami et al. [29]
Banana trunk fibres	Methyl red	96	Jandas et al. [30]
Sugarcane fibre	Congo red	56	Raymundo et al. [31]
Cotton shell and neem bark	Reactive red 120 (RR120) and reactive blue 15 (RB15)	79	Cholake et al. [32]
Chitin shell	Synthetic dyes	96	Lu et al. [33]
Tamarind peel	Methyl orange, methylene blue, rhodamine B, congo red, methyl violet and amido black	99	Binoj et al. [34], Reddy [35]
Pineapple leaf powder	Rose Bengal, crystal violet and basic green 4	58	Chowdhury et al. [36]
Corncob	Methyl red, methyl orange	94.54, 99.25	Memon and Khan [37], Salih et al. [38]
Pomelo peels	Rhodamine B	98.7	Wu et al. [23]
Pea nut hulls	Methylene blue	86	Nikkhah et al. [39], Etorki and Massoudi [40]
Azadirachta indica leaf powder	Malachite green	95.49	Binoj et al. [34]
Rice husk activated	Crystal violet	99.24	Dunnigan et al. [41], Sivalingam and Sen [42]
Hazelnut shells	Methylene blue, acid blue, procion blue H-EGN 125 (PB) and sandolan brilliant red N-BG 125 (SBR)	98.6, 99.6	Castro and Swart [43], Buyukada et al. [44]

(continued)

Table 1 (continued)

Adsorbent	Material adsorbed	Dye removal efficiency (%)	References
Yellow passion fruit waste—exocarp of Pulasan and Rambutan	Rhodamine B and reactive orange	90, 93.13	Nhung et al. [45]
Coconut coir pith	Methylene blue	99.9	Arena et al. [46], Etim et al. [47]
Rubber seed coat	Basic blue 3	87	Chin et al. [48], Rajesh et al. [49]
Apple pomade and wheat straw	Red dyes	66	Weinwurm et al. [50]
Castor seed hull	Safranin	99.6	Guil-Guerrero et al. [51]
Cucumis sativus	Crystal violet (CV) and rhodamine B (RH B)	72.27, 81.69	Smitha et al. [52]
Papaya peel	Remazol black B, remazol brilliant blue and remazol brilliant red	58.48–98.03, 46.15–93.47, 46.30–94.60	Weber et al. [53]
Inorganic adsorbents			
Natural clay bentonite	Malachite green	90	Burland [54]
Crushed brick	Methylene blue, basic dye	46	Kazmi et al. [55]
Fly ash (calcium-rich fly ash)	Methylene blue, congo red dyes	78–93	Capela et al. [56]
Bagasse fly ash and bottom fly ash	Methylene blue	80–85	Amin et al. [57]
Sepiolite clay	Reactive dyes	70–86	Hynes et al. [58]
Palygorskite clay	Ionic dyes	45–55	Yang et al. [59]
Red mud clay	Green colour dyes, congo red dye, methylene blue, rhodamine B, and fast green dyes	45, 82, 78, 92.5, 94.0, 75.0	Kismir and Aroguz [60]
Halloysite clay	Methylene blue on halloysite nanotubes	>80	Zhao and Liu [61]
Zeolite	Basic blue 41	73.2	Almasian et al. [62]
Pumice	Azo dyes-acid red 14, 18	83–89	Fiallos et al. [63]
Wood-shaving bottom ash	Azo reactive and red reactive 141 dyes	>75	Vital et al. [64]
Metal hydroxide sludge	Azo dyes, reactive dyes (CI reactive red 2, CI reactive red 120 and CI reactive red 141)	>90	Netpradit et al. [65]

the surface of a nanohybrid maintains its positive charge interacts strongly electrostatically with anionic dyes and has an extraordinary adsorption capacity when the pH is acidic (at a pH of three or below) [74]. Under alkaline pH, the surface remains negatively charged and exhibits weaker interaction with anionic dyes. In contrast, under alkaline pH, a strong interaction is achieved with cationic dyes [75]. In one research, Ibrahim et al. [76] explored the performance of modified burley straw in removal of ionic dyes. It was reported that complete removal of anionic dye was achieved at acidic pH (at pH 3 or less), whereas the removal efficiency decreases at higher alkaline conditions. This might be attributed to the fact that at acidic conditions, the surface of nanohybrid materials contains enormous positively charged ions and can attract more sulfonate group ions than that of alkaline environment.

8.2 Initial Concentration

A common method for visually controlling the amount of colour that is absorbed is to vary the amount of sorbent that is being utilised. Because the surface area of nanohybrid materials dosage grows in response to the growth in dosage amount, increasing the dosage may frequently make it simpler to acquire more active sites. When the dosage is raised, this immediately leads to an improvement in the efficiency with which dye molecules are eliminated. After the ideal dose amount has been attained, the findings of a few researches indicate that increasing the dosage in the future has very little impact on the substance's ability to be removed from the body. This might be due to the fact that the dye molecules have partly adhered to the surface of the hybrid material, which has the effect of reducing the surface area. When there are not enough dye molecules to cover all of the active sites of the hybrid material, the adsorption capacity of the dye molecules increases with an increase in dosage. This occurs when there are not enough dye molecules. This is what occurs when there are not enough dye molecules to fully cover the active sites of the enzyme [77].

8.3 Temperature

Temperature is an important process parameter of adsorption process and has a significant effect on the process [78]. For endothermic processes, the adsorption capacity increases with increase in temperature due to easy mobility of dye molecules. Bharali and his team investigated the adsorption of Congo red by NiAl-layered double hydroxide hybrid nanomaterials and reported that the removal efficiency of Congo red increases with increase in operating temperature [79]. However, for exothermic processes, the adsorption capacity decreases with increase in temperature. With increase in temperature, intermolecular force of attraction between dye molecules as well number of active sites gets reduced. Therefore adsorption capacity decreases.

8.4 Nanohybrid Materials Dosage

A common method for visually controlling the amount of colour that is absorbed is to vary the amount of sorbent that is being utilised. Because the surface area of nanohybrid materials dosage grows in response to the growth in dosage amount, increasing the dosage may frequently make it simpler to acquire more active sites. When the dosage is raised, this immediately leads to an improvement in the efficiency with which dye molecules are eliminated. After the ideal dose amount has been attained, the findings of a few researches indicate that increasing the dosage in the future has very little impact on the substance's ability to be removed from the body. This might be due to the fact that the dye molecules have partly adhered to the surface of the hybrid material, which has the effect of reducing the surface area. When there are not enough dye molecules to cover all of the active sites of the hybrid material, the adsorption capacity of the dye molecules increases with an increase in dosage. This occurs when there are not enough dye molecules. This is what occurs when there are not enough dye molecules to fully cover the active sites of the enzyme [80].

8.5 Contact Time

The length of time that dye molecules are in physical contact with a surface is the single most important component. Increasing the amount of time that the two substances are in contact often results in increased adsorption capability. The amount of interaction that occurs between the adsorbent and the adsorbate grows as the contact period between the two parties increases. The capacity of reactive sites to bind constituents of a system together also improves. After reaching equilibrium, increasing the contact period further does not result in a removal efficiency that is markedly different from what it was before the equilibrium was reached, as the results of the limited experiments that have been carried out have shown. It may be difficult to adsorb the dye molecule onto the surface of the hybrid material, which may occasionally result in the removal efficiency being reduced. This is something that may take place under particular conditions [81].

9 Efficacy of the Methods

The adsorption performance of nanohybrid materials for different dyes varies significantly, e.g. Mahouche-Chergui and his team synthesised graphene-like hybrid nanomaterials and investigated the performance in removal of anionic and cationic dyes. They reported that those material served as a better adsorbent with remarkable features. Cationic dye of rhodamine B and methylene blue and anionic dye of methyl orange were removed with adsorption capacity of 9.5–10 mg g^{-1} [82]. Lu and his friends explored the application of iron oxide coated layered double hydroxides in removal of Congo red and reported that the hybrid nanomaterial has a adsorption capacity of 813 mg g^{-1}. Further comparison was made in adsorption capacity of pristine material and hybrid nanomaterial and reported that hybrid nanomaterial has five times greater adsorption capacity than that of pristine compounds. It was worth mentioning that hybrid nanoparticles boosts the adsorption capacity [33].

Mohammadi et al. [83, 84] synthesised carboxylic group functionalised magnetic silica iron oxide nanoparticles and explored its performance ability as a nanoadsorbent in removal of malachite green. It was reported that the nanohybrid materials could remove 97.5% of malachite green. Narayani et al. synthesised magnetic amino-coated silica iron oxide nanoparticles and used it effectively in removal of acid red 114. It was found that nanohybrid material possess an adsorption capacity of 84.75 mg g^{-1}, and the separation was achieved within 2 min [85]. Liu et al. investigated the performance of α-Fe_2O_3/rGO in elimination of malachite green and reported that the hybrid possess adsorption capacity of 438.8 mg g^{-1}. The enhanced adsorption capacity is due to synergetic cooperation between α-Fe_2O_3 and reduced graphene oxide [86].

10 Recovery Through Adsorption

Regeneration of adsorbents remains a challenging issue as it concerns with environmental sustainability and economic feasibility. Very few reports are available concerned with recovery of nanohybrid materials. Recovery of the spent adsorbent is evaluated by means of adsorption/desorption process [87]. After dye adsorption, the aqueous solution was treated, and spent adsorbent was separated by means of filtration, magnetic separation and ultra-assisted filtration. Stable adsorption process should favour for the recovery of selective dyes from the aqueous phase [88]. The ability of the process to recover and reuse the adsorbent is a critical factor in determining whether or not the method is economically viable. The effectiveness of an adsorbent and the number of times it may be used before it begins to lose its capacity to carry out its intended purpose are two factors that contribute to an item's long-term stability. When things are chaotic, it might be difficult to apply selectivity effectively.

Processing by chemical reaction and processing via heat are the two procedures that may be used to recover the adsorbent. The chemical processing industry is by far the larger of the two. When selecting a method for recovery, it is essential to take into consideration the specific kind of adsorbent that will be put to use. Adsorbents manufactured from biomass are often subjected to heat treatment as standard procedure in the industry. Before heating the adsorbents to the requisite calcination temperature, which varies depending on the kind of absorbent that is being used, this process entails washing the adsorbents in distilled water. However, in comparison with the other methods, chemical treatment is the one that is used to reactivate adsorbents at a much higher frequency. During the process of chemical treatment, a wide variety of different chemicals may be used at any one time. Adsorbents that have been utilised in the past may be purified by subjecting them to a washing process that involves the application of a particular reagent and continues for a certain amount of time. The mixture is then either filtered or centrifuged, depending on which method is preferred, in order to remove the adsorbent that was previously added.

11 Biosorbents and Dye Removal

Tahir et al. [89] and researchers have investigated the adsorption of methylene blue using both bioadsorbents and *Ulva lactuca* sargassum as part of their methodology. For the adsorption process to be effective, it was important for all of the system components to be investigated, including the temperature, the quantity of adsorbent, the pH of the solution, the dye concentration and the length of time that the two substances were in contact with one another. The adsorption procedure is governed by not one but two distinct adsorption isotherm models: the Freundlich model and the Langmuir model. It was found that these adsorbents are capable of effectively extracting around 96.0% of the dye from the solution. According to the findings of one piece of research, making use of a powder that was produced from ground-up hen eggshells would be able to assist you in removing the bright green colour [13].

In a study on batch adsorption, several different experimental variables were used, such as the amount of adsorbent used, the temperature, the length of time the two substances were in contact with one another, the starting dye concentration and the pH. According to the findings of the investigation into the equilibrium data, the Langmuir isotherm equation provided a better match than any of the other equations that were considered. Vijay Kumar et al. analysed the adsorption capacities of raw bagasse from sugarcane (referred to as RB) and CAB (raw bagasse that had been chemically activated), respectively. Both of these substances were utilised in order to absorb the bright green dye. According to the findings of an experimental parameter study that was carried out at a pH of neutral, the percentage of dye that was removed increased in proportion to the quantity of adsorbent that was used. When the concentration of the bright green dye was lowered, there was a significant increase in the clearance percentage. An investigation into the kinetics of the reaction found that chemically activated raw bagasse performed significantly better than raw bagasse on its own.

Algerian kaolin was used as an alternative adsorbent by Meroufel et al. [90] with a focus on Congo red in particular, for the removal of potentially harmful anionic dyes from aqueous solutions. An experiment was conducted utilising batch adsorption to investigate the impact of factors like contact length, initial dye concentration, pH and temperature. The results of the experiments showed that the conditions necessary for optimal adsorption were achieved when the concentration of dye was high and the pH was basic. The pseudo-second-order kinetic equation might be comprehended with the greatest ease, thanks to the use of the Langmuir isotherm. After determining a variety of thermodynamic factors, the researchers came to the conclusion that Congo red adsorbs exothermically and spontaneously onto Algerian kaolin. The results provide credence to the hypothesis that anionic pigments found in industrial wastewater may be removed using kaolin extracted at a reasonable cost in the surrounding area. It was possible to remove the malachite green colour from the aqueous solutions by using powdered neem bark and powdered mango bark as adsorbents [91].

Experiments on adsorption were carried out under a wide range of conditions, some of which included temperatures, pH levels, starting concentrations and doses of the adsorbent that ranged from low to high. Isotherms of the Langmuir and Freundlich types were used in order to analyse the equilibrium data. When the two models were compared, it was discovered that the Langmuir isotherm provided a better fit to the data than did the Freundlich model. In addition, thermodynamic factors were scrutinised for their effects. In order to conduct a kinetics investigation, first- and second-order models were used. It was found that the results from each adsorbent follows second-order kinetics.

One of the experimental studies involves the process of extracting methylene blue (MB) from an aqueous solution by utilising *Calotropis procera* leaf powder that has been treated with hydrochloric acid (CPLP) [92]. Eighty minutes was the total amount of time required to reach equilibrium. The adsorption capacity decreased from 1.1 to 0.1 mg g^{-1} when the acid modified CPLP dosage increased from 16 to 160 g L^{-1}. Due to the fact that it was discovered that the Freundlich isotherm was not sufficiently characterised, this discovery was made. In contrast, using thermodynamic calculations, it was discovered that the process is endothermic, and the change in adsorption free energy was found to be -17.23 kJ mol^{-1}.

In the interest of scientific investigation, alizarin yellow was removed from an aqueous solution by using *Casuarina equisetifolla* as an adsorbent. In order to get the *Causarina equisetifolla* organism prepared for the experiment, it was subjected to a variety of different pre-treatments. In order to assess the adsorption behaviour, a large number of factors, such as pH, dosage size and contact duration, were investigated and tested. A summary of the findings from a variety of studies revealed that the Freundlich model provided the most convincing argument in support of the existence of multilayer adsorption [93]. Calculations have also been made about the thermodynamic system's parameter values. Clay from the natural world was used in our research to assist in the removal of methylene blue (MB) from successive batches of aqueous solutions. In this particular neighbourhood, there was an abundance of this kind of clay, and it was not difficult to get [94]. Throughout the course of the experiment, a wide variety of factors, such as the beginning dye concentration, starting pH, starting contact duration and starting mass of adsorbent, were investigated. An investigation of the equilibrium adsorption of MB onto bentonite was carried out, and the well-known Freundlich and Langmuir isotherms equations were used in order to determine a number of different isotherm parameters. The findings of the kinetic experiments indicate that a pseudo-second-order fit is the most appropriate model to use when attempting to characterise the process.

In Rubin et al. [15], it has come to light, as well as been the subject of study, that the cells of *Sargassum muticum* have the potential to be used as biosorbents for the purpose of recovering MB from aqueous solutions. Numerous additional pre-treatments, such as protonation and chemical cross-linking with $CaCl_2$ or H_2CO_3, have been applied in an effort to investigate the algal biomass's potential to keep its stability and adsorption capacities after being subjected to a variety of stressors. This has been done in an effort to learn more about the algal biomass's resilience. Several Langmuir and Freundlich isotherms have been used in order to provide an explanation for the equilibrium binding that occurred in accordance with the algal pre-treatment. The determination of the specific surface area of the algae was made possible by the use of measurements of the maximum adsorption capacity of the algae. For the purpose of characterising the findings, the first-order Langmuir model was used. Using this model, we were able to successfully calculate both the rate constant as well as the adsorption capacity.

In one of the study by Morosanu et al. [95] in order to remove the dyes from the aqueous effluents that were produced during the process of extracting textile colours, rapeseed was used as a biosorbent. In order to explore the impacts of the multiple distinct components, a batch mode sorption approach was developed and used in the laboratory. These factors included the temperature of the solution, the length of time it was in contact with the dye, the concentration of the dye at the beginning of the experiment and the pH of the solution. Based on the results of the experiment, it seemed that the pseudo-second-order model provided the clearest and most actionable insight into the kinetics of dye sorption.

Mature leaves of natural neem trees were employed as one of the adsorbent by using in the process of removing the dye known as BG from aqueous solutions [96]. Throughout the entire process of batch adsorption, the aqueous dye solution was utilised in a number of different concentrations. The process of adsorption was studied from the perspective of how different adsorbent doses, pH levels and temperatures influenced the outcome of the experiment. The findings demonstrate that the Freundlich and Langmuir adsorption models provide an accurate representation of the data obtained from the adsorption experiments. The researchers investigated whether or not it would be possible to remove rhodamine blow from an aqueous solution by utilising perlite, which is an adsorbent that occurs naturally. This is used in the process of removing the dye known as brilliant green from aqueous solutions [14]. In batch adsorption investigations, the dye adsorption equilibrium was swiftly attained after 50 min of contact time. Both the Langmuir and Freundlich adsorption isotherms properly described the process across the whole concentration range, which went from 20 mg L^{-1} all the way up to 100 mg L^{-1}. The modelling process made use of adsorption data derived from first- and second-order kinetic equations as well as intraparticle diffusion models. According to the thermodynamic features, the adsorption was of an exothermic nature and occurred spontaneously.

In this experiment, the removal of malachite green and safranin (SF) colour from aqueous solution was investigated using banana peduncle, which is a naturally occurring agricultural waste product [97]. In order to discover how the presence of various variables might influence the outcomes of the experiment, a batch experiment was conducted to study the adsorption potential. Studies on the effects of variations in starting dye concentration (between 10 and 150 mg L^{-1}), pH (ranging from 4 to 9), stirring speed (ranging from 90 to 180 revolutions per minute), contact time (ranging from 5 to 120 min), temperature (ranging from 25 to 35 °C) and adsorbent dose are included. Banana peduncle was used as one of the adsorbents, and further analysis revealed that it was effective in removing 97.6% of the malachite green and 85.2% of the SF from the sample. It was found that the optimal conditions for malachite green removal were a temperature of 300 °C, a pH of 7.0, an adsorbent dosage of 0.4 g per 100 millilitres (g/100 ml), a swirling speed of 150 revolutions per minute (rpm) and a contact time of 30 min for an initial dye concentration of 50 mg per litre (mg/l). All of these parameters were combined with a contact time of 30 min In addition, it was demonstrated that the optimal conditions for SF dye are a pH of 7.0, 0.1 g/100 ml of adsorbent, 150 rpm of stirring speed and 60 min of contact time for an initial dye concentration of 50 mg L^{-1} at a temperature of 30 °C. These are the conditions that produce the best results when using SF dye.

12 Potential in Large-Scale Applications

In order to find an effective alternative to organic dyes, researchers are continuously looking for novel nanomaterials that are not only inexpensive but also recyclable and kind to the environment. There have been a number of studies that have been published that investigate the possibility of using hybrid nanomaterials to act as dye adsorbents. The realm of academics has seen a great number of important developments in recent years. It is essential to keep in mind that cutting-edge hybrid nanomaterials provide appropriate adsorption capabilities of around 70–230 mg g^{-1} and better removal efficiency for textile dyes. This is something that should be kept in mind at all times. In spite of the considerable progress that has been made in this area, it is still fairly difficult to synthesise hybrid nanomaterials on a large scale. In the process of dye removal, the hybrid nanomaterial that was just developed might potentially be utilised. It is still quite difficult to isolate a specific dye from a mixture of several colours. The overwhelming majority of companies make an effort to collect and recycle a variety of colouring components [98]. It is of the utmost importance that people be made aware of the fact that a large number of nanohybrid materials are developed in laboratory settings in order to remove colour from synthetic aqueous solutions. At this time, however, it is not possible to make nanohybrid materials in sufficient numbers and then utilise them in a manner that is effective for level remediation on a large scale. This is due to the fact that the production of nanohybrid materials is now prohibitively expensive. In order to carry out the adsorption process in an effective manner, it is essential to first design and then constructs a sufficient number of columns and beds. In addition to this, the beds and columns have to be arranged in accordance with the design, which is of equal significance. The stability of synthetic nanohybrid materials and the lack of an individual's identity in nanohybrids of varied materials are two prevalent types of reporting failures. The hydrothermal stability of nanohybrid materials as well as their regeneration is significant aspects. In spite of the creation of a number of one-of-a-kind hybrid nanoparticles that have exceptionally high rates of effectiveness in terms of removal, there is still a problem that has to be solved in order for scientific progress to achieve its zenith.

13 Conclusion

Textile dyes are the major pollutants that need to be removed or recovered from industrial effluents because of their environmental pollution. The ecological effects of these dyes in polluted water remain a challenge for sustainability. Among the many methods used for dyes removal or recovery, Adsorption is found to be a versatile and effective method due to its simplicity in operation and economic feasibility. It remains to be a simple, physical method but proved to have maximum efficiency in dyes removal. The most remarkable feature in adsorption technique is the potential in recyclability of the adsorbents used in the method. The mechanism of adsorption is explained by surface attachment or interaction with the chemicals present on the surface. The different interactions between the absorbents and the dyes like electrostatic, hydrophobic, ion exchange, acid–base reaction, hydrogen bonding and Van der Waals forces decide the absorption maxima of the dyes and thereby the efficacy of the adsorbents.

The varieties of adsorbents are being used by researchers like organic, inorganic, agricultural, industrial wastes, different blends of nanohybrid materials and so on. The development of nanohybrid materials as adsorbents is an important milestone in the process of adsorption of textile dyes removal. The presence of numerous nanohybrid molecules are promising adsorbents; however, cost-effective synthesis of nanoparticle and their characterisation is still a concern. Even though low-cost adsorbents are used, the feasibility of using these methods in large scale remains a mainstay. The research gap lies in the optimisation conditions for large scale applications.

Further studies are warranted in the area of optimisation of process parameters with the well-studied isotherms and identification of low-cost nanohybrids for efficient recovery of textile dyes from bulk liquid. However, adsorption method proves to a promising technique to resolve the problem of textile dyes removal and recovery from polluted effluents.

References

1. Krithika T, Kavitha R, Dinesh M, Angayarkanni J (2021) Assessment of ligninolytic bacterial consortium for the degradation of azo dye with electricity generation in a dual-chambered microbial fuel cell. Environ Challenges 4:100093. https://doi.org/10.1016/j.envc.2021.100093
2. Chequer FD, De Oliveira GR, Ferraz EA, Cardoso JC, Zanoni MB, De Oliveira DP (2013) Textile dyes: dyeing process and environmental impact. Eco-Friendly Text Dyeing Finish 6(6):151–176. https://doi.org/10.5772/53659

3. Thiruppathi K, Rangasamy K, Ramasamy M, Muthu D (2021) Evaluation of textile dye degrading potential of ligninolytic bacterial consortia. Environ Challenges 4:100078. https://doi.org/10.1016/j.envc.2021.100078
4. Gambhir RS, Kapoor V, Nirola A, Sohi R, Bansal V (2012) Water pollution: impact of pollutants and new promising techniques in purification process. J Hum Ecol 37(2):103–109. https://doi.org/10.1080/09709274.2012.11906453
5. Hassan MM, Carr CM (2018) A critical review on recent advancements of the removal of reactive dyes from dyehouse effluent by ion-exchange adsorbents. Chemosphere 209:201–219
6. Mary JE, Krithika T, Kavitha R (2020) Biodegradation of textile dye by ligninolytic bacteria isolated from western Ghats. Int J Res Rev 7:22–29. https://www.ijrrjournal.com/IJRR_Vol.7_Issue.4_April2020/Abstract_IJRR004.html
7. Bhatia SC, Devraj S (2017) Pollution control in textile industry. WPI publishing
8. Wang CX, Chen SL (2005) Aromachology and its application in the textile field. Fibres Text East Eur 13(6):41–44
9. Aquino JM, Rocha-Filho RC, Ruotolo LA, Bocchi N, Biaggio SR (2014) Electrochemical degradation of a real textile wastewater using β-PbO2 and DSA® anodes. Chem Eng J 251:138–145
10. Khatri J, Nidheesh PV, Singh TA, Kumar MS (2018) Advanced oxidation processes based on zero-valent aluminium for treating textile wastewater. Chem Eng J 348:67–73
11. Sandhya S (2010) Biodegradation of azo dyes under anaerobic condition: role of azoreductase. Biodegradation Azo Dyes 39–57
12. Alaqarbeh M (2021) Adsorption phenomena: definition, mechanisms, and adsorption types: short review. RHAZES Green Appl Chem 13:43–51. https://doi.org/10.48419/IMIST.PRSM/rhazes-v13.28283
13. Kobiraj R, Gupta N, Kushwaha AK, Chattopadhyaya MC (2012) Determination of equilibrium, kinetic and thermodynamic parameters for the adsorption of Brilliant Green dye from aqueous solutions onto eggshell powder
14. Vijayakumar G, Tamilarasan R, Dharmendirakumar M (2012) Adsorption, kinetic, equilibrium and thermodynamic studies on the removal of basic dye Rhodamine-B from aqueous solution by the use of natural adsorbent perlite. J Mater Environ Sci 3(1):157–170
15. Rubín E, Rodríguez P, Herrero R, Sastre de Vicente ME (2010) Adsorption of methylene blue on chemically modified algal biomass: equilibrium, dynamic, and surface data. J Chem Eng Data 55(12):5707–5714
16. Shen K, Gondal MA (2017) Removal of hazardous Rhodamine dye from water by adsorption onto exhausted coffee ground. J Saudi Chem Soc 21:S120-S127
17. Zheng Y, Yang Y, Zhang Y, Zou W, Luo Y, Dong L, Gao B (2019) Facile one-step synthesis of graphitic carbon nitride-modified biochar for the removal of reactive red 120 through adsorption and photocatalytic degradation. Biochar 1:89–96
18. Dutta S, Gupta B, Srivastava SK, Gupta AK (2021) Recent advances on the removal of dyes from wastewater using various adsorbents: A critical review. Mater Adv 2(14):4497–4531
19. Pirbazari AE, Saberikhah E, Badrouh M, Emami MS (2014) Alkali treated Foumanat tea waste as an efficient adsorbent for methylene blue adsorption from aqueous solution. Water Resour Ind 6: 64–80
20. Cojocaru C, Samoila P, Pascariu P (2019) Chitosan-based magnetic adsorbent for removal of water-soluble anionic dye: Artificial neural network modeling and molecular docking insights. Int J Biol Macromol 123:587–599
21. Siddiqui SI, Manzoor O, Mohsin M, Chaudhry SA (2019) Nigella sativa seed based nanocomposite-MnO2/BC: An antibacterial material for photocatalytic degradation, and adsorptive removal of Methylene blue from water. Environ Res 171:328–340
22. Rápó E, Tonk S (2021) Factors affecting synthetic dye adsorption; desorption studies: a review of results from the last five years (2017–2021). Molecules 26(17):5419. https://doi.org/10.3390/molecules26175419
23. Wu J, Zhao Y, Qi H, Zhao X, Yang T, Du Y, Wei Z et al (2017) Identifying the key factors that affect the formation of humic substance during different materials composting. Bioresour Technol 244:1193–1196

24. Mittal A, Mittal J, Malviya A, Jaspal D, Gupta V (2010) Adsorption of hazardous dye crystal violet from wastewater by waste materials. J Colloid Interface Sci 343:463–473. https://doi.org/10.1016/j.jcis.2009.11.060
25. Jain R, Mathur M, Sikarwar S, Mittal A (2007) Removal of the hazardous dye rhodamine B through photocatalytic and adsorption treatments. J Environ Manag 85(4):956–964
26. Abbaszadeh S, Alwi SRW, Webb C, Ghasemi N, Muhamad II (2016) Treatment of lead-contaminated water using activated carbon adsorbent from locally available papaya peel biowaste. J Cleaner Prod 118:210–222
27. Binoj JS, Raj RE, Sreenivasan VS, Thusnavis GR (2016) Morphological, physical, mechanical, chemical and thermal characterization of sustainable Indian Areca fruit husk fibers (Areca Catechu L.) as potential alternate for hazardous synthetic fibers. J Bionic Eng 13(1):156–165
28. Careddu N, Marras G, Siotto G (2014) Recovery of sawdust resulting from marble processing plants for future uses in high value-added products. J Cleaner Prod 84:533–539
29. Arami M, Limaee NY, Mahmoodi NM, Tabrizi NS (2005) Removal of dyes from colored textile wastewater by orange peel adsorbent: equilibrium and kinetic studies. J Colloid Interface Sci 288(2):371–376
30. Jandas PJ, Mohanty S, Nayak SK (2013) Surface treated banana fiber reinforced poly (lactic acid) nanocomposites for disposable applications. J Cleaner Prod 52:392–401
31. Raymundo AS, Zanarotto R, Belisário M, Pereira MDG, Ribeiro JN, Ribeiro AVFN (2010) Evaluation of sugar-cane bagasse as bioadsorbent in the textile wastewater treatment contaminated with carcinogenic congo red dye. Braz Arch Biol Technol 53:931–938
32. Cholake ST, Rajarao R, Henderson P, Rajagopal RR, Sahajwalla V (2017) Composite panels obtained from automotive waste plastics and agricultural macadamia shell waste. J Cleaner Prod 151:163–171
33. Lu L, Li J, Ng DH, Yang P, Song P, Zuo M (2017) Synthesis of novel hierarchically porous Fe_3O_4@MgAl–LDH magnetic microspheres and its superb adsorption properties of dye from water. J Ind Eng Chem 46:315–323. https://doi.org/10.1016/j.jiec.2016.10.045
34. Binoj JS, Raj RE, Daniel BSS (2017) Comprehensive characterization of industrially discarded fruit fiber, *Tamarindus indica* L. as a potential eco-friendly bio-reinforcement for polymer composite. J Cleaner Prod 142:1321–1331
35. Reddy MC (2006) Removal of direct dye from aqueous solutions with an adsorbent made from tamarind fruit shell, an agricultural solid waste
36. Chowdhury S, Mishra R, Saha P, Kushwaha P (2011) Adsorption thermodynamics, kinetics and isosteric heat of adsorption of malachite green onto chemically modified rice husk. Desalination 265(1–3):159–168
37. Memon SA, Khan MK (2018) Ash blended cement composites: Eco-friendly and sustainable option for utilization of corncob ash. J Cleaner Prod 175:442–455
38. Salih SJ, Kareem ASA, Anwer SS (2022) Adsorption of anionic dyes from textile wastewater utilizing raw corncob. Heliyon 8(8):e10092
39. Nikkhah A, Khojastehpour M, Emadi B, Taheri-Rad A, Khorramdel S (2015) Environmental impacts of peanut production system using life cycle assessment methodology. J Cleaner Prod 92:84–90
40. Etorki AM, Massoudi FM (2011) The use of peanut hull for the adsorption of colour from aqueous dye solutions and dye textile effluent. Orient J Chem 27(3):875
41. Dunnigan L, Ashman PJ, Zhang X, Kwong CW (2018) Production of biochar from rice husk: particulate emissions from the combustion of raw pyrolysis volatiles. J Cleaner Prod 172:1639–1645
42. Sivalingam S, Sen S (2020) Rice husk ash derived nanocrystalline ZSM-5 for highly efficient removal of a toxic textile dye. J Mater Res Technol 9(6):14853–14864
43. Castro NR, Swart J (2017) Building a roundtable for a sustainable hazelnut supply chain. J Cleaner Prod 168:1398–1412
44. Buyukada M, Uzuner S, Evrendilek F (2018) Utilization of (modified-) ground hazelnut shells for adsorption of azo-metal toxic dyes: empirical and ANFIS modeling and optimization. Chiang Mai J Sci

45. Nhung NTH, Quynh, BTP, Thao PTT, Bich HN, Giang BL (2018) Pretreated fruit peels as adsorbents for removal of dyes from water. In: IOP conference series: earth and environmental science, vol 159, no. 1. IOP Publishing, p 012015
46. Arena N, Lee J, Clift R (2016) Life cycle assessment of activated carbon production from coconut shells. J Cleaner Prod 125:68–77
47. Etim UJ, Umoren SA, Eduok UM (2016) Coconut coir dust as a low cost adsorbent for the removal of cationic dye from aqueous solution. J Saudi Chem Soc 20:S67–S76
48. Chin BLF, Yusup S, Al Shoaibi A, Kannan P, Srinivasakannan C, Sulaiman SA (2014) Comparative studies on catalytic and non-catalytic co-gasification of rubber seed shell and high density polyethylene mixtures. J Cleaner Prod 70:303–314
49. Rajesh M, Jayakrishna K, Sultan MTH, Manikandan M, Mugeshkannan V, Shah AUM, Safri SNA (2020) The hydroscopic effect on dynamic and thermal properties of woven jute, banana, and intra-ply hybrid natural fiber composites. J Mater Res Technol 9(5):10305–10315
50. Weinwurm F, Turk T, Denner J, Whitmore K, Friedl A (2017) Combined liquid hot water and ethanol organosolv treatment of wheat straw for extraction and reaction modeling. J Cleaner Prod 165:1473–1484
51. Guil-Guerrero JL, Guil-Laynez JL, Guil-Laynez Á (2017) Bioprospecting for seed oils from wild plants in the Mediterranean Basin for biodiesel production. J Cleaner Prod 159:180–193
52. Smitha T, Santhi T, Prasad AL, Manonmani S (2017) Cucumis sativus used as adsorbent for the removal of dyes from aqueous solution. Arab J Chem 10:S244–S251
53. Weber C, Foletto E, Meili L (2013) Removal of tannery dye from aqueous solution using papaya seed as an efficient natural biosorbent. Water Air Soil Pollut 224. https://doi.org/10.1007/s11270-012-1427-7
54. Burland JB (1990) On the compressibility and shear strength of natural clays. Géotechnique 40(3):329–378
55. Kazmi SMS, Munir MJ, Wu YF, Hanif A, Patnaikuni I (2018) Thermal performance evaluation of eco-friendly bricks incorporating waste glass sludge. J Cleaner Prod 172:1867–1880
56. Capela MN, Cesconeto FR, Pinto PC, Tarelho LA, Seabra MP, Labrincha JA (2022) Biomass fly ash self-hardened adsorbent monoliths for methylene blue removal from aqueous solutions. Appl Sci 12(10):5134
57. Amin N, Faisal M, Muhammad Khan, Gul S (2016) Synthesis and characterization of geopolymer from bagasse bottom ash, waste of sugar industries and naturally available china clay. J Cleaner Prod 129. https://doi.org/10.1016/j.jclepro.2016.04.024
58. Hynes NRJ, Kumar JS, Kamyab H, Sujana JAJ, Al-Khashman OA, Kuslu Y, Kumar BS et al (2020) Modern enabling techniques and adsorbents based dye removal with sustainability concerns in textile industrial sector—a comprehensive review. J Cleaner Prod 272:122636
59. Yang R, Li D, Li A, Yang H (2018) Adsorption properties and mechanisms of palygorskite for removal of various ionic dyes from water. Appl Clay Sci 151:20–28
60. Kismir Y, Aroguz AZ (2011) Adsorption characteristics of the hazardous dye Brilliant Green on Saklıkent mud. Chem Eng J 172(1):199–206
61. Zhao M, Liu P (2008) Adsorption behavior of methylene blue on halloysite nanotubes. Microporous Mesoporous Mater 112(1–3):419–424
62. Almasian A, Olya ME, Mahmoodi NM (2015) Synthesis of polyacrylonitrile/polyamidoamine composite nanofibers using electrospinning technique and their dye removal capacity. J Taiwan Inst Chem Eng 49:119–128
63. Fiallos FRM, Robalino WSM, Beltrán BEP (2017) Biofiltración sobre Cama de Turba, para el Tratamiento de Aguas Residuales Provenientes del Lavado de Jeans. Revista Publicando 4(10):579–590
64. Vital RK, Saibaba KVN, Shaik KB, Gopinath R (2016) Dye removal by adsorption: a review. J Bioremediat Biodegrad 7:371. https://doi.org/10.4172/2155-6199.1000371
65. Netpradit S, Thiravetyan P, Towprayoon S (2003) Application of 'waste' metal hydroxide sludge for adsorption of azo reactive dyes. Water Res 37(4):763–772
66. Hossain MR, Rashid TU, Lata NP, Dey SC, Sarker M, Shamsuddin SM (2022) Fabrication of novel nanohybrid material for the removal of azo dyes from wastewater. J Compos Sci 6(10):304. https://doi.org/10.3390/jcs6100304

67. Farhadi S, Amini MM, Dusek M, Kucerakova M, Mahmoudi F (2017) A new nanohybrid material constructed from Keggin-type polyoxometalate and Cd(II) semicarbazone Schiff base complex with excellent adsorption properties for the removal of cationic dye pollutants. J Mol Struct 1130:592–602
68. Zhao Y, Qamar SA, Qamar M, Bilal M, Iqbal HMN (2021) Sustainable remediation of hazardous environmental pollutants using biochar-based nanohybrid materials. J Environ Manag 300:113762
69. Mahmoodi NM, Najafi F (2012) Synthesis, amine functionalization and dye removal ability of titania/silica nano-hybrid. Microporous Mesoporous Mater 156:153–160
70. Archana S, Jayanna BK, Ananda A, Ananth MS, Ali AM, Muralidhara HB, Kumar KY (2022) Numerical investigations of response surface methodology for organic dye adsorption onto Mg-Al LDH-GO nano hybrid: an optimization, kinetics and isothermal studies. J Indian Chem Soc 99(1):100249
71. Han J, Jun B-M, Heo J, Lee G, Yoon Y, Park CM (2019) Highly efficient organic dye removal from waters by magnetically recoverable La2O2CO3/ZnFe2O4-reduced graphene oxide nanohybrid. Ceram Int 45(15):19247–19256
72. Liu J, Yu H, Liang Q, Liu Y, Shen J, Bai Q (2017) Preparation of polyhedral oligomeric silsesquioxane based cross-linked inorganic-organic nanohybrid as adsorbent for selective removal of acidic dyes from aqueous solution. J Colloid Interface Sci 497:402–412
73. González MA, Pavlovic I, Rojas-Delgado R, Barriga C (2014) Removal of Cu^{2+}, Pb^{2+} and Cd^{2+} by layered double hydroxide–humate hybrid. Sorbate and sorbent comparative studies. Chem Eng J 254:605–611. https://doi.org/10.1016/j.cej.2014.05.132
74. Sarojini G, Venkateshbabu S, Rajasimman M (2022) Adsorptive potential of iron oxide based nanocomposite for the sequestration of Congo red from aqueous solution. Chemosphere 287:132371
75. Pal S, Ghorai S, Das C, Samrat S, Ghosh A, Panda AB (2012) Carboxymethyl tamarind-gpoly(acrylamide)/silica: a high performance hybrid nanocomposite for adsorption of methylene blue dye. Ind Eng Chem Res 51:15546–15556. https://doi.org/10.1021/ie301134a
76. Ibrahim S, Fatimah I, Ang H-M, Wang S (2010) Adsorption of anionic dyes in aqueous solution using chemically modified barley straw. Water Sci Technol 62:1177–1185. https://doi.org/10.2166/wst.2010.388
77. Guo X, Yin P, Yang H (2018) Superb adsorption of organic dyes from aqueous solution on hierarchically porous composites constructed by ZnAl-LDH/$Al(OH)_3$ nanosheets. Microporous Mesoporous Mater 259:123–133. https://doi.org/10.1016/j.micromeso.2017.10.003
78. Ahmed IM, Gasser MS (2012) Adsorption study of anionic reactive dye from aqueous solution to Mg–Fe–CO3 layered double hydroxide (LDH). Appl Surf Sci 259:650–656. https://doi.org/10.1016/j.apsusc.2012.07.092
79. Bharali D, Deka RC (2017) Adsorptive removal of Congo red from aqueous solution bysonochemically synthesized NiAl layered double hydroxide. J Environ Chem Eng 5:2056–2067. https://doi.org/10.1016/j.jece.2017.04.012
80. Lizundia E, Puglia D, Nguyen TD, Armentano I (2020) Cellulose nanocrystal based multifunctional nanohybrids. Prog Mater Sci 112:100668. https://doi.org/10.1016/j.pmatsci.2020.100668
81. Nazir MA, Khan NA, Cheng C, Shah SSA, Najam T, Arshad M, ur Rehman A et al (2020) Surface induced growth of ZIF-67 at Co-layered double hydroxide: removal of methylene blue and methyl orange from water. Appl Clay Sci 190:105564. https://doi.org/10.1016/j.clay.2020.105564
82. Mahouche-Chergui S, Boussaboun Z, Oun A, Kazembeyki M, Christian G, Carbonnier B, Ouellet-Plamondon CM (2021) Sustainable preparation of graphene-like hybrid nanomaterials and their application for organic dyes removal. Chem Eng Sci 236:116482
83. Mohammadi Galangash M, Mohaghegh Montazeri M, Ghavidast A, Shirzad-Siboni M (2018) Synthesis of carboxyl-functionalized magnetic nanoparticles for adsorption of malachite green from water: kinetics and thermodynamics studies. J Chin Chem Soc 65(8):940-950. https://doi.org/10.1002/jccs.201700361

84. Mohammadi Galangash M, Mohaghegh Montazeri M, Ghavidast A, Shirzad Siboni M (2018) Synthesis of carboxyl-functionalized magnetic nanoparticles for adsorption of malachite green from water: kinetics and thermodynamics studies. J Chin Chem Soc 65:940–950. https://doi.org/10.1002/jccs.201700361
85. Narayani H, Jose M, Sriram K, Shukla S (2018) Hydrothermal synthesized magnetically separable mesostructured $H_2Ti_3O_7/\gamma\text{-}Fe_2O_3$ nanocomposite for organic dye removal via adsorption and its regeneration/reuse through synergistic non-radiation driven H_2O_2 activation. Environ Sci Pollut Res 25:20304–20319. https://doi.org/10.1007/s11356-017-8381-2
86. Liu A, Zhou W, Shen K, Liu J, Zhang X (2015) One-pot hydrothermal synthesis of hematite-reduced graphene oxide composites for efficient removal of malachite green from aqueous solution. RSC Adv 5:17336–17342
87. Gagliano E, Sgroi M, Falciglia PP, Vagliasindi FGA, Roccaro P (2020) Removal of poly- and perfluoroalkyl substances (PFAS) from water by adsorption: role of PFAS chain length, effect of organic matter and challenges in adsorbent regeneration. Water Res 171:115381
88. Zubair M, Jarrah N, Manzar MS, Al-Harthi M, Daud M, Mu'azu ND, Haladu SA (2017) Adsorption of eriochrome black T from aqueous phase on MgAl-, CoAl- and NiFecalcined layered double hydroxides: kinetic, equilibrium and thermodynamic studies. J Mol Liq 230:344–352. https://doi.org/10.1016/j.molliq.2017.01
89. Tahir H, Sultan M, Jahanzeb Q (2008) Removal of basic dye methylene blue by using bioabsorbents Ulva lactuca and Sargassum. Afr J Biotechnol 7(15)
90. Meroufel B, Benali O, Benyahia M, Benmoussa Y, Zenasni MA (2013) Adsorptive removal of anionic dye from aqueous solutions by Algerian kaolin: characteristics, isotherm, kinetic and thermodynamic studies. J Mater Environ Sci 4(3):482–491
91. Srivastava R, Rupainwar DC (2011) A comparative evaluation for adsorption of dye on Neem bark and mango bark powder
92. Oyelude EO, Owusu UR (2011) Adsorption of methylene blue from aqueous solution using acid modified *Clotropis procer* leaf powder. J Appl Sci Environ Sanit 2011(6):477
93. Diraki A, Mackey HR, McKay G, Abdala A (2019) Removal of emulsified and dissolved diesel oil from high salinity wastewater by adsorption onto graphene oxide. J Environ Chem Eng 7(3):103106. https://doi.org/10.1016/j.jece.2019.103106
94. Karima O, Riazi G, Yousefi R, Movahedi AAM (2010) The enhancement effect of beta-boswellic acid on hippocampal neurites outgrowth and branching (an in vitro study). Neurol Sci 31(3):315–320. https://doi.org/10.1007/s10072-010-0220-x
95. Morosanu I, Gilca AF, Paduraru C, Fighir D, Peptu CA, Teodosiu C (2017) Valorisation of rapeseed as biosorbent for the removal of textile dyes from aqueous effluents. Cellul Chem Technol 51:175–184
96. Bhattacharyya J, Santhoshkumar P, Sharma KK (2003) A peptide sequence—YSGVCHTDL-HAWHGDWPLPVK [40–60]—in yeast alcohol dehydrogenase prevents the aggregation of denatured substrate proteins. Biochem Biophys Res Commun 307(1):1–7. https://doi.org/10.1016/S0006-291X(03)01116-1
97. Chukki J, Abinandan S, Shanthakumar S (2018) *Chrysanthemum indicum* microparticles on removal of hazardous Congo red dye using response surface methodology. Int J Ind Chem 9(4):305–316. https://doi.org/10.1007/s40090-018-0160-5
98. Salleh MAM, Mahmoud DK, Karim WAWA, Idris A (2011) Cationic and anionic dye adsorption by agricultural solid wastes: a comprehensive review. Desalination 280:1–13. https://doi.org/10.1016/j.desal.2011.07.019
99. Mittal A, Kurup L, Mittal J (2007) Freundlich and Langmuir adsorption isotherms and kinetics for the removal of Tartrazine from aqueous solutions using hen feathers. J Hazard Mater 146(1–2):243–248

Photocatalytic Degradation of Textile Dyes Using Nanohybrid Materials

A. A. A. Mutalib and N. F. Jaafar

Abstract The manufacturing activity in the textile industry has initiated environmental issue since dyes are one of the main materials and the discharge of dyes which being hazardous, carcinogenic, and poisonous would harm human health, environment, and aquatic ecosystem. Dyes not only resistant to degradation and can remain in the environment for a long time, but it also increases the biological oxygen demand (BOD) and chemical oxygen demand (COD) of water by modifying the pH and adjusting the organic–inorganic chemical contents which affect the aquatic ecosystem. There are various treatment methods have been reported for removal of dyes from the water stream like chemical oxidation, adsorption, coagulation, and microorganisms but some of them are eco-unfriendly and costly. Among them, advanced oxidation processes (AOPs) are identified as the most efficient in treating dyes wastewater. Photocatalytic degradation is one of the most popular AOPs methods where this process will produce hydroxyl radicals as a strong oxidizing agent for degradation of dye, and there is no production of secondary pollutant at the end of the process. To utilize this method, the selection of materials as a photocatalyst is crucial. There are various materials can used as photocatalysts. Hence, this chapter will give an overview on nanohybrid materials as photocatalyst for degradation of textile dyes. The chapter will cover on design of the nanohybrid materials which lead to the enhancement of the performance of photocatalysts towards textile dyes removal. The understanding will be valuable for the removal of dyes in the various contaminants in numerous media.

Keywords Photocatalyst · Photocatalytic degradation · Nanohybrid materials · Dyes · Textile industry

A. A. A. Mutalib · N. F. Jaafar (✉)
School of Chemical Sciences, Universiti Sains Malaysia, Gelugor 11800 USM, Penang, Malaysia
e-mail: nurfarhana@usm.my

© The Author(s), under exclusive license to Springer Nature Singapore Pte Ltd. 2023

A. Ahmad et al. (eds.), *Nanohybrid Materials for Treatment of Textiles Dyes*,
Smart Nanomaterials Technology, https://doi.org/10.1007/978-981-99-3901-5_10

1 Overview of Photocatalysis in Degradation of Pollutants in Wastewater

Water is essential to living life and it is very crucial to ensure the easily accessible of clean water. According to World Health Organization (WHO), there are 122 million people still using untreated surface water from various water bodies, while 368 million only get access to water from unprotected wells and springs. The rapid growth of manufacturing and industrial activities has brought along unwanted water pollution as one of their consequences. It is worrisome when at least 10% of the human population potentially consumes exposed food irrigated by wastewater. This situation is worrying since it has been known that manufacturing and industrial wastewater contain high toxic and hazardous pollutants from various sources such as chemicals and biologicals, it is continuously being released from their sites [1]. The treatment of wastewater is challenging even with the strict implementation of legislations and regulations to ensure the quality of water discharged due to the persistent nature and low biodegradability of pollutants found in wastewater.

There are various methods widely used to treat wastewater and can be categorized as chemical, physical, and biological as shown in Fig. 1. Commonly, the treatment is selected based on the type and composition of the wastewater. However, every treatment has their own advantages and disadvantages in terms of performance, efficiency as well as cost, and due to economic factors, only a few methods were used by the industries [2]. The chemical method usually involved the addition of chemicals or chemical reaction to convert and/or remove the pollutants. In contrast, a physical method uses physical forces that naturally take place to treat the pollutant without the changes in chemical and biological. While in biological method, the treatment will apply with chemical and physical methods before using its own method because this method focusses on the removal of biodegradable organic substances which it failed to treat by chemical and physical methods.

Among all the methods, advanced oxidation processes (AOPs) is one of the famous methods in elimination of pollutants from wastewater due to their effectiveness and the ability to convert the pollutants into less hazardous intermediates and products such as water (H_2O) and carbon dioxide (CO_2) [3]. In AOPs, photocatalysis is one of the favourable methods in wastewater treatment compared to other AOPs because it commonly used inexpensive heterogeneous catalyst to ease the separation process and usually it does not consume the oxidant after the reaction. Various pollutants have been reported in the literature to be successfully degraded by photocatalysis such as dyes, herbicides, pesticides, polymers, surfactants, aromatics, haloalkanes, and aliphatic alcohols. In photocatalysis, reactive oxygen species (ROS) like hydroxyl radicals ($^{\cdot}OH$) will be produced through these processes when the suitable catalyst is exposed to the light (*hv*) also known as a photocatalyst. Generally, the catalyst will adsorb the light to generate electron–hole (e^-–h^+) pairs when the e^- at valance band (V_B) excited to conduction band (C_B) (Eq. 1). The h^+ will be generated at V_B to react with H_2O molecules or a hydroxyl group (OH) to produce $^{\cdot}OH$ (Eqs. 2, 3).

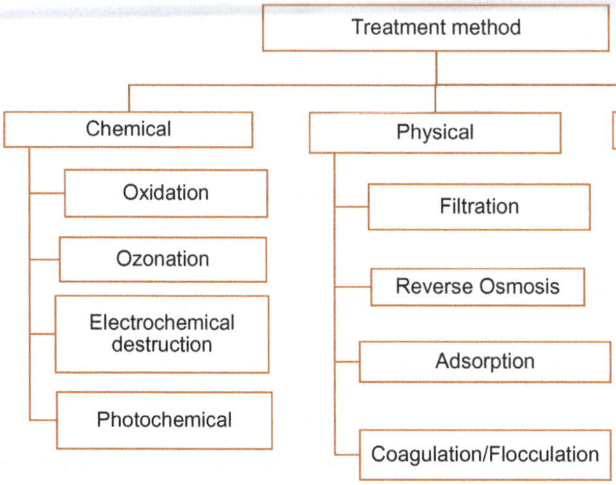

Fig. 1 Treatment method categories with examples

$$\text{catalyst} + h\upsilon \rightarrow \text{catalyst} + h_{VB} + e^{-}_{CB} \tag{1}$$

$$H_2O + h_{VB} \rightarrow {}^{\cdot}OH + H^{+} \tag{2}$$

$$OH^{-} + h_{VB} \rightarrow {}^{\cdot}OH \tag{3}$$

While the excited e^- at C_B will reduce to anion radicals, $^{\cdot}O_2{}^{-}$ (Eq. 4) resulting the production of fully generated $^{\cdot}OH$. The generated $^{\cdot}OH$ at V_B and C_B will partially (Eq. 5) or completely (Eq. 6) convert the pollutants into less harmful inorganic compounds such as CO_2 and H_2O. The generated $^{\cdot}OH$ will degrade the pollutants with different reactions such as addition to aromatic rings and double bonds, transfer of electrons and hydrogen transfer from aliphatic carbon are examples of radical attack, while the reaction with formation of intermediate may require additional reaction to form the final product [4]. Figure 2 illustrates the general mechanisms for degradation of pollutants.

$$O_2 + e_{CB} \rightarrow {}^{\cdot}O_2{}^{-} \tag{4}$$

$${}^{\cdot}O_2{}^{-} + H^{+} \rightarrow {}^{\cdot}HO_2 \rightarrow {}^{\cdot}OH \tag{5}$$

$${}^{\cdot}OH + \text{pollutant} \rightarrow \text{degraded pollutant} + CO_2 + H_2O \tag{6}$$

Other than suitable catalyst, light also one of the important things in photocatalysis process. Hence, the selection of the type of light is very critical to ensure it can activate

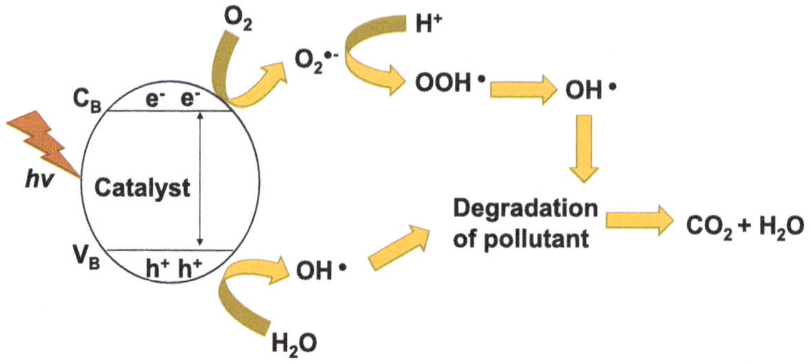

Fig. 2 General mechanism for photocatalytic degradation of pollutant

the catalyst to allow the generation of e^{-}–h^{+} pairs. The region of UV and visible light is 100 to 400 nm and 400 to 700 nm, respectively. It is very important to identify the band gap energy (difference in energy between V_B and C_B) of the selected catalyst to ensure the suitable light region is used for the photocatalytic process. The properties for suitable catalysts, specifically nanohybrid materials, will be further deliberated in the next section.

2 Nanohybrid Materials as Photocatalysts for the Degradation of Textile Dyes

Dyes are commonly used as a material for colouring the fibres or fabric. Nowadays, textile industries tend to use synthetic dyes compared with natural dyes because the colour of natural dyes can easily fade when exposed to sunlight, long extraction process and limited colour range. Table 1 lists the types of synthetic dyes and their characteristics. However, improper treatment of dyes wastewater from the textile industry gives negatives impact to environmental. Proper treatment of dyes wastewater is required to avoid these dyes permanently remaining as pollutants because it has high photo and thermal stability.

Nanohybrids is a term used to specify a combination of nanomaterials (more than one) linked by physiochemical or molecular forces despite having different dimensionalities due to their different chemical origins [5]. Due to their unique functionalities, nanohybrids have been useful in various fields, including energy storage, electronic materials, coating, sensing, catalysis, and biomaterials [6–11]. For wastewater remediation purposes, specifically photocatalytic dye degradation, different types of nanohybrids have been widely explored, including metal-based, metal–carbon, and clay-based nanohybrids. The synergetic effect between the different components incorporated to construct a highly active photocatalyst system has been proven to enhance the dye removal process and is further deliberated in this section.

Table 1 Types of synthetic dyes and its characteristics

Dyes	Examples of dyes	Characteristics
Acid dyes	Methyl orange, methyl red, congo red and orange (I, II)	Low-cost, water-soluble anionic dyes and dye absorption depend on amount of sulphuric acid
Basic dyes	Methylene blue, basic (red, brown, and blue), aniline yellow and crystal violet	Water-soluble cationic dyes and mostly added with acetic acid
Direct dyes	Martius yellow and direct (black, orange, blue, violet, and red)	Low cost, easy to use, time-consuming and mostly added with electrolyte (NaCl, Na_2SO_4, Na_2CO_3)
Vat dyes	Indigo, vat (blue, green) and benzanthrone	Insoluble in water, cannot directly dyeing the fibres, high cost and soluble by dissolving (NaHS) in (NaOH)
Reactive dyes	Reactive (red, blue, yellow, and black) and Remazol (blue, yellow, red, etc.)	Fast and most permanent dyes (covalent bond attach to natural fibres)
Sulphur dyes	Sulphur black and leuco sulphur black	Low cost and heating fabric in an organic solution to form dark colours
Azo dye	Mono azo dyes, diazo dyes, and triazo dyes	Insoluble azo dyes produce directly onto or within fibre, toxic chemicals, washable, bright, and high intensity colour

2.1 Metal–Metal Nanohybrids

Metal-based nanohybrids may be defined as nanomaterial systems composed of a metal mixture that includes metallic oxides, selenides, sulphides, etc. The development of metal-based nanohybrids is commonly useful when encountering certain semiconductor weaknesses. For example, the widely studied ZnO and TiO_2 are often correlated with wide band gaps (3.37 eV for ZnO and 3.2–3.35 eV for TiO_2), which generally limits their photoreactivity to the UV spectrum region only. Nevertheless, fast recombination of photogenerated electrons and high vulnerability to photo corrosion are also often reported regarding the application of ZnO and TiO_2.

Consequently, ZnO and TiO_2 with CuO combinations have been extensively studied to improve the reactivity of photocatalysts. The basis for CuO selection as the metal component of the nanohybrid system is attributed to its narrow band gap (1.35 eV), which was found to be effective in tuning the wide band gap of certain semiconductors. However, this p-type metal oxide catalytic activity itself is not appealing due to the rapid recombination rate of photogenerated electron–hole pairs. Regardless, this drawback does not hamper its applicability as a companion metal to form a visible light-active photocatalyst for dye degradation purposes. As proven, the research by Upadhyay et al. [12] and Bahnemann et al. [13] has managed to degrade approximately 90% methylene blue and 98% Congo red azo dyes, respectively, in a short amount of time (20–75 min) under visible light irradiation in the presence of Cu-based nanohybrids. Both studies stipulated that the incorporation of

Cu into the nanohybrids not only narrowed the band gap but also formed a substantial interaction with the metal component, which resulted in a prolonged lifetime of charge separation carriers of the catalyst.

Another typically selected metal nanoparticle to construct the highly active nanohybrid is Ag. The participation of Ag in a dye discoloration process can benefit the reaction rate due to its behaviour as an electron scavenger. Apart from that, the excellent capacity of visible light interaction through the resonance of free electron oscillations within its particle as well as the high thermal and electrical conductivity further make Ag an attractive metal companion for other semiconductor [14]. In addition, a study by Chakhtouna et al. [15] elucidated the two main advantages of incorporating Ag as the catalyst component for pollutant degradation, one of them is that Ag can trap the migrated electrons from the semiconductor conduction band and transfer them to oxygen for superoxide radical formation (Fig. 3). Second, the ability of Ag to create surface plasmon resonance can expand the light-harvesting capacity to the visible light region. This phenomenon has been proven by a study from Wang et al. [16], which successfully achieved near complete degradation of rhodamine B dye by exploiting ZnO–Ag nanohybrids under visible light irradiation. They agreed that the electronic and surface plasmon resonance impact from the interaction between the semiconductor material (ZnO) and Ag were responsible for the enhanced degradation reaction rate.

For practical applications, the ease of separation and reusability of the photocatalyst after a dye discolouration process is not less significant. Hence, to improve these two aspects of photocatalysts, the exploitation of the magnetic properties of Fe_3O_4 can be considered helpful, as the catalyst can be retrieved easily from the aqueous medium by the magnet as the separation tool. A study by Delnavaz et al. [17], who

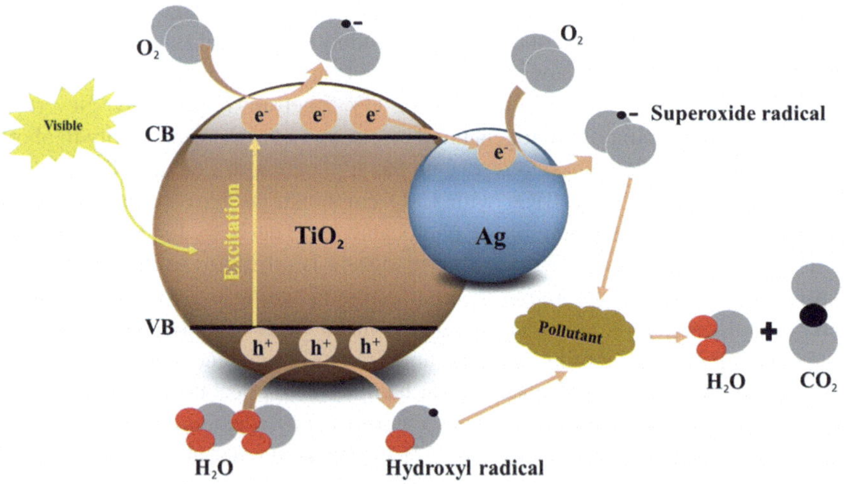

Fig. 3 The photocatalytic activity of Ag-based nanohybrids. Reproduced with permission from Chakhtouna et al. [15]. Copyright Springer

synthesized $CoFe_2O_4/SnO$, emphasized that the magnetic properties of the Fe-based oxide not only contribute to the high reusability and separability but are also discovered to be convenient in terms of impeding the rapid recombination of electron–hole pairs, contributing to a higher rate of dye removal. However, another recent report by Abouseada et al. [18] dictated contradictory findings, as the nanohybrid system constructed in their work ($CoFe_2O_4/SnO_2$) performed less efficiently than the SnO_2 catalyst during the degradation of Indigo carmine. It was believed that the active sites of SnO_2 were blocked by the presence of magnetic particles. Nevertheless, in addition to the report, the majority of metal-based nanohybrids demonstrated exceptional dye degradation capacity and are summarized in Table 2.

Table 2 Example of applications on metal-based nanohybrids in the dye degradation process

Nanohybrids	Type of dye	Degradation performance	Remarks	Ref.
ZnO/TiO_2	Methylene blue	90%, visible light, 5 g/L catalyst, pH 7.75 min	• Band gap energy decreased • Increased lifetime of charge separation carriers	Upadhyay et al. [12]
Cu–ZnO/ TiO_2	Congo red (75 mg/L)	98%, direct sunlight, 0.5 g/L catalyst, 20 min	• Cu contributed to the oxygen vacancies in the TiO_2 and ZnO band gap • Cu functioned as the electron mediator • Improved band gap	Bahnemann et al. [13]
α-Fe_2O_3/ WO_3	Reactive blue 19 (10 mg/L)	100%, visible light, 360 min, pH 2, 1 g/L catalyst	• Easy separation • More reusable • Integrating magnetic materials was useful in obstructing recombination of electron–hole pairs	Delnavaz et al. [17]
Cu–Fe/TiO_2	Methyl orange (50 mg/L)	97.06%, visible light, 1 g/L catalyst, 60 min	• The Cu and Fe metal did not generate new channels for electron de-excitation but modified the intensity and reduced the charge recombination	Khan et al. [53]
Ag–ZnO	Rhodamine B (2×10^{-5})	99.6%, 240 min, pH 7, 0.5 g/L catalyst, visible light	• The electronic effect of Ag nanoparticles enhanced the photodegradation rate • Strong surface plasmonic resonance of electrons for fast charge carriers between Ag and the semiconductor enhance the photoactivity of the nanohybrids	Wang et al. [16]

2.2 Carbon–Metal Nanohybrids

The main attractive points of carbon-based materials are their high mechanical strength, electron mobility, large surface area, and good thermal conductivity [19]. For wastewater treatment applications, nanohybrids comprising metal and nanocarbon materials, such as graphene (GNP), graphene oxide (GO), reduced graphene oxide (rGO), carbon nanotubes (CNTs), and graphitic carbon nitrides (g-C_3N_3), have become the focus of research due to their known wastewater removal capacity.

CNTs are the cylinder-shaped allotropic structure of carbon and are commonly generated via the chemical vapour deposition technique [20]. The two variations of CNTs are single-walled nanotubes and multiwalled nanotubes. Their advantages which have astounding chemical, electronic, mechanical, and optical properties, CNTs have been widely exploited in photocatalyst development. With their high surface area nature, semiconductor materials also can be well distributed on CNTs surfaces, hence augmenting the accessibility of pollutants, including dye molecules towards the active sites [21]. In addition, CNTs can serve as excellent electron acceptors and donors, facilitating the electron migration rate [22]. A report found that the enhanced catalytic activity of a CNT-based nanohybrid can also be accredited to the interaction that occurred between the CNT and the respective semiconductor, which creates separation potential (Schottky Barrier) that trapped and prolonged the electron life [21, 23].

In contrast to CNTs whose inner walls are unable to interact with pollutants, GNP, on the other hand, exhibits a unique two-plane structure that can serve as an interface for pollutant interaction, making it superior compared to CNTs [24]. However, at the same time, GNP also shares a similar property with CNT, specifically as a good electrical conductor, making it an efficient electron conduit that may enhance charge–carrier separation [25]. However, graphene oxide (GO) is an insulator, unlike semimetallic GNP, as the hybridization of sp^3 in the GO carbon network disturbs the sp^2-conjugated carbon existing in the original GNP [26]. Although the original properties of GNP can be partially re-established via GO sheet reduction (forming rGO), the removal of the functional groups by the oxidation process during the synthesis will trigger the reaggregation of rGO into graphite [27]. The formation of GO and rGO from GNPs is briefly summarized in Fig. 4. Though, regardless of the form, both GNP-based nanocarbon (GO and rGO) exhibited superior degradation efficiencies of dye pollutants, mainly methylene blue, which was assumed to be ascribed to the high surface area factor that eases the adsorption of dye pollutants onto the nanohybrid surface [28–30]. In some cases, the heterojunction formation between the rGO and the integrated metal interfaces also profits the dye discoloration process [29].

Meanwhile, graphitic carbon nitride (g-C_3N_3) is a valuable member of the graphitic compound family, which was constructed from nitrogen-linked triazine units [31]. Containing carbon and nitrogen atoms attached in an alternating manner, the g-C_3N_3-based nanohybrids have also been correlated with decent photocatalytic functionality. Wu et al. [32] tested the removal of different kinds of dye pollutants,

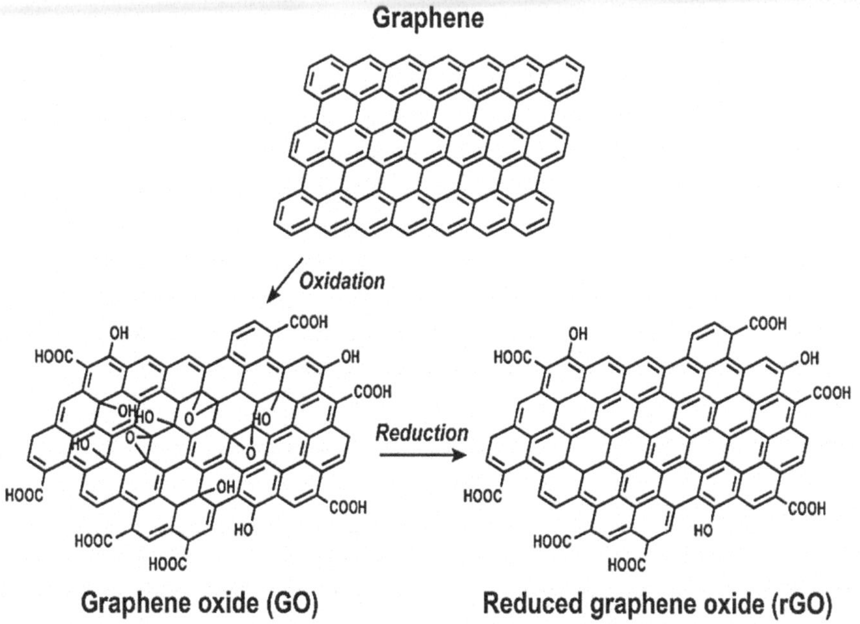

Fig. 4 The simplified formation of GO and rGO from GNP. Reproduced with permission from McCoy et al. [34]. Copyright Elsevier

such as methyl orange, Alizarin Yellow, and Orange G, under visible light irradiation in the presence of $ZnO/Fe_3O_4/g\text{-}C_3N_4$ nanohybrid. The attained high degradation rate of 83% (Alizarin Yellow) to 98% (Orange G) indicated the synergistic impact between each nanohybrid component, which increased the catalytic productivity. Some of the improvements observed were in terms of the redox capability, optical absorption, separation, and transportation behaviour of electrons [32, 33]. Table 3 lists the examples of reported applications of carbon-based nanohybrids for dye degradation.

2.3 Metal–Polymer Nanohybrids

In addition to metal-based and carbon-based nanohybrids, polymer-based nanohybrids have also been widely investigated mainly owing to their inimitable functionalities and flexible synthesis routes. The formulation of polymers can also be conducted based on their characteristics, such as surface morphology, architectural design, and porosity. Among the common approaches that have been established to construct polymer-based hybrids are electrospinning, sol–gel, phase inversion, and surface molecular imprinting [35].

Table 3 Examples of reported applications of carbon-based nanohybrids for dye degradation

Nanohybrids	Types of dye pollutants	Degradation performances	Remarks	Ref.
ZnO–GNP	Methylene blue (10 mg/L)	87%, 60 min, pH 5, 1.0 g/L catalyst, UV light	• Improved charge separation (due to high electron mobility in graphene) • The high surface area of graphene contributed to high pollutant adsorption	Mahmud et al. [25]
CdSe–rGO	Methylene blue (10 mg/L)	0.375 g/L, UV, 70%, 210 min	• rGO as an effective electron acceptor that improved photocatalytic activity • Improved separation of photogenerated charge carriers due to heterojunction formation between the rGO and CdSe interfaces	Patidar et al. [29]
ZnO–rGO	Methylene blue (10 mg/L)	Rate = 0.27 min^{-1}, UV, 0.3 g/L	• rGO tuned the band gap of the catalyst from UV to visible light region • Improved surface area	Prakash et al. [30]
GO–ZnO–Ag	Methylene blue (3.13×10^{-5} M)	100%, sunlight, 40 min, 0.5 g/L, pH 6.5	• The high surface area of GO enhanced the pollutant adsorption	Al-Rawashdeh et al. [28]
Sn–ZnO/GO	Methyl orange (50 mg/L)	96.2%, 2-h, 100 mg catalyst, visible light	• The resultant nanohybrid structure enhanced pollutant adsorption rate	Oyewo et al. [54]
BiSI/BiOI/CNT	Malachite green (10 mg/L)	93.1%, visible light, 210 min, 0.5 g/L	• The excellent conductivity of CNT eased the electron transfer from CB to the surface of CNT for reaction • CNT and BiSI/BiOI served as electron acceptors and donors, facilitating the electron's migration rate	Bargozideh et al. [22]

(continued)

Table 3 (continued)

Nanohybrids	Types of dye pollutants	Degradation performances	Remarks	Ref.
CNT/TiO_2	Methyl orange (15 mg/L)	87%, 180 min, 1 g/L, solar light	• CNT's presence narrowed the band gap • CNT increased the surface area and helped to disperse TiO_2 • Interaction between TiO_2 and CNT created separation potential (Schottky Barrier) • CNT served as a good electron acceptor and donor	Abega et al. [21]
Ag/$FeWO_4$/ g-C_3N_4	Rhodamine B (50 mg/L)	98%, pH 8, 0.5 g/L, 120 min, sunlight	• Efficient active species trapping and band energy potentials • g-C_3N_4 presence supported rapid electron dissociation and aided the migration rate of electrons and holes	Saher et al. [33]
ZnO/Fe_3O_4/ g-C_3N_4	Methyl orange (30 mg/L)	97.9%, 150 min, 0.2 g/L cat, visible light	• The heterojunction structure between g-C_3N_4 and ZnO improved light absorption properties	Wu et al. [32]
	Alizarin yellow (30 mg/L)	98.0%, 150 min, 0.2 g/L cat, visible light		
	Orange G (30 mg/L)	83.35%, 150 min, 0.2 g/L cat, visible light		
Sg-C_3N_4/ Bi_2O_3/Ag	Rhodamine blue dye (15 mg/L)	95%, sunlight, 90 min, 60 mg catalyst	• Improved stability and efficiency, • Better visible light-harvesting capacity • Efficient electron–hole pair separation • Heterojunction formation and synergetic interaction between nano-Ag, 2D Ag-C_3N_4 sheets, and irregular-shaped Bi_2O_3	Padhiari and Hota [55]

There are two classifications of polymers that have been explored in photocatalysis: conducting and coordination polymers. Conducting polymers are conjugated organic structures of alternating double and single bonds with respective unique optical and electrochemical properties [36]. Poly(3,4-ethylene dioxythiophene) (PEDOT), polyaniline (PANI), polyacetylene (PA), polyfuran (PF), and polypyrrole (PPy) are among the variations of conducting polymers. Even though the conducting polymer itself (without other material incorporation) has been linked with sufficient catalytic ability for the removal of dyes, the hybrid form of conducting polymer and metal oxides has demonstrated a further elevation in the photoreactivity, which caused by boosted photoresponse range and deferred charge carrier recombination effects [37–39].

The surface charge of the polymer itself also serves a vital role in attracting dye molecules towards the catalyst surfaces. In relation, Raveendran et al. [40] have proved that the PANI/TiO_2 hybrid presented a higher removal of Eosin yellow dye than TiO_2 alone. It was discovered that the incorporation of PANI, which was positively charged as a component of the nanohybrid system, was found to be beneficial in increasing the affinity of the anionic dye.

The PPy-based nanohybrid also displays profound photoreactivity as disclosed by research by Yang et al. [41]. They found that there was a synergistic impact from the interaction between PPy and Ag–TiO_2, based on the excellent visible-light catalytic activity, which was believed to be contributed by the improved electron–hole separation and photocurrent aspects. In agreement, Dimitrijevic et al. [42] also observed a similar scenario from the development of the nanohybrid TiO_2/PPy when enhanced absorptivity and catalytic activity were observed during the methylene blue degradation reaction under visible light irradiation.

Coordination polymers, on the other hand, are categorized as crystalline materials synthesized from structurally defined molecular building blocks connected by metal or nonmetal nodes [43]. The coordination polymers are further described in two main divisions, metal–organic frameworks (MOFs) and metal–organic complexes (MOCs), by which they exhibit distinguishable characteristics in terms of water solubility, porosity, and stability. Regardless of the differences, their functionality during a photocatalytic reaction is almost identical and is used as a chromophore for light absorption or as a reaction centre to initiate catalytic activity [44]. For example, a metal-integrated MOF hybrid system is correlated with desirable light-harvesting capability and thus has attracted much interest for photocatalytic purposes. Wang et al. [45] constructed a Ti-UiO-66 (Zr) MOF hybrid and proved that the enhancement of the photolysis of methylene blue dyes under sunlight was contributed by the harmonious properties of Ti as an electron donor that forms hetero-Zr oxo bridges, which raised the MOF optical properties. Figure 5 demonstrates how the optical properties of the nanohybrid were modified due to the interaction between the semiconductor material and MOF. Meanwhile, some examples of metal–polymer nanohybrids and their application in dye removal are presented in Table 4.

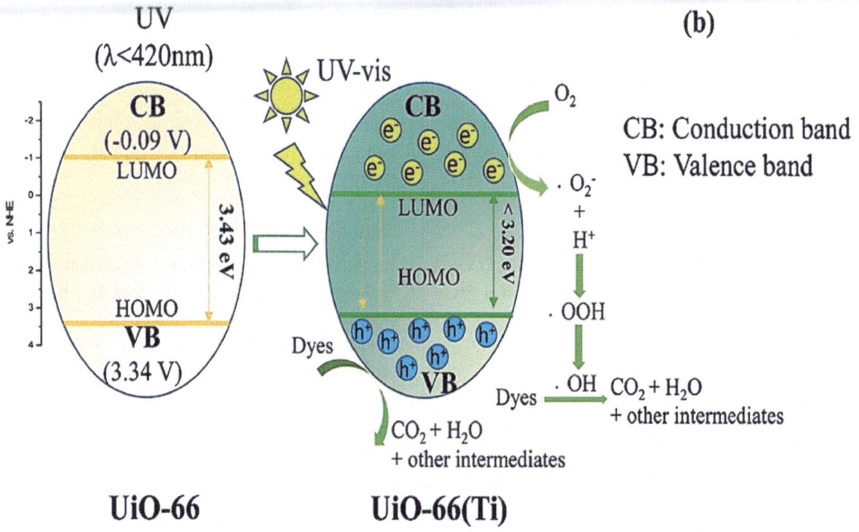

Fig. 5 The modification of the optical behaviour of MOF is due to the introduction of a semiconductor material. Reproduced with permission from Wang et al. [45]. Copyright Royal Society of Chemistry

2.4 Clay–Metal Nanohybrids

Clays, one of the most abundant earth resources, have important applications not only as materials for building construction and adsorbents but also as catalysts. Chemically, clay is composed of tetrahedra SiO_4 sheets that are linked with $Al(OH)_2$ octahedra [46]. The different types of clay are formed depending on the interaction between the silica sheets and the $Al(OH)_6$ plane. For instance, kaolinite is obtained if one silica sheet interacts with the $Al(OH)_2$ plane, forming two-tier $Al_2Si_2O_5(OH)_4$. Meanwhile, mica and smectite clay are composed of a three-tier sheet that occurs when an octahedral plane is sandwiched between two sheets of silica ($Al_2Si_4O_{10}(OH)_2$) [47]. These sheets are bonded together through a covalent bond between alumina and silica sheets, which then form a layer. The variations in how these layers are stacked result in many interesting properties and therefore extend the structural flexibility of clays.

For photocatalyst applications, there are two types of clay that have been explored: pillared clay and non-pillared clay. Sepiolite, bentonite, and montmorillonite (MT) are some of the clay examples that fall under the pillared clay group. The process of incorporating semiconducting materials into clay is usually accomplished by intercalating them within clay layers through the cation-exchange technique [46]. Interestingly, before the introduction of the semiconductor, the clay can be modified beforehand to increase the optical transparency to elevate the overall photoresponse aspect of the resultant nanohybrid. As an example, Yang et al. [48] introduced Mg^{2+}

Table 4 Examples of applications on polymer-based nanohybrids for dye degradation

Nanohybrid	Types of dye pollutants	Degradation performance	Remarks	Ref.
PANI/CdO	Malachite green (1.5 × 10^{-5} M)	99%, 0.4 mg/mL catalyst, sunlight, 4 h	• Interaction between PANI and CdO affected bond energy and electron density hence increasing photodegradation efficiency	Gülce et al. [56]
PANI/TiO_2	Eosin yellow	80.2%, UV light, 1.4 g/L, 8 h	• Enhanced light adsorption • Demonstrated greater degradation rate for anionic dye due to the PANI positive charge	Raveendran et al. [40]
PEDOT/ZnO	Methylene blue (1 × 10^{-5} M)	UV light (98.7%), sunlight (96.6%), 5 h, 0.4 mg/mL catalyst	• The ability of PEDOT to adsorb visible light enhanced the visible region degradation efficiency • Optimized charge separation and formation of oxyradicals	Abdiryim et al. [57]
TiO_2-PPy	Methylene blue (15 μM)	99%, visible light, 80 min, 40 mg catalyst, 20 min	• PPy helps to enhance the adsorption of pollutant for the degradation process	Dimitrijevic et al. [42]
Ag/Ag_3PO_4-MIL (Fe)	Ponceau BS	91.1%, visible light, 50 min, 1 g/L	• Improved charge separation • Reduced catalyst erosion leads to high stability and photoreactivity	Sofi and Majid [58]

and Fe^{2+} metal ions into vacant octahedral defect sites via a partial ion-exchange route before pillaring with TiO_2 nanoparticles (Fig. 6).

In terms of its photoactivity, several studies have emphasized the enhancement of the photocatalytic activity of clay-based nanohybrids for dye degradation applications. One of the cases includes the application of N- and S-doped TiO_2 that was intercalated into MT clay in the degradation process of 4BS dye under visible light irradiation [49].

Different from pillared clay, non-pillared clay does not allow the intercalation of foreign materials between the layers. Hence, the combination with the other materials is usually conducted by surface deposition. Among non-pillared clays, such as kaolinite, halloysite, saponite, and palygorskite, halloysite is drawing substantial attention owing to its tubular structure.

As a medium to hold semiconductor nanoparticles specifically through adhesion, non-pillared clay might as well prevent the nanoparticles from seeping into the liquid stream after a photodegradation process [50, 51]. This will benefit the separation process, as the catalyst can be retrieved much more easily from the water body by the sediment process and reused for the next cycle. UV-active semiconductors, such

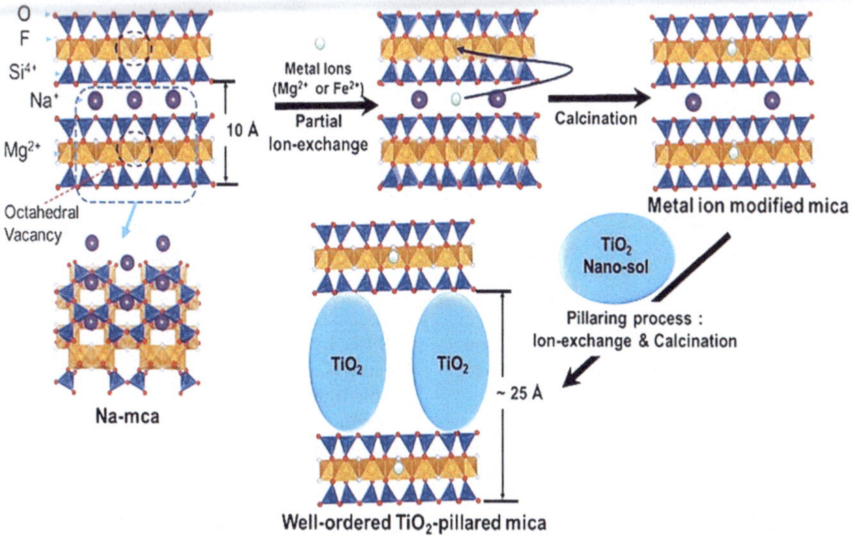

Fig. 6 Synthesis of TiO-pillared clay photocatalyst. Reproduced with permission from Yang et al. [48]. Copyright Royal Society of Chemistry

as TiO_2 and ZnO, have mostly been studied to develop nanohybrids with clay from this category, and most of the experimental results suggest the enhancement of photo reactivity during dye degradation. Since non-pillared clay offers less structural and molecular modification flexibility, the semiconductor material itself may be modified first before incorporation into the clay material. For instance, the immobilization of ZnO nanorods (NRs) on kaolin, by which their high photocatalytic performances were improved due to the high aspect ratio of the NR structure [52]. Table 5 presents the metal–clay nanohybrids and their applications in the dye removal process.

3 Conclusion and Future Prospect

In conclusion, the application of nanohybrid materials as photocatalyst in photocatalytic degradation of textile dyes unveiled good potential since the unique functionalities in the nanohybrid materials prove their high potential to be manipulated especially in modifications of their structure to adjust with the requirements on the materials for selected application. Photocatalytic degradation as one of the methods for wastewater treatment which required the suitable material as catalyst opens opportunities in exploring different types of nanohybrids such as metal-based, metal–carbon, metal–polymer and clay-based nanohybrids. The exploration in modification of nanohybrids is reliant on the target properties of the photocatalyst. In terms of application, nanohybrid materials have shown a promising result in degradation of various dyes. To realise a wider success of practical applications using nanohybrid

Table 5 The metal–clay nanohybrids and their applications in the dye removal process

Nanohybrids	Dye pollutant	Degradation efficiency	Remarks	Ref.
Bi_2O_3–Montmorillonite	Congo Red (80 mg/L)	86.4%, pH 9, 2 g/L catalyst, 30 min, visible light	• Recombination of photo electron–hole pair decreased • The combined effect of adsorption and photocatalytic reactivity	Patil et al. [59]
Fe_2O_3–Bentonite–TiO_2	Methylene blue (10 mg/L)	~100%, 1 g/L catalyst, visible light, 120 min	• High surface area aids the pollutant access on the photocatalyst surface • Extended optical adsorption to the visible region	Cao et al. [60]
Fe-ZnO/Montmorillonite	Malachite green (20 mg/L)	~100%, 0.5 g/L, 20 mg/L, UV, 20 min	–	Afifah et al. [61]
Fe_3O_4–SiO_2–$BiFeO_3$–sepiolite	A mixed solution of Rhodamine B, methyl orange, and methylene blue (20 mg/L for each dye)	~100%, visible light, 60 min, 1 g/L	• Sepiolite increased the adsorption capacity of pollutants • Catalyst structure consisted of the mesoporous channel, high surface area which responsible for increased light-harvesting capacity	Su et al. [62]
AgAgClTiO_2/Rectorite	Acid orange (50 mg/L)	100%, 20 min, visible light, 1.5 g/L	• Enhanced catalytic activity in the visible light region	Yang et al. [63]

materials in degradation of various dyes, following research directions are discovered: (1) The limitation of photocatalyst to be used only in UV spectrum region can be improved using metal-based modification by introduction of narrow band gap metal and/or noble metal because it potentially decreased the band gap to allow the application in visible spectrum by act as surface plasmon resonance as well as enhance the e^-–h^+ separation, (2) nanocarbon materials is a good option for nanohybrid materials if the aim of the photocatalyst properties is focus on the surface area, mechanical strength, and e^- mobility, (3) polymer-based nanohybrids is an option if the study requires flexible synthesis routes since the formulation of polymer can be conducted according to design characteristics, and (4) besides nanocarbon, clay which has been known rich with silica and alumina sources is also one of the potential materials to be used when the modification emphasis on the surface area, mechanical strength, and e^- mobility as well as increase the optical transparency to improve the photoresponse of nanohybrid. Other than the research directions listed above, there is a lot of room for research on the nanohybrid materials probably by combining more than one sources with the metal to study the potential changes in the catalytic properties. In addition, the nanohybrid materials applicability not only can focus on photocatalytic degradation of various dyes and other pollutants, but also in other applications such as energy storage, electronic materials, coating, sensing, catalysis, and biomaterials. Overall, the development of nanohybrid materials is not only expected to be valuable to photocatalytic researchers but also to the materials synthesis' community.

References

1. Crini G, Lichtfouse E (2019) Advantages and disadvantages of techniques used for wastewater treatment. Environ Chem Lett 17:145–155. https://doi.org/10.1007/s10311-018-0785-9
2. Hynes NRJ, Kumar JS, Kamyab H et al (2020) Modern enabling techniques and adsorbents based dye removal with sustainability concerns in textile industrial sector—a comprehensive review. J Clean Prod 272:122636. https://doi.org/10.1016/j.jclepro.2020.122636
3. Jaafar NF, Marfur NA (2021) Mesoporous titania nanoparticles as photocatalyst for degradation of 2-chlorophenol. J Chem Eng Ind Biotechnol 7:32–36. https://doi.org/10.15282/jceib.v7i1.6135
4. Machulek A, Oliveira, SC, Osugi ME et al (2013) Application of different advanced oxidation processes for the degradation of organic pollutants. Org Pollut Monit Risk Treat. https://doi.org/10.5772/53188
5. Liu LH, Métivier R, Wang S, Wang H (2012) Advanced nanohybrid materials: surface modification and applications. J Nanomater 2012. https://doi.org/10.1155/2012/536405
6. Anandakumar S, Savitha R (2011) Protective coatings: novel nanohybrid coatings for corrosion and fouling prevention. Green Corros Chem Eng Oppor Challenges 355–392. https://doi.org/10.1002/9783527641789.CH13
7. Deshmukh MA, Park SJ, Kang BC, Ha TJ (2020) Carbon nanohybrids for advanced electronic applications. Phys Status Solidi 217:2000199. https://doi.org/10.1002/PSSA.202000199
8. Mutalib AAA, Jaafar NF (2022) Potential of deep eutectic solvent in photocatalyst fabrication methods for water pollutant degradation: a review. J Environ Chem Eng 10:107422. https://doi.org/10.1016/J.JECE.2022.107422
9. Sun J, Ma Q, Xue D et al (2021) Polymer/inorganic nanohybrids: an attractive materials for analysis and sensing. TrAC Trends Anal Chem 140:116273. https://doi.org/10.1016/J.TRAC.2021.116273
10. Teranishi M, Naya SI, Tada H (2019) Nanohybrid catalysts for efficient synthesis of hydrogen peroxide at ambient temperature and pressure. J Phys Chem C 123:9831–9837. https://doi.org/10.1021/ACS.JPCC.9B00381/SUPPL_FILE/JP9B00381_SI_001.PDF
11. Wohlbier T (2021) Nanohybrids. Future materials for biomedical applications. Materials Research Forum LLC
12. Upadhyay GK, Rajput JK, Pathak TK et al (2019) Synthesis of ZnO:TiO2 nanocomposites for photocatalyst application in visible light. Vacuum 160:154–163. https://doi.org/10.1016/J.VACUUM.2018.11.026

13. Bahnemann W, Landge VK, Huang C-M et al (2022) Solar-energy-driven Cu-ZnO/TiO2 nanocomposite photocatalyst for the rapid degradation of Congo Red azo dye. Catalysts 12:605. https://doi.org/10.3390/CATAL12060605
14. Azam M, Adnan M, Julkapli NM et al (2016) Review on ZnO hybrid photocatalyst: impact on photocatalytic activities of water pollutant degradation. Rev Inorg Chem 2016:3–27. https://doi.org/10.1515/revic-2015-0015
15. Chakhtouna H, Benzeid H, Zari N et al (2021) Recent progress on Ag/TiO2 photocatalysts: photocatalytic and bactericidal behaviors. Environ Sci Pollut Res 28:44638–44666 (2833). https://doi.org/10.1007/S11356-021-14996-Y
16. Wang H, Liu X, Han S (2016a) The synthesis of a Ag–ZnO nanohybrid with plasmonic photocatalytic activity under visible-light irradiation: the relationship between tunable optical absorption, defect chemistry and photocatalytic activity. CrystEngComm 18:1933–1943. https://doi.org/10.1039/C5CE02381E
17. Delnavaz M, Farahbakhsh J, Mahdian SS (2021) Photodegradation of reactive blue 19 dye using magnetic nanophotocatalyst α-Fe2O3/WO3: a comparison study of α-Fe2O3/WO3 and WO3/NaOH. Water Sci Eng 14:119–128. https://doi.org/10.1016/J.WSE.2021.06.007
18. AbouSeada N, Ahmed MA, Elmahgary MG (2022) Synthesis and characterization of novel magnetic nanoparticles for photocatalytic degradation of indigo carmine dye. Mater Sci Energy Technol 5:116–124. https://doi.org/10.1016/J.MSET.2022.01.001
19. Gusain R, Kumar N, Ray SS (2020) Recent advances in carbon nanomaterial-based adsorbents for water purification. Coord Chem Rev 405:213111. https://doi.org/10.1016/J.CCR.2019.213111
20. Gomez-Gualdrón DA, Burgos JC, Yu J, Balbuena PB (2011) Carbon nanotubes: engineering biomedical applications. Prog Mol Biol Transl Sci 104:175–245. https://doi.org/10.1016/B978-0-12-416020-0.00005-X
21. Abega AV, Ngomo HM, Nongwe I et al (2019) Easy and convenient synthesis of CNT/TiO2 nanohybrid by in-surface oxidation of Ti3+ ions and application in the photocatalytic degradation of organic contaminants in water. Synth Met 251:1–14. https://doi.org/10.1016/J.SYNTHMET.2019.03.012
22. Bargozideh S, Tasviri M, Ghabraei M (2020) Effect of carbon nanotubes loading on the photocatalytic activity of BiSI/BiOI as a novel photocatalyst. Environ Sci Pollut Res Int 27:36754–36764. https://doi.org/10.1007/S11356-020-09759-0
23. Khan MR, Chuan TW, Yousuf A et al (2015) Schottky barrier and surface plasmonic resonance phenomena towards the photocatalytic reaction: study of their mechanisms to enhance photocatalytic activity. Catal Sci Technol 5:2522–2531. https://doi.org/10.1039/C4CY01545B
24. Zhang Y, Wu B, Xu H et al (2016a) Nanomaterials-enabled water and wastewater treatment. NanoImpact 3–4:22–39. https://doi.org/10.1016/J.IMPACT.2016.09.004
25. Mahmud RA, Shafawi AN, Ali KA et al (2020) Graphene nanoplatelets with low defect density as a synergetic adsorbent and electron sink for ZnO in the photocatalytic degradation of methylene blue under UV–vis irradiation. Mater Res Bull 128:110876. https://doi.org/10.1016/J.MATERRESBULL.2020.110876
26. Albiter E, Barrera-Andrade JM, Rojas-García E, Valenzuela MA (2019) Recent advances of nanocarbon-inorganic hybrids in photocatalysis. Nanocarbon its Compos Prep Prop Appl 521–588. https://doi.org/10.1016/B978-0-08-102509-3.00017-1
27. Katsukis G, Romero-Nieto C, Malig J et al (2012) Interfacing nanocarbons with organic and inorganic semiconductors: from nanocrystals/quantum dots to extended tetrathiafulvalenes. Langmuir 28:11662–11675. https://doi.org/10.1021/LA301152S/ASSET/IMAGES/MEDIUM/LA-2012-01152S_0001.GIF
28. Al-Rawashdeh NAF, Allabadi O, Aljarrah MT (2020) Photocatalytic activity of graphene oxide/zinc oxide nanocomposites with embedded metal nanoparticles for the degradation of organic dyes. ACS Omega 5:28046–28055. https://doi.org/10.1021/ACSOMEGA.0C03608/ASSET/IMAGES/LARGE/AO0C03608_0010.JPEG
29. Patidar D, Yadav A, Paul DR et al (2019) Nanohybrids cadmium selenide-reduced graphene oxide for improving photo-degradation of methylene blue. Phys E Low Dimensional Syst Nanostruct 114:113560. https://doi.org/10.1016/J.PHYSE.2019.113560

30. Prakash A, Singh S, Gupta J, Bahadur D (2019) Highly efficient zinc oxide-reduced graphene oxide nanohybrids for photocatalytic degradation of dye under dark and UV light. Mater Res Express 6. https://doi.org/10.1088/2053-1591/AB6256
31. Javaid A, Latif S, Imran M et al (2022) Nanohybrids-assisted photocatalytic removal of pharmaceutical pollutants to abate their toxicological effects—a review. Chemosphere 291:133056. https://doi.org/10.1016/j.chemosphere.2021.133056
32. Wu Z, Chen X, Liu X et al (2019) A ternary magnetic recyclable ZnO/Fe3O4/g-C3N4 composite photocatalyst for efficient photodegradation of monoazo dye. Nanoscale Res Lett 14:1–14. https://doi.org/10.1186/S11671-019-2974-2/FIGURES/14
33. Saher R, Hanif MA, Mansha A et al (2020) Sunlight-driven photocatalytic degradation of rhodamine B dye by Ag/FeWO4/g-C3N4 composites. Int J Environ Sci Technol 18:927–938. https://doi.org/10.1007/S13762-020-02888-6
34. McCoy TM, Turpin G, Teo BM, Tabor RF (2019) Graphene oxide: a surfactant or particle? Curr Opin Colloid Interface Sci 39:98–109. https://doi.org/10.1016/J.COCIS.2019.01.010
35. Shukla SK (2019) Polymer-based catalysts for water purification: fundamentals to applications. Polym Mater Clean Water 121–148. https://doi.org/10.1007/978-3-030-00743-0_6
36. Zhou Q, Shi G (2016) Conducting polymer-based catalysts. J Am Chem Soc 138:2868–2876. https://doi.org/10.1021/JACS.5B12474/ASSET/IMAGES/MEDIUM/JA-2015-12474U_0006.GIF
37. Floresyona D, Goubard F, Aubert PH et al (2017) Highly active poly(3-hexylthiophene) nanostructures for photocatalysis under solar light. Appl Catal B Environ 209:23–32. https://doi.org/10.1016/J.APCATB.2017.02.069
38. Ghosh S, Kouamé NA, Ramos L et al (2015) Conducting polymer nanostructures for photocatalysis under visible light. Nat Mater 14:505–511. https://doi.org/10.1038/NMAT4220
39. Yuan X, Floresyona D, Aubert PH et al (2019) Photocatalytic degradation of organic pollutant with polypyrrole nanostructures under UV and visible light. Appl Catal B Environ 242:284–292. https://doi.org/10.1016/J.APCATB.2018.10.002
40. Raveendran L, Veerapandy V, Mohan B et al (2019) Effect of morphology on the photocatalytic property of PANI/TiO2 on some synthetic dyes. Mater Res Express 6:125040. https://doi.org/10.1088/2053-1591/AB559E
41. Yang Y, Wen J, Wei J et al (2013) Polypyrrole-decorated Ag-TiO2 nanofibers exhibiting enhanced photocatalytic activity under visible-light illumination. ACS Appl Mater Interfaces 5:6201–6207. https://doi.org/10.1021/AM401167Y/SUPPL_FILE/AM401167Y_SI_001.PDF
42. Dimitrijevic NM, Tepavcevic S, Liu Y et al (2013) Nanostructured TiO2/polypyrrole for visible light photocatalysis. J Phys Chem C 117:15540–15544. https://doi.org/10.1021/JP405562B/SUPPL_FILE/JP405562B_SI_001.PDF
43. Zhang H, Li N, Tian C et al (2012) Unusual high thermal stability within a series of novel lanthanide TATB frameworks: synthesis, structure, and properties (TATB = 4,4',4"-s-Triazine-2,4,6-triyl-tribenzoate). Cryst Growth Des 12:670–678. https://doi.org/10.1021/CG2008655/SUPPL_FILE/CG2008655_SI_003.CIF
44. Zhang H, Liu G, Shi L et al (2016b) Engineering coordination polymers for photocatalysis. Nano Energy 22:149–168. https://doi.org/10.1016/J.NANOEN.2016.01.029
45. Wang A, Zhou Y, Wang Z et al (2016b) Titanium incorporated with UiO-66(Zr)-type metal-organic framework (MOF) for photocatalytic application. RSC Adv 6:3671–3679. https://doi.org/10.1039/C5RA24135A
46. Praneeth NVS, Paria S (2017) Clay-semiconductor nanocomposites for photocatalytic applications. In: Clay minerals: properties, occurrence and uses, pp 144–151
47. Dedzo GK, Detellier C (2016) Functional nanohybrid materials derived from kaolinite. Appl Clay Sci 130:33–39. https://doi.org/10.1016/J.CLAY.2016.01.010
48. Yang JH, Piao H, Vinu A et al (2015) TiO2-pillared clays with well-ordered porous structure and excellent photocatalytic activity. RSC Adv 5:8210–8215. https://doi.org/10.1039/c4ra12880j

49. Zhang G, Ding X, Hu Y et al (2008) Photocatalytic degradation of 4BS dye by N, S-codoped TiO2 pillared montmorillonite photocatalysts under visible-light irradiation. J Phys Chem C 112:17994–17997. https://doi.org/10.1021/JP803939Z
50. Abduljauwad SN, Ahmed HUR (2019) Enhancing cancer cell adhesion with clay nanoparticles for countering metastasis. Sci Rep 9. https://doi.org/10.1038/S41598-019-42498-Y
51. Aranda P, Ruiz-Hitzky E (2018) Immobilization of nanoparticles on fibrous clay surfaces: towards promising nanoplatforms for advanced functional applications. Chem Rec 18:1125–1137. https://doi.org/10.1002/TCR.201700113
52. Shirzad-Siboni M, Farrokhi M, Darvishi Cheshmeh Soltani R et al (2014) Photocatalytic Reduction of hexavalent chromium over ZnO nanorods immobilized on kaolin. Ind Eng Chem Res 53:1079–1087. https://doi.org/10.1021/IE4032583
53. Khan MS, García MF, Javed M, Kubacka A, Caudillo-Flores U, Halim SA, Khan A, Al-Harrasi A, Riaz N (2021) Synthesis, characterization, and photocatalytic, bactericidal, and molecular docking analysis of Cu–Fe/TiO_2 photocatalysts: influence of metallic impurities and calcination temperature on charge recombination. ACS Omega. https://doi.org/10.1021/acsomega.1c03102
54. Oyewo OA, Ramaila S, Mavuru L, Onwudiwe DC (2022) Enhanced photocatalytic degradation of methyl orange using Sn-ZnO/GO nanocomposite. J Photochem Photobio. https://doi.org/10.1016/j.jpap.2022.100131
55. Padhiari S, Hota G (2019) A Ag nanoparticle functionalized Sg-C_3N_4/$Bi_2O_3$2D nanohybrid: a promising visible light harnessing photocatalyst towards degradation of rhodamine B and tetracycline. Nanoscale Adv. https://doi.org/10.1039/C9NA00172G
56. Gülce H, Eskizeybek V, Haspulat B, Sarı F, Gülce A, Avcı A (2013) Preparation of a new polyaniline/CdO nanocomposite and investigation of its photocatalytic activity: comparative study under UV light and natural sunlight irradiation. Ind Eng Chem Res. https://doi.org/10.1021/ie401389e
57. Abdiryim T, Ali A, Jamal R, Osman Y, Zhang Y (2014) A facile solid-state heating method for preparation of poly (3, 4-ethelenedioxythiophene)/ZnO nanocomposite and photocatalytic activity. Nanoscale Res Lett. https://doi.org/10.1186/1556-276X-9-89
58. Sofi FA, Majid K (2018) Enhancement of the photocatalytic performance and thermal stability of an iron based metal–organic-framework functionalised by Ag/Ag_3PO_4. Mater Chem Front. https://doi.org/10.1039/C8QM00051D
59. Patil SP, Shrivastava VS, Sonawane GH, Sonawane SH (2015) Synthesis of novel Bi_2O_3–montmorillonite nanocomposite with enhanced photocatalytic performance in dye degradation. J Environ Chem Eng. https://doi.org/10.1016/j.jece.2015.09.005
60. Cao X, Luo S, Liu C, Chen J (2017) Synthesis of bentonite-supported Fe_2O_3-doped TiO_2 superstructures for highly promoted photocatalytic activity and recyclability. Adv Powder Technol. https://doi.org/10.1016/j.apt.2017.01.003
61. Afifah N, Djaja NF, Saleh R (2015) Photocatalytic activity of Fe-doped ZnO/montmorillonite nanocomposite for degradation of malachite green. Mater. Sci Forum. https://doi.org/10.4028/www.scientific.net/MSF.827.19
62. Su G, Liu L, Kuang Q, Liu X, Dong W, Niu M, Tang A, Xue J (2021) Enhanced visible-light photocatalytic activity and recyclability of magnetic core-shell Fe_3O_4@SiO_2@$BiFeO_3$–sepiolite microspheres for organic pollutants degradation. J Molecular Liq. https://doi.org/10.1016/j.molliq.2021.116566
63. Yang Y, Zhang G, Xu W (2012) Facile synthesis and photocatalytic properties of AgAgClTiO_2/rectorite composite. J Colloid Interface Sci. https://doi.org/10.1016/j.jcis.2012.03.003

Montmorillonite (MMt) Clay-Based Hybrid Materials for Textile Dyes' Removal

Babak Jaleh, Ensiye Shabanlou, Atefeh Nasri, and Mahtab Eslamipanah

Abstract Montmorillonite (MMt) clay mineral is one of the most available and affordable nature-inspired materials in the worldwide which is a member of the layered silicate family. In addition to the reasonably priced and abundance of MMt in the nature, its mechanical and thermal stability and fire resistance features make it as a high promising compound for large-scale utilization. These features together with porous structure and great surface area can improve the performance of composite materials which are synthesized for the different environmental applications such as removal of dyes by the photocatalytic, adsorption, and degradation methods. To overcome the challenge of water remediation and removal of dyes, the design and fabrication of (nano)composites involving the various effective compounds, namely hybrid materials with symmetrical two-/three-dimensional structures, single molecular weight, adjustable size, presence of many cavities, high biocompatibility, and low toxicity, have been more interested in research works. Modified MMt or MMt-based (nano)composites provide more active sites, higher surface energy, high porosity, and stronger physical properties for the removal of dye applications. This chapter deals with the identification and qualification of MMt material modified with various polymers and monomers as hybrid/hydrogel (nano)composites for cationic and anionic textile dyes' removal approaches.

Keywords Montmorillonite (MMt) · Removal · Adsorption · Wastewater · Cationic dyes · Anionic dyes

1 Introduction

Today's purification of water from different pollutants, namely organic dyes and toxic heavy metal ions, has drawn special attention. Therefore, a wide range of various approaches have been used for removal of contaminants from wastewater. The main component of water polluting is waste products fabricated in chemicals,

B. Jaleh (✉) · E. Shabanlou · A. Nasri · M. Eslamipanah
Department of Physics, Faculty of Science, Bu-Ali Sina University, 65174 Hamedan, Iran
e-mail: jaleh@basu.ac.ir

© The Author(s), under exclusive license to Springer Nature Singapore Pte Ltd. 2023
A. Ahmad et al. (eds.), *Nanohybrid Materials for Treatment of Textiles Dyes*,
Smart Nanomaterials Technology, https://doi.org/10.1007/978-981-99-3901-5_11

textiles, metallurgical, and mining industries [1, 2]. Textile wastes are one of the most hazardous pollutant types in water pollution which have been treated by conventional purification methods utilizing different materials. Recently, nanotechnology has provided a basis for generating desirable nanohybrid materials with significant physical and chemical properties to treat textile pollutants of water [2, 3]. With decades of research efforts, nanohybrid materials based on montmorillonite (MMt) clay have attained as one the most potential materials toward contaminants' removal fields of desirable properties, namely, naturally occurring, low-toxicity, widespread availability, and reasonably priced [4, 5]. Common methods which have been employed for the wastewater treatment using MMt clay mineral include degradation [6], adsorption [5], photo degradation [7], oxidation [8], and membrane filtration [9]. For this regard, numerous adsorbents, membrane, and catalysts have been successfully derived as modified MMt clays such as acid-washed MMt, organic MMt (OMMt) clays, inorganic pillared MMt, thermal-treated MMt [8]. Adsorption was introduced as one of the efficient treatment techniques for heavy metals and dyes' removal owing to good properties such as environmentally friendly, simple design, flexible, and cost-effective operation [4, 8, 10]. MMt adsorbents'-based hybrid materials have been extensively applied and represented high adsorption performance for various molecules and ions because of its physicochemical properties such as chemical and mechanical stabilities, high specific surface area, surface hydrophilicity, surface electronegativity, extraordinary, cation exchange capacity (CEC), and cation exchange selectivity [8, 11, 12]. Clay minerals (such as MMt) are mainly classified in hydrated phyllosilicates with particle size of less than 2 μm in soils, deposits, rocks and sediments [13, 14]. They have layered structure made up of one or two silica tetrahedral (T) sheets attached with an aluminum octahedral (O) sheet, which are mainly composed in 1:1 (T-O) and 2:1 (T-O-T) bonds [15]. The schematic of the top and side views of 2:1 layer structure of clay minerals is indicated in Fig. 1a [16]. MMt is chemically formulated in $(Na, Ca)_{0.33}(Al, Mg)_2Si_4O_{10}(OH)_2{\cdot}nH_2O$ and is composed in the 2:1 layered structure such that an alumina octahedral sheet sandwiches between two silica tetrahedral sheets as seen in Fig. 1b [8, 17]. The silica tetrahedral and aluminum octahedral sheets are generated of $Si_2O_6(OH)_4$ and $Al_2(OH)_6$ units, respectively. Because of the fact that lower-valent cations such as Al^{3+} and Mg^{2+} substitute for Si^{4+} and Al^{3+} in tetrahedral and octahedral sheets, respectively, the MMt layer structure has negative charge of about ~0.2–0.6, and thus the tendency to attract cationic species such as cationic dyes increases for balance the negative charge [17, 18].

In the adsorption method, the process starts by mixing an optimum amount of an appropriate adsorbent with a steady solution volume comprising the dye pollutants and it shakes in a water using bath thermostat shaker machine at definite time intervals. Then, the pollutant species (adsorbate) transfer to the adsorbent surface through the solution and bound by physical and chemical forces [5]. The solution is then centrifuged and its concentration is studied. On the other word, the adsorption mechanism of dyes follows by some steps including dye particles' migration from the solution to the adsorbent surface, their diffusion via the boundary layer, and subsequently the adsorption of dye particles on the adsorbent active sites [19,

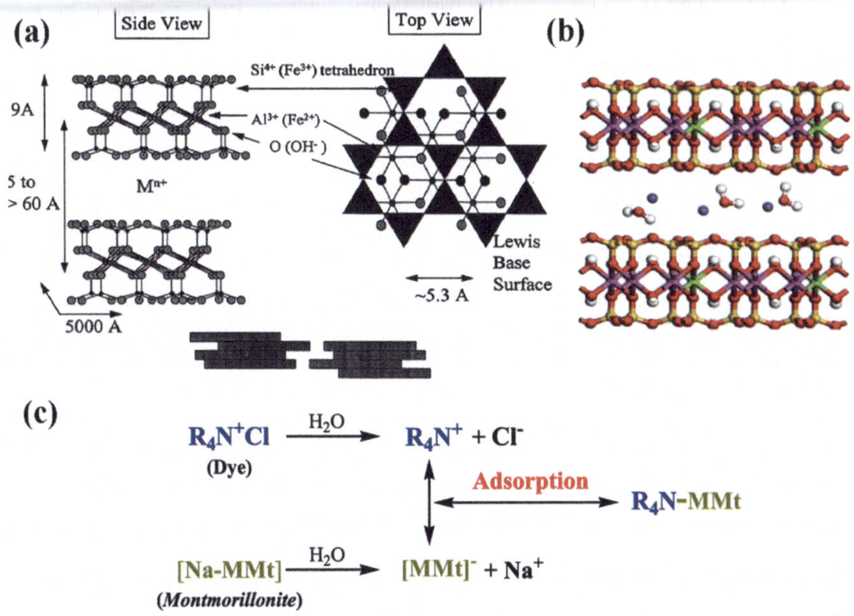

Fig. 1 **a** Side and top views' illustration of 2:1 layer clay minerals. Reproduced with permission from Ref. [16]. Copyright 2003, Elsevier. **b** Schematic of MMt geometry structure. Reproduced with permission from Ref. [8]. Copyright 2020, Elsevier. **c** Presentation of cationic dyes' adsorption on the surface of MMt material adsorbents. Reproduced with permission from Ref. [4]. Copyright 2022, Elsevier

20]. During adsorption process in aqueous solution, MMt liberates exchangeable cations, as example, H^+, Ca^{2+}, Na^+, resulting in the negatively charged surface of the MMt attracted the cationic dye particles [21]. The schematic representation in Fig. 1c indicates the cationic dye adsorption on the surface of MMt [4]. Although the negative charge on the surface of MMt displays well interaction with cationic dyes, but it also can gain the ability to be a suitable adsorbent for anionic dyes by the MMt modification processes [22, 23]. For example, in the case of the cationic dye (Basic Yellow 28), the adsorption is performed owing to interaction between negatively charged groups (Al–O^- and Si–O^-) and positively charged dye (N^+). However, in the case of the anionic dye (Acid Brown 75), it is occurred by negatively charged sulfonate SO_3 groups' interaction with the dye particles and the positive charges' interfoliar space of the adsorbent materials [24].

The adsorption of dyes on MMt adsorbent is performed through electrostatic interaction. The adsorption capacity of the MMt-based hybrids generally depends on adsorption parameters of adsorbent dose, the pH of solution, temperature, contact time, and initial dye concentration [4, 25]. In addition, it is largely affected by the pore size and the surface area of the adsorbent. In batch adsorption technique, after optimization adsorption parameters, the removal of dyes is often evaluated by a

UV–Vis spectrophotometer with the specific wavelength of dye absorbency [9]. The capacity of adsorption (q_e, mg/g) and its removal efficiency (R%) are determined as following equations:

$$q_e = \frac{(C_i - C_e)}{m} \times v, \tag{1}$$

$$R(\%) = \frac{(C_i - C_e)}{C_i} \times 100, \tag{2}$$

where C_i, C_e, V, and m are assigned to the initial concentration (mg/L), the equilibrium concentration (mg/L), the dye solutions volume (L), and the adsorbent dosage (g), respectively [26–28].

Theoretically mechanism of batch adsorption is investigated by isotherms and kinetics models [4, 29]. Functional equilibrium adsorption distribution is described by isotherm model with adsorbate concentration at stable temperature of solution. The adsorbent capacity for adsorption can also be quantitatively described by this model.

In isotherm study, the most widely used models are Langmuir equation,

$$\frac{C_e}{q_e} = \frac{1}{bq_m} + \frac{1}{q_m}c_e, \tag{3}$$

and Freundlich equations,

$$q_e = k_f C_e^n, \tag{4}$$

or

$$\log q_e = \log k_f + n \log C_e, \tag{5}$$

where q_e and C_e are attributed the equilibrium concentrations of the adsorbate in solid and liquid phases, respectively [4]). The Freundlich coefficients of n and K_f are the adsorption intensity and adsorption capacity, respectively. q_m and b as the Langmuir coefficients show the Langmuir monolayer adsorption capacity and its equilibrium constant, respectively [4, 30]. On the other hand, adsorption kinetic models determine using Lagergren pseudo-first-order and pseudo second-order models such that pseudo-first-order model describes the adsorption rate in the liquid phase by following equation [4, 29, 31]:

$$\frac{dq_t}{dt} = k_1(q_e - q_t) \tag{6}$$

and second-order kinetics is used when the first order becomes untenable to explain experimental adsorption results and it is determined by [4, 32]:

$$\frac{dq_t}{dt} = k_2(q_e - q_t)^2, \quad (7)$$

where q_e and q_t are assigned to mass density of adsorbed materials at stable condition and anytime (t), k_1 and k_2 (min^{-1}) are the coefficients of first-order and second-order kinetics adsorption rate, respectively.

In addition, MMt acted as a filler for synthesis of membranes as an absorbent, due to its desired properties comprising high hydrophilicity, good mechanical strength, and great thermal stability, indicating development of membrane-based separation of dyes [9, 33]. On the other, MMt has also been employed as a catalyst for degradation process with desired catalytic performance. In this method, the catalytic activity is evaluated by the dye decomposition in aqueous solution in the presence of reducing agent such as $NaBH_4$. The MMt-based compounds as a catalyst can facilitate the dye decomposition thorough the reduction process [34, 35]. This chapter attempts to focus on the new insights of dye removal techniques with the low-cost, reusable, novel, and efficient hybrid materials based on the MMt for treating the wastewater. In the next section, textile dye types are introduced, and then, latest works focusing on the application of MMt-based materials to improve the performance of removal technologies will be reviewed. The effect of modified MMt including hydrogel-based MMt, OMMt clays modified with polymers/monomers, inorganic pillared MMt, MMt-based chitosan (CS), or other chemical materials on the efficiency of the dye removal was investigated which have been applied in degradation, membrane, and especially in adsorption methods. Their dye removal properties have been clearly studied and development perspectives of new researches have also been suggested for the textile dye removal.

2 Textile Dye Removal with MMt-Based Materials

The wide chemical classification was introduced for textile dyes based on their functional groups such as phthalocyanine, anthraquinone, nitroso, sulfur, nitro, azo, and indigo via various physical and chemical properties [36]. They are often classified as three types of non-ionic (disperse), cationic (basic), and anionic (reactive, acid, and direct) dyes [37]. They can absorb light at a specific wavelength in visible region, caused their different colors [36]. Textile wastewater pollutants, especially water-soluble dyes with complex chemical structures, need severe treatment before discharging into aquatic environment. MMt materials can be appropriate choice for dye removal in adsorption, degradation, and membrane filtration processes [4]. Accordingly, the hybrid materials based on MMt as the dye remover and also the dye removal efficiency of various MMt-based compounds reported on research works are discussed in the following sections.

2.1 Cationic Dye Removal with MMt-Based Materials

Cationic (basic) dyes as a water-soluble dye contain the positive charge located on amine groups as the integral part of their formula. The cations of these dyes become colored in solutions and they are tended to attract matters with a negative charge through electrostatic interactions [38]. Hence, the negative surface charge of MMt can lead to electrostatically attracting of the cationic dyes, namely toluidine blue (TB), methylene blue (MB), crystal violet (CV), malachite green (MG), Safranin (ST), azur B (AB), etc. [8]. The structure of some cationic dyes is indicated in Fig. 2a–d [27].

In recent years, MMt hydrogels have been extensively applied for removal of textile dye from wastewater because of their special structure, chemical, and physical properties [39]. In 2014, the polysaccharide/clay polyelectrolyte hydrogels were prepared by acrylamide/sodium methacrylate (AAm/SMA), carrageenan (CR), and MMt with a radical solution polymerization method and they were introduced as a novel sorbent for cationic dyes [40]. The hydrophilic group involving chemicals, namely CR, SMA, and MMt, successfully was incorporated in acrylamide hydrogels. These hydrogel systems indicated high cationic ST dye absorbency. The charged properties of sulfonate groups and the incorporated MMt and CR into the copolymer network were the effective factors for increasing of the adsorption percentages to high values. It varied from 77.03% to 90.89% for AAm/SMA/CR/MMt, AAm/SMA/MMt, and AAm/SMA/CR hydrogel hybrid composites. Using ultrasound radical

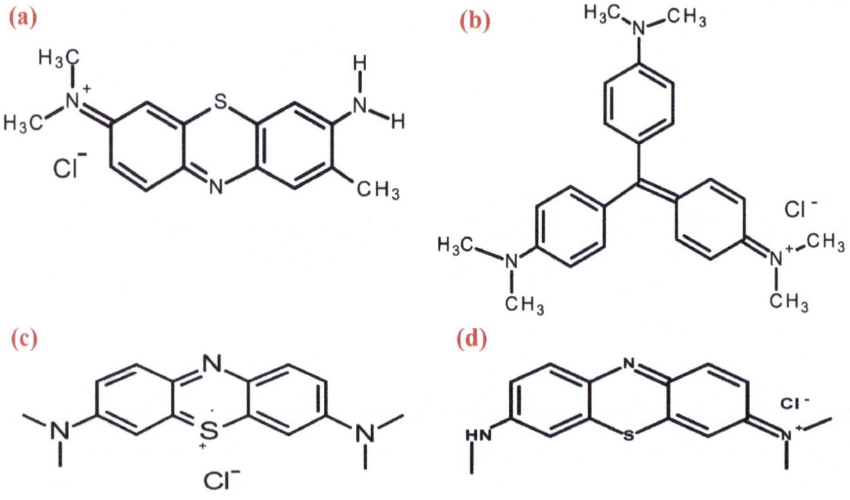

Fig. 2 Cationic dyes' structures of **a** TB, **b** CV, **c** MB, and **d** AB used for adsorption test. Reproduced with permission from Ref. [27]. Copyright 2020, Elsevier

polymerization, another hydrogel-based MMt was synthesized by grafting acrylamide and *N*-isopropyl acrylamide onto lignin hybriding with MMt (lignin-g-p(AM-*co*-NIPAM)/MMt) as seen in Fig. 3a [41]. After MMt hybriding, this hybrid hydrogel showed thin pore walls with high thermal stability and good mechanical strength, resulting in high efficiency for MB removal at neutral pH with adsorption capacity value of 9646.92 mg/g. The dye adsorption was extremely depended on temperature and pH. As seen in Fig. 3b, c, the special advantage of lignin-g-p(AM-*co*-NIPAM)/MMt adsorbent attributed to easily regeneration with excellent adsorption and also ability of utilization even for five cycles of adsorption–desorption process. Therefore, this hybrid hydrogel assisted by MMt was introduced as an efficient and environmental friendly adsorbent for dye adsorption.

For further investigation, cellulose nanofibrils (CNFs), alkali lignin (AL), and MMt were incorporated to polyvinyl alcohol (PVA) to synthesize composite hydrogels with different viscoelasticities for adsorption of MB dye [42]. Dispersing MMt in the polymer matrix acted as a multi-functional cross-linking agent of the PVA/CNF network and enhanced the adsorption performance. When the MMt was 40 wt%, the hydrogel efficiency of dye removal was increased over 98% toward MB adsorption, whereas maximum adsorption capacity and adsorption equilibrium time were 67.2 mg/g and 360 min, respectively. At about pH of 10.5, the adsorbents outperformed for dye removal due to electrostatic repulsion of same ionic charges and attendance of H^+ and MB cations, in the acidic condition. It is worth mentioning that an extremely high contents of MMt decreased the MB adsorption rate. It was incorporated in polymer hydrogels, decreasing of the swelling ratio of the hydrogel.

In 2016, the MMt nanoparticles were purposed to fabricate cationic dye adsorbent. In this regard, the MMt nanoparticles were doped in bridged silsesquioxane (SSO) hybrid films [43]. Different dyes, namely eosin y, rose bengal, fluorescein, and MB, were used in adsorption tests. The SSO-MMt film illustrated more efficiently for adsorption of cationic MB dye with initial concentration of 2.6 mg/L. The advantage of this hybrid film is the ability of nanoparticles dispersion in the SSO as a stable and chemically inert matrix which provide easy separation of absorbed dye from the medium without any requirement of complicated processes such as induced precipitation, centrifugation, and decantation. MMt nanoparticles are modified with special part of the critical micelle concentration (CMC) of cetyltrimethylammonium bromide (CTAB) and it was employed for synthesis of MMt/xanthan gum (XG)-sodium alginate (Alg) hybrid hydrogel [44]. The negative part of this nanocomposite could adsorb cationic MG dye. Different prepared adsorbents of MMt/XG-Alg (with any modify process on MMt nanoparticles), MMt(0.5CMC)/XG-Alg, MMt(1CMC)/XG-Alg, and MMt(2CMC)/XG-Alg were studied for adsorption test. Results showed that the adsorption capacity was reduced by increasing dosage of adsorbent and improved with increasing of temperature and initial pH values. The specific surface area of MMt(0.5CMC)/XG-Alg was increased up to 72.48 m^2/g, which was significantly more than that of MMt/XG-Alg (6.38 m^2/g). For this sample, maximum adsorption capacity was 769.23 mg/g in the Langmuir isotherm model under optimal contact time of 240 min, temperature of 318 K, pH of 6, and 0.02 g/50 mL of the adsorbent mass. The presence of the functional groups of carboxylates and hydroxyl

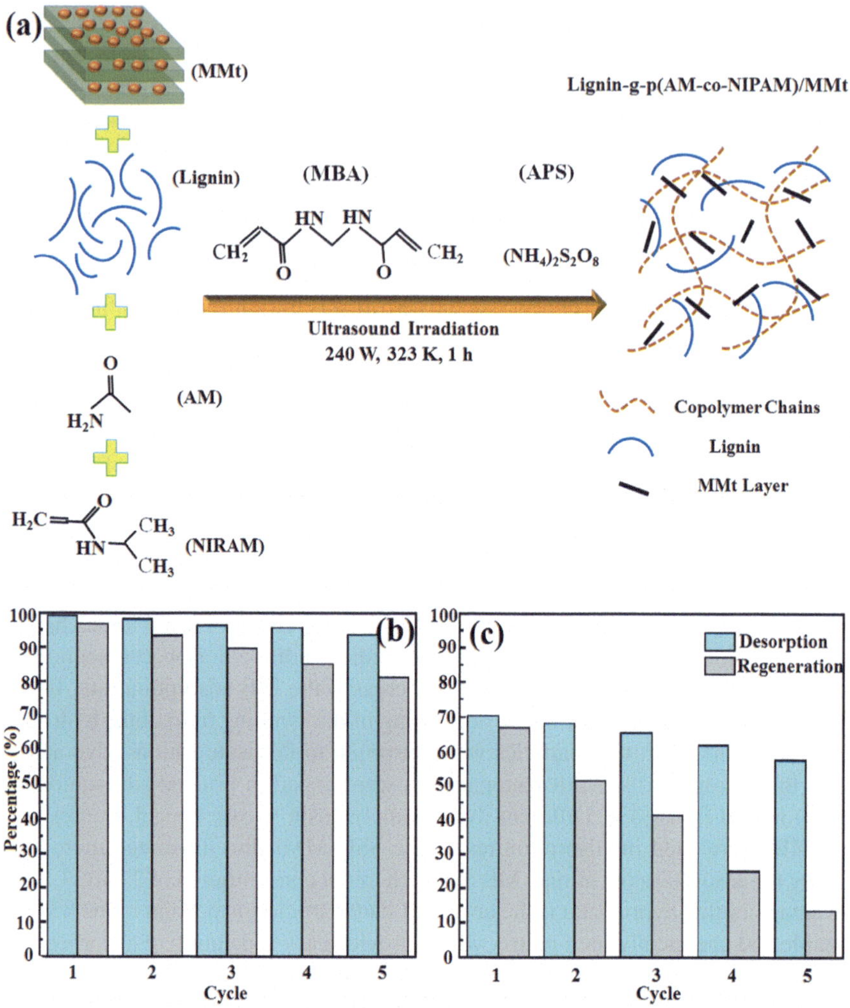

Fig. 3 **a** Schematic representation of lignin-g-p(AM-*co*-NIPAM)/MMt preparation. Regeneration and desorption efficiencies of MB adsorption on AN-L/M with **b** acid and **c** thermal desorption methods. Reproduced with permission from Ref. [41]. Copyright 2017, Elsevier

groups in the nanocomposite structure was the main factors in the improvement of adsorption process. The structural control of organo-clays and their surface charge after the adsorption of dyes can be acted as the critical factor on the dye adsorption process. In 2014, organic MMt (OMMt) clay was utilized for synthesis of styrene (St)/acrylic acid (AA)/OMMt nanocomposites with different wt % of OMMt, via cationic exchange reaction method [45]. These hybrid nanocomposites were employed for removing both anionic and cationic dyes and they presented best performance for the adsorption of cationic Maxilon C.I. Basic dye than that of anionic one (Acid Green B).

It was probably because of the cationic groups of Maxilon C.I. Basic dye interacted with carboxyl groups of nanocomposites. The batch experimental data of Maxilon C.I. Basic dye adsorption had the best fit with Freundlich isotherm model, indicating the best performance as regards capacity of 165.84 mg/g at pH of 6 and temperature of 30 °C using St/AA/OMMT (3 wt% OMMT) adsorbent. In order to produce desired absorber for cationic dyes, the MMt modification has been applied using different polymers such as polysaccharide, polyoxyethylene (PEO), poly(2-methacryloyloxy ethyl) trimethyl and PVA. Compared to the pure polymers, its nanocomposites with MMt indicated so good improvement in ionic conductivity [46], storage modulus [47], gas permeability [48], and thermal expansion coefficients [39, 49]. Recently, polymer/MMt hybrid (nano)composites have widely used for wastewater treatment because of their combined structure and excellent chemical and physical properties of organic and inorganic compounds [39]. In one report, sodium MMt/ureasil-PEO (U-PEO) nanocomposite was fabricated by sol–gel reaction of a U-PEO hybrid. Due to the combined properties of U-PEO hybrid and MMt, this adsorber was capable to recover easily after adsorption process [50]. The adsorption of the cationic dye was performed at the MMt nanoparticle surfaces which dispersed in the matrix of nanocomposite with negatively charged. The data of experimental adsorption were confirmed by the Langmuir isotherm and pseudo-first-order kinetic models. The UV–Vis absorption measurements for this nanocomposite indicated about 69% removal of the MB dye. The adsorption capacity was increased by increasing of the contact time and initial concentration of MB dye and the highest value was reported to be of 37–38 mg/g. In one more report, poly(2-methacryloyloxy ethyl) trimethyl was suggested for preparation of a hybrid adsorbent by grafting karaya gum onto poly(2-methacryloyloxy ethyl) trimethyl in the presence of MMt (KG-g-PMETAC/MMt) using microwave irradiation method [27]. The incorporation of MMt particles could improve the graft copolymer gel adsorption capacities by formation of hydrogen bond between molecules of cationic dye and hydrated MMt, resulting in the higher dyes' adsorption on the KG-g-PMETAC/MMt surface. This hybrid nanocomposite was applied for adsorption cationic dyes of AB, CV, TB, and MB with significant adsorption capacity of 128.78, 137.77, 149.64, and 155.85 mg/g, respectively, through hydrogen bonding and electrostatic interactions. Contact time effect on dyes adsorption is given in Fig. 4a. The dye adsorption was fitted with Freundlich isotherm and pseudo-first-order kinetic models, which demonstrated a two-step adsorption process. According to Fig. 4b, in the first cycle, the adsorption was nearly 70%, but it was decreased in second run, because during the first desorption cycle, some reactive sites are occupied with cationic groups, which blocked the adsorption sites, resulting in lower adsorption.

In another creative research, a reinforced hybrid porous beads (RHPBs) were fabricated using sodium alginate (SA), aluminum-pillared MMt (Al-PILMMt), PVA, and calcium carbonate ($CaCO_3$) by an extrusion-dripping technique [51]. These RHPBs were structured in a pores' network and a number of functional groups were investigated in different percentages of PVA, $CaCO_3$, and Al-PILMMT toward MG adsorption from aqueous solution. The results confirmed that the two important factors (i.e., stiffness and porosity) have great influence on the adsorption performance. RHPBs

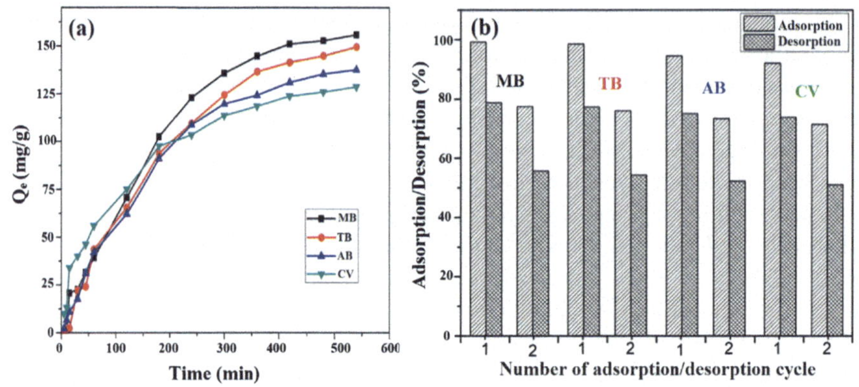

Fig. 4 **a** Effect of contact time on adsorption of different cationic dyes on KG-g-PMETAC/MMT, and **b** two cycles of cationic dyes' adsorption–desorption process. Reproduced with permission from Ref. [27]. Copyright 2020, Elsevier

with the higher content of PVA illustrated greater stiffness. The removal of MG was about 8% with SA beads, which significantly was improved to 27% and 46% with RHPBs and HPBs in the absence of PVA, respectively. The MG adsorption isotherm onto beads was matched with Freundlich model, which emphasized on the heterogeneous nature of the MG multilayer adsorption on the RHPB. A novel MMt/graphene oxide (MMt/GO) was introduced as multi-functional 2D/2D membranes for MB filtration [9]. It was prepared by restacking exfoliated MMt and GO nanosheets through their individual dispersion and employing simple solution casting method with a small amount linkage of PVA. MMt/GO membranes were designed by four ratios of 1/4, 2/3, 3/2, and 4/1, named as M1G4, M2G3, M3G2, and M4G1, respectively. MMt and GO nanopowders are individually known as the popular adsorbents for dyes. They difficulty separated from the water, which leaded to the second pollution. Mechanism for MB filtration is illustrated in Fig. 5a. The M2/G3 cross-section morphology is given in Fig. 5b, in which it can be seen the stacking of multilayered with internal porosity. The MB dye separation from wastewater was obtained by MMt/GO (M2/G3) hybrid membrane with maximum removal ratio and removal capacity of 87% and 304 mg/g, respectively, through filtration manner, higher than that of MMt membrane (280 mg/g) (seen Fig. 5c).

Photocatalytic oxidation in the presence of a catalyst is a significant technique to treat sewage, in which the catalyst causes to produce electron–hole pairs by absorption of light and degrades the pollutants in the wastewater. Hybrid nanofluids are well-known as new catalysts with improved photocatalytic activity. Hence, aqueous-based 10% w/w copper (Cu)–nickel (Ni)-doped titania (TiO_2)-supported MMt-k 10(MMt-k 10) clay-based hybrid nanofluid has been shown remarkable photocatalytic degradation of Rhodamine B [7]. The maximal degradation was achieved in the presence of Cu–Ni (75:25) metal doping at 2 vol.% nanofluid concentration due to improving high absorption and surface area. In addition, less reduction in catalytic activity was observed for MMt-supported Cu–Ni (75:25)–TiO_2/water compared with

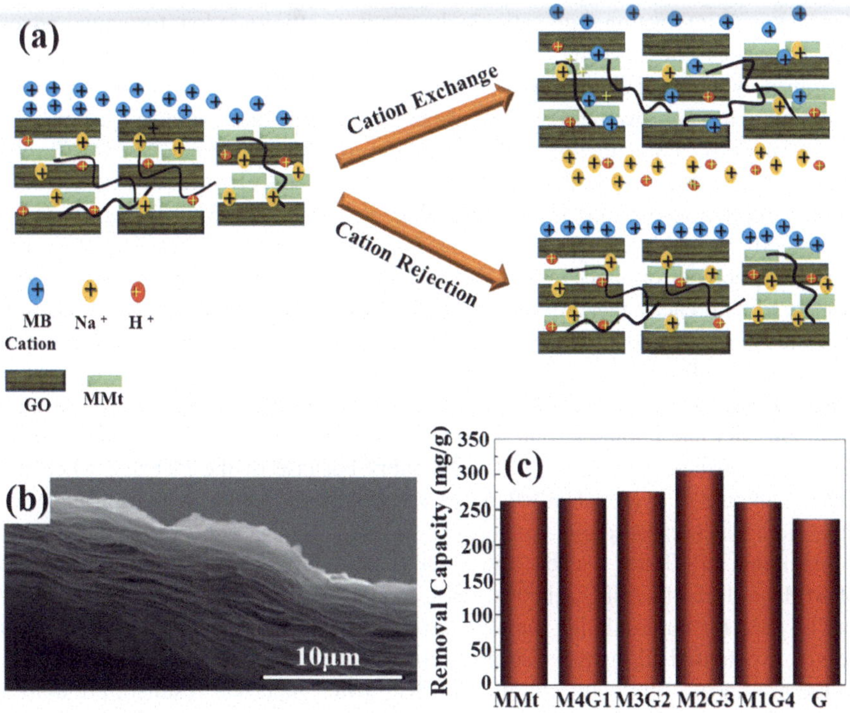

Fig. 5 **a** Schematic representation of the MB filtration mechanism, **b** SEM cross-sections image of M2G3 membrane, and **c** removal capacity of the different MMt/Go hybrid membranes. Reproduced with permission from Refs. [9]. Copyright 2020, Elsevier

Cu–Ni (75:25)–TiO_2/water. Therefore, the presence of MMt led to increasing the stability of Cu–Ni (75:25)–TiO_2 activity.

2.2 *Anionic Dye Removal with MMt-Based Materials*

Anionic (reactive, acid, and direct) dyes in the textile industries are mostly used because of their simple dyeing processes. Among them, azo dyes as reactive dyes have the most usage of textile production dyestuff (about 60–70%), so they are considered to be the main components in the textile industry [36, 52, 53]. The characterization of this kind of dyes is azo bonds (N=N) and its highly colored pollutant is because of associated chromophores [52, 53]. Another anionic dyes are acid dye whose molecules are composed of acid functions (SO_3H, COOH, and OH). The total usage of acid dyes is about 30–40% in the textile industry [36]. Therefore, the removal of anionic dyes is of great importance, which has been done with various methods and materials. Among them, the MMt clay material has been highly regarded. It displays well interaction with cationic dyes, but its affinity is low for anionic dyes [22, 23]. Thus, the MMt modification has been using different materials such as various polymers and chitosan (CS) for changing physical and chemical properties, resulting in the improvement of MMt usage for the removal of anionic species [23].

Assisted by a conducting polymer, starch-MMt/polyaniline (St-MMt/PANI) nanocomposite was prepared as an efficient and eco-friendly hybrid adsorbent toward the reactive dye adsorption [52]. All these three materials (St, PANI, and MMt) due to their good properties gained more interest in adsorption applications. Therefore, the dye adsorption performance of St, MMt, St-MMt, and St-MMt/PANI adsorbents was compared with each other. Among them, the St-MMt/PANI hybrid nanocomposite indicated maximum adsorption capacity because the polyaniline-containing nitrogen atoms grew well over the St-MMt nanocomposite surface, which created a rough surface for dye adsorption. The St-MMt/PANI nanocomposite removed reactive navy blue (SP BR) dye completely in a so short contact time (150 s), which could be remained constant even up to 200 s. Probably because of the fact that when the defined amount of dye is adsorbed on the nanocomposite for a certain time and after that, there was no absorption anymore.

Although MMT has a stable and nontoxic structure, keeping the fine particles of MMt is difficult in the aqueous phase [54]. Therefore, bulk polymer of PANI can be used for immobilizing the alginate (Alg)-MMt nanocomposite to remain its complete potential during the adsorption process of Reactive Orange 13 (RO13) dye. In the presence of high concentration of dye, lower yield was resulted, possibility due to satiation of the adsorptive sites on the nanocomposite surface. The high mass of sorbent could provide more adsorption sites with high surface area, leading to increase dye removal. pH was another affecting factor during removal process. At low pH, a positively charged was created on the catalyst surface with increasing concentration of H^+, resulted in a remarkable interaction between RO13 and catalyst. Therefore, an excellent dye removal was achieved in low pH. According to kinetics studies, the maximum dye removal obtained as high value of 99.9% in the optimal adsorption condition (64.4 mg/L of initial dye concentration, 72.39 mg of sorbent dosage, and pH of 2.59). In another report, by incorporation of MMt particles in PANI/polyvinyl alcohol (PANI/PVAL) aerogel matrix, an efficient hybrid aerogel was produced and enhanced the adsorption performance toward both Reactive Black 5 (RB5) and methyl orange (MO) dyes (seen Fig. 6a) [55]. This inorganic–organic compound aerogel exhibited favorable adsorption of anionic dye and improved adsorption capacity for cationic dyes compared to PANI/PVAL aerogel. For RB5 dye removal, the best rate was achieved at low pH and it was reduced by increasing pH, while for MO, the maximal removal rate was in pH range of 4–11. For cationic dye, a high adsorption capacity was obtained at pH of 11.5. These results can be explained by measuring zeta potential which showed positively charge for aerogels at $pH < 10$. Therefore, MO and RB5 can remove through electrostatic interaction with protonated NH^+ from PANI and hydrogen bonding with OH-Si of MMt, while electrostatic interaction of NH^+ and Si–O was major adsorption route in composite aerogel. As shown in Fig. 6b, the RB5 adsorption capacity slightly decreased by increasing the MMt content. However, the removal of all three dyes was enhanced when the aerogel dosage was increased.

KSF-based hybrid material can be introduced as highly efficient adsorbent for anionic dyes' removal compared with conventional adsorbents. By the reaction between a commercial clay (KSF-MMt) and protonated dodecylamine in an acidic medium, the dodecylammonium-functionalized MMt (KSF-DP) was successfully synthesized as a new hybrid adsorbent of Remazol brilliant blue R (RBBR) from aqueous solutions [53]. The KSF-DP had the maximum absorbance of RBBR dye at pH of 4–6, resulting electrostatic attraction of protonated surface with anionic dye. Besides, high temperature causes to increase mobility of dye molecules, leading to influence and availability to active sites of porous structure of KSF-DP. The dye

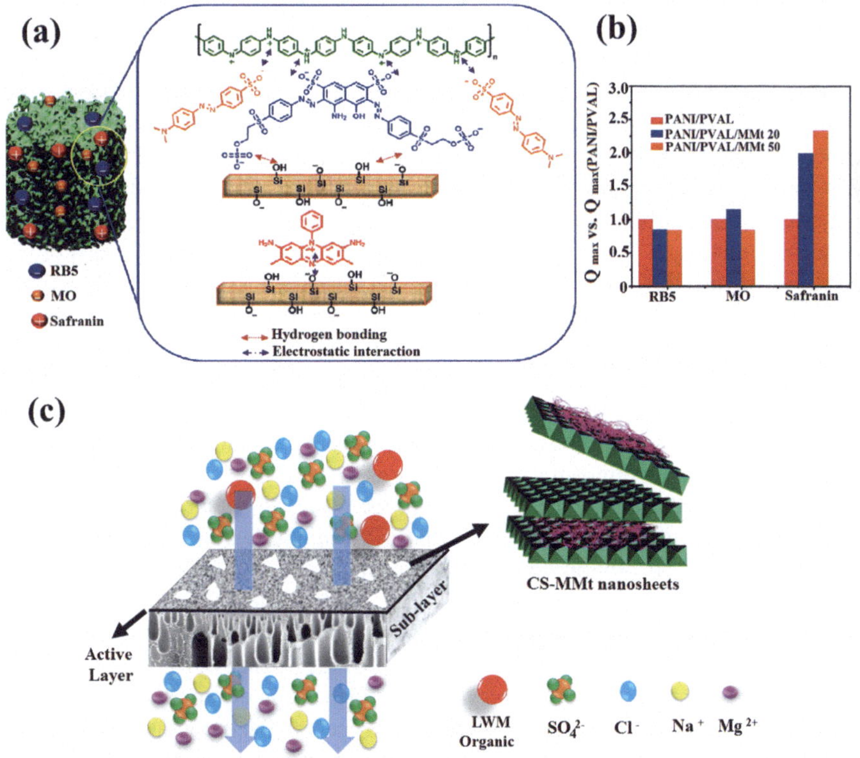

Fig. 6 **a** Schematic of adsorption mechanism of RB5, MO, and safranin onto PANI/PVAL/MMt-50 aerogel, **b** the maximum adsorption capacity of RB5, MO, and safranin on PANI/PVAL, PANI/PVAL/MMt-20, and PANI/PVAL/MMt-50 aerogels. Reproduced with permission from Ref. [55]. Copyright 2022, Elsevier. **c** Schematic representation of rejection LWM dyes (reactive red 49, reactive black 5) using NFM membrane. Reproduced with permission from Ref. [33]. Copyright 2015, Elsevier

adsorption was firstly increased by increasing adsorbent dosage owing to the adsorbent sites' presence and then slightly reduced. The Langmuir and pseudo-second-order models demonstrated better fitted with the experimental adsorption data with adsorption capacity of 38.99 mg/g at 323 K.

CS has been widely applied as the most abundant natural aminopolysaccharides for water purification because of its specific properties, namely good biodegradability, antimicrobial activity, and biocompatibility [39]. CS-MMt nanosheets were successfully employed for preparing a novel "loose" nanofiltration membrane (NFM) using phase inversion technique, which was a good selection for rejection of Reactive Black 5 and Reactive Red 49 (Fig. 6c) [33]. It has good properties such as hydrophilic nature, enhanced thermal stability, and high mechanical strength with finger-like structure. Under low pressure (0.4 MPa), NFM indicated rejection of Reactive Black 5 and Reactive Red 49 dyes with good overall flux of 65 kg/m^2 h, which was higher than conventional nanofiltration membranes such as the mixed matrix membranes (MMMs).

The attempts have been performed for enhancing the properties of MMt. The synthesis of CS/KSF-MMt (KSF-CTS) beads was successfully performed by sodium tripolyphosphate (TPP) as cross-linking agent in various MMt ratios of 1, 5, 15, and 25% w/w [22]. These green biocomposite beads were applied as adsorbents for removing Remazol blue RN. The Remazol blue RN structure is given in Fig. 7a. A positively charged was appeared on the bead surface due to amino groups of CS and KSF-MMt at low pH. Hence, a strong electrostatic interaction was prevailing for Remazol blue RN adsorption at pH 3. As shown in Fig. 7b, the gelatinous beads were formed due to electrostatic interaction of protonated amino groups of CS with negative TTP. The KSF-CTS beads' composite with 25% of MMt represented the best performance toward Remazol blue RN adsorption by the 0.5 kg/L of dye concentration and contact time of 480 min. Followed by pseudo-second-order kinetics evaluation, the maximum removal dye capacity was 310 mg/g, which was significantly independent of MMt clay mineral content in the beads. In another report, a hybrid membrane was prepared using CS/PVA and MMt in physical blending process for Reactive Red dye adsorption [26]. Because of the strong polymers interaction, the more stable hybrid membranes were created. This hybrid nanocomposite exhibited rapid response to liquid absorption and the hydrophilicity feature of nanocomposite increased through increasing PVA value. In the presence of MMt, the swelling behavior was significantly reduced owing to an increase in elastic force. According to batch adsorption analysis and isotherms experimental, these polymer membranes had the ability to stabilize organic dyes with a spontaneous process and were as L2 type, where the adsorption of the dye on the membrane surface is extendedly formed as a monolayer. The membrane composition of CS/PVA/MMt indicated the maximum adsorption capacity of 81.34 mg/g in Langmuir model compared to the neat ones (CS/PVA, pure PVA, and CS) due to its amorphous structure with more porous.

Organophilic montmorillonite (OMMt) can be prepared via cationic exchange between Na-MMt and vinyl benzyl triphenyl phosphonium chloride in an aqueous

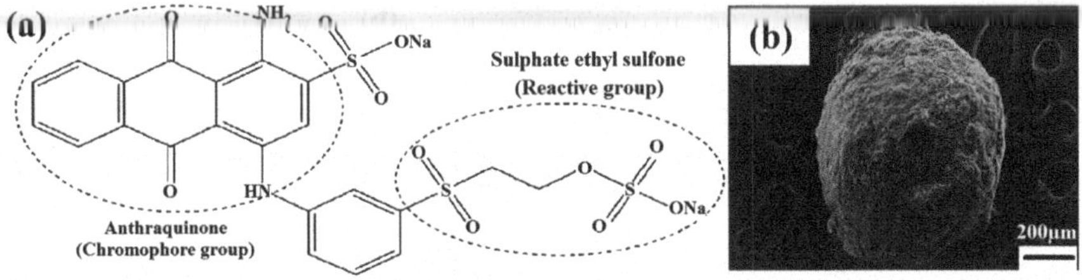

Fig. 7 **a** Illustration of RB dye structure, and **b** SEM image of KSF-CTS-25%. Reproduced with permission from Ref. [22]. Copyright 2022, Elsevier

solution. The presence of high content of OMMt leads to reduce water uptake, possibility due to reduce intervolume space and form high density of cross-linking sites. The composition of OMMt with CS and 4-acryloyl morpholine (ACMO) through γ-ray irradiation polymerization produced a hybrid nanocomposite for adsorption acid dyes of fast yellow G and green B [39]. CS/ACMO/OMMt nanocomposites expressed higher partition ratio and sorption capacities (74.9 mg/g) for Acid Green B than that of acid fast yellow G (54/7 mg/g), because Acid Green B contains more sulfonate group and charged chlorine ions in its anionic character, causing more interaction with polar cationic CS/ACMO/OMMt. It is worth mentioning that the acid dyes' adsorption increased with increasing OMMt weight from 0 to 3 wt% due to improve polymeric network. Although OMMt structure has been shown considerable adsorption ability owing to its high surface area, the modification of natural MMt or OMMt can enhance the negatively charged dyes. It was reported that the surfactant-modified OMMt by cetyltrimethylammonium bromide (CTAB) is effective for improving adsorption of hydrolyzed reactive dyes (Remazol Black B, Reactive Red 195, and Reactive Blue 19) [56]. According to kinetics adsorption and characterization studies, it was found that bilayer OMMt intercalation triggered expansion of large-enough intergallery spaces, which make the accommodation of large dye molecules possible and also disordering of crystalline direction of the clay platelets was occurred leading to facilitate access to dye. On the other hand, positive surface charge of this bilayer OMMt enhanced interaction with the anionic dyes. This new modified OMMt (with CTAB loading of 3.0 CEC) had the fastest and largest Remazol Black B uptake (77 $\pm$ 2 mg/g), whereas it was slow and poor with organo-clays with monolayer structure. Non-covalent, dye-mediated interactivities of OMMt with a CS-based polyelectrolyte (HTCC) can employ for highly impressive and rapid adsorption of anionic dyes (Reactive Blue 19, Reactive Red 195, and Reactive Black 5) from aqueous solutions [57]. The Black dye adsorption onto the OMMt occurred due to electrostatic and partition mechanisms. For HTCC, at low concentration, electrostatic repulsion of the positive charges along the polymer backbone provides an expanding out of the polymer chain, resulting in great availability to the binding sites. However, the high dye removal (96%) was achieved by composition of OMMt and HTCC. The adsorption mechanism of anionic dyes on cationic HTCC/OMMt hybrid is represented in Fig. 8a. For all three dyes, the dye diffusion and its adsorption onto surface of particle were more rapid compared with dye intercalation

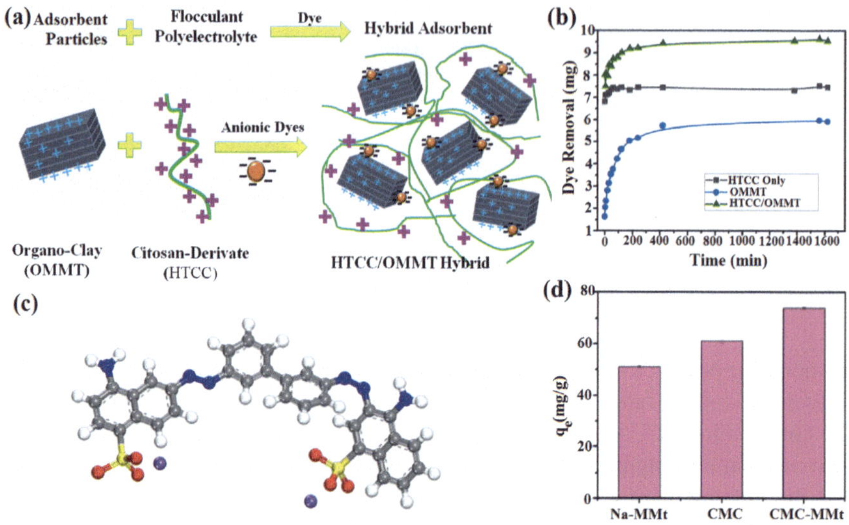

Fig. 8 **a** Schematic illustration of anionic dye removal by cationic adsorbent (HTCC/OMMt hybrid), **b** removal of reactive black 5 dye over time. Reproduced with permission from Ref. [57]. Copyright 2019, American Chemical Society. **c** The structure of CR, and **d** the adsorption capacities of CR of CMC-MMT, CMC, and Na-MMT. Reproduced with permission from Ref. [58]. Copyright 2020, Elsevier

and intergallery adsorption, suggesting limitation of the dye diffusion into the interlamellar clay spaces. By the intraparticle diffusion assessment of kinetic model, the uptake capacity of red and blue dyes mixture was achieved about 50% more than individual dyes and also indicated nearly 100% removal of different anionic dyes (Fig. 8b).

Inserting carboxymethyl CS (CMC) material (its geometry structure is shown in Fig. 8c) in MMt structure caused the interlayer space of MMT became larger resulting more surface active sites and enhanced the Congo Red (CR) adsorption [58]. CMC material containing chains of the COOH, OH, and NH_2 groups, which acted as coordination and interaction active sites, could improve the performance of CMC-MMt organic–inorganic composite as CR adsorbent. Isotherm and kinetic studies exhibited with the Langmuir and pseudo-second-order models determining the well adsorption of CR on CMC-MMt. According to Fig. 8d, removal capacity of CR by CMC-MMt adsorbent was as high as 82.4 mg/g, which was much more than that of CMC or MMt adsorbents. The speed of adsorption was fast at beginning of adsorption process and then rapidly became equilibrium. Also, it was resulted that the pH is the effective factor on the adsorption of CR by CMC-MMt hybrid adsorbent than temperature and reaction time. At low pH, the electrostatic interaction of negatively charged CR anions with positively charged adsorption sites was enhanced, increasing CR removal.

A hybrid MMt-Rich/CS (MCC) composite has been exhibited highly adsorption of Reactive Red 120 (RR120) dye [59]. The CS and MCC had positive zeta potential which led to electrostatic interaction of dye with adsorbents. However, the dye removal was completed in the presence of MCC compared with CS, which was due to combination of electrostatic attraction and hydrophobic interactions. At pH of 5.5 and dye concentration of 100 mg/L, the removal of RR120 was faster and had excellent removal rate with contact time of 90 min. In the presence of high concentration of dye, more dye molecules were transmitted to adsorbent surface, increasing adsorption of dye. Interestingly, this composite possessed complete dye adsorption in both acidic and basic solutions but with different required contact time intervals. However, the discrepancy in the magnitude of the zeta potential of the MCC and dye in acidic condition was higher than basic condition, causing to better dye removal at low pH. According to Koble–Corrigan isotherm model, it was expressed that the MCC hybrid material had various active sites in its heterogeneous surface, which could be a suitable adsorbent for the anionic dye molecules owing to more hydrophobic interaction and synergistic electrostatic attraction between the MMC composite and anionic dyes. Bentonite is well-known clay consisting of the smectite phase, namely MMt. This natural clay can be employed in biopolymer nanocomposites. For example, the composition of bentonite with CS through microwave irradiation indicated great adsorption capability for Reactive Violet 5R removal. The analyses showed that CS incorporated well in interlayer space of MMt and a flake-like morphology was preserved [60]. Some different adsorption factors (adsorbent dosage, pH, contact time, and initial dye concentration) were studied and resulted that the optimal condition of Reactive Violet 5R removal was at contact time of 20 min and a pH of 1. At low pH, the electrostatic attraction of nanocomposite with $-SO_3^-$ of dye was increased, leading to a high rate of dye removal. The adsorption capacity was increased by increasing dosage of nanocomposite, which was dependent on the higher content of CS. The data had better fit with the Langmuir isotherm and pseudo-second-order kinetic models via good adsorption capacity value of 282.01 mg/g.

Dye adsorption from aqueous media has been improved in the presence of the modified adsorbents. Cationic exchange or grafting is well-known approach for functionalization of clays. The grafted MMt using 3-aminopropyl-trietoxisilane (APTES) improved hydrophilicity of MMt and introduced new organic–inorganic hybrid (MMt-APTES) for Acid Red 1 (AR-1) and Acid Green 25 (AG-25) removal from aqueous solutions [23]. The presence of APTES leads more interaction between active silanol (Si–OH) and hydroxyl groups which occurred in the broken edges and interlayer spaces of MMt surface. In addition, the grafting APTES increased the positive charge of MMt. Since the AR-1 and AG-25 have negatively charged sites ($R-SO_3^-$), the dye adsorption occurred due to electrostatic attraction between these negatively charged and positively charged sites of adsorbent ($-NH_3^+$). Liu isotherm model has the best fit with experimental data, and it represents the maximum adsorption capacity as 364.1 and 397 mg/g for AR-1 and AG-25, respectively. Moreover, this hybrid structure suggested outstanding performance toward removal of synthetic dye effluents. As indicated in Figs. 9a and b, the adsorption percentage was obtained about 98 and 96% for effluent A and effluent B, respectively. In addition, the chemical modification of two bentonite samples (Bent1 and Bent2) using the CEC process by reaction with ethylenediamine (en) and ethylene sulfide (S) showed significant improvement in the adsorption capability of bentonites [28]. The modified bentonites as hydrophobic material were used for anionic azo-dye (Remazol blue RN) removal from aqueous solution, as depicted in Fig. 9c. Adsorption was evaluated at optimized condition of adsorbent dosage of 100 mg, pH value of 6.8, reaction time of 240 min. Equilibrium isotherms indicated that the maximum Remazol blue RN removal by solid modified adsorbent was ~40 mg/g, which was better than the pristine bentonites. The modification of MMT through acid leaching process is a facile method to produce active silica from clays. This process causes removal of octahedral sheets in clays, remaining a silicon tetrahedral sheet with a high SiO_2 value. Compared to direct synthesis of silica, acid leaching provides the silica materials with many pores and active surface groups. However, this process releases some acidic wastewater including meal ions of Fe, Mg, and Al which can form nanomaterials due to react with OH^-, CO_3^{2-}, and SiO_3^{2-} in solution. Hence, Dong et al. introduced an new 2D-layered double hydroxide (LDH) material using the metal ions of Fe(III), Fe(II), Al(III), and Mg(II) from acidic leachates of clay minerals (i.e., MMt, nontronite (NNt), and palygorskite (Pal)) [11]. These prepared MMt-LDH, NNt-LDH, and Pal-LDH adsorbents could adsorb CR dye with the adsorption capacities of 134.56 ± 3.49 mg/g, 123.20 ± 4.11 mg/g, and 127.16 ± 1.49 mg/g, respectively. After it, in order to make novel recyclable adsorbents, the calcination treatment of LDH-adsorbed Cr was done and new layered double oxides (LDOs) was generated with a 2D-nanolayered shape and more active adsorption sites for adsorbing dye pollutants from wastewater so that after reuse 3 times, a good adsorption capacity of CR was still indicated. It is worth mentioning that the Mg value has important role to specifying adsorption capability of LDO-based adsorbents. Moreover, the anion exchange, the electrostatic attraction, and chemisorption of Mg, Al, and C species on the adsorbents with dye enhance the adsorption of CR.

It has been reported that the immobilization of magnetite (M) NPs (with 2:1 molar ratio of Fe^{3+}/Fe^{2+}) into the poly (amidoamine) dendrimer/MMt (PAMAM/MMt) hybrids was improved the removal performance of xylenol orange (XO) and malachite green (MG) dyes [35]. The prepared M@PAMAM/MMt nanocomposite demonstrated an excellent degradation efficiency for catalytic oxidative of xylenol orange (XO) (about 85% mineralization) with the catalyst dosage of 1 g/L and the initial dyes' contents of 50 mg/L, while it showed lower activity toward removal of MG. However, the catalytic activity for removal of MG and XO was enhanced with increasing M content. It was due to the interaction of M NPs with PAMAM amidic groups, where generation of reducing environment was occurred by the terminal amine groups.

The MMt nanosheets have been introduced as the suitable matrix for MoS_2 for MO degradation in the presence of $NaBH_4$ reductant [34]. The hybridized MoS_2 and MMt via in situ hydrothermal method provided a hydrophilic surface, increasing dispersibility in the solution. The degradation mechanism of MoS_2/MMt hybrid catalysis is schematically exhibited in Fig. 10 such that by adding the pure $NaBH_4$

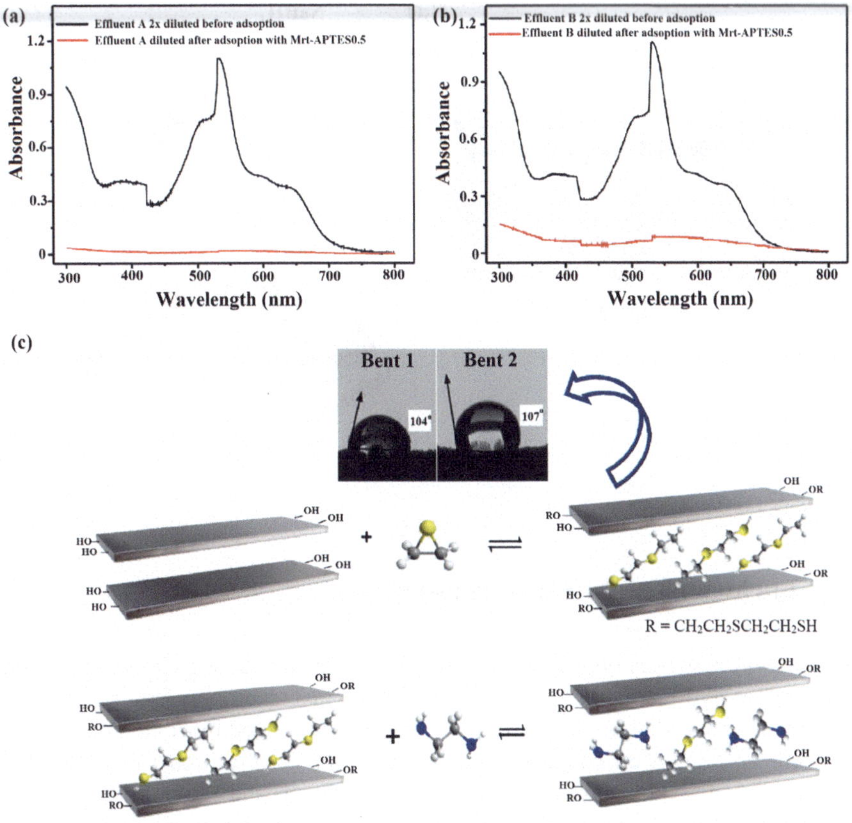

Fig. 9 Adsorption of effluent A **a** and effluent B **b** dyes. Reproduced with permission from Ref. [23]. Copyright 2017, Elsevier. And the presentation of organic–inorganic hybrid structures (**c**). Reproduced with permission from Ref. [28]. Copyright 2019, Elsevier

in to MO solution, the molecules $NaBH_4$ interacted with water molecules, and after hydrolyzing, a few hydrogen atoms were released. When the quinone imine and azo groups of MO meet the hydrogen atoms, the decomposition of MO starts. The MoS_2/MMt catalyst with more active sites and smaller particles was added into the solution for better hydrolysis of $NaBH_4$ and greatly improved the MO decomposition. The presence of MoS_2 enhances the hydrolysis of $NaBH_4$ and hydrogen atoms are diffusion in solution, improving MO degradation. The catalytic decomposition of MO was obtained 98% for MoS_2/MMt which was better than MoS_2 (48.6%). Also, reaction rate of MoS_2/MMt was 8 times higher than that of MoS_2. Notably, the MoS_2/MMt hybrid carried out an excellent catalytic reusability and stability, even up to five regeneration cycles.

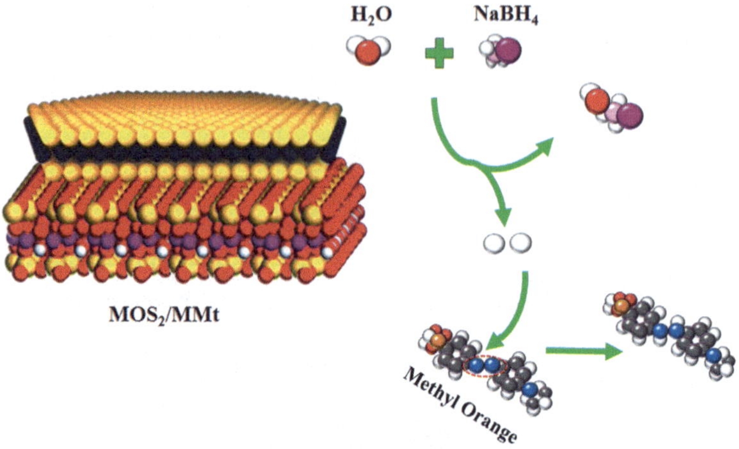

Fig. 10 Schematic of catalytic mechanism of MoS_2/MMt hybrid. Reproduced with permission from Refs. [34]. Copyright 2020, Elsevier

3 Conclusions and Future Perspectives

MMt hybrid materials due to its special advantages such as high specific surface area, surface hydrophilic nature, chemical and thermal stabilities, high mechanical strength, negative charge surface, easy access in nature, good biocompatibility, and extraordinary cation exchange selectivity have been widely used for removal of various kind of textile anionic (reactive, acid, and direct) and cationic (basic) dyes. This chapter focuses on the developments of dye removers based on MMt clay minerals formulating in (nano)organic and inorganic MMt, multi-functional 2D/2D MMt membranes, MMt-biocomposite beads, polymer-MMt nanocomposites, MMt-based hybrids/hydrogels. The presence of MMt clay could improve the porosity, specific surface area, active sites, pore-sizes, surface charge, elastic forces, and physical properties of modified hybrid materials toward dyes' removal processes such as degradation, membrane separation, and adsorption. The MMt structure has negative charge; therefore during removal process, in particular in adsorption, it has the tendency to attract cationic dyes for balancing the negative charge. It also can be a suitable adsorbent for anionic dyes by the MMt modification procedures using different polymers (such as PVA, PANI, PAMAM, and TPP), metal ions (Mg, Fe, Mo, and Al), and other effective molecules, namely CS, starch, ACMO, CTAB, and APTES. Some of these hybrid materials could increase the adsorption of anionic dyes due to the having hydroxyl groups and carboxylates' functional groups in the nanohybrid structures and also effecting on the rough and surface wettability of nanocomposites. The pH solution, adsorbent dose, contact time, temperature, and initial dye concentration play important roles in dye adsorption by addressing issues of electrostatic attraction, ion exchange, hydrogen bonding, surface wettability, as well as regeneration of MMt material adsorbents. For example, at low pH, the positively charge surface of modified MMt-hybrid materials caused the facilitated adsorptions of the anionic dyes or in the high mass of sorbent, more adsorption sites provided leading to enhance the removal of cationic and anionic dyes. Although the various mechanisms were proposed in the scientific researches for achieving superior removal performance of dyes by MMt-clay-based hybrid materials, there are still some momentous issues relating to MMt improvements which are suggested as follows:

- The vast parts of future research can be assigned to synthesize novel MMt hybrid adsorbents to remove different wastewaters such as mixed inorganic/organic dyes, of non-ionic (disperse), phenolic compounds, herbicides, pesticides, etc.
- The low expense and clean syntheses of MMt modification with different chemical materials should be given more attention with considering the higher removal efficiency.
- There is further possibility of improvements in the future studies about the regeneration/reuse of the dye-loaded MMt, which can be give more attention to have the economic wastewater treatment.
- In creative approaches, it can be fabricated three-dimensional (instead of low-dimensional) porous structures of MMt clay coupled with functional groups of various polymers, which provide more adsorption sites and large specific surface area enhancing the adsorption performance.
- More dye removal studies of different manners including bio/photodegradation, degradation, oxidation should be found more favor using cost-effective MMt material.
- The high-quality purification of MMt from clay minerals and ceramics should be improved by applying easy and cost-effective techniques.

References

1. Rajkumar D, Kim JG (2006) Oxidation of various reactive dyes with in situ electro-generated active chlorine for textile dyeing industry wastewater treatment. J Hazard Mater 136(2):203–212
2. Singh S, Barick K, Bahadur D (2013) Functional oxide nanomaterials and nanocomposites for the removal of heavy metals and dyes. Nanomater Nanotechnol 3:3–20
3. Hillie T, Munasinghe M, Hlope M, Deraniyagala Y (2006) Nanotechnology, water and development. New Haven
4. Al Kausor M, Gupta S, Bhattacharyya KG, Chakrabortty D (2022) Montmorillonite and modified montmorillonite as adsorbents for removal of water soluble organic dyes: a review on current status of the art. Inorg Chem Commun 143:109686
5. Liu P, Zhang L (2007) Adsorption of dyes from aqueous solutions or suspensions with clay nano-adsorbents. Sep Purif Technol 58(1):32–39
6. Zhao Y, Kang S, Qin L et al (2020) Self-assembled gels of fe-chitosan/montmorillonite nanosheets: dye degradation by the synergistic effect of adsorption and photo-fenton reaction. Chem Eng J 379:122322
7. Malika M, Sonawane SS (2021) Statistical modelling for the ultrasonic photodegradation of rhodamine B dye using aqueous based Bi-metal doped TiO_2 supported montmorillonite hybrid nanofluid via RSM. Sustain Energy Technol Assess 44:100980

8. Zhang T, Wang W, Zhao Y et al (2021) Removal of heavy metals and dyes by clay-based adsorbents: from natural clays to 1D and 2D nano-composites. Chem Eng J 420:127574
9. Yang Z, Yuan Z, Shang Z, Ye S (2020) Multi-functional membrane based on montmorillonite/graphene oxide nanocomposites with high actuating performance and wastewater purification. Appl Clay Sci 197:105781
10. Peng W, Li H, Liu Y, Song S (2017) A review on heavy metal ions adsorption from water by graphene oxide and its composites. J Mol Liq 230:496–504
11. Dong W, Ding J, Wang W et al (2020) Magnetic nano-hybrids adsorbents formulated from acidic leachates of clay minerals. J Clean Prod 256:120383
12. Han H, Rafiq MK, Zhou T et al (2019) A critical review of clay-based composites with enhanced adsorption performance for metal and organic pollutants. J Hazard Mater 369:780–796
13. Bergaya F, Lagaly G (2006) General introduction: clays, clay minerals, and clay science. Dev Clay Sci 1:1–18
14. Guggenheim S, Martin R (1995) Definition of clay and clay mineral: joint report of the AIPEA nomenclature and CMS nomenclature committees. Clays Clay Miner 43(2):255–256
15. Kumar BS, Dhakshinamoorthy A, Pitchumani K (2014) K10 montmorillonite clays as environmentally benign catalysts for organic reactions. Catal Sci Technol 4(8):2378–2396
16. Swearingen C, Macha S, Fitch A (2003) Leashed ferrocenes at clay surfaces: potential applications for environmental catalysis. J Mol Catal A Chem 199(1–2):149–160
17. Frenkel M (1974) Surface acidity of montmorillonites. Clays Clay Miner 22(5):435–441
18. Miller RW, Gardiner DT (1958) Soils in our environment: Upper Saddle River, 10th edn. Prentice Hall PTR
19. Kannan N, Sundaram MM (2001) Kinetics and mechanism of removal of methylene blue by adsorption on various carbons—a comparative study. Dyes Pigm 51(1):25–40
20. Shahadat M, Isamil S (2018) Regeneration performance of clay-based adsorbents for the removal of industrial dyes: a review. RSC Adv 8(43):24571–24587
21. Hu Q, Qiao S, Haghseresht F et al (2006) Adsorption study for removal of basic red dye using bentonite. Ind Eng Chem Res 45(2):733–738
22. Pereira FA, Sousa KS, Cavalcanti GR et al (2017) Green biosorbents based on chitosan-montmorillonite beads for anionic dye removal. J Environ Chem Eng 5(4):3309–3318
23. Thue PS, Sophia AC, Lima EC et al (2018) Synthesis and characterization of a novel organic-inorganic hybrid clay adsorbent for the removal of acid red 1 and acid green 25 from aqueous solutions. J Clean Prod 171:30–44
24. Chaari I, Fakhfakh E, Medhioub M, Jamoussi F (2019) Comparative study on adsorption of cationic and anionic dyes by smectite rich natural clays. J Mol Struct 1179:672–677
25. Kalpaklı Y, Toygun Ş, Köneçoğlu G, Akgün M (2014) Equilibrium and kinetic study on the adsorption of basic dye (BY28) onto raw Ca-bentonite. Desalin Water Treat 52(37–39):7389–7399
26. Amri N, Radji S, Ghemati D, Djamel A (2019) Studies on equilibrium swelling, dye adsorption, and dynamic shear rheology of polymer systems based on chitosan-poly(vinyl alcohol) and montmorillonite. Chem Eng Commun 206(6):716–730
27. Preetha BK, Vishalakshi B (2020) Microwave assisted synthesis of karaya gum based montmorillonite nanocomposite: characterisation, swelling and dye adsorption studies. Int J Biol Macromol 154:739–750
28. Santos S, França DB, Castellano LR et al (2020) Novel modified bentonites applied to the removal of an anionic azo-dye from aqueous solution. Colloids Surf A Physicochem Eng Asp 585:124152
29. Lagergren S (1898) Zur theorie der sogenannten adsorption geloster stoffe. Kungliga svenska vetenskapsakademiens. Handlingar 24:1–39
30. Freundlich H (1906) Over the adsorption in solution. J Phys chem 57(385471):1100–1107
31. Yuh-Shan H (2004) Citation review of Lagergren kinetic rate equation on adsorption reactions. Scientometrics 59(1):171–177
32. Ho YS, McKay G (1999) Batch lead (II) removal from aqueous solution by peat: equilibrium and kinetics. Process Saf Environ Prot 77(3):165–173

33. Zhu J, Tian M, Zhang Y et al (2015) Fabrication of a novel "loose" nanofiltration membrane by facile blending with chitosan-montmorillonite nanosheets for dyes purification. Chem Eng J 265:184–193
34. Yang L, Wang Q, Rangel-Mendez JR et al (2020) Self-assembly montmorillonite nanosheets supported hierarchical MoS2 as enhanced catalyst toward methyl orange degradation. Mater Chem Phys 246:122829
35. Hassan SA, Darwish AS, Gobara HM et al (2017) Interaction profiles in poly(amidoamine) dendrimer/montmorillonite or rice straw ash hybrids-immobilized magnetite nanoparticles governing their removal efficiencies of various pollutants in wastewater. J Mol Liq 230:353–369
36. Benkhaya S, M'rabet S, El Harfi A (2020) A review on classifications, recent synthesis and applications of textile dyes. Inorg Chem Commun 115:107891
37. Hassan MM, Carr CM (2018) A critical review on recent advancements of the removal of reactive dyes from dyehouse effluent by ion-exchange adsorbents. Chemosphere 209:201–219
38. Broadbent AD (2001) Basic principles of textile coloration, 1st edn. Society of Dyers and Colorists Bradford
39. Al-Sigeny S, Abou Taleb MF, El-Kelesh NA (2008) Hybrid nanocomposite prepared by graft copolymerization of 4-acryloyl morpholine onto chitosan in the presence of organophilic montmorillonite. J Macromol Sci A 46(1):74–82
40. Karadağ E, Hasgül B, Kundakci S, Üzüm ÖB (2014) A study of polymer/clay hybrid composite sorbent-based AAm/SMA hydrogels and semi-IPNs composed of ι-carrageenan and montmorillonite for water and dye sorption. Adv Polym Technol 33(4)
41. Wang Y, Xiong Y, Wang J, Zhang X (2017) Ultrasonic-assisted fabrication of montmorillonite-lignin hybrid hydrogel: highly efficient swelling behaviors and super-sorbent for dye removal from wastewater. Colloids Surf A Physicochem Eng Asp 520:903–913
42. Luo J, Ma X, Zhou X, Xu Y (2021) Construction of physically crosslinked cellulose nanofibrils/alkali lignin/montmorillonoite/polyvinyl alcohol network hydrogel and its application in methylene blue removal. Cellulose 28(9):5531–5543
43. Waiman CV, Chesta CA, Gómez ML (2016) Hybrid films based on a bridged silsesquioxane doped with goethite and montmorillonite nanoparticles as sorbents of wastewater contaminants. J Nanomater
44. Rahmani M, Dadvand Koohi A (2021) Adsorption of malachite green on the modified montmorillonite/xanthan gum-sodium alginate hybrid nanocomposite. Polym Bull 79:1–27
45. El-Sigeny S, Mohamed SK, Abou Taleb MF (2014) Radiation synthesis and characterization of styrene/acrylic acid/organophilic montmorillonite hybrid nanocomposite for sorption of dyes from aqueous solutions. Polym Compos 35(12):2353–2364
46. Wu J, Lerner M (1993) Structural, thermal, and electrical characterization of layered nanocomposites derived from sodium-montmorillonite and polyethers. Chem Mater 5(6):835–838
47. Yao K, Song M, Hourston DJ, Luo DZ (2002) Polymer/layered clay nanocomposites: 2 polyurethane nanocomposites. Polymer 43(3):1017–1020
48. Usuki A, Tukigase A, Kato M-J (2002) Preparation and properties of EPDM–clay hybrids. Polymer 43(8):2185–2189
49. Sun T, Garces JM (2002) High-performance polypropylene-clay nanocomposites by in-situ polymerization with metallocene/clay catalysts. Adv Mater 14(2):128–130
50. Bermudez YH, Truffault L, Pulcinelli SH, Santilli CV (2018) Sodium montmorillonite/ureasil-poly(oxyethylene) nanocomposite as potential adsorbent of cationic dye. Appl Clay Sci 152:158–165
51. Chabane L, Cheknane B, Zermane F et al (2017) Synthesis and characterization of reinforced hybrid porous beads: application to the adsorption of malachite green in aqueous solution. Chem Eng Res Des 120:291–302
52. Olad A, Azhar FF (2014) Eco-friendly biopolymer/clay/conducting polymer nanocomposite: characterization and its application in reactive dye removal. Fibers Polym 15(6):1321–1329
53. Silva M, Oliveira M, Avelino MC et al (2012) Adsorption of an industrial anionic dye by modified-KSF-montmorillonite: evaluation of the kinetic, thermodynamic and equilibrium data. Chem Eng J 203:259–268

54. Ayazi Z, Khoshhesab ZM, Azhar FF, Mohajeri Z (2017) Modeling and optimization of adsorption removal of reactive orange 13 on the alginate–montmorillonite–polyaniline nanocomposite via response surface methodology. J Chin Chem Soc 64(6):627–639
55. Lyu W, Li J, Trchová M et al (2022) Fabrication of polyaniline/poly (vinyl alcohol)/ montmorillonite hybrid aerogels toward efficient adsorption of organic dye pollutants. J Hazard Mater 435:129004
56. Huang P, Kazlauciunas A, Menzel R, Lin L (2017) Determining the mechanism and efficiency of industrial dye adsorption through facile structural control of organo-montmorillonite adsorbents. ACS Appl Mater Interfaces 9(31):26383–26391
57. Huang P, Xia D, Kazlauciunas A et al (2019) Dye-mediated interactions in chitosan-based polyelectrolyte/organoclay hybrids for enhanced adsorption of industrial dyes. ACS Appl Mater Interfaces 11(12):11961–11969
58. Zhang H, Ma J, Wang F et al (2020) Mechanism of carboxymethyl chitosan hybrid montmorillonite and adsorption of Pb(II) and congo red by CMC-MMT organic-inorganic hybrid composite. Int J Biol Macromol 149:1161–1169
59. Muangrak W, Thouchprasitchai N, Phongboonchoo Y, Pongstabodee S (2020) Dual functional composite of montmorillonite-rich/chitosan (MCC) for decolorizing the water used in joss paper process: thermodynamic, isotherm, and kinetic studies. Appl Sci 10(21):7493
60. Da Silva JC, França D, Rodrigues F et al (2021) What happens when chitosan meets bentonite under microwave-assisted conditions? Clay-based hybrid nanocomposites for dye adsorption. Colloids Surf A Physicochem Eng Asp 609:125584

Metal-Decorated Nanohybrid Materials for Textile Dyes' Removal from Wastewater

Babak Jaleh, Ensiye Shabanlou, Mahtab Eslamipanah, and Atefeh Nasri

Abstract Water is a vital material for human life, and its pollution by textile dyes is a significant problem the world has encountered today. The textile dyes can enter the food chain and provide mutagenicity and carcinogenicity, destroy photosynthesis, and prevent plant growth. Recently, various approaches such as photocatalytic degradation, electrocatalytic degradation, and adsorption employ to removal of dyes from water. However, identification of affordable materials with high removal efficiency is also important. Metallic hybrid materials, a compound of different materials with metals, have been noticed as catalyst dyes' removal due to their large surface area and specific optical, physical, and chemical features. In this chapter, utilization of metallic hybrid materials with different structures was investigated toward the removal of cationic and anionic dyes. The effect of temperature and pH on the removal efficiency was also perused.

Keywords Metal (nano)hybrid · Removal · Adsorption · Catalysts · Wastewater · Cationic dyes · Anionic dyes

1 Introduction

Water contamination is complex and the demand for the supply of clean water increases every day. Therefore, removing textile dyestuffs from aqueous solution is considered to be the main issue in scientific society [10, 44, 45, 47]. Cationic and anionic dyes from the textile industry discharge, agricultural effluents, and other human activities are well-known as significant sources of water pollution. Due to the mutagenesis, carcinogenesis, chronic toxicity, and teratogenesis properties of textile dyes, they have harmful health effects on both human and wildlife [30, 55]. Accordingly, some separation and treatment technologies, including photocatalytic degradation, adsorption, electrochemical oxidation, membrane separation, gravity separation, ion exchange, evaporation, and flocculation coagulation, have been

B. Jaleh (✉) · E. Shabanlou · M. Eslamipanah · A. Nasri
Department of Physics, Faculty of Science, Bu-Ali Sina University, 65174 Hamedan, Iran
e-mail: jaleh@basu.ac.ir

© The Author(s), under exclusive license to Springer Nature Singapore Pte Ltd. 2023
A. Ahmad et al. (eds.), *Nanohybrid Materials for Treatment of Textiles Dyes*,
Smart Nanomaterials Technology, https://doi.org/10.1007/978-981-99-3901-5_12

widely used for the decolorization of textile dyes from wastewater [3, 11, 66]. The requirement of these methods is material with excellent properties leading to low cost, high effectiveness, easy operation, and superior performance of dyes' removal processes [4, 29, 62]. Compared to single materials, the designed and fabricated composites based on various effective materials such as (nano)hybrid materials with two/three-dimensional structures with adjustable size, single molecular weight, presence of many cavities, high biocompatibility, and low toxicity are more interesting in research works [39, 44, 45, 54, 64]. Nanotechnology approaches in wastewater treatment suggest effective nanomaterials such as nanoadsorbents, nanocatalysts, nanoparticles (NPs), and nanostructured membranes, which improve different properties like the surface area, catalytic activity, photosensitivity, electrochemical, magnetic, and optical features [29, 31, 38, 43, 46]. Among them, the (nano)metal products such as metal NPs, metal nanoclusters, metal nanotubes, and metal nanorods of different kinds of mono/bimetal/trimetal, noble and transition metals have attracted a lot of attention for synthesizing nanohybrid materials in order wastewater treatment [14, 38]. The effect of metals on water purification extremely depends on their shape, particle size, surrounding the dielectric condition, and assembly state, which can improve the chemical, physical, magnetic, optical, biocompatibility, and wettability properties of adsorbents/catalysts [38]. The most conventional methods which have been used for dye removal by metal-based nanohybrids include absorption, (photo)degradation, and oxidation, in which metals are often used as catalysts, membranes, precursors, and adsorbents. The adsorption technique was introduced as one of the practical, easy, and feasible dye removal methods owing to the great benefits such as easy operation, availability, environmentally friendly, simple, and affordable design. Moreover, this method can be used for any inorganic, insoluble and soluble organic, biological contaminants with the different types of available and reusable absorbents [3, 66]. Metal-based adsorbents can provide excellent adsorption conditions, namely a good specific surface area, special pores' structures, and tunable and high porosities, which lead to increased absorption efficiency [38, 51, 70]. For example, by the combination of metal–organic framework (MOFs) or metal NPs with suitable material, an effective adsorbent can be made with specific pores' structures, enhanced synthesis kinetics, improved physical and chemical properties, and high stability [3]. The (photo)degradation is another dye removal method, which has been extensively employed in wastewater treatment. In the case of the degradation method, the dye decomposition performs in an aqueous solution containing reducing agent such as $NaBH_4$. The metal-based compounds as a catalyst can facilitate the dye decomposition through the reduction process. In photocatalytic degradation, the photoactivated reactions are determined through the aid of the unfastened radical mechanism initiated by the interplay of photons of a definite energy level under UV or visible irradiation, resulting in the degradation of dye molecules 14. The (nano)metal-based photocatalysts indicate better photocatalytic activity due to improving light absorption depending on their band gap energy, electronic configurations, the property of charge carrier transport, and excited-state lifetimes 14. Also, the advanced oxidation process (AOP) as the photocatalytic technique can provide hydroxyl radicals as strong oxidizing agents through oxidation processes with H_2O_2 under UV irradiation. In other words, by adsorbing the light, the photo-generated electrons can be produced and it caused the reduction of adsorbed oxygen (O_2) and the creation of photogenerated holes and superoxide radicals, which react with H_2O to generate hydroxyl radicals. Subsequently, the hydroxyl radicals oxidize the dye pollutants leading to degrading dye into smaller and nontoxic molecules like H_2O, CO_2, and mineral acids (seen Fig. 1a) [38]. Due to the fact that total solar energy contains 40% visible light and 5% UV light, so the improving photocatalysis technology in the visible range has attracted enormous interest [38]. In recent years, the NPs of noble metals, namely gold (Au), platinum (Pt), silver (Ag), copper (Cu), and palladium (Pd), because of their localized surface plasmon resonance (LSPR) and high optical light absorption property have been considered to be suitable materials for adsorbing both visible and UV lights energy [38]. The plasmons of surfaces are determined as collective oscillations of free electrons on the noble metal surfaces, which can improve adsorption, and cross-section scattering, as well as the field enhancement near the surface of particles. According to Fig. 1b and c, two mechanisms are enhanced the photocatalytic reactions of nanohybrid photocatalysts using semiconductors and noble metal NPs. First, the light excites the LSPR in the metal NPs and electron–hole pairs in the semiconductor in nanohybrid structures including metal NPs and a semiconductor with the energies of plasmon resonance and overlapping absorption (Fig. 1b). Therefore, the plasmonic integration can improve the photocatalytic performance under visible irradiation with the sensitizing of the surface plasmon through the LSPR excitation such that electron transition occurred at the metal–semiconductor (Fig. 1c). Up to date, various nanohybrid materials based on metal oxides have been synthesized like Ag/ZnO, Ag/Ag_3PO_4, TiO_2–Ag, GO–Ag, Ag/rGO, AgCl/Ag, Ag_3PO_4/TiO_2, Sn–ZnO, Cs–Ag, Sn–Fe–CO, Sn–rGO, Sn–Fe_2O_3, $AgNO_3$–Sn Au–Sn, AgCl–Sn, for improving photosensitivity properties [49].

This chapter attempts to focus on the new insights of dye removal techniques and fabricating of the novel, low-cost, reusable, high electronic, and thermal conductivity metal-based nanohybrid materials using various ceramics, polymers, metal–organic frameworks (MOFs), and carbon-based materials like graphene, reduced graphene oxide (rGO), carbon nanotubes (CNTs)1, and activated carbon (AC).

2 Textile Dyes' Removal by Hybrid-Based Materials

The textile dyes' industry has been being for more than 4000 years. Recently, the wastewater consisting of organics dyes produced from industrials process such as textile, paper-making, leather and food additives has negative effect on human life and limits the water purification. Hence, the removal of dyes has been well-known as an important environmental problem because even a low content of dyes can toxic [25, 57]. There are three types of dyes including nonionic, anionic, and cationic dyes, in which cationic dyes are well-known as basic dyes and anionic dyes consist of reactive

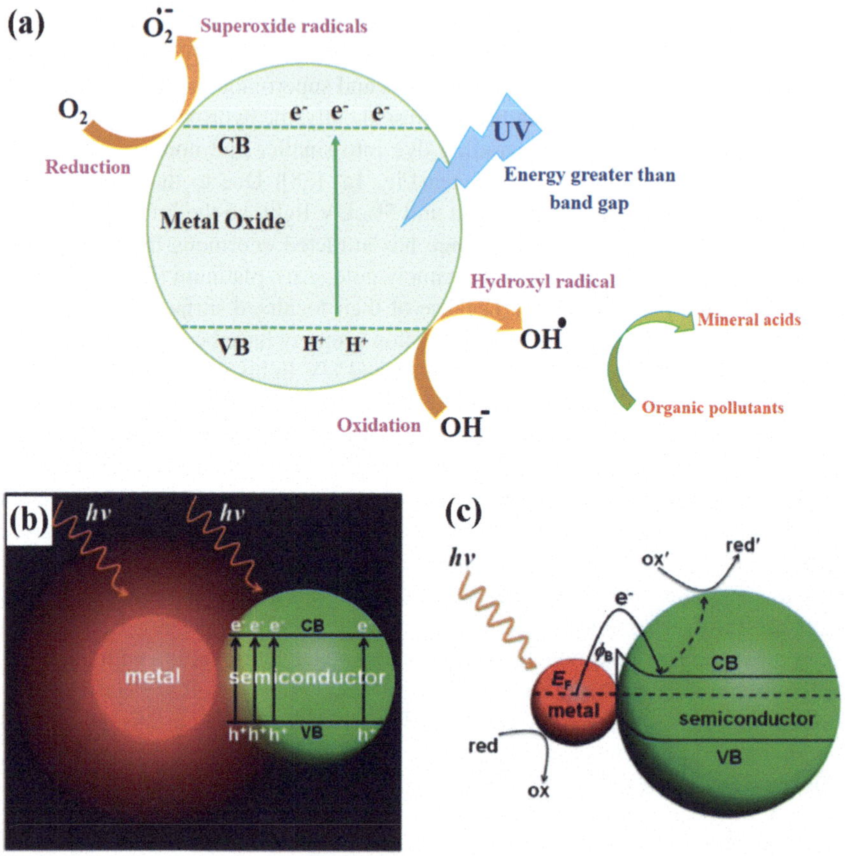

Fig. 1 **a** Representation of the photodegradation procedure, the illustration of two plasmons in chemical reactions mechanisms: **b** the improvement of plasmonic light absorption **c** plasmonic sensitization. Reproduced with permission from Ref. [38]. Copyright 2020, Elsevier

and direct acid dyes. Various materials such as polymeric materials, clays, carbon-based compounds, metallic and metal oxide nanoparticles have been introduced as effective candidates for both anionic and cationic dyes' removal. Utilization of hybrid-based materials, a mixture of different types of materials, leads to improve stability, increase removal efficiency, and reduce the environmental effect during dyes' removal process [53].

2.1 Cationic Dyes' Removal

One of the most widely used toxic dyes is cationic dyes, consisting of chemical structures of the successor aromatic groups. This classification of dyes can provide detrimental effects of allergic dermatitis, mutations, cancer, and skin irritation. Cationic dyes are also named as basic dyes, depending on a positive ion. Cationic dyes carry positively charged and generate colored cations in solutions. Basic dyes are observable and possess high severity of color. One of the important and useful cationic dyes in the textile industry is methylene blue (MB) [57]. Different techniques with highly efficient material are widely employed for the purification of textile cationic dyes, which will be discussed in the following.

Graphene or GO can be used as the dye adsorbent due to their great specific surface area and adsorption sites. Still, the adsorption process for removal of dyes from wastewater requires a long reaction time, even hundreds of minutes. Recently, the properties of GO have been improved using (nano)metal ions. Hence, different GO multifunctional metal-based hybrids were synthesized for wastewater purification. In 2012, the synthesis of Fe nanoparticles (NPs)@graphene composites (FGC) was performed with GO as a supporting matrix. GO could prevent the agglomeration of Fe NPs and reduce the Fe NPs' size to about 5 nm which enhanced dye decolorization efficiency [24]. GO was reduced to graphene by borohydride at the same time as Fe NPs' formation leading to FGC. FGC hybrids could remove MB dye in a short time of 15 min (Fig. 2a, b). The reaction between Fe and H^+ or H_2O caused atomic H generation, which can destroy the dye chromophore groups with the Fe intermediate products through an unstable and active thermodynamic process. Furthermore, the passive iron oxide shells, such as FeOOH, Fe_2O_3, and Fe_3O_4, fabricated on the surface of Fe NPs, helped the MB molecules adsorption. In another new approach, one-pot solvothermal technique was employed for synthesizing rGO-supported ferrite (MFe_2O_4, M=Co, Ni, Mn, and Zn) hybrids using metal ions (M^{2+} and Fe^{3+}) and GO as starting materials (Fig. 2c). The monodispersed MFe_2O_4 microspheres homogeneously grew on the flake-like rGO nanosheets as seen in Fig. 2d [13]. The good adsorption property of rGO and the photocatalytic and magnetic separation properties of MFe_2O_4 NPs made the rGO–MFe_2O_4 hybrids a potential dye remover. The rGO–$MnFe_2O_4$ hybrids illustrated high-performance absorbance of rhodamine B (RhB) (92%) and MB (100%). The graphitized basal plane of rGO structure, its high specific surface area, the electrostatic interactions, and π–π stacking between dyes (RhB and MB) and the rGO nanosheets could prepare great conditions for adsorption of dyes. Also, the large-size microspheres of $MnFe_2O_4$ could act as spacers and reduce the rGO nanosheet agglomeration. On the other hand, the $MnFe_2O_4$ microspheres firmly stacked with rGO nanosheets resulting a desirable structure with numerous pores which improved the adsorption performance. The rGO–$MnFe_2O_4$ magnetic hybrid was able to be separated by a magnet and subsequently washed and reused in the adsorption process in five cycles. Moreover, rGO-ferrite hybrids indicated an enhanced photodegradation activity of MB and RhB dyes compared to pure rGO

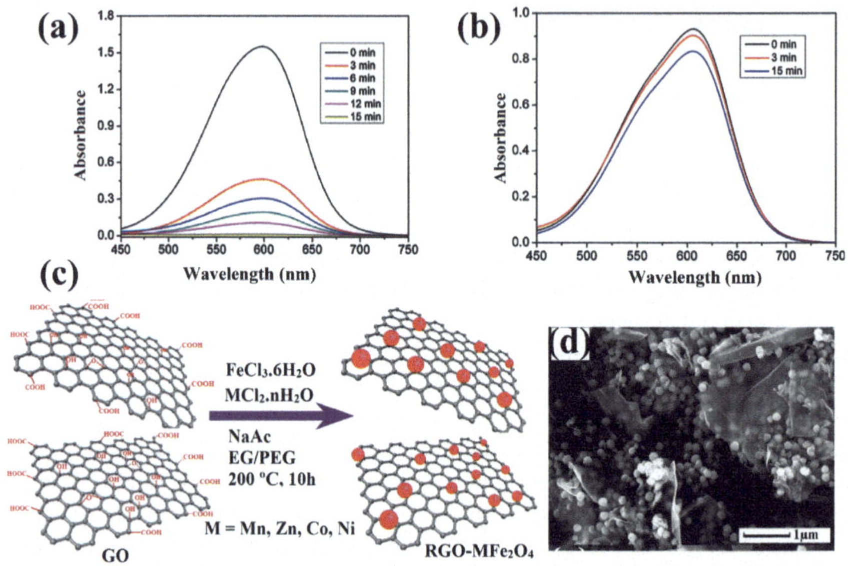

Fig. 2 UV–vis spectra of MB absorbance with FGC **a** and GO **b** in time interval of 15 min. Reproduced with permission from Ref. [24]. Copyright 2012, Elsevier. **c** The representation of rGO–MFe_2O_4 hybrids' synthesis (M=Mn, Ni, Co, Zn), **d** FESEM of the $MnFe_2O_4$ hybrid. Reproduced with permission from Ref. [13]. Copyright 2012, Elsevier

nanosheets and $MnFe_2O_4$ because their combined synergistic effect resulted in the separation of photocarriers in the coupled system of $MnFe_2O_4$ and rGO.

In another study, by combination of one-dimensional (1D) ZnO nanorod (NR) and two-dimensional (2D) rGO, the multifunctional ZnO NR-rGO hybrids' nanocomposites were fabricated via hydrothermal process [52]. The morphological analysis confirmed that the rGO sheets grafted well on the ZnO NRs, which provided good adsorption condition for degrading MB, RhB, and methyl orange (MO) dye molecules under visible-light irradiation. It was mainly because of the large surface area of both ZnO NRs and rGO and the easy transportation of electrons by rGO connected parallel with elongated 1D ZnO NRs. The photodegradation interactions of MB and RhB dyes with the hybrid system were faster than MO dye. It was owing to the negative charges in hybrid nanocomposites, resulting more cationic dye absorbance rate. Several hybrid nanocomposites based on zeolitic imidazolate framework (ZIF-8) as a metal–organic framework (MOF), GO, and CNT with high adsorption capacity were applied for cationic dye (malachite green (MG)) removal [3]. The generation of new pores between the MOF and GO or CNT was very useful for dye molecules' adsorption by impeding aggregate formation, enhancing dispersive forces, good thermal stability, and appropriate pore size. The adsorption capacities of ZIF-8, ZIF-8@CNT and ZIF-8@GO hybrid nanocomposites archived 1667, 2034, and 3300 mg g^{-1}, respectively. The adsorption test was studied in different pH solutions in the range of 2.2–7 for investigating the protonation and deprotonation effects of MG adsorbents. It was concluded that in low pH solution (acidic conditions), the proton accumulation or positive charges on the surface of adsorbents increased, reducing electrostatic interaction with positively charged MG molecules. Thereby, the high pH solution was more efficient for enhancing adsorption capacity by these hybrid adsorbent-based ZIF-8 MOFs.

The polymerization of octavinyl polyhedral oligomeric silsesquioxane (POSS) on the surface of Fe_3O_4 NPs was successfully performed to fabricate Fe_3O_4@POSS material with high specific surface area of 653.59 m^2 g^{-1} and an average particle size of 20 nm [26]. After treatment of Fe_3O_4@POSS with the thiol-ene addition, the specific surface area of magnetic Fe_3O_4@POSS-SH functionalizable hybrid was achieved 224.20 m^2 g^{-1} with a good porous morphology. It was used as an efficient single adsorbent for multiple pollutants containing organic textile dyes and heavy metal ions even for five cycles. Due to the strong interactions between the porous hybrid coating stemming from cross-linked POSS and cationic dyes, the Fe_3O_4@POSS-SH NPs demonstrated good absorbance capacity of 100 and 96% for MG and RhB, respectively. The adsorption ability of dyes increased by increasing pH and reached to its maximum value at pH of ~7. At low pH, the strong –SH groups' protonation on the material provided electrostatic repulsion by cationic dyes, resulting to reduce adsorption capacity. On the other hand, by increasing pH, the adsorption capacity decreased, which attributed to the weak hydrophobic interaction of dyes and material. By photopolymerization method, polymer nanocomposites containing in situ photogenerated Au NPs were fabricated using methacrylated glycomonomers with a-D-glucofuranose or D-mannitol structural units, other mono(di)methacrylates, and $AuCl_3$ precursor [18]. The generation of Au NPs inside the matrix of polymer was confirmed with different analyses. When the methacrylates functionalized with quaternary ammonium, carboxylic or thiol groups (MA-SH, MA-COOH, MA-N^+) added in each formulation, the average size of Au NPs decreased to 1.5 nm. These hybrid films were used in catalysis applications for the photodegradation of MO and MB dyes under the action of visible irradiation. The performance of the catalyst was improved by decreasing the Au NPs' size. The small sizes of Au NPs can help to catalyze different reactions of the organic compounds even at ambient temperature and also absorb the visible light because of its SPR effect. The hybrid film preparing 30% MA-SH could be decomposed 98 and 94% of MB and MO, respectively, after 100 min of visible irradiation. Moreover, the Au NPs' holes could interact with water hydroxyl groups, and then, the conduction band electrons could be reacted with dissolved oxygen molecules resulting in superoxide radical anions O_2^-. Then, by generation of hydroxyl radicals HO_2, the dye degradation was completed through a superoxide anion activity. In new approach, a novel core@double-shell-structured magnetic halloysite nanotubes (HNTs) was prepared by co-precipitation and modified mussel-inspired co-modification approach [66]. The HNTs' skeleton was structured with Fe_3O_4 NPs and poly(DA + KH550) as inner and outside shells, respectively, which provided an excellent condition for adsorption process such that the intermediate magnetite Fe_3O_4 NPs created the magnetic separation property and high surface area. In addition, poly(DA + KH550) provided numerous active adsorption sites (Fig. 3). By the contribution of inner and outside shell properties, the HNTs/Fe_3O_4/poly-(DA + KH550) nanohybrids have acted as potential recyclable adsorbent for MB removal with high adsorption capacity of 714.29 mg g^{-1} and good cycling stability. The solution pH had the strong effect on the ionic state of functional groups on the surface of HNTs/Fe_3O_4/poly-(DA + KH550). The MB removal was slowly changed in pH 3 to 7 and by increasing pH to 10, and the adsorption capacity was rapidly increased due to negative charge surface of the HNTs/Fe_3O_4/poly(DA + KH550). In the low pH, the –NH– protonation was occurred to generate $–NH^+$ cation and reduced –NH– sites on the catalyst surface, providing the electrostatic repulsion of cationic dye with catalyst.

The anatase TiO_2, Au and Pt–TiO_2 thin films were successfully deposited on glass and silicon substrates through plasma-enhanced chemical vapor deposition/

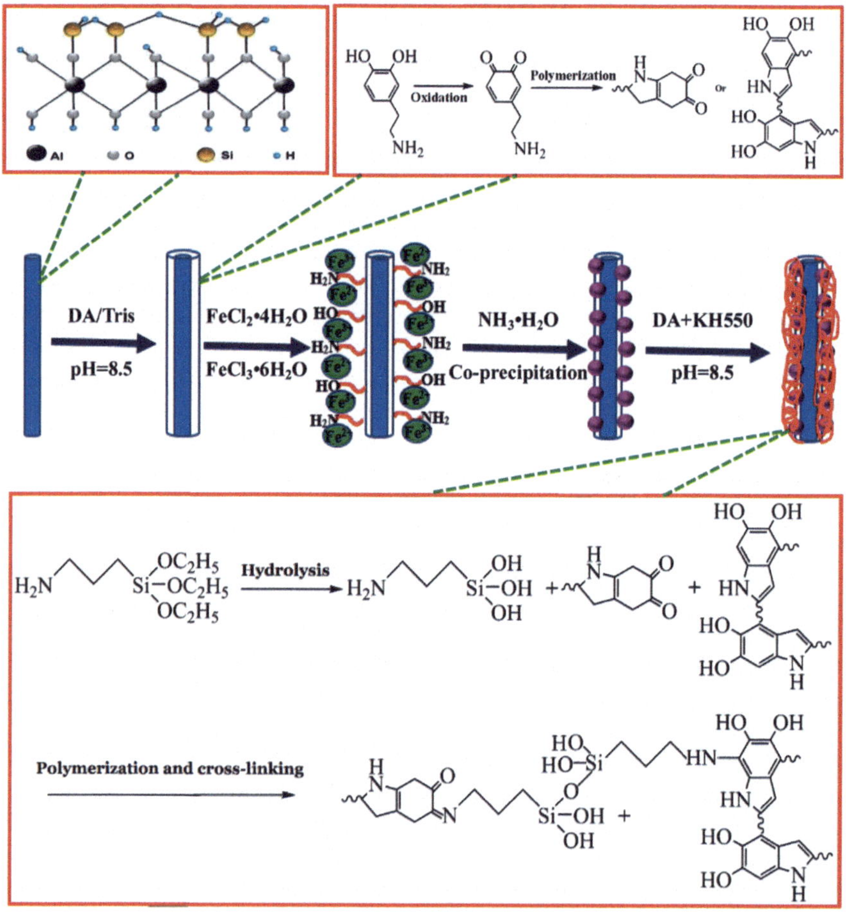

Fig. 3 Illustration of HNTs/Fe_3O_4/poly(DA + KH550) nanohybrids' fabrication and its chemical reactions. Reproduced with permission from Ref. [66]. Copyright 2017, Elsevier

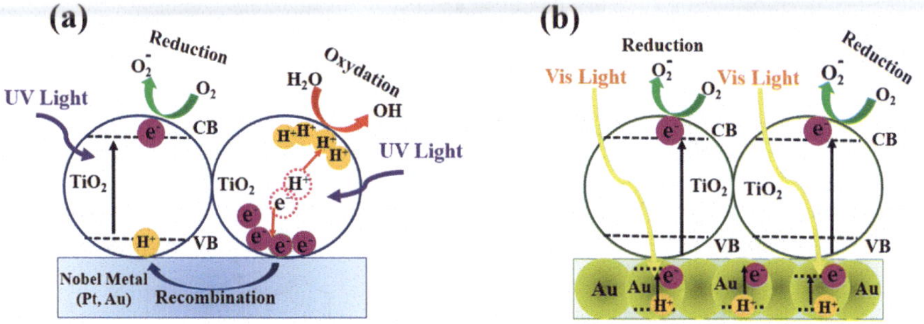

Fig. 4 **a** Effect of noble Pt and Au metal layer on the charge recombination reduction **b** and the effect of SPR of TiO_2–Au layer on the photocatalytic properties. Reproduced with permission from Ref. [12]. Copyright 2017, American Chemical Society

physical vapor deposition (PECVD/PVD) approach [12]. The MB was applied as dye pollutants to study the Pt and Au nanolayer efficiency for water decontamination, indicating 80 and 100% photodegradation of MB dye by Au–TiO_2 and Pt–TiO_2 catalysts, respectively. The UV photoactivity of Pt and Au–TiO_2 nanocomposites was two to three folds greater than that of TiO_2 owing to the electrons trapping in the noble metal layer, which prevent the recombination of charge carrier (Fig. 4a). In contrast, the Au–TiO_2 illustrated more visible photoactivity toward to MB dye because of its more SPR effect as shown in Fig. 4b.

In 2018, a series of α-MoO_3 and navel transition metal molybdates' ($MMoO_4$) hybrid catalysts were fabricated in the presence of metal (M=Fe, Cr, Zn, Mn, Co, Cu, and Ni) by a facile hydrothermal method [76]. They were applied in the catalytic degradation process for different cationic dyes (red GTL, safranine T, MB, RhB, and rhodamine 6G (Rh6G)) via a catalytic wet air oxidation (CWAO) system. The pathway of MB and degradation intermediates is exhibited in Fig. 5a. It was resulted that catalytic degradation activities were significantly enhanced compared with pure α-MoO_3, pure transition metal molybdates, and even the physical mixture of these two bulks. Among the hybrids, $NiMoO_4$/α-MoO_3 and $MnMoO_4$/α-MoO_3 indicated the best degradation performance with 100% removal of dye within 30 min, although it was only 20% for pure transition metal molybdates or pure α-MoO_3 (Fig. 5b). According to electron spin resonance (ESR) analysis, the %OH radicals were generated during the oxidation process due to the co-existence of α-MoO_3 and transition metal molybdates, which played an important role in high catalytic activity. Also the O^{2-} ions of the oxygen deficient regions can easily contribute in the oxidation process owing to its higher mobility than lattice oxygen. As indicted in Fig. 5c, in the CWAO system, increasing temperature caused generation more active radicals and increased the activities of $NiMoO_4$/α-MoO_3 and $MnMoO_4$/α-MoO_3 catalysts.

Applying loose nanofiltration (NF) membranes for textile wastewater purification can provide a adequate permeate flux at low operating pressure, leading to reduction of the energy use. Creating loose layered double hydroxides/polymer (LDHs/PEI)

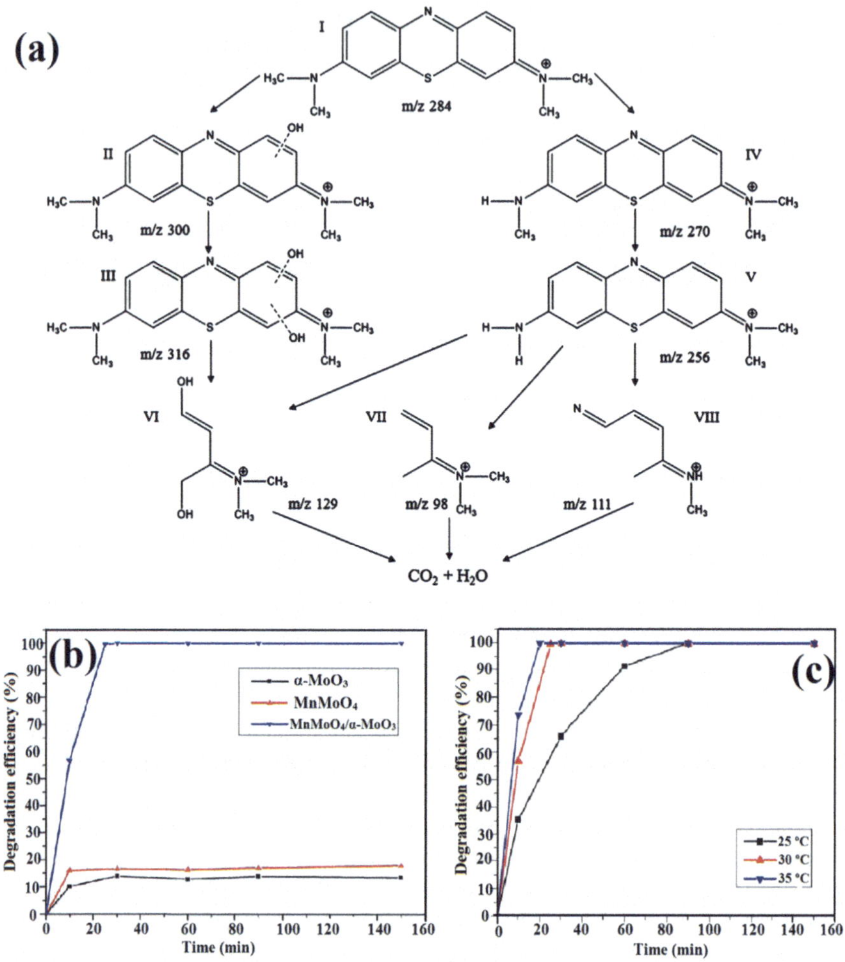

Fig. 5 **a** Identification of MB degradation pathway with mass spectroscopy, **b** the performance of $MnMoO_4/\alpha\text{-}MoO_3$ hybrid, pure $\alpha\text{-}MoO_3$, and pure $MnMoO_4$ catalysts in degradation process, **c** the effect of temperature on MB degradation. Reproduced with permission from Ref. [76]. Copyright 2018, Elsevier

hybrid on an organic support through chelating-assisted in-situ growth led to optimize rejection of MB (97.9%) and acid fuchsin (97.5%) [77]. The presence of LDHs reduced contact angles and enhanced hydrophilicity of the membrane surface, leading to low membrane fouling and high permeate flux. In addition, the synthesized hybrid membrane consisted positively charged which was useful for remove cationic dyes. The LDHs/PEI exhibited durable dye rejection during 30 h filtration test. Doping transition metals can enhance photo-induced electron–hole separation yield. However, particulate photocatalysts have some disadvantages in recycling capability and fast electron–hole recombination. Sakthivel et al. reported that the supporting g-C_3N_4

on Cu by pyrolysis could provide more photocatalytic activity than Cu_2O, glass–g-C_3N_4, and physically coated Cu–g-C_3N_4 film, because it reduced the electron–hole recombination [56]. In fact, the abundant Cu metal with high conductivity has been provided hybrid bridge which acts as a charge carrier separation The existence of Cu–g-C_3N_4 led to the removal of 85% of initial MB.

Geopolymers are an affordable inorganic matter that usually acquired from metakaolin or low-calcium alumino-silicate wastes via geopolymerization process over heat-curing conditions. After dyes' removal, they can easily be modified due to their excellent thermostability and corrosion resistance. The geopolymer-based membranes can separate Cr(III) from wastewater. Recovering Cr(III) to Cr_2O_3 causes to formation of a suitable semiconductor on geopolymers (Cr_2O_3/GPCM) for photocatalytic degradation of dye from wastewaters (Fig. 6a) [16]. The photo-induced decolorization of basic green molecules via Cr_2O_3/GPCM indicated that the photo-oxidation rate of the sample enhanced with increasing Cr_2O_3 values on the membrane surface because the Cr_2O_3 is strongly nucleophilic and can adsorb the green molecules. It was found that the dye was 100% destroyed after 100 min, as shown in Fig. 6b. In this sample, the membrane acted as a separator for H_2O/basic green molecules, passing only H_2O molecules. The Cr_2O_3 provided electron-hole pairs (Fig. 6c). Electrocoagulation, in which the electrocatalytic oxidation at the sacrificial electrode forms coagulants of metal hydroxides and oxyhydroxides, is well-known as a significant way of dye adsorption. In this removal method, the atomic layer deposition (ALD)-enabled TiO_2 ultrathin on the stainless-steel cathode (Ti/SSM) indicated promising performance toward MB removal [21]. The catalytic performance of Ti/SSM was improved when the coating cycle increased to 100 cycles (100Ti/SSM), the MB removal was achieved to 73.2% by 100Ti/SSM after 90 min because the increasing coating cycles causes to increase the impedance for electron transfer. The pH was an important factor to MB removal because the removal rate was changed from 81.5% to 88.8% when the pH was reduced from 5.5 to 3.4. In less pH, more protons were accessible for the electroproduction of H_2O_2 at the cathode.

Metal–organic frameworks (MOFs) with a mesoporous/microporous networks are well-known filler for preparation of mixed matrix membranes (MMMs). Among them, Material of Institute Lavoisier (MIL)-100(Fe) (Fe-MOF), a sub-family of MOFs, has significant catalytic performance toward dye removal. The prepared hybrid flat sheet through blending polyethersulfone (PES) with various percentages of Fe-MOF indicated excellent rejection of > 98.5% for cationic and anionic dyes and optimal permeation flux as high as 165.68 L m^{-2} h^{-1} [33]. The presence of high values of Fe-MOF on the membrane increased hydrophilicity. Moreover, the membrane had small pore size which led to positive impacts of increased hydrophilicity. Therefore, the presence of MOF in the membrane matrix enhanced the water transfer. Photocatalytic membrane reactor consisting of polysulfone (PSU) and MOF (NH_2-MIL125(Ti)) was also introduced as an efficient hybrid system for successful decomposition of MB. Adding NH_2-MIL125(Ti) to the polymer increased hydrophilicity of the membrane owing to the natural hydrophilicity of the MOF particles [5]. In addition, the porosity of the membrane was increased from 0.338 to

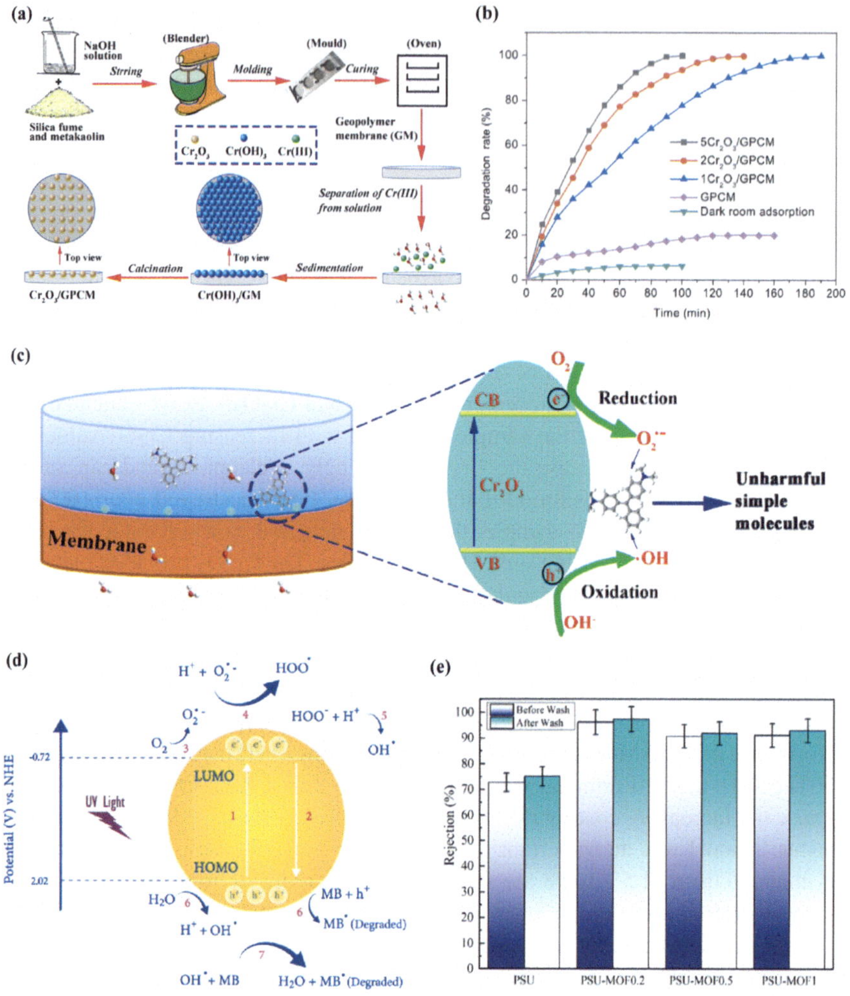

Fig. 6 **a** Schematic of Cr_2O_3/GPCM fabrication, **b** photocatalytic degradation activity of the membranes under static condition, **c** the mechanism of the degradation of dyes using Cr_2O_3/GPCM. Reproduced with permission from Ref. [16]. Copyright 2020, Elsevier. **d** Photocatalytic removal of MB dye with NH_2-MIL125(Ti) mechanism and **e** dye rejection of the photocatalytic membrane reactor system before and after washing. Reproduced with permission from Ref. [5]. Copyright 2022, Elsevier

0.432 after adding 0.2 wt% of MOF, increasing water flux. According to zeta potential analysis, the membrane demonstrated negatively charged which increased with adding MOF nanoparticles. As depicted in Fig. 6d, e, the MB removal was obtained 97% after absorption of UVC light by NH_2-MIL125(Ti). This sample showed the antifouling ability of the membrane, preventing the formation of a dye-foulant layer on the membrane surface. Application of four different MOFs (Ti-MOF, Zr-MOF,

Al-MOFs, and Fe-MOF)-enclosed functionalized cotton (O_x-cotton) hybrids has been exhibited efficient catalytic activity for removal of MB and RhB [2]. The maximal capacity for dyes adsorption was acquired for the Zr-MOF@O_x-cotton hybrid. However, the MB adsorption was more than RhB due to high reactivity and low molecular weight of MB. The Zr-MOF@O_x-cotton catalyst possessed high content of MOF, which led to great active sites in porous structure and functional groups. After five regenerated cycles, the dye adsorption ability was reduced by 10–12% in the presence of the Zr-MOF@O_x-cotton.

Composition of cellulose nanocrystal (CNC) with metal oxide has provided a hybrid material with good absorption, recyclability, and excellent degradation features. However, CNC-based catalysts possess low absorption ability due to their small porosity and specific surface area. Application of freeze-drying the air-bubble templated emulsion (Fig. 7a) was reported as an easy strategy to fabricate CNC/MnO_2 with porous microsphere surface, leading to high porosity of 98.23% and low density of 0.027 g cm^{-3} [37]. The rejection efficiency of MB was achieved at 95.4% during 10 min in the presence of CNC/MnO_2 (Fig. 7b), in which the MB molecules were firstly adsorbed by carboxyl groups of CNC and sodium alginate with negative charges and then oxidized and degraded to CO_2 or other molecules by MnO_2. Therefore, the decolorization of MB was performed during electrostatic adsorption and oxidative degradation by CNC/MnO_2. In another report, the MnO_2 NPs were supported onto the poly(amidoxime-hydroxamic acid)-modified microcrystalline cellulose (pAHA-MCC@MnO_2) using an oximation reaction followed by in-situ growth (Fig. 7c), forming flower-like MnO_2 NPs on hierarchical porous-structured cellulose microrods [32]. After five cycles, the maximal removal of MB was acquired 97.6% in 25 min. In acidic condition, the desire for MB oxidation was increased, causing the optimal catalytic performance at pH 3. Figure 7d presents a schematic of MB degradation process. However, the good catalytic activity of single metal oxide for wastewater treatment was limited by electron–hole recombination. Therefore, a CNC/manganese dioxide/titanium dioxide (CNC/MnO_2/TiO_2) with porous microspheres was prepared through a bubble template and ionic cross-linking of sodium alginate with Ca^{2+} ions, as depicted in Fig. 7e [63]. This catalyst showed significant removal efficiency of 97% for MB as well as fast regeneration and facile recovery process. Compared with CNC-based adsorbent, it also possessed more activity for MB degradation. As mentioned above, the CNC had great adsorption ability due to its abundant carboxylic group. Although TiO_2 possesses a wide band gap, it can rapidly degrade MB by generation of OH radicals after light irradiation. Adding the MnO_2 reduced the band gap of hybrid. Hence, the hybridization of TiO_2 and MnO_2 provided a synergetic effect which increased the absorbable wavelength of light for TiO_2 and prevented recombination of electron–hole pairs.

Bimetallic nanocomposite (Ag–Sn) prepared using easily hydrothermal approach was reported remarkable for photocatalytic degradation of MG, Methyl Red (MR), and MB (Fig. 8a) [49]. As a result, the Ag-Sn nanocomposite indicated better performance of MG photocatalytic degradation compared with Sn and Ag due to the electrons' generation by Ag as active species for degradation process and the Sn transition metal stimulating. (Table 1).

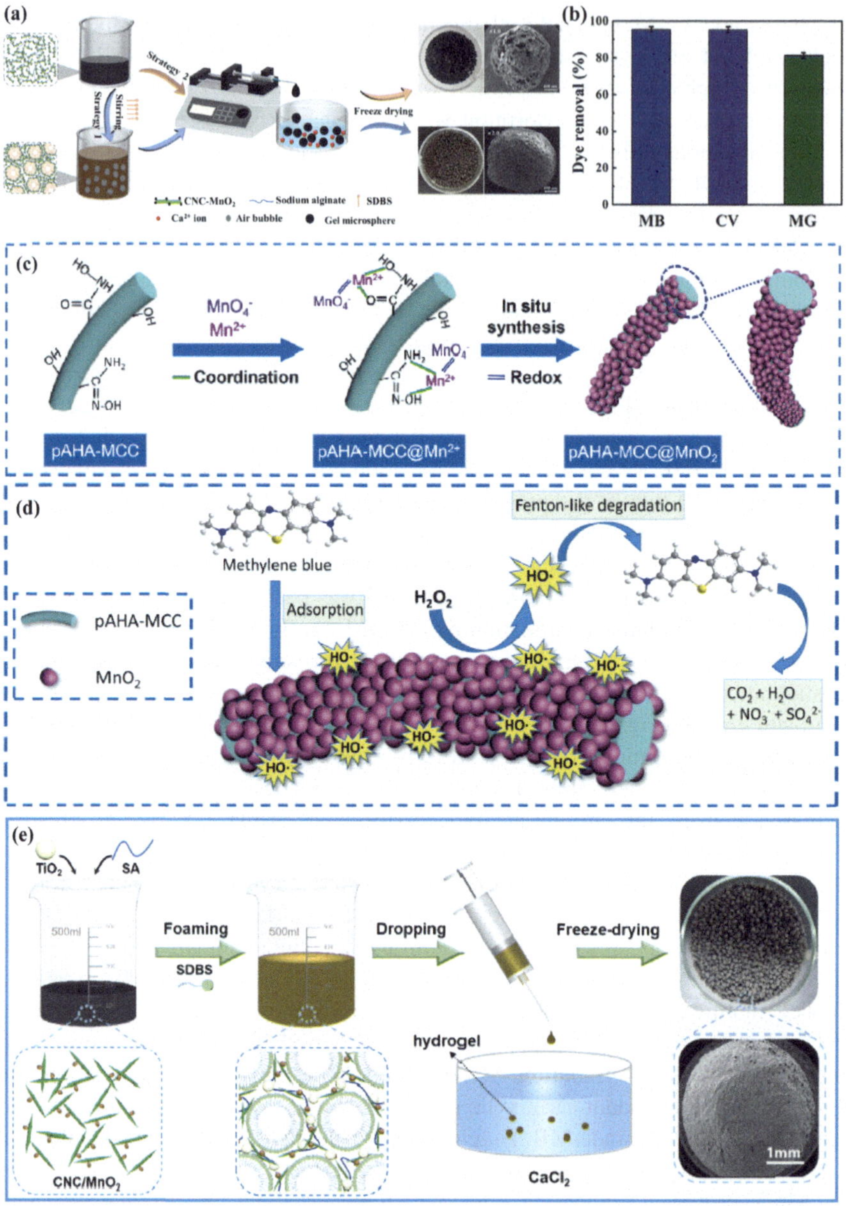

Fig. 7 **a** CNC/MnO_2/SA porous microspheres synthesis route, **b** assessment of rejection capability of various dyes. Reproduced with permission from Ref. [37]. Copyright 2021, Elsevier. **c** Schematic of the pAHA-MCC@MnO_2 synthesis, **d** a possible MB degradation mechanism in the presence of pAHA-MCC@MnO_2 microrods. Reproduced with permission from Ref. [32]. Copyright 2021, Elsevier. **e** The synthesis of CNC/MnO_2/TiO_2 porous microspheres. Reproduced with permission from Ref. [63]. Copyright 2021, Elsevier

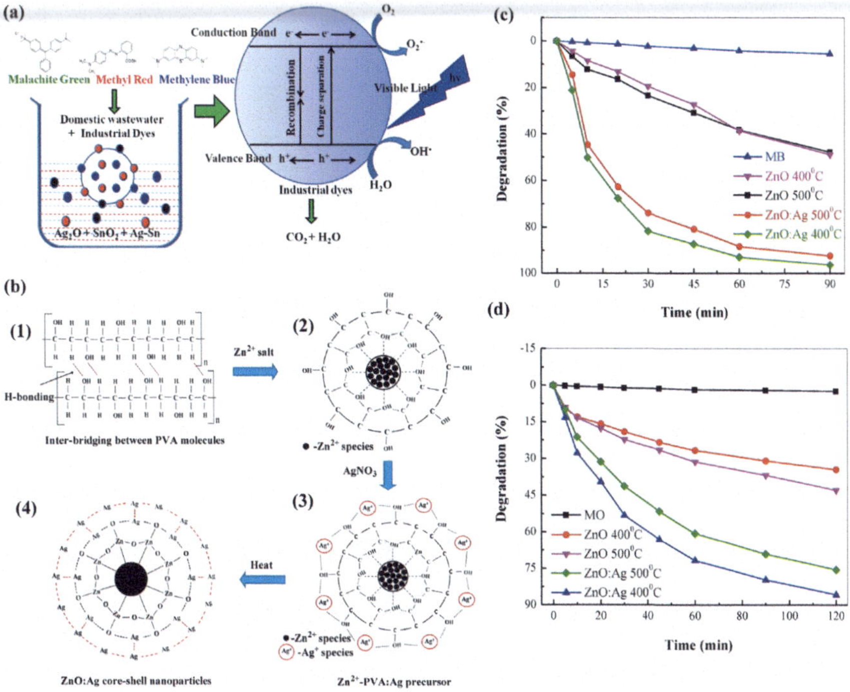

Fig. 8 **a** Dye degradation process using Ag_2O/SnO_2. Reproduced with permission from Ref. [49]. Copyright 2022, Elsevier. **b** A schematic of the growth mechanism of ZnO:Ag core–shell NPs. The photodegradation of **c** MB and **d** MO dyes using pristine and Ag-coated ZnO NPs. Reproduced with permission from Ref. [28]. Copyright 2018, Elsevier

2.2 Anionic Dye's Removal

Anionic dyes include negative ions that compose from the most diverse category of dyes with different specifications in their structure (azoic, anthraquinone, triphenylmethane, and nitro dyes). The anionic dyes include direct dyes, anionic azo dyes, and reactive dyes. Reactive dyes can create high covalent bonds with other materials, while acid dyes exhibit high solubility in water which have harmful effect on human [57].

By further investigation, magnetic $MnFe_2O_4$–rGO and $MnFe_2O_4$ NPs hybrid were synthesized with chemical deposition and reduction processes [72]. They were applied as catalysts in peroxymonosulfate (PMS) oxidation and generating active sulfate radicals for organic dyes (methyl violet, MB, MO, orange II, and RhB) degradation. They could separate with a suitable external magnet owing to the magnetic property and fenton-like activities and be used after four repeated cycles with high activities. The PMS activation by $MnFe_2O_4$ NPs includes both redox pairs of Fe(III)/Fe(II) and Mn(II)/Mn(III) species in $MnFe_2O_4$ NPs, leading to high catalytic activity. The investigation revealed that chemisorbed oxygen on the $MnFe_2O_4$–rGO surface

Table 1 Other metallic hybrid catalysts for cationic dyes' removal

Catalyst	The dye removal method	Dye	The removal performance	References
Polydopamine–chitosan–Fe_3O_4 (PDA/CS/Fe_3O_4)	Adsorption	MB	96.9%	Wang et al. [67]
$M_2Mo_4O_{13}$/α-MoO_3 (M = Li, Na ad K)	Advanced oxidation	Red GTL	99.8%	Zhang et al. [75]
HKUST-1@PVA-co-PE/PVA hydrogel	Selective adsorption	MB	96.1%	Zhu et al. [78, 79]
Alpha-Fe_2O_3 Nanodisk/bacterial cellulose (α-Fe_2O_3 nanodisk/ BFO)	Photodegradation	RhB	100%	Zhu et al. [78]
TiO_2-MnO_2	Oxidation	Methyl violet 2B (MV 2B)	96.8%	Mohod et al. [40]
NiCo@NCNTs	Advanced oxidation	MB	99%	Kang et al. [34]
Fe_3O_4/CuO NPs	Degrdation	MB	94%	Ghasemi et al. [23]
Cellulose acetate (CA)/Au/ZnO	Photodegradation	Eosin Y	95.6%	Abad et al. [1]
PtTAOPP-silica	Adsorption	MB	7.3 mg/g	Anghel et al. [9]
Kappa carrageenan-graft-poly(Nhydroxyethylacrylamide) (κC-g-PHEAA/Fe_3O_4)	Adsorption	MB and Rh6G	94.2% (MB) and 96.4% (Rh6G)	Kulal and Badalamoole [35]
Tungsten disulfide/tungsten trioxide (WS_2/WO_3)	Adsorption	RhB	237.1 mg/g	Li et al. [36]
Copper oxide/zinc oxide-tetrapods (CuO/ZnO-T)	Photodegradation	Yellow-145 (RY-145) and Basic violet-3 (BV-3)	80% (RY-145) and 86% (BV-3)	Sharma et al. [59]
Nanodiamond/UiO-66 (OND-UiO)	Adsorption	MG	43%	Molavi et al. [41]
Chitosan/poly(vinyl alcohol)/cationic Fe(III)-porphyrin (Cht/PVA-FeTMPyP)	Oxidation	MR	92%	Mota et al. [42]
Fe-BTC MOF	Adsorption	MG	177 mg/g	Delpiano et al. [19]
ZnS–WO_3–$CoFe_2O_4$	Photodegradation	MB	95.97	Palanisamy et al. [48]

(continued)

Table 1 (continued)

Catalyst	The dye removal method	Dye	The removal performance	References
g-C_3N_4/Mn–ZnO	Photodegradation	MB	98%	Qamar et al. [50]
Ag/g-C_3N_4/$LaFeO_3$	Photodegradation	MB	98.9%	Zhang et al. [74]
Silver-embedded porous TiO_2 (Ag–TiO_2)	Photodegradation	RhB	100%	Heng et al. [27]
AgZnO/polyoxometalates (AgZnO/POMs)	Adsorption	Basic magenta (BM)	94.1%	Tian et al. [65]
Polyarylene ether nitrile (PEN)-Fe^{3+}@TiO_2	Photodegradation	MB	99%	Duan et al. [20]
Fe–Al-1,4-benzene-dicarboxylic acid (FeAl(BDC)) MOF	Adsorption	RhB	48.59 mg/g	Singh et al. [61]
Tea leaf biochar-based nanocomposite (nAg-TC)	Adsorption	RhB	95.89%	Shaikh et al. [58]
Durian shell fiber–Fe_3O_4 MOF	Adsorption	MB	53.31 mg/g	Cai et al. [15]

was the most active oxygen species, which showed substantial role in the oxidation reaction. The $MnFe_2O_4$–rGO hybrid exhibited the higher catalytic degradation of orange II (90%) within 120 min compared with $MnFe_2O_4$ (55%). The synergistic function of $MnFe_2O_4$ and graphene enhanced the surface area, chemical reaction sites, and electron transport ability, so the $MnFe_2O_4$–rGO/PMS system had the ability to destroy the dye chromophoric structure. A controllable surface coating of Ag around the ZnO NPs led to form a hybrid ZnO:Ag core-shell nanostructure with improved photocatalytic dyes' degradation compared with ZnO (Fig. 8b) [28]. The photocatalytic reduction of MB was obtained about 96 and 92% with ZnO:Ag processed at 400 and 500 °C, respectively (Fig. 8c). Under similar experimental conditions, the photocatalytic removal of MO efficiency was 86 and 75% for nanostructure processed at 400 °C and 500 °C, respectively (Fig. 8d). The great performance at lower processing temperature was due to form high surface area and more active sites for interaction with dye molecules. In addition, the production of electron–hole pairs and oxygen species was facilitated due to Schottky barrier junctions between the ZnO core and Ag shell.

Hybrid cation exchanger with chain-like structure, poly-o-toluidine–thorium(IV)molybdophosphate (POT-TMP), has produced through sol–gel approach [6]. This hybrid matter showed significant chemical stability in acidic, basic, and organic solvents, possibility due to the attendance binding polymer. It was reported that the POT-TMP can act as a photocatalyst for Congo red (CR) dye degradation and 74% of dye was degraded over 8 h irradiation. Thorium and molybdenum disrupted dye molecules and produced electron–hole pairs through charge separation. Graphitic carbon nitride (g-C_3N_4), a two-dimensional metal-free polymer, is a rich in carbon and nitrogen and has a small band gap for visible-light absorption. Hence, it is a beneficial candidate for application in photocatalytic reaction. Magnetic solid-phase extraction (MSPE) is well-known as a noble mode of solid-phase extraction using magnetic adsorbents. Magnetic MOFs including Cu(II)-BTC (BTC = 1, 3, 5-benzenetricarboxylate) (Fe_3O_4@SiO_2–Cu-BTC) has been indicated remarkable performance toward quick MSPE of CR (97%) [69]. The Cu-BTC has been provided high surface area, high chemical stability, and large pore windows. Besides, the presence of Fe_3O_4@SiO_2 improved dispersion of the catalyst and segregation of its from water. In addition, the MSPE mode and the Fe_3O_4@SiO_2 can enhance the superficial area between the catalyst and liquid, leading to quick mass transfer and high CR adsorption ability. At pH of 5, the structure of CR was positive, while the surface of MOF Cu-BTC had negative charge, resulting great adsorption at pH 5. This hybrid absorbent can also adsorb MB (97.7%), Basic Red 2 (69.3%) and Crystal Violet (92.5%), which was due to its negative surface charge. Laccase (benzenediol oxygen oxidoreductase) has copper atoms in its active sites which can carry four electrons to oxygen as an electron acceptor and form water. Capsulizing laccase into a hybrid giant vesicles (AuNPs@vesicles), denoted as laccase $\subset$ AuNPs@vesicles, indicated high activity for decoloration of CR [68]. The decoloration efficiency was achieved 99% which was 2.3 times better than free laccase in 3 h. This biocatalyst was also durable and did not present remarkable reduction in decoloration efficiency (from 99.5 to 98.5%) during the fifth cycles.

Three-dimensional graphene structure can rapidly pass water due to its high specific surface area and porous structure. Random distribution of graphene sheets can seriously reduce the water flux. Therefore, the aerogel graphene with aligned sheets can be more beneficial for catalytic degradation of pollutants (Fig. 9a) [73]. The anisotropic hybrid aerogel based on $CoFe_2O_4$ and graphene (CFO@GA-A) was synthesized through hydrothermal synthesis followed by freeze-drying for dyes' removal. The long and vertical aligned pores of the anisotropic aerogel provided flux of 1100 L m^{-2} h^{-1} which was 450% higher than rough and zigzag pores in the isotropic aerogel. The CFO@GA-A had excellent efficiency for removal of indigo carmine, MO, orange II, and MG and showed fast degradation within 7.5–12.5 min. The rejection of indigo carmine was better than other dyes. In this aerogel, the dyes can adsorb by graphene due to π-π interaction and auxiliary adsorption. Then, the PMS of dyes' solution connected to $CoFe_2O_4$ and subsequently activated, which improved dye molecules adsorption in the aerogel. It is also indicated more stability for wastewater treatment. As depicted in Fig. 9b, nitrogen-doped porous carbon-encapsulating iron NPs (CN-Fe) hybrid can synthesize by carbothermal reduction process in the presence of polyaniline (PANI) and α-Fe_2O_3 precursors [17]. Coupling CN-Fe with peroxymonosulfate (PMS) (CN-Fe/PMS) has been provided anionic dyes removal efficiency over 98% within 40 min. The presence of Fe into CN led to activating PMS for pollutants' removal. It was also durable, with approximately no loss of activity after three months. A hybrid nanofibrous aerogel is based on aramid nanofibers (ANFs), carbon nanotubes (CNTs), and gold NPs (AuNPs) (Fig. 9c) [60]. The AuNPs on the surface have enhanced light absorption, while the presence of CNTs caused through-body light-to-heat activity. The AuNP@ANF/CNT aerogels with high porosity and open-cell cellular structure indicated purification capability of > 99% for sewage water with MB, MO, CR, and RhB (Table 2).

3 Conclusion

This chapter has presented a view about the application of metallic hybrid materials toward the removal of cationic and anionic dyes by different methods. Hybrid-based materials are introduced as stable and effective candidate for removal of dyes. The modification and functionalization of hybrid materials can improve their efficiency of dye removal. The presence of magnetic materials such as Fe_2O_4 and Fe_3O_4 in hybrid structures not only causes to collect the catalyst easier but also improve their stability during catalytic process. It is also improving the interaction of dyes with catalysts. The hybrid compounds including metal oxide such as TiO_2, ZnO, Cu_2O, and Cr_2O_3 have been indicated to have more photo-induced degradation of dyes due to low electron–hole recombination and their high dye molecules' adsorption ability. The MOFs were widely used to produce hybrid materials due to their functional groups and porous structure. Furthermore, carbon-based materials such as g-C_3N_4, CNTs, graphene, and GO were widely employed to enhance pore size, adsorption ability, and interaction of dyes with catalyst. It is worth mentioning that the pH values

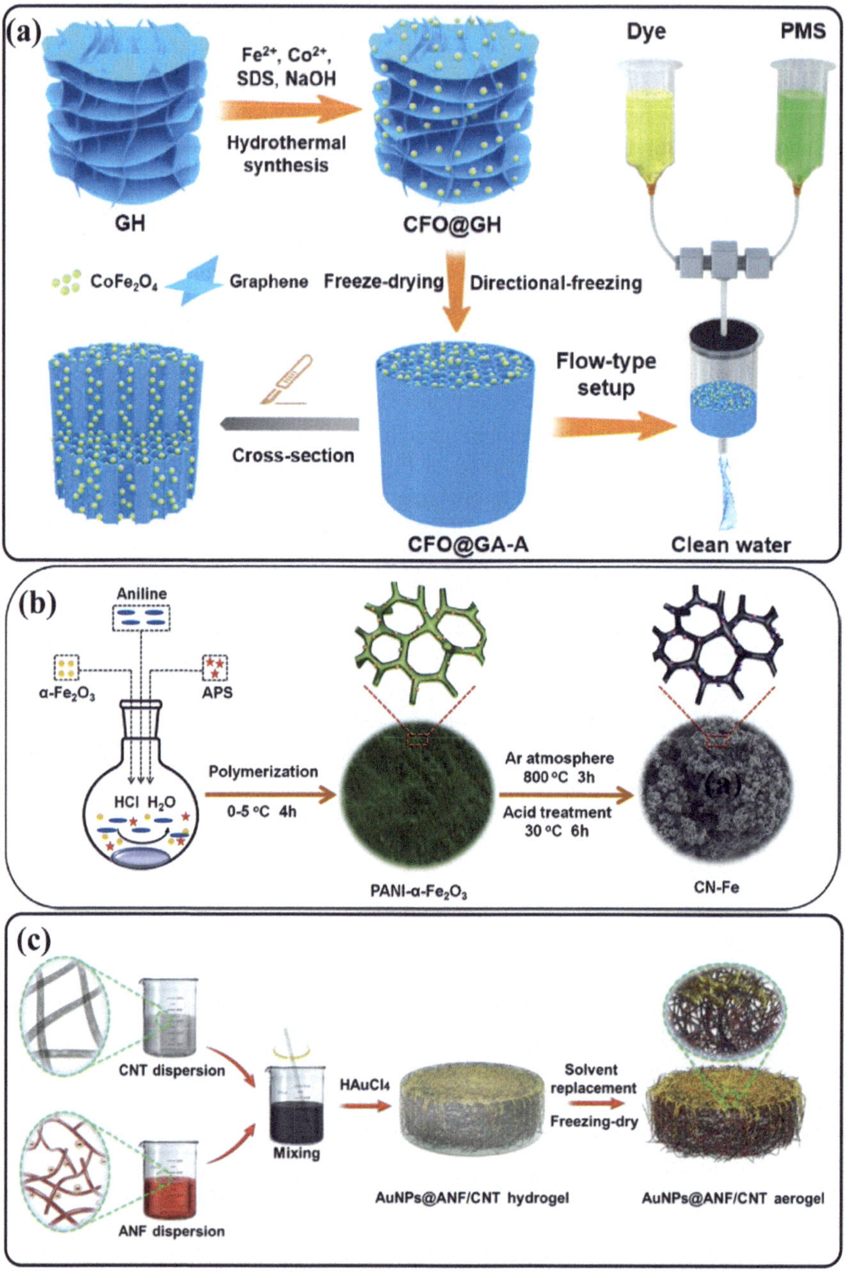

Fig. 9 **a** Schematic of the $CoFe_2O_4$@graphene hybrid aerogel synthesis. Reproduced with permission from Ref. [73]. Copyright 2019, American Chemical Society. **b** The CN-Fe synthesis process. Reproduced with permission from Ref. [17]. Copyright 2020, Elsevier. **c** Fabrication of AuNPs@ANF/CNT aerogel. Reproduced with permission from Ref. [60]. Copyright 2022, Elsevier

Table 2 Other metallic hybrid catalysts for anionic dyes' removal

Catalyst	The dye removal method	Dye	The removal performance	References
NiCo@NCNTs	Advanced oxidation	MO	100%	Kang et al. [34]
Carbon fibers-assisted iron carbide (Fe_3C@CFs)	Advanced oxidation	Acid Red 1 (AR1)	97.3%	Gao et al. [22]
Chitosan/poly(vinyl alcohol)/cationic Fe(III)-porphyrin (Cht/PVA-FeTMPyP)	Oxidation	MO	100%	Mota et al. [42]
Chitosan-bismuth cobalt selenide hybrid microspheres	Photodegradation	CR	85%	Ali et al. [8]
Solanum tuberosu peel–silver nanoparticle (STpe-Ag NP)	Adsorption	Bromophenol blue (BB)	88.5%	Akpomie and Conradie [7]
Nanodiamond/UiO-66 (OND-UiO)	Adsorption	MR	59%	Molavi et al. [41]
Fe-BTC MOF	Adsorption	Alizarin Red S (ARS)	80 mg/g	Delpiano et al. [19]
Fe/N-codoped carbon	Advanced oxidation	MO	99%	Yang et al. [71]
Silver-embedded porous TiO_2 (Ag–TiO_2)	Photodegradation	MO	99.7%	Heng et al. [27]
Tea leaf biochar-based nanocomposite (nAg-TC)	Adsorption	CR	94.10%	Shaikh et al. [58]

of solution were an effective factor because the charge surface of hybrids was change in different pH, leading to electrostatic repulsion/attraction between dye molecules and hybrid surface. However, there are some important issues to investigate in future about application of metal-based hybrid materials for dyes removal:

- Application of natural and affordable sources for preparation of hybrid materials for dye removal.
- Modification of hybrid materials by bimetallic NPs.
- More investigation about synthesis effective metallic hybrid electrodes for electrochemical degradation of dyes.
- Evaluation of metal-based hybrid performance for removal of more dyes.
- Investigation the possibility of removing mixed dyes.

References

1. Abad SNK, Mozammel M, Moghaddam J et al (2020) Highly porous, flexible and robust cellulose acetate/Au/ZnO as a hybrid photocatalyst. Appl Surf Sci 526:146237
2. Abdelhameed RM, Emam HE (2022) Modulation of metal organic framework hybrid cotton for efficient sweeping of dyes and pesticides from wastewater. SM&T 31:e00366
3. Abdi J, Vossoughi M, Mahmoodi NM, Alemzadeh I (2017) Synthesis of metal-organic framework hybrid nanocomposites based on GO and CNT with high adsorption capacity for dye removal. J Chem Eng 326:1145–1158
4. Ahmad PT, Jaleh B, Nasrollahzadeh M, Issaabadi Z (2019) Efficient reduction of waste water pollution using GO/γMnO_2/Pd nanocomposite as a highly stable and recoverable catalyst. Sep Purif Technol 225:33–40
5. Ahmadi A, Sarrafzadeh MH, Hosseinian A, Ghaffari SB (2022) Foulant layer degradation of dye in Photocatalytic Membrane Reactor (PMR) containing immobilized and suspended NH_2-MIL125 (Ti) MOF led to water flux recovery. J Environ Chem Eng 10(1):106999
6. Akhtar A, Khan MDA, Nabi SA (2015) Synthesis, characterization and photolytic degradation activity of poly-o-toluidine–thorium (IV) molybdophosphate cation exchanger: analytical application in metal ion treatment. Desalination 361:1–12
7. Akpomie KG, Conradie J (2020) Biogenic and chemically synthesized Solanum tuberosum peel–silver nanoparticle hybrid for the ultrasonic aided adsorption of bromophenol blue dye. Sci Rep 10(1):1–18
8. Ali N, Uddin S, Khan A et al (2020) Regenerable chitosan-bismuth cobalt selenide hybrid microspheres for mitigation of organic pollutants in an aqueous environment. Int J Biol Macromol 161:1305–1317
9. Anghel D, Lascu A, Epuran C et al (2020) Hybrid materials based on silica matrices impregnated with Pt-porphyrin or PtNPs destined for CO_2 gas detection or for wastewaters color removal. Int J Mol Sci 21(12):4262
10. Anis SM, Hashemi SH, Nasri A et al (2022) Decorated ZrO_2 by Au nanoparticles as a potential nanocatalyst for the reduction of organic dyes in water. Inorg Chem Commun 141:109489
11. Arnawtee WH, Jaleh B, Nasrollahzadeh M et al (2022) Lignin valorization: facile synthesis, characterization and catalytic activity of multiwalled carbon nanotubes/kraft lignin/Pd nanocomposite for environmental remediation. Sep Purif Technol 290:120793
12. Baba K, Bulou S, Quesada-Gonzalez M et al (2017) Significance of a noble metal nanolayer on the UV and visible light photocatalytic activity of anatase TiO_2 thin films grown from a scalable PECVD/PVD approach. ACS Appl Mater Interfaces 9(47):41200–41209

13. Bai S, Shen X, Zhong X et al (2012) One-pot solvothermal preparation of magnetic reduced graphene oxide-ferrite hybrids for organic dye removal. Carbon 50(6):2337–2346
14. Bhaviya Raj R, Umadevi M, Parimaladevi R (2021) Synergy photodegradation of basic dyes by ZnO/Bi_2O_3 nanocomposites under visible light irradiation. IJCCE 40(4):1121–1131
15. Cai Z, Liu Q, Li H et al (2022) Waste-to-resource strategy to fabricate functionalized MOFs composite material based on durian shell biomass carbon fiber and Fe_3O_4 for highly efficient and recyclable dye adsorption. Int J Mol Sci 23(11):5900
16. Chen H, Zhang YJ, He PY et al (2020) Coupling of self-supporting geopolymer membrane with intercepted Cr (III) for dye wastewater treatment by hybrid photocatalysis and membrane separation. Appl Surf Sci 515:146024
17. Chen L, Huang Y, Zhou M et al (2020) Nitrogen-doped porous carbon encapsulating iron nanoparticles for enhanced sulfathiazole removal via peroxymonosulfate activation. Chemosphere 250:126300
18. Chibac AL, Buruiana T, Melinte V et al (2015) Tuning the size and the photocatalytic performance of gold nanoparticles in situ generated in photopolymerizable glycomonomers. RSC Adv 5(110):90922–90931
19. Delpiano GR, Tocco D, Medda L et al (2021) Adsorption of malachite green and alizarin red s dyes using Fe-BTC metal organic framework as adsorbent. Int J Mol Sci 22(2):788
20. Duan F, Liu C, Liu X et al (2021) TiO_2 immobilized on polyarylene ether nitrile/Fe^{3+} complex for efficient adsorption and photocatalytic degradation towards methylene blue. J Alloys Compd 875:159951
21. Fan T, Deng W, Feng X et al (2020) An integrated electrocoagulation–electrocatalysis water treatment process using stainless steel cathodes coated with ultrathin TiO_2 nanofilms. Chemosphere 254:126776
22. Gao S, Zhou H, Xia Y et al (2019) Carbon fiber-assisted iron carbide nanoparticles as an efficient catalyst via peroxymonosulfate activation for organic contaminant removal. Catal Sci Technol 9(16):4365–4373
23. Ghasemi H, Aghabarari B, Alizadeh M et al (2020) High efficiency decolorization of wastewater by Fenton catalyst: magnetic iron-copper hybrid oxides. J Water Process Eng 37:101540
24. Guo J, Wang R, Tjiu WW et al (2012) Synthesis of Fe nanoparticles@ graphene composites for environmental applications. J Hazard Mater 225:63–73
25. Hassan MM, Carr CM (2021) Biomass-derived porous carbonaceous materials and their composites as adsorbents for cationic and anionic dyes: a review. Chemosphere 265:129087
26. He HB, Li B, Dong JP et al (2013) Mesostructured nanomagnetic polyhedral oligomeric silsesquioxanes (POSS) incorporated with dithiol organic anchors for multiple pollutants capturing in wastewater. ACS Appl Mater Interfaces 5(16):8058–8066
27. Heng JZX, Tang KY, Regulacio MD et al (2021) Solar-powered photodegradation of pollutant dyes using silver-embedded porous TiO_2 nanofibers. Nanomaterials 11(4):856
28. Jadhav J, Biswas S (2018) Hybrid ZnO: Ag core-shell nanoparticles for wastewater treatment: growth mechanism and plasmonically enhanced photocatalytic activity. Appl Surf Sci 456:49–58
29. Jaleh B, Karami S, Sajjadi M, et al (2020a) Laser-assisted preparation of Pd nanoparticles on carbon cloth for the degradation of environmental pollutants in aqueous medium. Chemosphere 246:125755
30. Jaleh B, Nasrollahzadeh M, Mohazzab B et al (2021) State-of-the-art technology: Recent investigations on laser-mediated synthesis of nanocomposites for environmental remediation. Ceram Int 47(8):10389–10425
31. Jaleh B, Zare E, Azizian S, et al (2020b) Preparation and characterization of polyvinylpyrrolidone/polysulfone ultrafiltration membrane modified by grapheme oxide and titanium dioxide for enhancing hydrophilicity and antifouling properties. J Inorg Organomet Polym 30(6):2213–2223
32. Jiao C, Wei N, Liu D et al (2021) Sustainable Fenton-like degradation of methylene blue over MnO_2-loaded poly (amidoxime-hydroxamic acid) cellulose microrods. Int J Biol Macromol 193:1952–1961

33. Johari NA, Yusof N, Lau WJ et al (2021) Polyethersulfone ultrafiltration membrane incorporated with ferric-based metal-organic framework for textile wastewater treatment. Sep Purif Technol 270:118819
34. Kang J, Zhang H, Duan X et al (2019) Magnetic Ni-Co alloy encapsulated N-doped carbon nanotubes for catalytic membrane degradation of emerging contaminants. J Chem Eng 362:251–261
35. Kulal P, Badalamoole V (2020) Hybrid nanocomposite of kappa-carrageenan and magnetite as adsorbent material for water purification. Int J Biol Macromol 165:542–553
36. Li G, Wang Y, Bi J et al (2020) Partial oxidation strategy to synthesize WS_2/WO_3 heterostructure with enhanced adsorption performance for organic dyes: synthesis, modelling, and mechanism. Nanomaterials 10(2):278
37. Li J, Zhou L, Song Y et al (2021) Green fabrication of porous microspheres containing cellulose nanocrystal/MnO_2 nanohybrid for efficient dye removal. Carbohydr Polym 270:118340
38. Mansor MSAC, Amir MNI, Julkapli NM, Ma'amor A (2020) Gold hybrid nanomaterials: prospective on photocatalytic activities for wastewater treatment application. Mater Chem Phys 241:122415
39. Mohazzab BF, Jaleh B, Nasrollahzadeh M et al (2020) Upgraded valorization of biowaste: laser-assisted synthesis of Pd/calcium lignosulfonate nanocomposite for hydrogen storage and environmental remediation. ACS Omega 5(11):5888–5899
40. Mohod AV, Hinge SP, Raut RS et al (2018) Process intensified removal of methyl violet 2B using modified cavity-bubbles oxidation reactor. J Environ Chem Eng 6(1):574–582
41. Molavi H, Neshastehgar M, Shojaei A, Ghashghaeinejad H (2020) Ultrafast and simultaneous removal of anionic and cationic dyes by nanodiamond/UiO-66 hybrid nanocomposite. Chemosphere 247:125882
42. Mota HP, Quadrado RF, Burgo TA et al (2020) Polysaccharide/Fe (III)-porphyrin hybrid film as catalyst for oxidative decolorization of toxic azo dyes: an approach for wastewater treatment. Arab J Chem 13(7):5923–5938
43. Mousavi SS, Jaleh B, Nasrollahzadeh M et al (2021) Hardystonite/palladium nanocomposite as a high performance catalyst for electrochemical hydrogen storage and Cr (VI) reduction. Int J Hydrog Energy 46(49):25175–25188
44. Nasri A, Jaleh B, Nezafat Z et al (2021a) Fabrication of g-C_3N_4/Au nanocomposite using laser ablation and its application as an effective catalyst in the reduction of organic pollutants in water. Ceram Int 47(3):3565–3572
45. Nasri A, Nezafat Z, Jaleh B et al (2021b) Laser-assisted preparation of C_3N_4/Fe_2O_3/Au nanocomposite: a magnetic reusable catalyst for pollutant degradation. Clean Technol Environ Policy 23(6):1797–1806
46. Nasrollahzadeh M, Shafiei N, Eslamipanah M et al (2020) Preparation of Au nanoparticles by Q switched laser ablation and their application in 4-nitrophenol reduction. Clean Technol Environ Policy 22(8):1715–1724
47. Nezafat Z, Mohazzab BF, Jaleh B et al (2021) A promising nanocatalyst: upgraded Kraft lignin by titania and palladium nanoparticles for organic dyes reduction. Inorg Chem Commun 130:108746
48. Palanisamy G, Bhuvaneswari K, Bharathi G et al (2021) Construction of magnetically recoverable ZnS–WO_3–$CoFe_2O_4$ nanohybrid enriched photocatalyst for the degradation of MB dye under visible light irradiation. Chemosphere 273:129687
49. Pandiyan R, Dharmaraj S, Ayyaru S et al (2022) Ameliorative photocatalytic dye degradation of hydrothermally synthesized bimetallic Ag-Sn hybrid nanocomposite treated upon domestic wastewater under visible light irradiation. J Hazard Mater 421:126734
50. Qamar MA, Shahid S, Javed M et al (2021) Fabricated novel g-C_3N_4/Mn doped ZnO nanocomposite as highly active photocatalyst for the disinfection of pathogens and degradation of the organic pollutants from wastewater under sunlight radiations. Colloids Surf, A Physicochem Eng Asp 611:125863
51. Ramalingam G, Pachaiappan R, Kumar PS et al (2022) Hybrid metal organic frameworks as an exotic material for the photocatalytic degradation of pollutants present in wastewater: a review. Chemosphere 288:132448

52. Ranjith KS, Manivel P, Rajendrakumar RT, Uyar T (2017) Multifunctional ZnO nanorod-reduced graphene oxide hybrids nanocomposites for effective water remediation: Effective sunlight driven degradation of organic dyes and rapid heavy metal adsorption. Chem Eng J 325:588–600
53. Rigoletto M, Calza P, Gaggero E, Laurenti E (2022) Hybrid materials for the removal of emerging pollutants in water: classification, synthesis, and properties. Chem Eng J Adv 10:100252
54. Ruiz-Santoyo V, Andrade-Espinoza BA, Romero-Toledo R et al (2022) Use of nanostructured photocatalysts for dye degradation: a review. Period Polytech Chem Eng 66(3):367–393
55. Sajesh K, Jayakumar R, Nair SV, Chennazhi K (2013) Biocompatible conducting chitosan/polypyrrole–alginate composite scaffold for bone tissue engineering. Int J Biol Macromol 62:465–471
56. Sakthivel T, Ramachandran R, Kirubakaran K (2018) Photocatalytic properties of copper-two dimensional graphitic carbon nitride hybrid film synthesized by pyrolysis method. J Environ Chem Eng 6(2):2636–2642
57. Salleh MAM, Mahmoud DK, Karim WAWA, Idris A (2011) Cationic and anionic dye adsorption by agricultural solid wastes: a comprehensive review. Desalination 280(1–3):1–13
58. Shaikh WA, Kumar A, Chakraborty S et al (2022) Biochar-based nanocomposite from waste tea leaf for toxic dye removal: From facile fabrication to functional fitness. Chemosphere 291:132788
59. Sharma M, Poddar M, Gupta Y et al (2020) Solar light assisted degradation of dyes and adsorption of heavy metal ions from water by CuO–ZnO tetrapodal hybrid nanocomposite. Mater. Today Chem. 17:100336
60. Shi L, Sun K, Zhang G et al (2022) Hybrid nanofibrous aerogels for All-in-one solar-driven interfacial evaporation. J Colloid Interface Sci
61. Singh H, Raj S, Rathour RKS, Bhattacharya J (2022) Bimetallic Fe/Al-MOF for the adsorptive removal of multiple dyes: optimization and modeling of batch and hybrid adsorbent-river sand column study and its application in textile industry wastewater. Environ Sci Pollut Res:1–16
62. Singh S, Barick K, Bahadur D (2013) Functional oxide nanomaterials and nanocomposites for the removal of heavy metals and dyes. Nanomater Nanotechnol 3(Godište 2013):3–20
63. Song Y, Jiang M, Zhou L et al (2022) Rapidly regenerated CNC/TiO_2/MnO_2 porous microspheres for high-efficient dye removal. Carbohydr Polym 292:119644
64. Sriram G, Bendre A, Mariappan E et al (2021) Recent trends in the application of metal-organic frameworks (MOFs) for the removal of toxic dyes and their removal mechanism-a review. Sustain Mater Technol:e00378
65. Tian H, Luo J, Zhang K et al (2021) Synergistic photocatalytic-adsorption removal of basic magenta effect of AgZnO/polyoxometalates nanocomposites. Nanoscale Res Lett 16(1):1–12
66. Wan X, Zhan Y, Long Z et al (2017) Core@ double-shell structured magnetic halloysite nanotube nano-hybrid as efficient recyclable adsorbent for methylene blue removal. Chem Eng J 330:491–504
67. Wang Y, Zhang Y, Hou C, Liu M (2016) Mussel-inspired synthesis of magnetic polydopamine–chitosan nanoparticles as biosorbent for dyes and metals removal. J Taiwan Inst Chem Eng 61:292–298
68. Wu G, Liu X, Zhou P et al (2019) A facile approach for the reduction of 4-nitrophenol and degradation of congo red using gold nanoparticles or laccase decorated hybrid inorganic nanoparticles/polymer-biomacromolecules vesicles. Mater Sci Eng C 94:524–533
69. Xu Y, Jin J, Li X et al (2015) Magnetization of a Cu (II)-1, 3, 5-benzenetricarboxylate metal-organic framework for efficient solid-phase extraction of Congo Red. Mikrochim Acta 182(13):2313–2320
70. Yang F, Du M, Yin K et al (2022) Applications of metal-organic frameworks in water treatment: a review. Small 18(11):2105715
71. Yang L, Chen W, Sheng C et al (2021) Fe/N-codoped carbocatalysts loaded on carbon cloth (CC) for activating peroxymonosulfate (PMS) to degrade methyl orange dyes. Appl Surf Sci 549:149300

72. Yao Y, Cai Y, Lu F et al (2014) Magnetic recoverable $MnFe_2O_4$ and $MnFe_2O_4$-graphene hybrid as heterogeneous catalysts of peroxymonosulfate activation for efficient degradation of aqueous organic pollutants. J Hazard Mater 270:61–70
73. Yu XJ, Qu J, Yuan Z et al (2019) Anisotropic $CoFe_2O_4$@ graphene hybrid aerogels with high flux and excellent stability as building blocks for rapid catalytic degradation of organic contaminants in a flow-type setup. ACS Appl Mater Interfaces 11(37):34222–34231
74. Zhang W, Ma Y, Zhu X et al (2021) Fabrication of Ag decorated g-C_3N_4/$LaFeO_3$ Z-scheme heterojunction as highly efficient visible-light photocatalyst for degradation of methylene blue and tetracycline hydrochloride. J Alloys Compd 864:158914
75. Zhang Y, Zhang Z, Yan Q, Wang Q (2016) Synthesis, characterization, and catalytic activity of alkali metal molybdate/α-MoO_3 hybrids as highly efficient catalytic wet air oxidation catalysts. Appl Catal A: Gen 511:47–58
76. Zhang Z, Gao Y, Wang Q (2018) Fabrication, activity and mechanism studies of transition metal molybdate/molybdenum trioxide hybrids as novel CWAO catalysts. Sep Purif Technol 191:354–363
77. Zhao S, Zhu H, Wang Z et al (2018) A loose hybrid nanofiltration membrane fabricated via chelating-assisted in-situ growth of Co/Ni LDHs for dye wastewater treatment. Chem Eng J 353:460–471
78. Zhu Q, Li Y, Wang W et al (2018) High performance HKUST-1@ PVA-co-PE/PVA hybrid hydrogel with enhanced selective adsorption. Compos Commun 10:36–40
79. Zhu ZS, Qu J, Hao SM et al (2018) α-Fe_2O_3 nanodisk/bacterial cellulose hybrid membranes as high-performance sulfate-radical-based visible light photocatalysts under stirring/flowing states. ACS Appl Mater Interfaces 10(36):30670–30679

Nano-engineered Hybrid Materials for Cationic Dye Removal

Nana Aboagye Acheampong, Emmanuel Okoampah, Nana Kobea Bonso, and Abubakari Zarouk Imoro

Abstract Nano-engineered hybrid materials are materials that are made of nanoparticle sizes covalently or non-covalently bonded to each other. Their varied compositions (organic and inorganic materials) explain their hybrid nature. They are generally synthetic materials. Nano-engineered hybrid materials have several uses in industry including wastewater treatment. Many studies have reported their use for the treatment of dye wastewaters. For instance, cationic dyes including anthraquinone dyes and azo dyes can be removed from wastewater by the use of nano hybrid materials. This chapter presents discussions on cationic dyes, nano-engineered hybrids materials, the synthesis route for obtaining nanoparticles, and their use in the treatment of wastewater containing cationic dyes. Also, a teaser on the potential regeneration of nanoparticles after use is presented.

Keywords Nano-engineered materials · Dye · Wastewater treatment · Nanomaterials · Nano-photocatalysts

1 Introduction

Hybrid nanomaterials are derived from the combinations of at least two types of nanomaterials, whether they are inorganic or organic or a mixture of the two [51]. The final product is a substance that works well with good qualities as well as

N. A. Acheampong
Department of Microbiology, Faculty of Biosciences, University for Development Studies, Tamale, Ghana

E. Okoampah
Department of Biochemistry, Faculty of Biosciences, University for Development Studies, Tamale, Ghana

N. K. Bonso · A. Z. Imoro (✉)
Department of Environment, Water and Waste Engineering, School of Engineering, University for Development Studies, Tamale, Ghana
e-mail: zabubakari@uds.edu.gh

© The Author(s), under exclusive license to Springer Nature Singapore Pte Ltd. 2023
A. Ahmad et al. (eds.), *Nanohybrid Materials for Treatment of Textiles Dyes*,
Smart Nanomaterials Technology, https://doi.org/10.1007/978-981-99-3901-5_13

performance to create applications with distinct characteristics, controlled by the molecular or supramolecular level interaction of the combined elements.

Due to the challenge of creating materials with more potential multifunctional uses, the addition of nanoscale components has become necessary because it increases the unique qualities of hybrid materials even more. In this light, nano chemistry and molecular engineering are currently focusing more on nanoscale materials' atom- and molecule-level manipulation. The uniqueness and efficiency of nano hybrid materials have brought about their use in many industries including wastewater sector where they are used for dye removal [15].

Generally, dyes are complex organic molecules containing a benzene ring, a chromophore (chemical group that imparts color), and an auxochrome (chemical group that conveys ionization property to the chromogen) [23]. They impart color upon adsorption onto substrate surfaces. They have been widely employed in diverse industries such as the paper, food, plastic, cosmetic, textile, leather, pulp, and paint industries [92]. Dyes are primarily classified into two major groups, namely natural and synthetic dyes. The advent of synthetic dyes in the early nineteenth century saw a drastic shift in preference from natural dyes to synthetic alternatives due to their cost effectiveness, improved affinity, and substantivity for textile fiber [36]. Synthetic dyes can be broadly categorized into anionic, cationic, and non-ionic dyes. Although the present cationic dyes are the most technically excellent of all dyes, they have been reported to be more harmful than the others [34].

Over time, industries ranging from food, textile, paper, paints, oils, soaps, and so on have emerged that make use of various cationic dyes such as methylene blue (MB) [50], rhodamine B (RhB) [8], safranine T (T) [71], gentian violet (GV) [74], and fuchsin basic (FB) [65]. These dyes are highly colored and non-biodegradable. Their application generates huge volumes of wastewater, particularly in the textile industry, where about 40–50 L of dye wastewater is generated per 1 kg of the product formed [62]. The effluents of this industry in many cases end up in the sea via streams, affecting human and aquatic sustenance [55, 122, 124, 126].

According to the UN and WHO, approximately two (2) billion people live in countries with high levels of water stress [20, 121]. In 2010, the United Nations (UN) stated that access to quality and safe drinking water is a human right (SDG 6) [31, 108, 120]. Water is an essential element that saves lives, so we must protect it at all costs. Literature has revealed that approximately 700,000 tons of dyes are produced annually to meet the growing demands of industry and commerce [13, 54]. However, since regulatory bodies in developing countries fail to regulate the discharges of dye waste, the disposal of effluent into the sea from these dye industries has become a topic of public discourse, as there are concerns about cancer-related diseases caused by drinking polluted water and eating aquatic animals. This warrants sustainable purification technologies to make polluted water pure and clean for human consumption.

Over time, existing purification technologies such as physical processes (filtration, sedimentation, etc.); chemical processes (flocculation and chlorination); and biological processes (activated sludge) have been shown to have low efficiency in decontaminating dyes from water. Some even produce by-products that have been linked to cancer [22, 118]. This necessitates improvements in the efficiency of traditional water purification technologies to make them more efficient and environmentally friendly. The field of nanotechnology has emerged as promising for augmenting conventional water purification technologies in order to mitigate the negative consequences (cancer causing by-products) some of these conventional technologies bring about [15, 22, 70].

Nanotechnology is the manipulation of matter on a near-atomic scale to create new structures ranging in size from 1 to 100 nm, with nanoparticles (NPs) serving as their fundamental unit. The intrinsic physicochemical properties of NPs, such as size, shape, surface chemistries, hydrophilicity, hydrophobicity, and so on [128, 130], greatly improve their performance as nano-adsorbents [100], nano-photocatalysts [94], and nano-membranes [127]. With these NP applications, new strategies in dye-contaminated water treatment technologies are introduced, with limited inefficiencies [15].

Researchers have successfully synthesized several NPs such as gold nanoparticles (AuNPs), silver nanoparticles (AgNPs), graphene, carbon dots (CDs), copper nanoparticles (NPs), zinc oxide nanoparticles (ZnO NPs) and many more as nano-adsorbents, nano-membranes, and nano-photocatalysts for removing dye from wastewater [107, 123]. Against this backdrop, we focus in this book chapter on cationic dyes as a pollutant in water, NPs in perspective, and the synthesis route for obtaining NPs. The various NPs approaches that can be used to degrade dyes from water as a sustainable strategy are discussed.

2 Cationic Dyes

Cationic (also known as basic/alkaline) dyes represent a group of water-soluble synthetic dyes that carry localized or delocalized positive charge in their molecules that can easily dissociate into colored positively charged ions in aqueous solution [141]. The main chemical classes are acridine, triarylmethane, cyanine, diazahemicyanine, hemicyanine, oxazine, and thiazine [141]. Cationic dyes possess aromatic and functional groups that render stability and recalcitrance on the dyes [141]. They are the brightest class of soluble dyes among all commercial synthetic dyes available. They have profound tinctorial strength and easily bind to negatively charged surface of cells, fiber or adsorbents by electrostatic force [117, 141].

2.1 Types of Cationic Dyes

Cationic dyes can be categorized into two main types: localized or pendant cationic dyes (where the positive charge is localized on a single atom, often nitrogen) and delocalized cationic dyes (where the positive charge is delocalized over all the atoms in the molecule) [141]. The current forms of cationic dyes for acrylic combine remarkably with high tinctorial strength, brightness, and light fastness, which make them the most technically excellent dyes [34].

2.2 *Properties of Cationic Dyes*

Cationic dyes contain an auxochrome in addition to their chromophore, a quaternary group ($-N(CH_3)_3$) tied with an alkilic chain that exists as a guide anion which aids the solubility and isolation of the dye [109]. Generally, cationic dyes impart color onto acrylic fiber by binding to acidic groups in the third monomer of acrylon, thus ensuring remarkable fastness [106]. The increased mobility of electrons as a result of the dissociation of cationic dye molecules results in the generation of attractive cationic groups, which have bathochromic as well as light and industrial gases resistance ameliorating effects [109]. The positively charged colored cation confers sufficient solubility in aqueous solution in order to impart color onto the substrate [109].

2.3 *Uses of Cationic Dyes*

Some cationic dyes have been highlighted below to expatiate on the merits and demerits of cationic dye applications:

(a) **Malachite Green**

Malachite green (also known as Basic green 4) is a cationic dye of the triphenylmethane group that has the chemical formula $C_{23}H_{25}N_2Cl$ as depicted in Fig. 1. It is the most common colorant spanning several industrial applications [51]. It is extensively used in the textile, paper and leather industries for dyeing cotton, silk, wool, jute, paper, pulp, and leather [70, 131]. It serves as a key input in the manufacture of some pharmaceuticals, anthelminthic and medical disinfectants [115]. In the aquaculture industry worldwide, it is widely used as fungicide, bactericide, and parasiticide. In the food and distillery industries, it is used as a food additive and coloring agent [115].

However, it has been reported to be extremely harmful to mammalian cells [51]. It has also been implicated as teratogenic, carcinogenic, and mutagenic [70]. It causes liver, eye, lung, spleen, kidney, heart, brain, bone, and skin lesions [115].

(b) **Methylene Blue**

Methylene blue (also known as Swiss blue) is a basic dye belonging to the thiazine group and has the chemical formula $C_6H_{18}N_3SCl$ as depicted in Fig. 2. It has a wide application spanning the textile (for cotton and wool dyeing), paper, renewable energy (for solar cells), and cosmetic (for hair dyeing) industries due to its brightness even

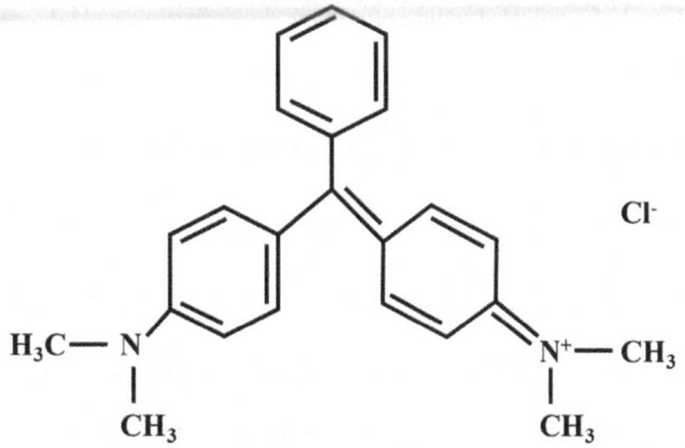

Fig. 1 Chemical structure of malachite green

at lower concentrations [51]. Also, it is used to stain and inactivate microorganisms [51]. It is also utilized as an antidote for carbon dioxide and cyanide poisoning as well as an antiseptic [51]. However, it is noted for several harmful effects on mankind and the environment [51]. It has been implicated as a causative agent of gastrointestinal disorders, headache, jaundice, quadriplegia, fever, dizziness, tissue necrosis, high blood pressure, nausea, shock, and cyanosis [57, 90]. Contact with the eye can cause eye burns that could lead to permanent eye damage [51]. Its inhalation could lead to acute breathing disorders, whilst its ingestion could lead to nausea, methemoglobinemia, profuse sweating, and mental disorder [82].

(c) **Crystal Violet**

Crystal violet (also known as Gentian violet) belongs to triarylmethane group of cationic dyes and has the chemical formula $C_{25}H_{30}N_3Cl$ as depicted in Fig. 3. It is commonly used to dye products such as paper, leather, detergent, and fertilizer [51]. Other critical areas of application include dermatology, textile, paint, veterinary and printing industries [91, 104]. Also, it is employed in poultry feed production to suppress the growth of molds and intestinal parasites [34, 81]. Additionally, crystal violet is also invaluable to gram staining, in terms of classifying bacteria, and has also demonstrated both antibacterial and antifungal abilities [89]. Nonetheless, it has been reported to be extremely toxic in nature and harmful to the environment [83]. It is known to be mutagenic and carcinogenic [26]. Also, it is extremely toxic to mammalian cells and can pose as an irritant to the skin and the gastrointestinal tract [77, 78]. Moreover, it can cause eye irritations, vomiting, jaundice, paralysis,

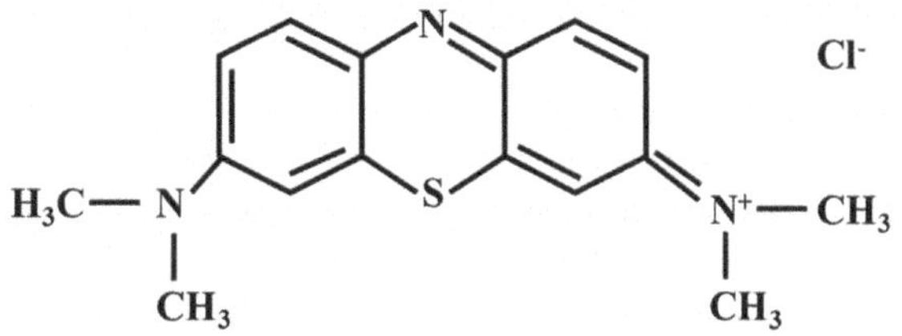

Fig. 2 Chemical structure of methylene blue

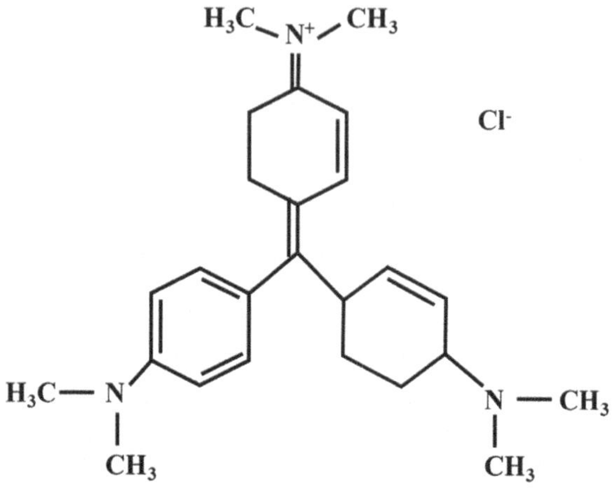

Fig. 3 Chemical structure of crystal violet

cyanosis, and heart diseases in humans [131]. In severe cases, it can lead to kidney and lung failure [5, 91]. Even more alarmingly, it is known to pose great threat to the environment (even at a low level of 1 ppb) and thus must be eradicated from the environment [42].

In spite of the potential harm cationic dyes pose to mankind and the environment, they are still being used in several industrial processes as mentioned earlier. Thus, it is imperative to develop effective technologies to remove cationic dyes from industrial effluents or wastewater streams before they are discharged into receiving waters or the environment. This will minimize the threat they pose to humans and the environment.

3 Nanomaterials

Nanomaterials are materials (matter) with nanoscale dimensions ranging from 1 to 100 nm. Engineered nanomaterials are materials that have been engineered into small sizes that fall within the scale of a nanoscale [29, 53, 130]. They can be found in nature and are made from bulk metals such as gold, silver, copper, carbon-containing materials, and so on. Nanomaterials obtained from bulk metals or precursors may have different physicochemical properties from the bulk metal or precursors. As a fundamental unit of nanotechnology, several (NPs) have been developed, including gold nanoparticles (AuNPs), silver nanoparticles (AgNPs), copper nanoparticles, carbon dots (CDs), graphene, fullerene, and others [17, 49, 61, 128]. Inorganic NPs (metal NPs and metal oxide), organic NPs (dendrimers, liposomes and micelles, and ferritin), ceramic NPs, and carbon base NPs are the various classes of these NPs (fullerene, carbon nanotube, graphene, and carbon nanofiber) [26, 67, 72]. Their distinct physicochemical properties, such as size, shape, surface chemistry, hydrophilicity, and hydrophobicity, have piqued the interest of researchers to study them for several applications including in the field of biomedical science [21], pharmaceutical [63], renewable energy generation [140], wastewater treatment [32], and many more. NPs emerging importance for the removal of dye from water has drawn the interest of many researchers these days thus its consideration as important subject matter for this book.

3.1 Synthesis of Metal Nanoparticles

Several synthesis routes for reducing bulk metals to their NPs form have existed for decades [101]. These metal NPs are obtained from their bulk metals in the presence of reducing and stabilizing agents such as sodium borohydrate, citrate, glutathione, plant materials, and others which reduce the oxidation number of the bulk metal from + 3 to + 0, thus diminishing microparticles to NPs [39, 93]. Some of the reducing agents also act as stabilizing agents, preventing NP aggregation after the reduction. The synthesis route of NPs has been divided into top-down and bottom-up approaches, which have been classified as chemical, physical, and biological synthesis approaches [38, 58]. The bottom-up method, also known as the "wet" method, employs the chemical synthesis approach which is based on nucleating atomic-sized matter into NPs.

The co-precipitation [12], polyol method [104], micro-emulsion [119], thermal decomposition [44], electrochemical [137], microbial [45], and plant synthesis [78] are all common methods. Top-down methods also employ the physical synthesis approach which necessitate the physical breakdown of bulk metals into smaller molecules through milling [30], laser ablation [98], spark ablation [35], and microwave [60]. It is worth mentioning that the biological synthesis approach follows the principles of top-down method. The various synthetic routes are discussed further below.

3.2 Chemical Synthesis Approach

The chemical synthesis approach which employs the principles of the bottom-up approach has gained much attention over time for reducing bulk metals into their NPs form. Some frequently used chemical synthesis methods are discussed below.

3.3 Co-precipitation

The co-precipitation technique involves the precipitation of metals in the form of hydroxide from a salt precursor in a solvent with the assistance of a base [112]. Its reactions entail the simultaneous occurrence of nucleation, growth, coarsening, and/or agglomeration processes in the reduction of bulk metals to NPs [3, 113]. To keep the NPs from aggregating, a stabilizing agent that adheres to the NPs' surface is usually added [10]. The co-precipitation method has the advantage of producing crystalline sizes in the small range when compared to other synthesis processes, depending on the precipitating agent used during the reaction.

Furthermore, by using capping agents, the crystalline size and morphology of the NPs prepared using this method can be controlled [10, 142]. For instance Yazid et al. [138] employing the co-precipitation techniques have been able to synthesize magnetic NPs (MNPs) from iron (II) and iron (III) metals in the presence of the base solutions NaOH, KOH, and NH_4OH at varying temperature (25, 50, and 70 °C). Farahmandjou and Soflaee [43] also synthesized Fe_2O_3 NPs from iron chloride hexahydrate ($FeCl_3 \cdot 6H_2O$) using NH_4OH solution as their base. Additionally, Mohamed et al. [91] synthesized gold nanoparticles (AuNPs) by co-precipitation method through the reduction of gold chloride hydrate ($HAuCl_4$) with sodium citrate tribasic dehydrate as the reduction and stabilizing agent under sunder conditions.

3.4 Polyol Method

The polyol method exists as a chemical method for the synthesis of NPs which employs non-aqueous liquid (polyol) as a solvent and reducing agent [24]. Their mode of synthesis involves the suspension of the metal precursor in a polymer, glycol solvent and heating the mixture to a refluxing temperature. With this method, there is flexibility in size control, texture, and shape of NPs [24, 57]. This has been made evident in a research study by Fereshteh et al. [46] where they used a polyol method to synthesis AgNPs using polyethylene glycol (PEG) as the capping agent, reductant, and media agent to reduce the metal precursor, $AgNO_3$. In controlling the size and shape of the as-synthesized AgNPs, they revealed that upon the addition of the polymer, PEG, the size changed with varying sizes 524 ± 87 nm (agglomerated), 41 ± 20.2 mm, 413 ± 62 nm, and 352 ± 40 nm.

Employing a microwave polyol method, Ider et al. [66] have synthesized AgNPs from $AgNO_3$ in the presence of ethanol and a copolymer agent with intrinsic physicochemical properties such as controlled size between 4 and 10 nm. Chieng and Loo [27] synthesized various zinc oxide NPs introducing the zinc precursor in ethylene glycol (EG), diethylene glycol (DEG), and tetraethylene glycol (TTEG) to obtain ZnO/EG, ZnO/DET, and ZnO/TTEG. They revealed that, with increased glycol, the average size of the ZnO increases. The sizes of ZnO particles determined from TEM were 19.62 nm, 38.84 nm, and 68.57 nm for ZnO/EG, ZnO/DET, and ZnO/TTEG, respectively. The shape of the ZnO nanoparticles changed from spherical (ZnO/EG), spherical and rod (ZnO/DEG) to "diamond" like structure (ZnO/TTEG) as observed in a field emission scanning electron microscope. This makes it okay to employ polyol method in the synthesis of NPs with varying sizes and shapes [135].

3.5 Micro-emulsion

Micro-emulsion employs liquid mixtures of oil, water, and surfactant, and in some instance, a co-surfactant is added [87]. Usually, when water and oil are mixed, they separate into two phases as they are immiscible which requires an energy input to mix the two to create a uniform mixture of water–oil. It has been revealed that the interfacial tension between bulk oil and water can be as high as 30–50 dynes/cm which can be avoided by using surfactants, like CTAB [14, 24]. These surfactants contain hydrophilic (water-loving) and lipophilic (oil-loving) groups. The interface can be aligned and established between oil and water by reducing the interfacial tension if there are enough surfactant molecules. This method has been used to synthesize inorganic NPs.

Primarily, the preparation procedure of metallic NPs in water-in-oil micro-emulsion commonly consists of mixing of two micro-emulsion containing metal salt and a reducing agent [24]. In a typical micro-emulsion method, the morphology and size of NPs are based on the size and shape of the nano-droplets and the type of the surfactant [19, 139]. The surfactant is used to stabilize the particle and protect them from proceeding to grow. Micro-emulsion methods could be advantageous due to the efficient control of the size, shape, and composition of the NPs obtained. Water-in-oil (W/O) micro-emulsions consist of aqueous microdomains dispersed in a continuous oil phase and stabilized by surfactant molecules [28, 96]. These work as nano-reactors where the synthesis of the desired nanoparticles takes place through a co-precipitation chemical reaction [119]. Asab et al. [14] from their study synthesized magnetite and silica-coated magnetite (Fe_3O_4) NPs by water-in-oil (W/O) micro-emulsion method from hydrated ferric nitrate, ferrous sulfate precursors, and ammonia a precipitating agent with the assistance of Tween-80 and SDS surfactants with a size range of 7.3 ± 0.05 to 10.83 ± 0.02 nm for uncoated Fe_3O_4 and 16 ± 0.14 nm for silica-coated Fe_3O_4 NP. They revealed that particle size of Fe_3O_4 NPs slightly increased with the temperature and precursor concentration.

Salvador et al. [119] synthesized superparamagnetic magnetite NPs (SPIONs) with average diameters between 5.4 and 7.2 nm and large monodispersity through precipitation in a water–oil (W/O) micro-emulsion, with Cetyl Trimethyl Ammonium Bromide (CTAB) as a main surfactant, 1-butanol as a co-surfactant, and 1-hexanol as the continuous oily phase. Beygi and Babakhani [18] investigated the synthesis of $Fe_xNi_{(1-x)}$ bimetallic NPs employing micro-emulsion method. They revealed that the NPs synthesis occurred by simultaneous reduction of metal ions and indicated the NP structure as homogeneous alloy $Fe_xNi_{(1-x)}$ NPs with different sizes, morphologies, and compositions. The synthesis proceeded with the changing of the micro-emulsion parameters such as water/surfactant/oil ratio, presence of co-surfactant, and $NiCl_2 \cdot 6H_2O$ to $FeCl_2 \cdot 4H_2O$ molar ratio. The results indicated that presence of butanol as co-surfactant led to chain-like arrangement of NPs. Also, finer NPs were synthesized by decreasing the amount of oil and water and increasing the amount of CTAB.

3.6 Thermal Decomposition

Thermal decomposition of NPs is a chemical reaction in which heat is the reactant. An endothermic reaction requires heat to break chemical bonds in compounds undergoing decomposition. However, if decomposition becomes sufficiently exothermic, a positive feedback loop is formed, resulting in thermal runaway and possibly an explosion or other chemical reactions [24, 112]. Thermal decomposition has emerged as a novel method for producing stable monodispersed NPs. This synthesis route has received a lot of attention, and it is thought to be more cost-effective than the traditional method. An earlier study by Whitesides et al. [134] where they investigated the thermal decomposition mechanism of n-Butyl (tri-n-butylphosphine) copper (I). It provided the first easily interpretable example of a metal hydride and its parent metal alkyl succeeding reactions.

Unni et al. [132] have shown how thermal decomposition synthesis of magnetic nanoparticles in the absence of oxygen yielded iron oxide NPs of size approximately 20 nm and magnetic diameters that are significantly smaller than their physical diameters. Also, Hufschmid et al. [64] investigated critical parameters for synthesis of monodisperse SPIONs by organic thermal decomposition. The study was done to present a comprehensive template for the design and synthesis of iron oxide nanoparticles with control over size, size distribution, phase, and resulting magnetic properties. Betancourt-Galindo et al. [16] have synthesized copper nanoparticles (CuNPs) via thermal decomposition using copper chloride, sodium oleate, and phenyl ether as solvent agents. The study revealed that NPs formed from the thermal decomposition route correspond to NPs synthesized from other methods. The formation of the CuNPs in their study was evidenced by XRD and TEM which revealed the sizes to be between 4 and 18 nm and spherical shapes.

3.7 Electrochemical Synthesis

The electrochemical synthesis method uses an electrochemical cell to synthesize chemical compounds. This method employs electrical energy to drive chemical change in order to replace toxic and expensive chemical reagents, allowing for cleaner and cheaper synthesis with higher production efficiency and at a lower cost. Islam and Islam [68] investigated the electrochemical deposition method for the synthesis of platinum NPs. Changing the electrolysis parameters led to the control of particle size, and changing the composition of the electrolytic solutions improved platinum particle homogeneity. Platinum NPs were deposited on electrode surfaces, with particle sizes larger than 10 nm and a wide particle size distribution. Kuntyi et al. [79] reported on the electrolysis of silver electrodes under conditions of altered current polarity in sodium polyacrylate (NaPA) solutions to produce silver sols containing AgNPs with sizes up to 10 nm. According to the findings, the values of the current of the anode dissolution of silver increased with increasing NaPA concentration, and the i_{anode}—E dependence was linear. They proposed a multistage scheme for the formation of silver nanoparticles stabilized by polyacrylate anion (PA) that includes (1) silver ionization and complexation of produced Ag+ ions with polyacrylate, (2) cathode reduction of silver ions from the complex, (3) formation of PA-stabilized nanoclusters, and (4) aggregation of nanoclusters and formation of nanoparticles. The rate constants of nucleation and growth of silver nanoparticles were estimated using UV–vis spectroscopy in relation to temperature and sodium polyacrylate concentration. It was discovered that the formation of silver nanoparticles occurs in electrode space and is governed by diffusion.

3.8 *Physical Synthesis Approach*

The top-down approach is typically used in the physical synthesis of NPs because it employs a large tube furnace to reduce bulk materials to the NPs while maintaining their original properties without atomic or subatomic level changes. The most common physical techniques are laser ablation, microwave, spark ablation, gamma radiation, and ultrasound [125]. This method of synthesis has benefits since it preserves the uniform distribution of NPs and prevents NP solvent contamination in the generated thin films [99]. Some of the commonly used physical synthesis methods are discussed in the following subsections.

3.8.1 Spark Ablation

The spark ablation process employs VSP-G1 NPs generator (VSP-G1)—thus a tabletop product that allows the user to produce NPs of the desired material in a controlled way. Spark ablation is a purely physical process that only requires electricity, a carrier gas, and electrode material to produce clean NPs. No additional chemicals are required for the production or to stabilize the NPs in the aerosol compared to the chemical synthesis which requires chemicals to stabilize the NPs to prevent aggregation. Spark ablation in gas technology has been reported to be green, fast quenching, and dynamic in producing metallic NPs. Using a spark ablation generator (VSP-G1 B.V., Delft,) with an operating voltage and current of 1.35 kV and 7 mA, Lu et al. [86] synthesized AgNPs and in situ loaded them on carbon fiber using nitrogen flow as the gas carrier. The electric spark was created when initial electrons collided and ionized with background gas atoms while moving to the anode under the influence of high voltage.

In the spark discharge process, Ag electrodes were partially vaporized by a high-temperature electric spark, and then, with the quenching effect of flowing carrier gas, AgNPs were formed and deposited on a carbon fiber. Li et al. [84] successfully synthesized $La(OH)_3$ MENPs by combining a green and on-step method with spark ablation aerosol. The study revealed the use of a spark ablation NPs generator (VSP-G1, VSPARTICLE B.V., Delft, The Netherlands) with a voltage of 1.38 kV and a current of 8.1 mA. One pair of cylindrical La electrodes (diameter = 8 mm, length = 30 mm) were placed in the holders inside the bottom bracket and separated to form a 1 mm gap through which a carrier gas (2% O_2, 98% N_2) flowed perpendicularly at 10 L/min. This lends support to the findings of Lu et al. [86] studies; however, due to the expected reaction of La and O_2, Li et al. [84] included O_2 gas in their study. Because of the electric gas breakdown, pulsed sparks formed in the gap between the two electrodes. Through the reaction of La and O_2, the sparks ablated the La electrodes, causing them to condense into La_2O_3 prime particles. Following the formation of the prime particles, bypassed water vapor (65 °C) was injected via the N_2 gas at a standard flow rate of 10 L/min (La_2O_3). To form $La(OH)_3$ prime particles, the La_2O_3 prime particles were homogeneously mixed and hydrated with water vapor. On a commercial filter holder, the $La(OH)_3$ prime particles coagulated and deposited on 0.01 g of the prepared PAN electrospun nanofiber membrane. After 15 min, the $La(OH)_3$ MENP-loaded nanofibers were collected.

3.8.2 Laser Ablation

Laser ablation, also known as photo-ablation or laser blasting, is the process of removing material from a solid or (occasionally) liquid surface using a beam of light. At low laser flux, the absorbed laser energy heats the material, causing it to evaporate or sublimate. When the flux is high, the material is typically converted to plasma. Over time, they have shown promise in synthesizing inorganic NPs. Nucleation and growth of laser-vaporized species in a background gas generate NPs in this method. Alhamid et al. [9] in their study synthesized AgNPs using pulse laser ablation method. During the experiment, a high purity silver plate (99.95%) was immersed in a 25 ml pure aquades placed in a petri dish with a diameter of 50 nm. The Nd: YAG 1064 nm laser was focused on the surface of the sample and shot for 13 h to produce AgNPs colloids.

During the shooting process, the petri dish containing silver samples is moved slowly and continuously to obtain evenly distributed nanoparticle colloids. The results show that colloidal silver nanoparticles had yellowish color. Furthermore, the produced AgNPs had spherical shape with an averaged diameter size of 10 nm. Employing this synthesis method tends to curb the menace such as tedious sample pre-treatment, contamination from additional agents, hazard, and among many others that comes with the chemical and biological synthesis of NPs. By focusing femtosecond laser pulses onto a silicon wafer immersed in an aqueous $KAuCl_4$ solution, John and Tibbetts [73] reported the synthesis of silica-AuNPs using a one-step femtosecond-reactive laser ablation in liquid (fs-RLAL) technique. Thus, the synthesis of silica-AuNPs occurred in two steps: first, the dense electron plasma formed within tens of femtoseconds of the laser pulse which initiated the reduction of the $[AuCl_4]$-complex, resulting in the formation of larger isolated AuNPs. Secondly, after hundreds of picoseconds or later, the initial laser pulse silicon species ejected from the wafer surface reduced the remaining $[AuCl_4]$- and encapsulate the growing clusters, forming ultrasmall AuNPs stabilized by the silica matrix.

3.8.3 Microwave

Microwave synthesis has been used in the synthesis of NPs because it is thought to be faster and more uniform in heating the metal precursor. Microwave irradiation has a penetration characteristic that allows for uniform heating of the reaction solution. As a result, uniform nucleation and rapid crystal growth occur, resulting in the formation of crystallites with a narrow size distribution. In comparison with other conventional methods, synthesis by microwave irradiation has the advantage of a short reaction time, which is attributed to the combined forces created by the microwave's electric and magnetic components, which generate friction and collisions of the molecules. Hebbar et al. [59] created $ZnGa_{2-x}Eu_xO_4$ NPs for the first time using a microwave-assisted synthesis at 200 °C from aqueous solutions of $Zn(NO_3)_2$, $Ga(NO_3)_3$, and Eu_2O_3 (99.9%) which were mixed thoroughly for uniformity and pH adjusted to 8 using ammonia. Interestingly, the study stated that 10 min is sufficient for NP synthesis, making it ideal for obtaining NPs in a short period of time. The as-synthesized $ZnGa_{2-x}Eu_xO_4$ NPs were found to be nearly spherical in shape and approximately 7 nm in size.

In another study by Rameshkumar et al. [111], a reduced graphene oxide-silver (rGO–Ag) nanocomposite was obtained from silver nitrate with the slow addition of ammonia before the addition of GO solution via a facile microwave-assisted synthesis. The study revealed the synthesis of the rGO–AgNPs at varying times, 30 s, 1 min, and 3 min which is essential for NPs size control. So, in the end, they got rGO–AgNPs (30 s), rGO–AgNPs (1 min), and rGO–AgNPs (3 min). This makes it novel to obtain NPs within a short period of time compared to the 10 min used by Hebbar et al. [59] in synthesizing $ZnGa_{2-x}Eu_xO_4$ NPs. The rGO–AgNPs from Rameshkumar et al. [111] study had a spherical shape with an average size of 20 nm as confirmed from the single surface plasmon absorption feature and the TEM analysis. Another study conducted by Tiwary et al. [131] also synthesized AgNPs in aqueous medium by a simple, efficient, and economic microwave-assisted synthetic route using hexamine as the reducing agent and the biopolymer pectin as stabilizer. The characterization (TEM, UV/Vis, and SEM) of the as-synthesized AgNPs showed the shape of the NPs to be spherical shape with an average diameter of 18.84 nm.

3.8.4 Gamma Radiation

Gamma radiation has emerged as promising in the synthesis of metallic NPs because it is reproducible, shape can be controlled, it is relatively easier to use, cheap, and uses fewer toxic precursors. In water or solvents such as ethanol, gamma radiation NPs synthesis method uses least the number of reagents and operate at a reaction temperature close to room temperature. For instance, Nguyen et al. [102] created platinum nanoparticles (PtNPs) with gamma-ray irradiation as a reducing factor and chitosan as a stabilizer. This was accomplished by dissolving CTS in lactic acid to create a CTS solution in which K_2PtCl_6 was dissolved to create Pt^{4+}, which is expected to be reduced to Pt^0 to form NPs. The conversion dose to reduce Pt^{4+}

to Pt^0 was found to be approximately 14 kGy for an initial Pt^{4+} concentration of 1 mM in the study. The TEM analysis revealed that the size of the as-synthesized PtNPs was 1.4–1.6 nm. The interaction of ionizing radiation, such as gamma rays, with aqueous solutions results in the generation of randomly distributed reducing and oxidizing agents with large redox potentials, where the use of additives such as alcohols or NO_2 are capable of modifying the environment to generate a high concentration of reducing or oxidizing agents. The inherent reactivity of solvent radiolysis products triggers chemical follows up reactions that involve other species present in the system, such as metal ions, metal complexes, monomers, and natural or synthetic polymers. Irradiating precursors can be introduced in the presence of templating agents to confine the locus of reaction into virtual "nano-reactors" or "nanomolds," which impart the desired particle size and shape, or in the absence of templates [47]. Metallic Ni NPs have been synthesized from aqueous solution of $NiCl_2$ using γ-radiation induced reduction. However, a post-irradiation treatment was needed to avoid re-oxidation of Ni, which is pertinent for the stability of the Ni NPs [136].

3.8.5 Ultrasound

Ultrasound is a one-of-a-kind technique related to cavitation (extremely high local temperatures (around 5000 K) and pressures (over 1000 atmospheres) generated in a liquid phase) that has gained popularity in the synthesis of NPs. Ultrasonic cavitation forms when liquids are irradiated with ultrasonic irradiation. Ultrasonic cavitation causes a number of physical and chemical effects, including high temperature, pressure, and cooling rate, creating a unique environment for chemical reactions under extreme conditions. Ultrasound has been proposed as a good method for preparing NPs with controllable morphologies. Polythlene fibers containing AgNPs have been prepared from silver nitrate solution through chemical reduction using n-propanol, ethylene glycol, and ethanolamine as a reducing agent under ultrasound irradiation. The study revealed the effect of reducing reagent, power of ultrasound irradiation, reaction time, and temperature in the growth of the AgNPs. The particle sizes and morphology of the NPs were found to be dependent on the power of ultrasound irradiation. The particle size decreased with increasing power of the ultrasound irradiation, and also, increasing the temperature caused an increase in the AgNPs size [1]. Karthik et al. [75] used an ultrasound-assisted method to synthesize V_2O_5 NPs from benzyl alcohol (C_6H_5 CH_2 OH; 99%) and vanadium oxytrichloride ($VOCl_3$; 99%) precursors. The TEM analysis revealed that the V_2O_5 NPs were 23 nm in size.

4 Biological Synthesis Approach

Biological synthesis of NPs is a green chemistry approach that interconnects nanotechnology and biotechnology. Biological materials such as plants, microorganisms, enzymes, and many more have been utilized for synthesizing NPs which follows the bottom-up approach. Researchers assert that the biological material are simple, clean, and effective in the synthesis process which has the same physicochemical properties as NPs synthesized from chemical and physical approaches [125]. Among the biological synthesis approach, plant materials have gained dominance.

4.1 Plants

Plants are among the biomaterials available for NP synthesis; however, the potency of a plant to reduce bulk metal to their NPs form is dependent on the presence of phytochemicals such as flavonoids, ketones, phenols, saponins, terpenes, amino acid, and many more that act as reducing and stabilizing agents. The reaction is easily scalable, as the shape and size of the NPs can be fine-tuned by changing the reaction conditions compared to other methods. Any part of the plant can be used in the synthesis of NPs. Evidently, Lakshmanan et al. [83] synthesized AgNPs by reducing silver nitrate ions with fruit extracts of *Cleome viscosa* L. The synthesis was carried out at room temperature, and a change in color from light green to brown indicated that the AgNPs had been successfully synthesized. UV/Vis, FTIR, FESEM-EDAX, and TEM were used to characterize the as-biosynthesized AgNPs. The AgNPs were 20–50 nm in size, as determined by TEM, and had the shape of a face-centered cubic silver crystal, as revealed by XRD analysis.

Elumalai et al. [40] reported the synthesis of zinc oxide NPs using *Vitex trifolia* L. leaf extract to reduce zinc nitrate hexahydrate crystals. This was possible because of the intrinsic physicochemical properties of *Vitex trifolia* leaves. UV/vis, PL, XRD, FTIR, SEM, FE-SEM, EDX, and TEM were used to characterize the as-biosynthesized ZnO NPs, which revealed that the NPs were 30 nm in size and spherical and hexagonal in shape. Liu et al. [85] also discovered that not all phytochemicals found in plant material can reduce bulk metals. They classified the phytochemicals in *Cacumen platycladi* leaf extract into three groups using column chromatography, Fourier transform infrared spectroscopy, and high-performance liquid chromatography. The synthesis of AuNPs using the three types of phytochemicals to reduce chloroauric acid ($HAuCl_4$) was successful. Bio-reduction, on the other hand, revealed a strong-to-weak sequence of reducing power of polyphenols, flavonoids, and sugars. Sugars provided no protection strength for the biosynthesis of AuNPs, while polyphenols were discovered to be involved in the isotropous stabilization of the AuNPs and flavonoids in the anisotropic growth of the AuNPs. This study gives researchers some insight into quantifying the presence of polyphenols and flavonoids in plant components during bio-reduction of bulk metals.

4.2 Microorganism

Microorganisms have been valuable in the synthesis of NPs since the inception of the biological synthesis approach. Several microorganisms such as bacteria and fungi have been explored in recent times. Synthesis of NP by microorganisms can be classified into intracellular synthesis and extracellular synthesis depending on the NP formation site. When ions are transported into the microbial cell to form NPs in the presence of enzymes, it is indicated as the intracellular method, and the extracellular method is the formation of NPs by trapping metal ions on the cell membrane with enzymes. Extracellular synthesis of AgNPs using native *Bacillus* sp. strain AW1-2 [7] has been reported in literature with size range from 22.33 to 41.95 nm and spherical shapes. The synthesis of NPs has also been investigated in both eukaryotic and prokaryotic organisms. MubarakAli et al. [95] recently synthesized AuNPs using a eukaryote, *Coelastrella* sp., and a prokaryote, *Phormidium* sp., to reduce chloroauric acid (gold precursor). The AuNPs were 30 nm in size and spherical in shape, according to TEM analysis (Table 1).

5 Nanomaterials for Dye Removal

Metal nanoparticles due to their intrinsic physicochemical properties have emerged as promising in degrading dyes from water. The high adsorption, large surface area, and localized plasmon surface resonance (LPSR) of NPs make them a great candidate to be explored as adsorbent, catalyst, and filtration membranes to remove dyes from water. In the section below, we discuss the various applications of NPs for dye removal in wastewater.

5.1 Nano-adsorbents

Adsorption is the process of separating a substance from one phase accompanied by its accumulation or concentration at the surface of another. The adsorbing phase is the adsorbent, and material concentrated or adsorbed at the surface is the adsorbate. The adsorbate can be organic or inorganic, small, and light or a heavy material. Most adsorbents can be regenerated.

This treatment method makes use of organic or inorganic nanomaterials with a high affinity for adsorption. These adsorbents are extremely effective at removing a wide range of contaminants. The ideal absorbent is small, has a large surface area, has a high catalytic potential, and is highly reactive. The usage of nanomaterials in adsorption can be grouped into diverse groups including metallic NPs, magnetic NPs, metal oxide, nanostructured mixed oxides, carbonaceous nanomaterials (carbon nanotubes, carbon nanoparticles, and carbon nanosheets), silicon

Table 1 Summary of some nanoparticles synthesis methods

S/N	Nanoparticles	Size (nm)	Shape	Synthesis method	References
1	α-Fe_2O_3 NPs	30	Spherical	Co-precipitation	[43]
2	AuNPs	25	Cubic crystal structure	Co-precipitation	[91]
3	AgNP	~3	Spherical	Polyol	[136]
4	AgNPs	524 ± 87 413 ± 62 352 ± 40		Polyol	[47]
5	AgNPs	4–10	Spherical	Polyol	[66]
6	(1) ZnO/EG, (2) ZnO/DET, (3)ZnO/TTEG	(1) 19.62, (2) 38.84, (3) 68.57	(1) Spherical, (2) Spherical and rod, (3) Diamond-like	Polyol	[28]
7	Fe_3O_4 MNPs	11–338	–	Polyol	[104]
8	(1) Uncoated-Fe_3O_4 NPs, (2) Silica-coated-Fe_3O_4 NPs	(1) 7.3 ± 0.05 to 10.83 ± 0.02, (2) 16 ± 0.14	Flower-like	Micro-emulsion	[15]
9	Superparamagnetic iron Oxide Nanoparticles (SPIONs)	5.4 to 7.2	–	Micro-emulsion	[120]
10	Fe_3O_4 NPs	~20		Thermal decomposition	[119]
11	CuNPs	4–18	Spherical	Thermal decomposition	[17]
12	AgNPs	10	–	Electrochemical	[79]
13	AgNPs	4.73, 3.79, and 2.41	Face-centered cubic	Spark ablation	[86]

(continued)

Table 1 (continued)

S/N	Nanoparticles	Size (nm)	Shape	Synthesis method	References
14	AgNPs	10	Spherical	Laser ablation	[10]
15	$ZnGa_{2-x}Eu_xO_4$ NPs	~7	Spherical	Microwave	[59]
17	PtNPs	1.4–1.6	–	Gamma irradiation	[102]
18	Ni NPs	3.47 ± 0.71 nm	–	Gamma irradiation	[137]
19	AgNPs	20–50	Cubic	Plant-mediated synthesis	[83]
20	ZnO NPs	30	Spherical and hexagonal	Plant-mediated synthesis	[41]
21	AgNPs	22.33–41.95	Spherical	Microorganism	[8]
22	AuNPs	30	Spherical	Microorganism	[95]

nanomaterials (silicon nanotubes, silicon nanoparticles, and silicon nanosheets), nanoclays, nanofibers, polymer-based nanomaterials, and xerogels and aeroge [11].

The mechanism of adsorption involves physical and chemical interaction between an adsorbent and an adsorbate which is dependent on the reacting environment. The mechanism of the nanoadsorption can be looked as an isothermal mechanism that has to do with the quantification of the quantity of adsorbate that an adsorbent can absorb at equilibrium conditions and constant temperature. This adsorption equilibrium relationship known as an isotherm allows the adsorption capacity of the adsorbent to be calculated at any given liquid-phase adsorbate concentration. The equilibrium adsorption phase is achieved when the concentration of the adsorbate remains unchanged due to zero net transfer of pollutant that is adsorbed and desorbed from the surface of the adsorbent. This equilibrium phenomenon is vital for adsorption mechanism pathways optimization, surface properties indication, and adsorbents capacities as they can describe the relationship of contaminants with the adsorbents.

Another mechanism is the adsorption kinetics which kinetic models assist in understanding the transfer rate of dye onto the adsorbent surface and how this rate is affected and controlled by the change in adsorption parameters [105]. Using the Taguchi design, Ghasemi et al. [48] synthesized a novel Fe_3O_4/AC nanocomposite that was used as an adsorbent for the removal of methylene blue (MB) from water. The ANOVA results showed that, in descending order, pH (66.81%) > adsorbent dose (25.54%) > temperature (4.83%) > initial MB concentration (1.23%) > contact time (0.32%). The kinetic data were fitted to pseudo-first-order, pseudo-second-order, and intra-particle diffusion models, with MB dye adsorption following pseudo-second-order kinetics. The regression coefficient values for Langmuir (0.98), Freundlich (0.93), and Dubinin-Radushkevich (0.94) indicated that the adsorption process fits the Langmuir isotherm and that the maximum adsorption capacity is 384.6 mg/g.

Furthermore, thermodynamic studies suggested that the adsorption process was spontaneous. During an adsorption–desorption study, Ghasemi et al. [48] indicated that HCl can be used to desorb MB dye from Fe_3O_4/AC nanocomposite. This conclusion was reached after the desorption of MB dye from the nanocomposites (Fe_3O_4/AC), and thus, the nanocomposites can be reused for about three times. Dutta et al. [37] reported that dyes SnO_2 quantum dots decorated silica nanoparticles (QDs-MSN) can be used as an adsorbent material for the removal of methylene blue dye from water due to their large BET surface area and uniform pore size distribution with large pore volumes. The results showed that QDs-MSN could adsorb 100% of methylene blue in 5 min at room temperature. This means that the NPs had high absorbance rate for the removal of dye from polluted water. Gong et al. [52] created a multi-wall carbon nanotube nanocomposite that can be used as an adsorbent for the removal of methylene blue (MB), neutral red (NR), and brilliant cresyl blue (BCB) from water. The results indicated that the three cationic dyes' adsorption capacity on MMWCNT adsorbent increased with temperature. The cationic dyes adsorption increased with pH due to electrostatic attraction between the negatively charged MMWCNT adsorbent surface and the positively charged cationic dyes. In MB, NR, and BCB, the dye removal ratio increased from 30.1 to 99.16%, 17.11 to 98.33%, and 17.6 to 98.8%, respectively. The researchers then investigated the regeneration potential of the MMWCNT nanocomposite using ethanol and found that ethanol favored the adsorbent desorption. The regeneration potential using ethanol will be prudent in obtaining NPs back after initial removal of dyes from wastewater.

5.2 *Nano-photocatalysts*

NPs have been explored as photocatalyst in degrading dyes from water in order to make the water pure and clean for use. Photocatalytic activities, such as the interaction of light energy with metallic nanoparticles, are used for these treatments. Through the reaction with hydroxyl radicals, photocatalytic activities destroy microorganisms (bacteria) and organic substances. Nanoparticles dye degradation through photocatalysis is based on advanced oxidation method which employs the NPs as semiconductor photocatalyst. The degradation mechanism of nano-photocatalyst is initiated by the absorption of irradiated light with energy higher than or equal to the band gap of the semiconductor photocatalyst. From a thermodynamic perspective, the bottom of the CB must be located at a more negative potential than the reduction potential of H^+ to H_2 (−0.41 vs. NHE, at pH 7), while the top of the VB must exceed the oxidation potential of H_2O to O_2 (0.82 V vs. NHE).

When illuminated by light energy, the semiconducting nano-photocatalyst enters its excited state by absorbing the light energy and producing a pair of electrons-and-holes (e−/h+). This h+ acts as a strong oxidizer, reacting with water molecules to produce a super reactive hydroxyl radical (·OH) and superoxide anion radical (O_2^-), which then reacts with the dye to degrade them to smaller intermediates, carbon dioxide and water. In a previous study by our team, we synthesized AuNPs using *Parkia biglobosa* leaves as a reducing agent and investigated their ability to function as a photocatalyst to degrade rhodamine b dye from water when exposed to visible light. The results showed that rhodamine b dye was degraded from water in the presence of the AuNPs photocatalyst within 75 min [123]. Rafique et al. [110] also studied the photocatalytic efficiency of CuO NPs in removing rhodamine b (RhB) dye from water. The study revealed that under visible light irradiation 98% of the dye were removed from water within 10 min. This is an interesting results and improvement in dye removal compared to the 75 min reported in our study [123]. Also, Roushani et al. [115] studied the photocatalytic behavior of a new graphene quantum dot (GQD) for the removal of a new fuchsin dye from water under visible light irradiation. The study found that the kinetic rate constants and decolorization efficiency of the photocatalyst, GQDs, were independent of dye concentration and slightly decreased as the initial dye concentration was increased. This tends to demonstrate the photocatalytic behavior of the photocatalyst because it was able to generate more electron–hole pairs to begin the degradation. Nonetheless, if the generated electrons return to the holes generated at the valence band (VB), a photocatalyst may be unable to function effectively. As a result, it is sometimes necessary to use sacrificial agents such as triethanolamine to arrest the generated holes at the VB in order to improve photocatalytic activity. Roushani et al. [115] discovered that the GQDs photocatalyst can be reused for at least 5 cycles. However, there is a decrease in photocatalyst activity after the sixth run.

5.3 *Nano-membranes*

Nano-membrane filtration technologies have emerged as prudent in wastewater treatment such as dye pollutant removal and reuse of treated water. They are operated at lower pressure rates and require less energy to operate. They bring about high separation efficiency and are capable of operating in a continuous mode [4]. Nano-membranes are widely used to remove heavy metals, dyes, and other contaminants based on the philosophy of reverse osmosis. Nano-membranes have membrane pore sizes ranging from 0.5 to 2 nm and operate with pressures ranging from 5 to 40 bar [2]. The pores are large enough to allow water molecules to pass through but small enough to retain dyes. The use of NPs in membranes can thus reduce energy demand, chemical use for membrane cleaning, and cost. For instance, metal oxide NPs are used in nano-enhanced membrane (NEM) to protect the membrane from fouling and to increase water permeability due to their hydrophilic nature. In recent years, the nano-membrane method has seen the compliment of nano-photocatalyst functioning as photocatalytic membranes in degrading dyes without the loss of the nano-photocatalyst [116]. Naresh Yadav et al. [97], for example, investigated the removal of rhodamine b dye from water with ZnO as a photocatalyst and ceramic nanoporous tubular membranes. The photocatalysis process alone showed 33% decolorization, whereas ceramic nanofiltration showed 50% decolorization, implying that integrating nano-photocatalyst and nano-membranes is prudent in decolorizing dye polluted water. Over the course of 90 min, the nano-photocatalyst and nano-membrane were able to decolorize 96% of the rhodamine b dye in water. Also, Wang et al. [133] developed polyethersulfone-sulfonated polyethersulfone (PES-SPES) membranes and graphene oxide-polydopamine (GO-PDA)/PES-SPES adsorptive membranes for the removal of rhodamine B (RhB) from aqueous solutions. The results showed that the Langmuir model and the pseudo-second-order kinetic model can well describe the adsorption of RhB on the membrane, and the maximum adsorption capacity of the GO-PDA layer was reported as 87.03 mg g^{-1}, indicating a chemisorption dominated adsorption process. According to the study, GO adsorbents and PDA crosslinker contributed to synergistic adsorption. The adsorptive membranes were also reusable in a dynamic adsorption/desorption cycle (Table 2).

6 Summary and Outlook

The number of dye manufacturing companies and industries such as textile, food, ink, and paper that use dyes has grown. Effluents from these industries pose a significant health risk because they affect water quality. The increasing pollution of water

Table 2 Summary of some cationic dyes degraded from water using nanoparticles (NPs)

S/N	Nanoparticles	Dyes degraded	Method	References
1	Fe_3O_4/AC nanocomposite	Methylene blue	Nano-adsorbent	[48]
2	QDs-MSN	Methylene blue	Nano-adsorbent	[37]
3	MMWCNT	(1) Methylene blue, (2) Neutral red (NR), (3) Brilliant cresyl blue (BCB)	Nano-adsorbent	[52]
4	AuNPs	Rhodamine b	Photocatalytic	[123]
5	QDs	New fuchsin	Photocatalytic	[115]
6	CuO NPs	Rhodamine b	Photocatalytic	[110]
7	ZnO–graphene composite	Methylene blue	Photocatalytic	[42]
8	BISe-CM	Crystal violet	Photocatalytic	[6]
9	ZnO ceramic nanoporous tubular membrane	Rhodamine b	Photocatalytic and nano-mebrane	[97]
10	Graphene oxide/polydopamine adsorptive membrane	Rhodamine b	Nano-membrane	[133]

bodies cannot be ignored, especially in an era where much public discourse is on raising environmental awareness and promoting environmentally friendly technologies. Over time, the market for nanotechnology in water pollution removal has grown significantly. Nanotechnology uses matter on the atomic, molecular, and supermolecular scales (1 to 100 nm) for industrial purposes including wastewater treatment. In wastewater treatment, nanoparticles (NPs) have been used in a variety of ways including their use as nano-membranes, nano-photocatalysts, and nano-adsorbents. Several scientists have investigated these NPs since their discovery. As shown in this chapter, there are several methods for producing engineered NPs which are capable of removing cationic dyes from wastewaters. Nanoparticles maybe used alone or combined with polymers or other molecules to improve degradation activity. The inherent results are seen with various nanotechnology methods in degrading dye in wastewater, and the reuse of such treated water is shifting the paradigm of wastewater treatment to nanotechnology. However, it is recommended that future research works look at the safe use of treated water by checking the water chemistry of treated water and also exploring some clinical trials in vivo to alleviate concern about deposits. By so doing, indigenous knowledge would be strengthened, and also, the acceptability and sustainability of nanotechnology as a purification technique will be realized.

References

1. Abbasi AR, Kalantary H, Yousefi M, Ramazani A, Morsali A (2012) Synthesis and characterization of Ag nanoparticles@ polyethylene fibers under ultrasound irradiation. Ultrason Sonochem 19(4):853–857
2. Abid MF, Zablouk MA, Abid-Alameer AM (2012) Experimental study of dye removal from industrial wastewater by membrane technologies of reverse osmosis and nanofiltration. Iran J Environ Health Sci Eng 9(1):1–9
3. Abid N, Khan AM, Shujait S, Chaudhary K, Ikram M, Imran M, Maqbool M (2021) Synthesis of nanomaterials using various top-down and bottom-up approaches, influencing factors, advantages, and disadvantages: a review. Adv Coll Interface Sci:102597
4. Ağtaş M, Ormancı-Acar T, Keskin B, Türken T, Koyuncu İ (2021) Nanofiltration membranes for salt and dye filtration: effect of membrane properties on performances. Water Science Technol 83(9):2146–2159
5. Ahmad R (2009) Studies on adsorption of crystal violet dye from aqueous solution onto coniferous pinus bark powder (CPBP). J Hazard Mater 171:767–773. https://doi.org/10.1016/j.jhazmat.2009.06.060
6. Ahmad W, Khan A, Ali N, Khan S, Uddin S, Malik S, Bilal M (2021) Photocatalytic degradation of crystal violet dye under sunlight by chitosan-encapsulated ternary metal selenide microspheres. Environ Sci Poll Res 28(7):8074–8087
7. Ajaz S, Ahmed T, Shahid M, Noman M, Shah AA, Mehmood MA, Li B (2021) Bioinspired green synthesis of silver nanoparticles by using a native Bacillus sp. strain AW1-2: characterization and antifungal activity against Colletotrichum falcatum Went. Enzyme Microbial Technol 144:109745
8. Al-Buriahi AK, Al-Gheethi AA, Kumar PS, Mohamed RMSR, Yusof H, Alshalif AF, Khalifa NA (2022) Elimination of rhodamine B from textile wastewater using nanoparticle photocatalysts: a review for sustainable approaches. Chemosphere 287(2):132162
9. Alhamid MZ, Hadi BS, Khumaeni A (2019) Synthesis of silver nanoparticles using laser ablation method utilizing Nd: YAG laser. Paper presented at the AIP conference proceedings
10. Ali A, Zafar H, Zia M, ul Haq I, Phull AR, Ali JS, Hussain A (2016) Synthesis, characterization, applications, and challenges of iron oxide nanoparticles. Nanotechnol Sci Appl 9:49
11. Anastopoulos I, Hosseini-Bandegharaei A, Fu J, Mitropoulos AC, Kyzas GZ (2018) Use of nanoparticles for dye adsorption. J Dispers Sci Technol 39(6):836–847
12. Anbarasu M, Anandan M, Chinnasamy E, Gopinath V, Balamurugan K (2015) Synthesis and characterization of polyethylene glycol (PEG) coated Fe_3O_4 nanoparticles by chemical co-precipitation method for biomedical applications. Spectrochim Acta Part A: Mol Biomol Spectrosc 135:536–539
13. Artifon W, Cesca K, de Andrade CJ, de Souza AAU, de Oliveira D (2021) Dyestuffs from textile industry wastewaters: trends and gaps in the use of bioflocculants. Process Biochem 111(2):181–190
14. Asab G, Zereffa EA, Abdo Seghne T (2020) Synthesis of silica-coated Fe3O4 nanoparticles by microemulsion method: characterization and evaluation of antimicrobial activity. Int J Biomater:1–11
15. Badawi AK, Abd Elkodous M, Ali GA (2021) Recent advances in dye and metal ion removal using efficient adsorbents and novel nano-based materials: an overview. RSC Adv 11(58):36528–36553
16. Betancourt-Galindo R, Reyes-Rodriguez P, Puente-Urbina B, Avila-Orta C, Rodríguez-Fernández O, Cadenas-Pliego G, García-Cerda L (2014) Synthesis of copper nanoparticles by thermal decomposition and their antimicrobial properties. J Nanomater
17. Beyene HD, Werkneh AA, Bezabh HK, Ambaye TG (2017) Synthesis paradigm and applications of silver nanoparticles (AgNPs), a review. Sustain Mater Technol 13:18–23
18. Beygi H, Babakhani A (2017) Microemulsion synthesis and magnetic properties of $Fe_xNi(1-x)$ alloy nanoparticles. J Magn Magn Mater 421:177–183

19. Bibi I, Maqbool H, Iqbal S, Majid F, Kamal S, Alwadai N, Iqbal M (2021) $La_{1-x}Gd_xCr_{1-y}Ni_yO_3$ perovskite nanoparticles synthesis by micro-emulsion route: dielectric, magnetic and photocatalytic properties evaluation. Ceram Int 47(4):5822–5831
20. Boretti A, Rosa L (2019) Reassessing the projections of the world water development report. NPJ Clean Water 2(1):1–6
21. Burduşel A-C, Gherasim O, Grumezescu AM, Mogoantă L, Ficai A, Andronescu E (2018) Biomedical applications of silver nanoparticles: an up-to-date overview. Nanomaterials 8(9):681
22. Cai Z, Sun Y, Liu W, Pan F, Sun P, Fu J (2017) An overview of nanomaterials applied for removing dyes from wastewater. Environ Sci Poll Res 24(19):15882–15904
23. Cappuccino JG, Welsh C (2019) Microbiology: a laboratory manual, 12th edn. Pearson, New York
24. Cele T (2020) Preparation of nanoparticles. In: Engineered nanomaterials-health safety
25. Chackraborty S, Chowdury S, Saha PD (2011) Adsorption of crystal violet from aqueous solution onto NaOH-modified rice husk. Carbohyd Polym 86:1533–1541. https://doi.org/10.1016/j.carbpol.2011.06.058
26. Chauhan V, Gautam P, Kanwar S (2022) Nanoparticle-associated lipopeptides: a new class of antimicrobials. In: Biomedical translational research. Springer, pp 483–494
27. Chieng BW, Loo YY (2012) Synthesis of ZnO nanoparticles by modified polyol method. Mater Lett 73:78–82
28. Cid A (2018) Synthesis of NPs by microemulsion method. In: Microemulsion-a chemical nanoreactor. IntechOpen
29. Colvin VL (2003) The potential environmental impact of engineered nanomaterials. Nat Biotechnol 21(10):1166–1170
30. De Carvalho J, De Medeiros S, Morales M, Dantas A, Carriço A (2013) Synthesis of magnetite nanoparticles by high energy ball milling. Appl Surf Sci 275:84–87
31. Dinka MO (2018) Safe drinking water: concepts, benefits, principles and standards. In: Water challenges of an urbanizing world. Intechopen, London, pp 163–181
32. Dodoo-Arhin D, Asiedu T, Agyei-Tuffour B, Nyankson E, Obada D, Mwabora JM (2021) Photocatalytic degradation of Rhodamine dyes using zinc oxide nanoparticles. Mater Today: Proc 38:809–815. https://doi.org/10.1016/j.matpr.2020.04.597
33. Doke KM, Yusufi M, Joseph RD, Khan EM (2016) Comparative adsorption of crystal violet and Congo red onto $ZnCl_2$ activated carbon. J Disper Sci Technol 37(11):1671–1681. https://doi.org/10.1080/01932691.2015.1124342
34. Doltabadi M, Alidadi H, Davoudi M (2016) Comparative study of cationic and anionic dye removal from aqueous solutions using sawdust-based adsorbent. Environ Prog Sustainable Energy 35(4):1078–1090. https://doi.org/10.1002/ep.12334
35. Domaschke M, Schmidt M, Peukert W (2018) A model for the particle mass yield in the aerosol synthesis of ultrafine monometallic nanoparticles by spark ablation. J Aerosol Sci 126:133–142
36. Dulo B, Phan K, Githaiga J, Raes K, De Meester S (2021) Natural quinone dyes: a review on structure, extraction1 techniques, analysis and application potential. Waste Biomass Valor 12:6339–6374. https://doi.org/10.1007/s12649-021-01443-9
37. Dutta D, Thakur D, Bahadur D (2015) SnO_2 quantum dots decorated silica nanoparticles for fast removal of cationic dye (methylene blue) from wastewater. Chem Eng J 281:482–490
38. Ealia SAM, Saravanakumar M (2017) A review on the classification, characterisation, synthesis of nanoparticles and their application. Paper presented at the IOP conference series: materials science and engineering
39. El Shafey AM (2020) Green synthesis of metal and metal oxide nanoparticles from plant leaf extracts and their applications: a review. Green Process Synth 9(1):304–339
40. Elumalai K, Velmurugan S, Ravi S, Kathiravan V, Raj GA (2015) Bio-approach: plant mediated synthesis of ZnO nanoparticles and their catalytic reduction of methylene blue and antimicrobial activity. Adv Powder Technol 26(6):1639–1651

41. Fabryanty R, Valencial C, Soetaredjo FE, Putro JN, Santoso SP, Kurniawan A, Ju Y, Ismadji S (2017) Removal of crystal violet dye by adsorption using bentonite–alginate composite. J Environ Chem Eng 5(6):5677–5687. https://doi.org/10.1016/j.jece.2017.10.057
42. Fan H, Zhao X, Yang J, Shan X, Yang L, Zhang Y, ... Gao M (2012) ZnO–graphene composite for photocatalytic degradation of methylene blue dye. Catalysis Commun 29:29–34
43. Farahmandjou M, Soflaee F (2015) Synthesis and characterization of α-Fe_2O_3 nanoparticles by simple co-precipitation method. Phys Chem Res 3:191–196
44. Farhadi S, Javanmard M, Nadri G (2016) Characterization of cobalt oxide nanoparticles prepared by the thermal decomposition. Acta Chim Slov 63(2):335–343
45. Fariq A, Khan T, Yasmin A (2017) Microbial synthesis of nanoparticles and their potential applications in biomedicine. J Appl Biomed 15(4):241–248
46. Fereshteh Z, Rojaee R, Sharifnabi A (2016) Effect of different polymers on morphology and particle size of silver nanoparticles synthesized by modified polyol method. Superlattices Microstruct 98:267–275
47. Flores-Rojas G, López-Saucedo F, Bucio E (2020) Gamma-irradiation applied in the synthesis of metallic and organic nanoparticles: a short review. Radiation Physics Chemistry 169:107962
48. Ghasemi M, Mashhadi S, Azimi-Amin J (2018) Fe_3O_4/AC nanocomposite as a novel nano adsorbent for effective removal of cationic dye: process optimization based on Taguchi design method, kinetics, equilibrium and thermodynamics. J Water Environ Nanotechnol 3(4):321–336
49. Gilbertson LM, Zimmerman JB, Plata DL, Hutchison JE, Anastas PT (2015) Designing nanomaterials to maximize performance and minimize undesirable implications guided by the principles of green chemistry. Chem Soc Rev 44(16):5758–5777
50. Ginimuge PR, Jyothi SD (2010) Methylene blue: revisited. J Anaesthesiol Clin Pharmacol 26(4):517
51. Gómez-Romero P, Sanchez C (2004) Hybrid materials, functional applications. An introduction. Funct Hybrid Mater pp 1–14
52. Gong J-L, Wang B, Zeng G-M, Yang C-P, Niu C-G, Niu Q-Y, Liang Y (2009) Removal of cationic dyes from aqueous solution using magnetic multi-wall carbon nanotube nanocomposite as adsorbent. J Hazard Mater 164(2–3):1517–1522
53. Gottschalk F, Sun T, Nowack B (2013) Environmental concentrations of engineered nanomaterials: review of modeling and analytical studies. Environ Pollut 181:287–300
54. Gou Z, Hopla GA, Yao M, Cui B, Su Y, Rinklebe J, Sun Y (2022) Removal of dye pollution by an oxidase derived from mutagenesis of the Deuteromycete Myrothecium with high potential in industrial applications. Environ Pollut 310:119726
55. Goutam Mukherjee A, Ramesh Wanjari U, Eladl MA, El-Sherbiny M, Elsherbini DMA, Sukumar A, Vellingiri B (2022) Mixed contaminants: occurrence, interactions, toxicity, detection, and remediation. Molecules 27(8):2577
56. Gupta TB, Lataye DH (2019) Removal of crystal violet and methylene blue dyes using Acacia nilotica sawdust activated carbon. Indian J Chem Technol 26:52–68
57. Hang NTN, Yang Y, Nam NQT, Nogami M, Phuc LH, Long NV (2022) Pt-based multi-metal electrocatalysts and potential applications: recent advancements in the synthesis of nanoparticles by modified polyol methods. Crystals 12(3):375
58. Hasan S (2015) A review on nanoparticles: their synthesis and types. Res J Recent Sci 2277:2502
59. Hebbar D, Choudhari K, Pathak N, Shivashankar S, Kulkarni SD (2018) $ZnGa_2$-xEuxO_4 nanoparticles: 10 minutes microwave synthesis, thermal tuning of Eu3+ site distribution and photophysical properties. J Alloys Compd 768:676–685
60. Henam SD, Ahmad F, Shah MA, Parveen S, Wani AH (2019) Microwave synthesis of nanoparticles and their antifungal activities. Spectrochim Acta Part A: Mol Biomol Spectrosc 213:337–341
61. Hu X, Zhang Y, Ding T, Liu J, Zhao H (2020) Multifunctional gold nanoparticles: a novel nanomaterial for various medical applications and biological activities. Front Bioeng Biotechnol 8:990

62. Hua Y, Xiao J, Zhang Q, Cui C, Wang C (2018) Facile synthesis of surface-functionalized magnetic nanocomposites for effectively selective adsorption of cationic dyes. Nanoscale Res Lett 13:1–9. https://doi.org/10.1186/s11671-018-2476-7
63. Huang J, Wang D, Huang L-H, Huang H (2020) Roles of reconstituted high-density lipoprotein nanoparticles in cardiovascular disease: a new paradigm for drug discovery. Int J Mol Sci 21(3):739
64. Hufschmid R, Arami H, Ferguson RM, Gonzales M, Teeman E, Brush LN, … Krishnan KM (2015) Synthesis of phase-pure and monodisperse iron oxide nanoparticles by thermal decomposition. Nanoscale 7(25):11142–11154
65. Ibrahim AG, Sayed AZ, Abd El-Wahab H, Sayah MM (2020) Synthesis of a hydrogel by grafting of acrylamide-co-sodium methacrylate onto chitosan for effective adsorption of Fuchsin basic dye. Int J Biol Macromol 159:422–432
66. Ider M, Abderrafi K, Eddahbi A, Ouaskit S, Kassiba A (2017) Rapid synthesis of silver nanoparticles by microwave-polyol method with the assistance of latex copolymer. J Cluster Sci 28(3):1025–1040
67. Ijaz I, Gilani E, Nazir A, Bukhari A (2020) Detail review on chemical, physical and green synthesis, classification, characterizations and applications of nanoparticles. Green Chem Lett Rev 13(3):223–245
68. Islam M, Islam MS (2013) Electro-deposition method for platinum nano-particles synthesis. Eng Int 1(2)
69. Islam MA, Ali I, Abdul Karim SM, Firoz MSH, Chowdhury A, Morton DW, Angove MJ (2019) Removal of dye from polluted water using novel nano manganese oxide-based materials. J Water Process Eng 32:1–21. https://doi.org/10.1016/j.jwpe.2019.100911
70. Jain K, Patel AS, Pardhi VP, Flora SJS (2021) Nanotechnology in wastewater management: a new paradigm towards wastewater treatment. Molecules 26(6):1797
71. Janaki V, Oh B-T, Shanthi K, Lee K-J, Ramasamy A, Kamala-Kannan S (2012) Efficiency of various semiconductor catalysts for photodegradation of Safranin-T. Res Chem Intermed 38(7):1431–1442
72. Jiang S, Chekini M, Qu Z-B, Wang Y, Yeltik A, Liu Y, … Demir HV (2017) Chiral ceramic nanoparticles and peptide catalysis. J Am Chem Soc 139(39):13701–13712
73. John MG, Tibbetts KM (2019) One-step femtosecond laser ablation synthesis of sub-3 nm gold nanoparticles stabilized by silica. Appl Surf Sci 475:1048–1057
74. Kanagamani K, Muthukrishnan P, Saravanakumar K, Shankar K, Kathiresan A (2019) Photocatalytic degradation of environmental perilous gentian violet dye using leucaena-mediated zinc oxide nanoparticle and its anticancer activity. Rare Met 38(4):277–286
75. Karthik K, Nikolova MP, Phuruangrat A, Pushpa S, Revathi V, Subbulakshmi M (2020) Ultrasound-assisted synthesis of V_2O_5 nanoparticles for photocatalytic and antibacterial studies. Mater Res Innovations 24(4):229–234
76. Kaur S, Rani S, Mahajan RK (2013) Adsorptive removal of dye crystal violet onto low-cost carbon produced from Eichhornia plant: kinetic, equilibrium and thermodynamic studies. Desalin Water Treat 53(7):1957–1969. https://doi.org/10.1080/19443994.2013.841109
77. Khan MMR, Rahman MW, Cheng CK (2015) Tea dust as potential low-cost adsorbent for the removal of crystal violet from aqueous solution. Desalin Water Treat 57(31):14728–14738. https://doi.org/10.1080/19443994.2015.1066272
78. Kharissova OV, Dias HR, Kharisov BI, Pérez BO, Pérez VMJ (2013) The greener synthesis of nanoparticles. Trends Biotechnol 31(4):240–248
79. Kuntyi O, Kytsya A, Mertsalo I, Mazur A, Zozula G, Bazylyak L, Topchak R (2019) Electrochemical synthesis of silver nanoparticles by reversible current in solutions of sodium polyacrylate. Colloid Polym Sci 297(5):689–695
80. Kumar R, Ahmad R (2011) Biosorption of hazardous crystal violet dye from aqueous solution onto treated ginger waste (TGW). Desalination 265:112–118. https://doi.org/10.1016/j.desal.2010.07.040
81. Kyzas GZ, Kostoglou M (2014) Green adsorbents for wastewaters: a critical review. Materials 7:333–364. https://doi.org/10.3390/ma7010333

82. Lascar N, Kumar U (2017) Adsorption of crystal violet from wastewater by modified Bambusa tulda. KSCE J Civ Eng 22(8):2755–2763. https://doi.org/10.1007/s12205-017-0473-5
83. Lakshmanan G, Sathiyaseelan A, Kalaichelvan P, Murugesan K (2018) Plant-mediated synthesis of silver nanoparticles using fruit extract of Cleome viscosa L.: assessment of their antibacterial and anticancer activity. Karbala Int J Modern Sci 4(1):61–68
84. Li S, Huang X, Wan Z, Liu J, Lu L, Peng K, … Bhattarai R (2020) Green synthesis of ultrapure La $(OH)_3$ nanoparticles by one-step method through spark ablation and electrospinning and its application to phosphate removal. Chem Eng J 388:124373
85. Liu H, Lian T, Liu Y, Hong Y, Sun D, Li Q (2017) Plant-mediated synthesis of Au nanoparticles: separation and identification of active biomolecule in the water extract of Cacumen Platycladi. Indus Eng Chem Res 56(18):5262–5270
86. Lu J, Guo J, Song S, Yu G, Liu H, Yang X, Lu Z (2020) Preparation of Ag nanoparticles by spark ablation in gas as catalysts for electrocatalytic hydrogen production. RSC Adv 10(63):38583–38587
87. Malik MA, Wani MY, Hashim MA (2012) Microemulsion method: a novel route to synthesize organic and inorganic nanomaterials: 1st nano update. Arab J Chem 5(4):397–417
88. Mashkoor F, Nasar A, Inamuddin AAM (2018) Exploring the reusability of synthetically contaminated wastewater containing crystal violet dye using Tectona grandis sawdust as a very low-cost adsorbent. Sci Reports 8(1):1–16. https://doi.org/10.1038/s41598-018-26655-3
89. Mashkoor F, Nasar A (2020) Magsorbents: potential candidates in wastewater treatment technology—a review on the removal of methylene blue dye. J Magn Magn Mater 500:1–19. https://doi.org/10.1016/j.jmmm.2020.166408
90. Mittal A, Mittal J, Kaur D, Gupta VK (2010) Adsorption of hazardous crystal violet from wastewater by waste materials. J Colloid Interface Sci 343:463–473. https://doi.org/10.1016/j.jcis.2009.11.060
91. Mohamed MM, Fouad SA, Elshoky HA, Mohammed GM, Salaheldin TA (2017) Antibacterial effect of gold nanoparticles against Corynebacterium pseudotuberculosis. Int J Veterinary Sci Med 5(1):23–29
92. Moharm AE, El Naeem GA, Soliman HMA, Abd-Elhamid AI, El-Bardan AA, Kassem TS, Nayl AA, Bräse S (2022) Fabrication and characterization of effective biochar biosorbent derived from agricultural waste to remove cationic dyes from wastewater. Polymers 14:1–14. https://doi.org/10.3390/polym14132587
93. Montes-García V, Squillaci MA, Diez-Castellnou M, Ong QK, Stellacci F, Samori P (2021) Chemical sensing with Au and Ag nanoparticles. Chem Soc Rev 50(2):1269–1304
94. Moosavi S, Li RYM, Lai CW, Yusof Y, Gan S, Akbarzadeh O, Johan MR (2020) Methylene blue dye photocatalytic degradation over synthesised Fe_3O_4/AC/TiO_2 nano-catalyst: degradation and reusability studies. Nanomaterials 10(12):2360
95. MubarakAli D, Arunkumar J, Nag KH, SheikSyedIshack K, Baldev E, Pandiaraj D, Thajuddin N (2013) Gold nanoparticles from Pro and eukaryotic photosynthetic microorganisms—comparative studies on synthesis and its application on biolabelling. Coll Surf B: Biointerfaces 103:166–173
96. Nakagawa T, Hibara A, Hinze WL, Takagai Y (2020) Nanoparticle induced formation of self-assembled zwitterionic surfactant microdomains which mimic microemulsions for the in situ fabrication and dispersion of silver nanoparticles. RSC Adv 10(56):34161–34166
97. Naresh Yadav D, Anand Kishore K, Bethi B, Sonawane SH, Bhagawan D (2018) ZnO nanophotocatalysts coupled with ceramic membrane method for treatment of Rhodamine-B dye waste water. Environ Dev Sustain 20(5):2065–2078
98. Naser H, Alghoul M, Hossain MK, Asim N, Abdullah M, Ali MS, Amin N (2019) The role of laser ablation technique parameters in synthesis of nanoparticles from different target types. J Nanopart Res 21(11):1–28
99. Nasrollahzadeh M, Sajadi SM, Issaabadi Z, Sajjadi M (2019) Biological sources used in green nanotechnology. In: Interface science and technology, vol 28. Elsevier, pp 81–111
100. Nassar MY, Khatab M (2016) Cobalt ferrite nanoparticles via a template-free hydrothermal route as an efficient nano-adsorbent for potential textile dye removal. RSC Adv 6(83):79688–79705

101. Natsuki J, Natsuki T, Hashimoto Y (2015) A review of silver nanoparticles: synthesis methods, properties and applications. Int J Mater Sci Appl 4(5):325–332
102. Nguyen TKL, Nguyen ND, Dang VP, Phan DT, Tran TH, Nguyen QH (2019) Synthesis of platinum nanoparticles by gamma co-60 ray irradiation method using chitosan as stabilizer. Adv Mater Sci Eng Int
103. Nazifa TA, Habba N, Aris A, Hadibarata T (2017) Adsorption of procion red MX-5B and crystal violet dyes from aqueous solution onto corncob activated carbon. J Chin Chem Soc 65(2):259–270. https://doi.org/10.1002/jccs.201700242
104. Oh AH, Park H-Y, Jung Y-G, Choi S-C, An GS (2020) Synthesis of Fe_3O_4 nanoparticles of various size via the polyol method. Ceram Int 46(8):10723–10728
105. Panda SK, Aggarwal I, Kumar H, Prasad L, Kumar A, Sharma A, Vo D-VN, Thuan DV, Mishra V (2021) Magnetite nanoparticles as sorbentsfor dye removal: a review. Environ Chem Lett 19(3):24872525
106. Patra A (2021) Cationic dyes. https://www.academia.edu/49326512/Cationic_Dyes. Accessed 19 Sep 2022
107. Phan CM, Nguyen HM (2017) Role of capping agent in wet synthesis of nanoparticles. J Phys Chem A 121(17):3213–3219
108. Price H, Adams E, Quilliam RS (2019) The difference a day can make: the temporal dynamics of drinking water access and quality in urban slums. Sci Total Environ 671:818–826
109. Rădulescu C, Hossu AM, Ioniţă I, Moater EI (2006) Synthesis and characterization of new cationic dyes for synthetic fibres. Dyes Pigm 76:366–371. https://doi.org/10.1016/j.dyepig.2006.08.047
110. Rafique M, Shafiq F, Gillani SSA, Shakil M, Tahir MB, Sadaf I (2020) Eco-friendly green and biosynthesis of copper oxide nanoparticles using Citrofortunella microcarpa leaves extract for efficient photocatalytic degradation of Rhodamin B dye form textile wastewater. Optik 208:164053
111. Rameshkumar P, Yusoff N, Ming HN, Sajab MS (2016) Microwave synthesis of reduced graphene oxide decorated with silver nanoparticles for electrochemical determination of 4-nitrophenol. Ceram Int 42(16):18813–18820
112. Rane AV, Kanny K, Abitha V, Thomas S (2018) Methods for synthesis of nanoparticles and fabrication of nanocomposites. In: Synthesis of inorganic nanomaterials. Elsevier, pp 121–139
113. Rao BS, Kumar BR, Reddy VR, Rao TS, Chalapathi G (2011) Preparation and characterization of CdS nanoparticles by chemical co-precipitation technique. Chalcogenide Lett 8(3):177–185
114. Raval NP, Shah PU, Shah NK (2017) Malachite green "'a cationic dye"' and its removal from aqueous solution by adsorption. Appl Water Sci 7:3407–3445. https://doi.org/10.1007/s13201-016-0512-2
115. Roushani M, Mavaei M, Rajabi HR (2015) Graphene quantum dots as novel and green nanomaterials for the visible-light-driven photocatalytic degradation of cationic dye. J Mol Catal A: Chem 409:102–109
116. Roy A, Bhattacharya J (2015) Nanotechnology in industrial wastewater treatment. IWA Publishing
117. Sakti SCW, Laily RN, Aliyah S, Indrasari N, Fahmi MZ, Lee HV, Akemoto Y, Tanaka S (2020) Re-collectable and recyclable epichlorohydrin-crosslinked humic acid with spinel cobalt ferrite core for simple magnetic removal of cationic triarylmethane dyes in polluted water. J Environ Chem Eng 8(4):1–14. https://doi.org/10.1016/j.jece.2020.104004
118. Saleh TA (2021) Protocols for synthesis of nanomaterials, polymers, and green materials as adsorbents for water treatment technologies. Environ Technol 24:101821
119. Salvador Fernández M, Gutiérrez Cervelló G, Noriega S, Moyano Artime A, Blanco López MdC, Matos González M (2021) Microemulsion synthesis of superparamagnetic nanoparticles for bioapplications. Int J Mol Sci 22(1):427
120. Schiel R, Wilson BM, Langford M (2021) The determinants of access to sanitation: the role of human rights and the challenges of measurement. Water 13(12):1676
121. Schlosser CA, Strzepek K, Gao X, Fant C, Blanc É, Paltsev S, … Gueneau A (2014) The future of global water stress: an integrated assessment. Earth's Future 2(8):341–361

122. Shah N, Dwivedi AH, Pillai SG (2021) Processes for the treatment of industrial wastewater. In: Advanced materials technologies for wastewater treatment, pp 285–316
123. Shine DJ, Shasha S, Emmanuel O, Fometu SS, Guohua W (2020) Biosynthesizing gold nanoparticles with Parkia Biglobosa leaf extract for antibacterial efficacy in vitro and photocatalytic degradation activities of Rhodamine B dye. Adv Sci Eng Med 12(7):970–981
124. Simate GS, Cluett J, Iyuke SE, Musapatika ET, Ndlovu S, Walubita LF, Alvarez AE (2011) The treatment of brewery wastewater for reuse: state of the art. Desalination 273(2–3):235–247
125. Singh JP, Kumar M, Sharma A, Pandey G, Chae KH, Lee S (2020) Bottom-up and top-down approaches for MgO. In: Sonochemical reactions. IntechOpen London, UK
126. Singh RP, Singh PK, Gupta R, Singh RL (2019) Treatment and recycling of wastewater from textile industry. In: Advances in biological treatment of industrial waste water and their recycling for a sustainable future. Springer, pp 225–266
127. Singha I, Mishrab PK (2020) Nano-membrane filtration a novel application of nanotechnology for waste water treatment. Mater Today: Proc 29:327–332
128. Srivastava A, Katiyar A (2022) Zinc oxide nanostructures. In: Ceramic science and engineering. Elsevier, pp 235–262
129. Sriram G, Kigga M, Uthappa UT, Rego RM et al (2020) Naturally available diatomite and their surface modification for the removal of hazardous dye and metal ions: a review. Adv Colloid Interface Sci. https://doi.org/10.1016/j.cis.2020.102198
130. Sudha PN, Sangeetha K, Vijayalakshmi K, Barhoum A (2018). Nanomaterials history, classification, unique properties, production and market. In: Emerging applications of nanoparticles and architecture nanostructures. Elsevier, pp 341–384
131. Tiwary K, Choubey SK, Sharma K (2013) Structural and optical properties of ZnS nanoparticles synthesized by microwave irradiation method. Chalcogenide Lett 10(9):319
132. Unni M, Uhl AM, Savliwala S, Savitzky BH, Dhavalikar R, Garraud N, Arnold DP, Kourkoutis LF, Andrew JS, Rinaldi C (2017) Thermaldecomposition synthesis of iron oxide nanoparticles with diminished magneticdead layer by controlled addition of oxygen. ACS Nano 11(2):2284–2303
133. Wang X, Guo Y, Jia Z, Ma H, Liu C, Liu Z, Shi Q, Ren B, Li L, Zhang X (2021) Fabrication of graphene oxide/polydopamine adsorptive membrane by stepwise in-situ growth for removal of rhodamine B from water. Desalination 516:115220
134. Whitesides GM, Stedronsky ER, Casey CP, San Filippo Jr J (1970) Mechanism of thermal decomposition of n-butyl (tri-n-butylphosphine) copper (I). J Am Chem Soc 92(5):1426–1427
135. Wolf JB, Stawski TM, Smales GJ, Thünemann AF, Emmerling F (2022) Towards automation of the polyol process for the synthesis of silver nanoparticles. Sci Rep 12(1):1–9
136. Yang Y, Johansson M, Wiorek A, Tarakina NV, Sayed F, Mathieu R, Johnson M, Soroka I (2021) Gamma-radiation induced synthesis of freestanding nickel nanoparticles. Dalton Trans 50(1):376–383
137. Yanilkin VV, Gul'naz RN, Kokorekin VA (2018) Mediated electrochemical synthesis of metal nanoparticles. Russian Chem Rev 87(11):1080
138. Yazid NA, Joon YC (2019) Co-precipitation synthesis of magnetic nanoparticles for efficient removal of heavy metal from synthetic wastewater. Paper presented at the AIP Conference Proceedings
139. Yousuf MA, Baig MM, Waseem M, Haider S, Shakir I, Khan SU-D, Warsi MF (2019) Low cost micro-emulsion route synthesis of Cr-substituted $MnFe_2O_4$ nanoparticles. Ceram Int 45(17):22316–22323
140. Yue D, Qian X, Kan M, Ren M, Zhu Y, Jiang L, Zhao Y (2017) Sulfurated [NiFe]-based layered double hydroxides nanoparticles as efficient co-catalysts for photocatalytic hydrogen evolution using CdTe/CdS quantum dots. Appl Catal B 209:155–160
141. Yunusa U, Usmana B, Ibrahim MB (2021) Cationic dyes removal from wastewater by adsorptive method: a systematic in-depth review. Algerian J Chem Eng 2:6–40. https://doi.org/10.5281/zenodo.5101197
142. Zhang D, You H, Zhang L, Fang J (2020) Facile surface modification of mesoporous Au nanoparticles for highly sensitive SERS detection. Anal Chem 92(23):15379–15387

Textile Dyes Removal Using Silica-Dendrimer Hybrid Materials

Akbar Esmaeili

Abstract Textile industries are one of the largest industries that consume water and produce colored wastewater. Textile dyes are the largest group of water-soluble synthetic dyes with the most diversity. Therefore, the removal of dyes from wastewater has attracted much attention recently due to their potential toxicity. Polluting dyes are widely used in various industries such as textiles, papermaking, color pictures, pharmaceuticals, food, and cosmetics. More than 70 million tons of artificial dyes are produced annually worldwide. Recently, mesoporous compounds have had good performance and efficiency in the separation process due to their advantages such as high specific surface area, lack of internal penetration resistance, and tendency to target species in solution. The structure of magnetic mesoporous silica includes a magnetic core surrounded by silica cavities. In other words, magnetic iron oxide nanoparticles sit inside the pores of mesoporous silica. The intermediate compound and the chitosan-dendrimer compound were used to decolorize the effluent of textiles dyed with reactive dyes. The polymer-dendrimer absorbent can be an absorbent with high potential and biodegradable to remove anionic compounds such as reactive dyes from textile industry wastewater. The unique feature of these absorbents is high absorption power, biodegradability, biocompatibility, and nontoxicity. The prepared biopolymer-dendrimer combination can potentially produce various industrial applications, including water purification, purification of organic and metal pollutants harmful to the environment, or preparation of nanofibers for medical applications.

Keywords Smart textiles · Natural resources · Clothing technology · Clothing industry

A. Esmaeili (✉)
Department of Chemical Engineering, Islamic Azad University, North Tehran Branch, PO Box 1651153311, Tehran, Iran
e-mail: akbaresmaeili@yahoo.com

© The Author(s), under exclusive license to Springer Nature Singapore Pte Ltd. 2023

A. Ahmad et al. (eds.), *Nanohybrid Materials for Treatment of Textiles Dyes*, Smart Nanomaterials Technology, https://doi.org/10.1007/978-981-99-3901-5_14

1 Introduction

The application of nanomaterials in the removal of color from wastewater has attracted wide attention in recent years. This chapter highlights current advances in using nanomaterials and special dendrimers for the adsorption of dyes from textile wastewater. Specific adsorption mechanisms and recent advances, especially in increasing the adsorption capacity and reducing the toxicity of textile wastewater, are discussed for each dendrimer. The collected data show that nano-dendrimer can be effectively used to treat sewage color; nano chitosan has a high potential for commercial applications. This nanomaterial's stability, nontoxicity, and cheapness make it perform well in the absorption process. The use of nanomaterials for various purposes has developed rapidly in recent years. However, this technology has a long way to go before reaching its ultimate commercialization goal. Other fields, such as methods of recovery and purification of commercial textile dyes from textile wastewater have not been thoroughly studied.

Textile dyes have been widely used in textiles, dyeing, pigments, and many other applications for thousands of years. Today, paints play a vital role in the textile, dye, and pigment industries, and currently, at least one hundred thousand different types of dyes are commercially available. The amount of paint produced annually is estimated to be approximately 1.6 million tons, and 10–15% of this volume is disposed of as wastewater. As a result, dyes are one of the primary sources of water pollution. Excessive exposure to color causes skin irritation and respiratory problems, and exposure to specific colors increases the risk of cancer in humans. In addition, the presence of dyes in wastewater increases chemical oxidation and, as a result, creates a foul smell. For these reasons, it is essential to effectively remove stains from wastewater and ensure the safe discharge of treated liquid to the bed of running waters.

Under normal conditions, colored wastewater is treated using flocculation, aerobic or anaerobic treatment, electrochemical treatment, membrane filtration, and absorption methods. The absorption method is considered the most famous purification method due to its fruitfulness and simplicity. Dye factories generally use commercial activated carbon for dye removal due to its high porosity and large surface area (500–2000 m^2 per gram). The use of commercial activated carbon is relatively expensive due to its high production cost. In addition, activated carbon regeneration requires high-pressure steam, which increases the operating costs of this treatment system. This high cost has become an incentive to discover alternative economical and efficient adsorbents for color removal [65].

During the last decade, low-cost adsorbents obtained from agricultural and solid wastes have attracted the attention of researchers. The effectiveness of most of these wastes in removing colors and heavy metals has been proven. For example, the adsorbent from waste palm oil can remove copper, zinc, and reactive dyes. Such adsorbents with possibly low cost have been extensively investigated by Gupta [94]. However, most of these inexpensive adsorbents consist of microparticles, and the small contact surface of these particles requires a considerable amount of time to achieve maximum removal of pollutants. Since most industries need rapid removal to keep pace with the increasing capacity of pollutants, developing these adsorbents for industrial applications is impossible. As a result, there is an urgent need to develop economically sustainable adsorbents with high removal rates and absorption capacity.

Nanomaterials, especially dendrimers, which are also called nanoparticles, are particles whose dimensions are in the range of 1–100 nm. In general, known nanomaterials are valuable due to their high strength, highly active areas, and low mass. In addition to wastewater treatment, this research study focuses on developing nanomaterials for optical data storage, the construction of sensors, and light and durable materials. Both nanomaterials and activated carbon have a considerable contact surface. But some nanomaterials have two other main advantages as adsorbents compared with activated carbon: first, they can be synthesized quickly and cheaply. Secondly, smaller amounts of these nanomaterials are required to remove pollutants effectively. For this reason, it is expected that nanomaterials in absorption applications will be more economical in the future than activated carbon [67].

This article discusses the various applications of nanomaterials and dendrimers to absorb dyes. Generally, for this purpose, multiple parameters related to adsorption, isothermal and kinetic are also reported. But in this evaluation, each nanomaterial's absorption mechanism has been emphasized. So, the fundamental understanding of how these nanomaterials and dendrimers become more accessible works as absorbents for color removal. Recent advances and development of these nanomaterials and dendrimers to increase their absorption efficiency have also been discussed.[1]

Dendrimers are new-generation materials and catalysts. Approaches to grow dendrimers are functionalization of dendrimers in the outer environment or inner core and incorporation of various nanoparticles. The activity of derived dendrimers is higher in the outer layer of functionalized dendrimers with nanoparticles. The modified dendrimers trap nanoparticles and create covalent and non-covalent immobilized dendrimers. The contribution of ionic liquid and ion gel in catalysis, as a new topic, has recently received public attention. Swelling and shrinking polymer and quaternary ammonium salt immobilized on a silica support the control of textile dye removal activities. This chapter describes these aspects [10].

Many industries, such as textile, papermaking, printing, leather, tanning, ink, cosmetics, etc., enter a large amount of colored waste into the environment [15]. It is reported that more than 700,000 tons and 10,000 types of commercial dyes are produced worldwide due to the lack of proper stabilization of color molecules on the fibers and the inefficiency of dyeing units; about 20% of these dyes enter the industrial effluent [55]. Rimazol Black Bay (RBB) dye is one of the types of reactive dye. These dyes are widely used due to their excellent color fastness, easy application techniques, and low energy consumption. Approximately 45% of all textile dyes produced annually belong to the reactive group. Due to their high solubility in water, reactive dyes are found in higher concentrations in textile wastewater than other stains. They are not easily removed by conventional treatment systems [11, 96]. In

[1] https://nano.ir/index.php?ctrl=paper&actn=paper_view&id=3888&lang=1.

addition, the colors prevent the complete passage of light into the water entered. As a result, the photosynthesis process was reduced, followed by oxygen reduction in receiving waters [105]. Also, various studies have shown that most colors are toxic, allergenic, carcinogenic, and mutagenic [86]. These compounds also negatively affect water quality for drinking and other purposes from an aesthetic point of view [21, 110]. For this reason, this wastewater has been considered one of the most critical factors threatening public health and the environment worldwide and should be adequately treated before discharging into the atmosphere [29].

Absorption is one of the good essential techniques to reduce the concentration of dyes dissolved in aqueous solutions [2]. Compared with other separation processes, the advantages of this process are their simplicity and cheapness [6]. The most common adsorbent material is activated highly effective carbon, but its high cost and high cost of regeneration have made it not desirable to use [9]. In addition, various other adsorbents, such as eggshell [43], pumice mineral adsorbent, bentonite, and fish bone powder, have been used to remove colored wastewater. However, researchers are developing the use of low-cost adsorbents with high absorption potential that have high absorption power with low consumption and do not harm the environment. Recently, most of the researchers' investigations have focused on adsorbents that have a natural base, are available in high amounts in nature, and are not harmful to the environment. Also, economic problems and the recovery of adsorbents have made researchers focus on adsorbents at a cheaper price [31, 32].

2 Commonly Used Colors

2.1 Azo Dyes

Commercially, azo dyes are the most critical dyes, making up more than 5% of the total amount of dyes produced. The most significant number of dyes have been studied, and the enormous volume of articles is related to this class. This term is one of the production methods of insoluble dyes. Azo in the material, like the fibers, is applied to the interlayer, and the intermediate azo (coupling component) that has a high tendency to be absorbed in cellulose fibers is used. Then it passes through the second intermediary (diazo component). This group of dyes for dyeing and printing textile fibers, especially cotton, has an exceptional use because they create stable colors against light and moisture. It is one of the most basic azo dyes widely used [45, 51].

Azo dyes are known with a double nitrogen bond attached to aromatic groups [79]. At least 3000 types of azo dyes are used in the textile, paper, food, cosmetic and pharmaceutical industries. Among the various uses of synthetic dyes, more than 30,000 tons of different shades are produced annually worldwide, among which azo dyes make up more than 60% of stains in factories [87]. Paints are classified based on their application and chemical structure.

Daily, large amounts of these dyes enter the wastewater due to various industrial activities (especially textile factories), which destroy water sources, soil, and organisms (4). During the dyeing process, not all colors are attached to the fibers, and some of the colors remain in the dye bath; So 20% of acid dyes, 30% of direct dyes, and 50% of reactive dyes are released in wastewater [82]. The amount of color used is small, but the main problem is the stability of these dyes in sewage; due to having one or more sulfonic acid groups in their aromatic ring, they inhibit the growth or synthesis of nucleotide acids [22]. The resistance of azo dyes is due to the strongly electron-withdrawing bond of azo to oxygenases and the presence of large aromatic groups attached to the azo bond [79]. The degree of decolorization depends on the type, molecular weight, and substituent groups of the dye. Azo compounds with hydroxy and amino groups tend to degrade more than methyl, methoxy, sulfur, and nitro [74].

Enzymatic and microbial treatment methods to remove dyes and break down stains have advantages over physical and chemical processes. Among them, they can be mentioned: (1) Low price; (2) Sludge production with a small amount; (3) Production of nontoxic products; (4) Compatible with nature; (5) Less water is needed in cleaning than physical and chemical methods [101]. Bacteria, fungi, algae, actinomycetes, yeasts, and plants are the microorganisms in which color removal has been studied so far [87]. Various salts are used in the textile industry; textile factories generally use 60–100 g of salt per liter in reactive dyes. As a result, microorganisms must be able to tolerate these salty conditions for the dyeing process [7]. The first studies have been conducted on the ability to remove color in salt-loving and tolerant species of the *Halomonas* genus [38, 77, 90, 102].

2.2 Reactive Dyes

Reactive dyes have excellent washing stability due to covalent bonding with anionic agents of cotton fiber and low molecular weight; for this reason, this group is known as the bride of chemical dyes. These dyes are classified into three groups, three chlorine, two chlorine, and one chlorine, depending on how many chlorine agents they have to create covalent bonds. The main limitation of reactive dyes is that a significant part of the dyes always remains inside the fiber. This issue, especially in the case of dark colors, is more intense. This paints category is widely used as ink in textile printing industries, especially cotton. Reactive dyes with azo structure, which have a nitrogen double bond color agent, are among our textile industries' most widely used shades. Due to these dyes' low toxicity and degradability, they are classified as dangerous substances for the environment that must be purified before discharge [56, 57, 77].

2.3 Acid Dyes

Dyeing protein fibers with acid dyes are challenging because the color fixation method is different in this process. When we come to the fact that water acts as a lubricant for fibers like nylon, the problem becomes even more complicated. Anyway, in this article, we intend to discuss acid dyes briefly. So, stay with us. Acid dyes are mainly salting sulfuric acid or carboxylic acid for their use. For this reason, they are called acid dyes. Since there are different acid dyes; so, this is a logical question. The simple answer is that the choice of color should be based on the prerequisites of color stability and adjustment. Vital acidic dyes leave a uniform and homogeneous shade (such as Luling dye), but their color stability against washing is weak. At the same time, the super colors produce more durable color shades. If the color stability of textiles is to be improved after the dyeing stage by treatment with metal salts, colors with chelating backgrounds are preferred. Electrolytes act as a moderating or retarding agent in acid dyeing. Here, acid dye absorption takes place faster because of the different electrical natures of the dye and fibers and the rate of dye uptake by the fibers. At the same time, electrolytes slow down the initial dyeing speed by creating temporary bonds with dyes or fibers when separated due to increased temperature. Therefore, electrolytes prevent the unevenness of the dyeing process. Not only does the total amount of dye absorbed by the fibers depend on the volume of acid, but the amount of color discharge also depends on the acidity or alkalinity of the bath (Fig. 1).

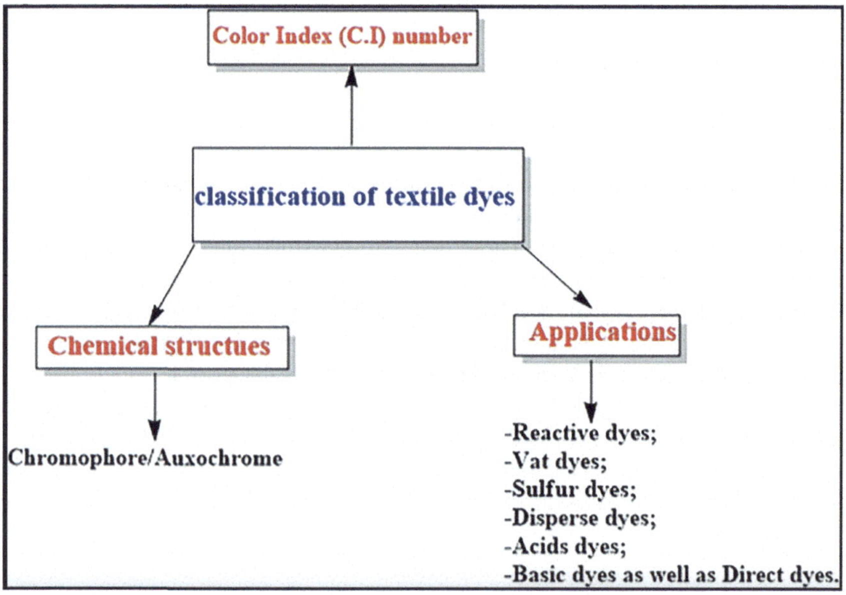

Fig. 1 Classification structure of the textile [14]

Acidic dyes added to the bath can attach themselves to the cationic site of the fiber by replacing the anion released by the acid. The subject is not possible at room temperature and only when the bath is heated; the molecular acceleration of the dye creates the required displacement. If the dyeing starts at a temperature of 40 °C, the dye solution boils slowly, with most of the dyeing during the boiling time; the dyeing results will be excellent. Acidic dyes are not transferred to the fibers in a bath whose temperature is below 39 °C. Milling acid dyes require a minimum temperature of 60 °C to discharge, but the paint is quickly transferred to the fibers at a temperature of 70 °C. Super milling colors can produce a uniform color only when boiling [77].

The color stability of nylons dyed with acid dyes, the fabric is treated with tannin teeth (back tanning). This processing method is necessary for swimwear or multi-colored fabrics. In the process, a film-like layer is created on the surface of the fibers, which prevents the color from separating. The treatment of fabric with tannin teeth is done in the following two ways: general treatment of material with tannin teeth: Here, the fabric is treated with a solution containing 2% tannin and 2% acetic acid for 20 min at 70 °C. Half-half fabric treatment with tannin teeth: If the fabric is treated with syntax, it is called half-half treatment; because, in this case, the color stability against washing does not increase as much as in the previous point[35].

Wool fibers contain a large number of amino groups. The number of amino groups in wool fibers is approximately 20 times the amino groups of nylon fibers and five times the amino groups of silk fibers. Since wool has a very amorphous nature; therefore, it is easy to apply the shadow of dark colors to it. The characteristics of the nylon dyeing process with acid dyes are similar to wool. The shades of colors produced on nylon fibers are identical to the shadows of wool colors but have a lower saturation point. Therefore, it is not possible to use darker shades on them. Silk has a great tendency to acid stains, but its colors have less color stability than wool fibers. The movement of silk to acid dyes occurs at a lower temperature than wool fibers; therefore, silk fiber dyeing is usually done at 40 °C, and the maximum temperature of the dyeing environment is limited to 85 °C. Sodium sulfate is used in silk dyeing, because it reduces its brightness. Because solid acids such as sulfuric acid damage silk, then acetic acid is used.[2]

2.4 Cationic Dyes (Basic)

Cationic dyes are chromatographically complete and highly chromatic. They are water-soluble dyes with solid color and can ionize into colored organic cations and simple anions in water. Since the cationic part in the molecular structure of the dye has a primary group, cationic dyes and basic dyes are in the list of dyes. In China, it is customary to call them cationic dyes, and some of the old main types are still called basic dyes. When cationic dyes are dyed, it is usually done in an acidic environment, in which case both the dye and the fiber are in an ionized state that combines with

[2] https://nasaji.com.

the anions of the thread by absorbing charge. Cationic dyes have low light and soap fastness on polyacrylonitrile, "acrylic fiber." They are still unique colors for acrylic yarn and can also be dyed with modified polyester and nylon. Because the size of the cationic dye molecule is close to the gap of acrylic macromolecules in the static state, in this state, the dye molecule is difficult or almost impossible to enter the fiber space. When the dyeing temperature is higher than the acrylic fiber's glass transition temperature, the thread's molecular fragment undergoes serious movement. Because the dependence of the cationic dye on the acrylic fiber is high, the dye absorption rate increases suddenly, and the cationic dye has poor transmittance, so the dyeing is not uniform [53].

To improve the unevenness of dyeing caused by too fast dyeing speed the first thing is to control the heating speed when dyeing seriously and appropriately extend the dye absorption time or add suitable additives simultaneously. The second way is to choose the right color when matching colors. Because the content of the dyeing seat in acrylic fiber is limited, if the color compatibility is not inconsistent, the dyeing amount will be different, and the phenomenon of competitive dyeing will occur, which will affect the color and uniformity, therefore, to select the right color for the right color[3] [35].

2.5 Sulfur Colors

In general, sulfur dyes can be classified into two parts, soluble and insoluble, which differ in their solubility in water. In this way, soluble sulfur dyes can dissolve in water, but they do not have an attraction to cellulose before regeneration. Insoluble sulfur dyes are insoluble in water or have very low solubility, which can be dissolved in water by combining with sodium sulfide solution (regenerative). One of the essential features of this dye is its relatively good optical stability. Even though these dyes do not have ideal wear stability, this undesirable feature can become a strong point. Of course, due to some of their excellent shades, it is possible to produce clothes with various colors and affordable. These dyes use primarily dark shades such as black, brown, navy, olive, etc. Of course, they do not have a limited or incomplete color shade because sulfur dyes also have yellow, pink, etc. colors. For example, from sulfur dye, you can get a kind of beautiful green called hospital green, which is widely used in children's clothes. Still, sulfur is mainly used in dark shades, such as women's coats, incredibly dark red color, and polyester/cotton mix, which has good washing stability.

Sulfur dyes are mixed with sodium sulfur (sodium sulfide) and hot water to make a solution. Then, with the help of salt, the absorption between the product and the dye is carried out, and finally, it enters the oxidation bath so that the paint becomes insoluble on the fabric again. In the case of dyeing with sulfur dye in a black color shade, for each percentage of color, you can use 1–2.4 times 60% sodium sulfur dye

[3] https://af.sulphurblackcn.com/info/cationic-dyes-38610024.html.

and 70–80 g salt/l. If the sulfur used was weaker, more could be used. The duration of the interaction between paint and salt is about 45 min to 1 h, after which you can wash with cold water. Of course, in the case of using a jigger, washing is done by double cold overflow water, after which it enters the oxidation bath containing 5 g/l of hydrogen peroxide, 1 ml per liter of acetic acid at a temperature of 55 to 60 °C for 20 min.

One critical factor is paying attention to the amount of sulfur, provided that the consumed sulfur is healthy. The subject means that if sulfur is exposed to air, it starts to oxidize. Sulfur oxidizes itself, and as a result, its strength decreases. For this reason, to store sulfur, the lid of the barrel or bag must be closed well, and no air can enter it. The amount of sulfur is also low, the loco dye does not perform enough, and due to the decrease in the strength of the paint, some of it is lost.

Along with the amount of sodium sulfur mentioned, you can use the amount of 5 g/l of carbonate to help the formation of dye solution loco. If you don't use the right solution and oxide, the stability and the super dry stability will decrease. As a result, if the effluent is not colored after washing, if the dyed product is worn with another product, it will stain the product.

When using a sulfur dye, washing after the dyeing operation is more important than the process itself. Because if the washing operation is not done correctly, the remaining color will turn into sulfuric acid over time and cause the fabric to rot and, eventually, damage the body. For this purpose, in the best possible way after dyeing and oxidation with high-quality soap. This process is helpful. Also, using cationic softeners is unsuitable due to overshadowing the abrasion stability of the fabric.

Dyeing with sulfur dyes is usually done in a jigger machine; hence, when the heavy pants fabrics with thick fabric borders enter through the open door of the jigger, the air enters from the side of the material and causes the color to increase or decrease to some extent. One of the points that failure to comply with can cause numerous problems is not paying attention to the dyeing water level of the jigger machine, which should have a higher water level than other dyes. For example, suppose in cotton dyeing with reactive dyes, 60% of the device takes water, then in sulfur dyes. Another issue in using sulfur dyes is the sensitivity in the number of consumables. The lack of this causes unevenness in the environment and factories. It is necessary to pay attention to the alignment of the jigger, the parallelism of the roller to the horizon line, and the foundation surface to not cause a difference in the shade of the sides. When dyeing the woven fabric in the Haspel machine, the water level should be high, and the door should be slightly open so that complete absorption is done; of course, the amount of dye absorption in Haspel is less than that of jigger.[4]

[4] https://nasajimovafagh.com/interview/440.

2.6 Color Removal Technology of Textile Wastewater

Many methods have been used to remove colored pollutants from wastewater from different industries. These methods can be divided into biological, chemical, and physical categories. In the following, the practices of color removal will be reviewed [26, 77, 83].

2.6.1 Biological Methods of Textile Wastewater

Naturally, due to the activity of microorganisms, organic materials are decomposed into carbon dioxide, carbon, water, and a small part of them into stable mineral compounds.

Aerobic Treatment

This purification is done in the presence of oxygen: (1) The system of spreading organisms in wastewater and then treating them is an example of the activated sludge method. Other methods include the aeration pond oxidation method and the biological oxidation pond method. (2) Organism stabilization system in the material: This system contains a thin shell of microorganisms attached to the solid material. When they hit the wastewater, the water purification process takes place. This method is called the critical thin crust. This system uses drip filtration, rotating disk, contact aeration, ground irrigation, and intermittent sand filtration. Aerobic microorganisms can decompose wastewater organic materials in the following way: Sewage-purifying microorganisms can generally exist in two forms: (1) in the form of suspended substances sprayed in the form of clots in water (2) in the form of materials deposited on floating solid materials in water. Usually, in the treatment method, the first case is called the activated sludge method, and the second case is called the critical shell method.

Anaerobic Treatment

In anaerobic purification, the process occurs in oxygen-free places and is not much different from the aerobic type. A clear example of anaerobic treatment includes hot and cold washing. In addition, the mentioned system is effective for treating leachate of bulky waste materials. The working method is that in the first stage, the amount of organic matter in this system is reduced. Then the purification process is performed with an anaerobic purification device. In addition, the system is widely used to purify the excess sludge produced in the activated sludge process. The biological treatment method is exclusively used in urban wastewater treatment. Anaerobic treatment is done without air storage. In natural medicine, it is necessary to keep the environmental conditions stable, which causes the development of the maximum ability of microorganisms. This method's primary ecological conditions are temperature, inappropriate atmosphere, and pH. Different types of microorganisms participate in wastewater treatment. The group of microorganisms that play the leading role in wastewater treatment is bacteria, protozoa, algae, etc.—Loading of pollution and its elimination rate by organisms: Although biological treatment has different methods, a general principle (that the relationship between the amount of organism and the amount of pollutant separated) is established in all of them. This purification is done in the presence of oxygen. After wastewater treatment to control pollution due to biological, physical, and chemical modification methods, organic and inorganic sludge will remain. Usually, to dispose of sludge, it is dried through dewatering. It is essential to mention that the remaining sludge in the treatment plants contains different amounts of toxic and dangerous compounds.[5]

2.6.2 Chemical Methods of Textile Wastewater

Chemical treatment of sewage is one of the most critical steps in treating industrial and sanitary wastewater, which significantly impacts removing pollutants and increasing the quality of effluent from treatment plants. Despite the high efficiency of physical and biological purification methods, many colloidal particles and solid organic substances are not decomposed and removed. They are considered a severe threat to the environment. For this purpose, various wastewater treatment methods with chemicals are provided and used in different treatment plants.

Chemical wastewater treatment removes solid organic substances, colloidal particles, and pollutants in industrial and sanitary wastewater that many biological and active sludge methods cannot remove. The type of wastewater and the contaminants present in them, as well as the quality of the output of treatment plants, different ways are used for the chemical treatment of sewage. So that in some conditions, only the coagulation of fat and colloidal particles can bring the wastewater to the desired quality. Other methods and processes will be needed, such as neutralization, oxidation, flotation, etc.

The Most Common Chemical Wastewater Treatment Processes

Due to their high electricity, some solid particles suspended in sewage cannot settle on the floor of the treatment plant. For example, colloidal particles, fat, and oil in sewage and industrial and sanitary effluents contain many of these particles. These problems cannot be solved by biological treatment, and the only way is the chemical treatment of wastewater. For this purpose, with the help of coagulant materials, efforts are made to reduce the charge of these particles as much as possible. Aluminum and

[5] https://abnikco.com.

iron salts are used as coagulants. In the next step, the colloidal particles that have lost their electric charge will connect and form a clot-like layer. The layers are placed in the lower pool on the surface of the sewage. It can also be collected with the help of a spatula.

Sewage Neutralization

The bacteria in the wastewater are essential in the biological treatment process. To provide suitable conditions for the proper activity of bacteria for purification, the environment of the treatment plant and the wastewater in it must be neutral. In other words, wastewater should not have a pH lower than 6.5 and above 8.5. Using different chemicals to neutralize the wastewater and adjust its pH to a certain level is necessary. For example, wastewater may be passed through limestone beds or alkaline solid materials; strong caustic soda, magnesium hydroxide, sodium carbonate, sodium bicarbonate, solid sulfuric acid, and weak carbon dioxide are used.

Wastewater Oxidation

Aerobic and anaerobic bacteria decompose and remove large amounts of organic matter from wastewater in biological processes. But these methods cannot decompose vital organic substances in industrial and chemical wastewater treatment processes that are necessary. One of the best ways to remove these organic substances is the oxidation of wastewater. Wastewater oxidation processes are carried out in two ways: treatment with ultraviolet rays, oxygenated water, Gamma rays and the use of ozone, ultraviolet ozone, oxygenated water ozone, ultraviolet oxygenated water, etc. In this step, hydroxyl radical is produced. In this group of wastewater purification processes, ammonium, BOD, resistant organic compounds, and microorganisms are significantly removed.

Treatment of Industrial and Sanitary Wastewater

Sedimentation is another type of wastewater treatment using chemical methods, which by adding some substances, leads to settling suspended particles in the wastewater. By carrying out the sedimentation process, the quality of wastewater will increase to a great extent.

Ion Exchange

Ion exchange processes can remove solids dissolved in wastewater, nitrogen, and heavy metals. In this method, ion exchange resin is used, widely used not only in wastewater treatment but also for water hardness removal and municipal water treatment. Cationic resins lead to removing positive ions in wastewater, and anionic resins remove the negative charge.

Flotation

One of the most essential and practical methods in the chemical treatment of wastewater is flotation. In this process, air bubbles stick to suspended materials with the help of aeration equipment such as a diffuser. Wastewater causes them to be transferred to the upper layers of the sewage at a faster rate.

Chemical Treatment of Wastewater with the Disinfection

In the last step, to treat wastewater with chemical methods, it is necessary to disinfect it and remove the unpleasant smell of sanitary and industrial wastewater. For this purpose, with the help of a chlorination package, you can disinfect the wastewater. In addition to using chlorine, the disinfection process can also be done with the help of ultraviolet rays, which brings a lot of money.

2.6.3 Physical Methods of Textile Wastewater

Physical treatment is the first step in textile and other industries. Wastewater treatment is considered the easiest. This step eliminates suspended solid particles in the wastewater, such as thread, fluff, and similar items. For this purpose, different equipments, such as mechanical and manual garbage collectors can be used. In addition, by injecting coagulants it is used for mechanical and chemical flocculation, and after coagulating the suspended particles, they are given a chance to settle. By performing these processes, it is possible to prevent damage to the equipment of the following steps and increase their efficiency. The effluents from this stage are not allowed to enter nature due to their low quality[6] [44].

Various physical methods, such as membrane filtration processes 17, and surface absorption techniques 18, are also widely used to remove dyes. The limited life before sedimentation in the membrane and economic issues are the limitations of this method [77].

2.7 An Overview of Dendrimers

Dendrimer or tree is derived from the Greek word "dendron," meaning tree or branch, and "meros," meaning part. These compounds have many components, all of which

---

[6] https://ig7.ir/.

are connected to the central nucleus. These compounds are considered polymers, but unlike linear polymers, all the branches in dendrimers are connected to a metal that is the central core. Although these compounds are considered polymers, unlike polymers, they are never made by polymerization reactions. Different synthesis methods are used to make these compounds. Each layer that is added is known as a generation. Figure 2 shows a dendritic structure with a "layer" between each focal point called a generation. Some examples of commercial dendrimers available in the market are shown in this figure. For example, the third layer is known as the third generation, and so on.

Dendrimers are made reversibly due to their unique synthesis methods. Many dendrimers are entirely organic, either made purely of carbon and hydrogen or carbon, hydrogen, oxygen, and nitrogen (poly amidoamine). However, in addition to organic dendrimers, many dendrimers are inorganic, specifically when they have inorganic compounds as branching points in their structure. The central metal in

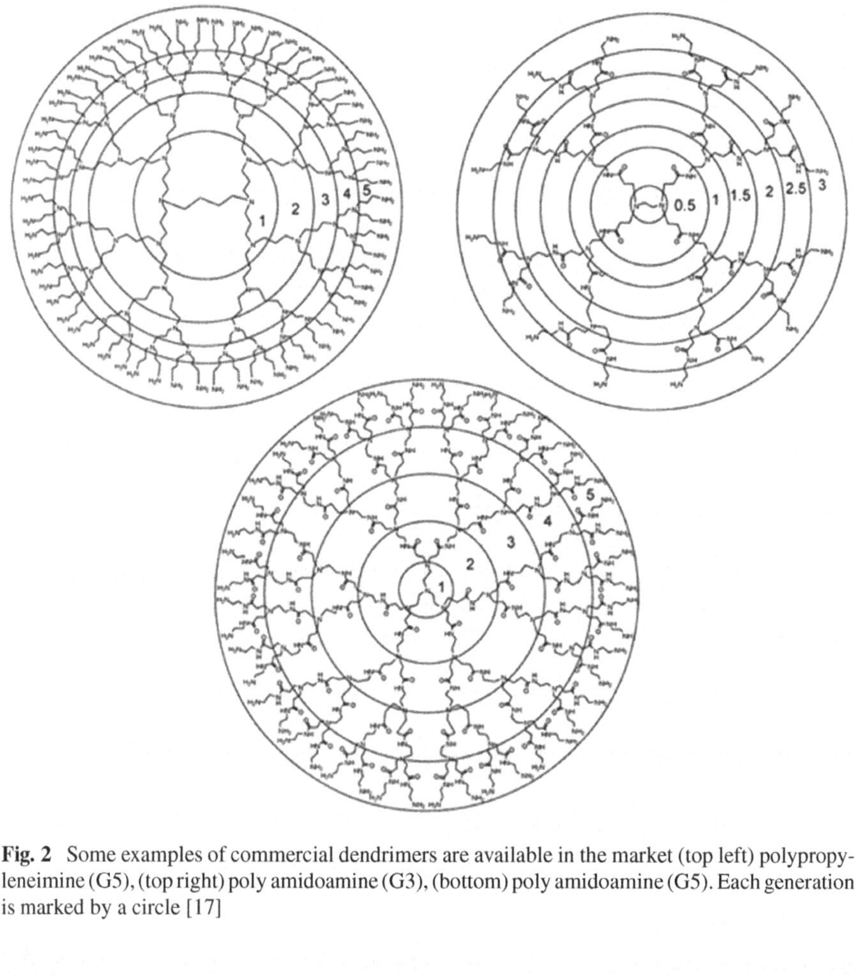

Fig. 2 Some examples of commercial dendrimers are available in the market (top left) polypropyleneimine (G5), (top right) poly amidoamine (G3), (bottom) poly amidoamine (G5). Each generation is marked by a circle [17]

mineral dendrimers is usually silicon, phosphorus, boron, germanium, or bismuth, and these metals are used as branching points. Ideal dendrimer models would have a "starburst" topology. [93].

At the molecular level, branching in dendrimers creates a spherical state, which is formed from hemispherical to spherical, depending on the amount of structural branching. The dendrimer core is a 1,4-di-amino-butane compound. The naming method in dendrimers has always been controversial [72]. In standard terminology, each dendrimer is named based on the layers between the core and the surface layer (known as focal points). For example, when moving from the center to the surface layer, generation five, or "G5," means there are five layers between the core and the surface layer. The body is called layer zero or "G0." Hydrogen substitutions are not considered layers.

In the dendrimer, ammonia is placed in the core. Also, in this dendrimer, the intermediate compounds with carboxylate groups in their surface layer are known as semi-generation dendrimers. Dendrimer design can be based on countless connections. For example, poly amidoamine dendrimer, a mixture of amides and amine, is a variety of hydrophobic compounds [47, 95]. The core of new dendrimers is hydrophobic [17, 47]. The nature of new dendrimers is based on the carbohydrates [97], macromolecules [85], or the cores of metals of the third period [3].

2.8 The History of Dendrimers

Dendrimers were noticed in polymer science in the 1980s as beautiful, star-like molecules with a low weight index. Still, the idea of making branch-like molecules dates back to 1941, when Flory came up with the idea of making these compounds. The first article related to the synthesis of dendrimers with features such as low molecular weight was published by Vogtel in 1978, which focused on the methods of dendrimer synthesis [19].

Three years later, Denkwalter reported the synthesis of polylysine dendrimers with about ten generations [28]. The method was based on divergence methods. The first person who used the term dendrimer was Denqualter.

In the same year, Newcome had synthesized polyether dendrimers, which he had not published. In this work, he synthesized this dendrimer with about three generations [73]. These results were accompanied by Hawker and Frecht's article in 1990, which presented a new method for synthesizing dendrimers called the convergent method [73]. Until the 1990s, many essays and writings about dendrimers were published. Dendrimers with less used components in their structure were synthesized in the following years.

Dendrimers with different functional groups in their surface layer were noticed in many practical aspects. Dendrimers with several other functional groups on their surface have been synthesized. Therefore, the concept that dendrimers are molecules with no active group on their character and are entirely inactive has wholly changed. The early decades of the twenty-first century for dendrimers were associated with

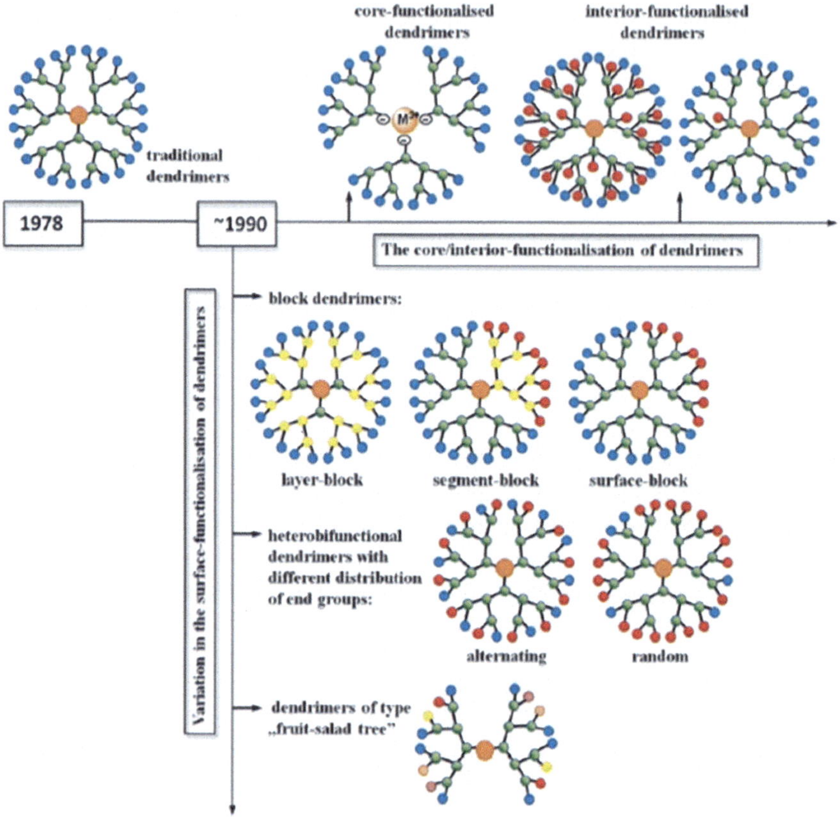

Fig. 3 Structural evolution of dendrimers [91]

changes such as that the core of dendrimers had properties such as catalytic, photocatalytic, electrochemical activity, etc. Also, the surface of dendrimers had functional groups [16]. Due to their interactive parties, dendrimers are used in many scientific fields, such as biology, medicine, physics, and material engineering [52]. Figure 3 shows the evolution of dendrimers from a compound. That is, without a functional group on its surface to a combination with an active group.

2.9 *Types of Dendrimers*

2.9.1 Poly Amidoamine Dendrimers

Polyamidoamine dendrimers are the first nanoscale structures chemically synthesized and have commercial properties.

The unique properties of sterols, such as controlled size, monodispersity, and variable surface groups, make these molecules desirable for biomedical applications. Different therapeutic and imaging agents can activate the end groups in the derivates in a specific and controlled manner, which is a potential for their use in targeted drug delivery. In addition, empty holes in the wood chips are used to encapsulate hydrophobic drug molecules. Third-type amine groups in poly amidoamine (PAMAM = Poly amidoamine) create acid–base interactions, hydrogen bonds, and non-covalent interactions with encapsulated host molecules. All these features make the tree saps suitable agents for the solubilization of hydrophobic drugs. The emergence of polyethylene glycol (PEG)-functionalized surfactants are considered to cause solubility in water and increase drug loading. G3 and G4 generation of "PAMAM," functionalized with amine has been used to encapsulate the Ibuprofen and check their absorption into cells. [58].

2.9.2 Silicone-Based Dendrimers

Nanosilica has been considered due to the increased mechanical strength of composite resins. Particle size affects surface strength and smoothness, and most studies use a range of 5–200 nm. The amount of transparency of the surface is also directly related to the size of the particles. The size of particles more than 100 nm causes the scattering of visible light and the reduction of transparency. Also, the high specific surface area of nanoparticles creates high surface energy and effective suspension of silica nanoparticles in aqueous and non-aqueous solutions. Nanoparticles should be noted that in their chemical nature, silica particles cannot dissolve in aqueous solvents and many non-aqueous conventional solvents. This feature of stable suspension of nano-silica particles is of interest in many industrial applications [69].

Nano-silica particles have found their place in many fields, including drug delivery systems, catalysts, biological treatment, photography of living organisms, color imaging, sensors, liquid armor, and as a filler in composite materials [42, 63, 69, 103]. Producing porous and suspended silica particles by hydrolysis of TEOS was reported for the first time in 1968 [92]. Later, the concentration of TEOS, alkaline environment, and water was about the size of the obtained silica nanoparticles [18]. Also, the shape and size of silica particles can be controlled by reaction parameters such as time, temperature, and solution concentration [66].

Rice husk, bran, slag ash, and partially burned straw ash are the disposable materials used to make sodium silicate solution [48, 61, 89, 104, 106]. Acids destroy sodium silicate particles. Like hydrochloric acid as a precipitant, they are precipitated from sodium silicate solution [48]. Carbon dioxide can also be a precipitant [25].

Another application of nanoparticles is in thermal insulation and electrical insulation. The porous structure has exciting applications; among other things, they can be used as cleaners. They have produced mechanical and chemical polishers from silica nanopowder with low particle size distribution. This method solved the problems in polishing surfaces using acids and other polishers.

2.10 An Introduction to Porous Materials (Prussian) and Mesoporous Silica Materials and Their Synthesis Methods

Today, more than 60% of different products are synthesized by chemical methods, mainly using catalysts to increase efficiency and productivity [99]. William Oswald stated that there is no chemical reaction that is not affected by stimuli. Spurs are essential in the preparation and synthesis processes of raw chemicals, drugs, and energy carriers. The advantages of heterogeneous catalysts, such as easier separation and recovery function and more excellent compatibility with the environment have made these materials always attract attention. Due to their high surface area, Prussian materials have attracted much attention from researchers for catalysts. The type of active compounds used in the structure of stimuli is the most crucial part of the design of heterogeneous catalysts. Microprocessed motivations are of great interest in industries. However, the pore size of these catalysts has prompted researchers to synthesize catalysts with larger pore sizes. All these finally led to the design of silica mesoporous materials [59]. The International Union of Pure and Applied Chemistry divided nanoporous materials into three categories [84].

(1) Microporous materials have pore sizes smaller than 2 nm. (2) Mesoporous materials: Mesoporous materials have pore sizes between 2 and 50 nm [78]. (3) Microporous materials have a pore size of more than 50 nm, and due to their small surface area, these materials are used less in heterogeneous catalysts.

The history of mesoporous materials dates back to the last decades of the twentieth century when MCM-like materials were synthesized by researchers at Mobil Company [107].

2.11 Mesoporous Silica

These materials have holes with dimensions of 20–30 nm. The high surface area of 700–21,500 m/g and high thermal stability have made these materials very interesting in the field of catalysts. In addition to high thermal stability, these materials are easily functionalized and used as substrates, absorbents, and other catalytic functions in biological devices. They are known among mesoporous silica materials, and their difference is in their structure and applications. Silica mesoporous materials have a stable structure, high surface area, hexagonal structure, and uniform pores. These materials are mainly used as absorbents and catalysts.

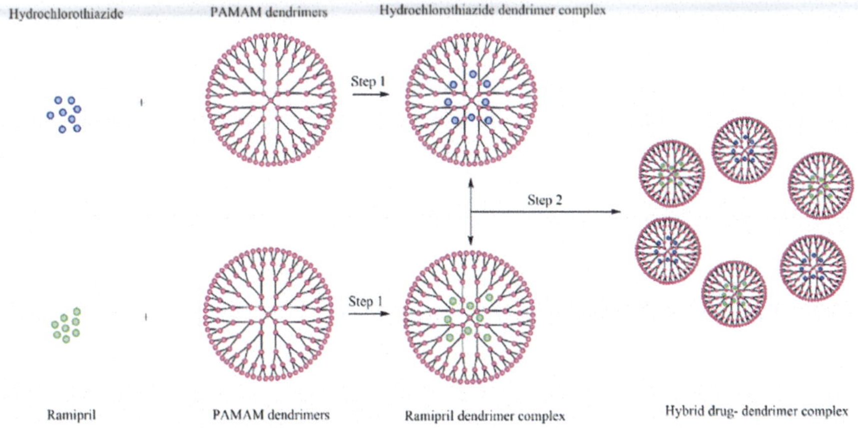

Fig. 4 Schematic of the hybrid drug-dendrimer complex [24]

From another point of view, mesoporous materials are divided into two categories: silica and non-silica. Unmodified silicas are known as silicates. Non-silica mesoporous materials include transition metal oxides such as phosphates, sulfates, and mesoporous carbons.

In general, mesoporous materials are synthesized by three available methods.

(1) Sol–gel method: This process is a wet chemical method widely used in materials science and ceramic engineering. In sol–gel, first, a colloidal solution known as sol is formed. Then, the conversion of sol to gel, an interconnected and three-dimensional network, is performed. The precursors of the sol–gel method are usually different. In the sol–gel process (Fig. 4), various forms, such as copolymers, organic molecules, and surfactants synthesize mesoporous materials.
(2) The method of using the mold: This method is exciting as a cheap method, divided into two methods [49].
(3) Soft mold method: It is also known as the end template method and does not use common complex molds such as silica. In this method, surfactants are usually used as templates. In the last step, with specific methods such as increasing acid or base or heating, the desired template, which is a surfactant, is pulled out.
(4) Hard molding method: They usually use a porous solid as a mold. The steps of this method are the same as the previous method. In the first stage, the precursors are placed in the mold's holes, then converted into the desired material with hydrothermal processes; in the last step, the mold is removed. It is tough to control the particle size with this method. This method is usually used to make mesoporous carbons.
(5) Microwave method: This method also has steps similar to the efforts of the previous two ways, with the difference that each step is much faster due to the use of microwave waves. Figure 5 shows an example of this method in which

organic matter is placed inside the cavities of mesoporous silica by microwave waves.

(6) Brushing method is based on the structural difference between the core and the shell (which is made of silica). Standard methods, such as soft and hard molding are not successful in controlling morphology like this. During this method, a substance such as acid or base is added, and the desired morphology can be obtained by retaining the conditions [49].

2.12 Effects of Pollutants on Dendrimer

The process of biological absorption is one of the methods of removing environmental pollution. Bioabsorption is an energy-independent process during which elements or materials are absorbed on surfaces and places. That can be linked to the surface of the cells of microorganisms or the absorbing agent. In bioaccumulation, the adsorbed material is absorbed into the cell's surface or transferred inside. Bioaccumulation is a process dependent on the cell's energy, vital mechanism, and biological metabolism.

Basically, in the process of biological absorption, microorganisms are used in the non-living form, and in the process of natural accumulation, they are used in the living form, each of which has advantages and disadvantages.

Bioabsorption is based on separating metal ions from aqueous solutions, so different microorganisms, such as bacteria, fungi, algae, and yeasts play a significant role in such processes. Recently, biological absorption has been developed as a solution for industrial wastewater treatment.

Some biosorptions can remove a wide range of heavy metals, while others only absorb certain types of metals. The complex structure of microorganisms enables them to absorb metals in different ways. These processes can be classified in two aspects, classification based on dependence on cellular metabolism (metabolism-dependent and metabolism-independent) and type based on the site of metal absorption from the solution (metal accumulation outside the cell, surface absorption, and collection of metal inside the cell).

The transfer of metal through the cell wall causes its accumulation inside the microorganism. This absorption type is related to the active defense system of the organism, which reacts in the presence of heavy metal. Heavy metals are transported through the microbial cell membranes by transporters of essential metabolic ions such as potassium, magnesium, and sodium. The balance of this system is disturbed in the presence of metals with the same charge and ionic radius as other ions. This absorption type is relatively fast and reversible [75, 108]. The proposed mechanism is shown in Fig. 5 [88].

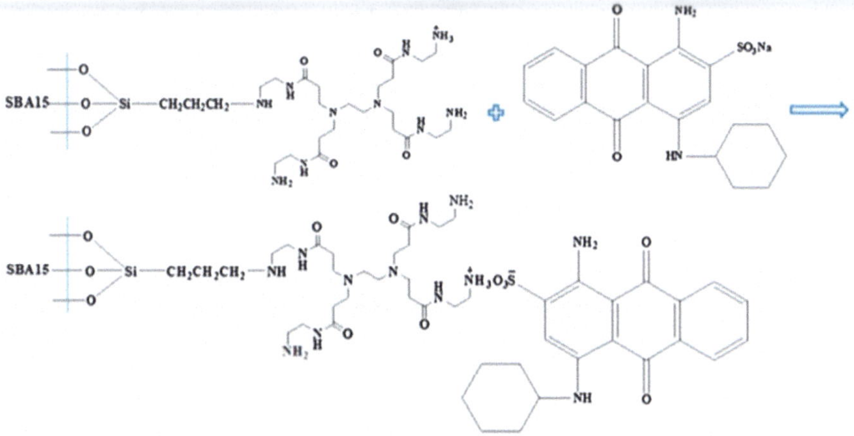

Fig. 5 Proposed mechanism of metal adsorption on silicate substrate with dendrimer [75]

2.13 *Dendrimers with Silica-Supported*

The term nano comes from the Greek word Nanos which means "dwarf." A nanometer is one billionth of a meter. This branch of science and technology related to materials has at least one spatial dimension in size range of 1 to 100 nm. The definitions of Richard Feynman (Nobel Prize Winner in Physics, 1965) are often used to introduce the concept of nanotechnology. Silica particles below 300 nm are called nanosilica. Silica size above this value has reached submicron.

Nanosilica has been considered due to the increased mechanical strength of composite resins. Particle size affects surface strength and smoothness, and most studies use a range of 5–200 nm. The amount of transparency of the surface is also directly related to the size of the particles. The size of particles more than 100 nm causes the scattering of visible light and the reduction of transparency. Also, the high specific surface area of nanoparticles creates high surface energy and effective suspension of silica nanoparticles in aqueous and non-aqueous solutions.

Nano-silica particles have found their place in many fields, including drug delivery systems, catalysts, biological treatment, photography of living organisms, color imaging, sensors, liquid armor, and as a filler in composite materials [42, 63, 69, 103]. Producing porous and suspended silica particles by hydrolysis of TEOS was reported for the first time in 1968 [92]. Later, the concentration of TEOS, alkaline environment, and water was about the size of the obtained silica nanoparticles. Also, the shape and size of silica particles can be controlled by reaction parameters such as time, temperature, and solution concentration. Rice husk, bran, slag ash, and partially burned straw ash are the disposable materials used to produce sodium silicate solution. Acids destroy sodium silicate particles. Like hydrochloric acid as a precipitant, they are precipitated from sodium silicate solution. Carbon dioxide can also be precipitant [25].

Nanosilica generally has an amorphous and hollow (mesoporous) structure. In some cases, the crystalline structure of nanosilica can be obtained, which is usually impossible to detect from x-ray tests. Due to Gibbs free energy, silica nanoparticles strongly tend to agglomerate (stick together). The best particle shape for silica nanoparticles for suspension is spherical. This spherical form is accessible by the synthesis in an ammonia medium. Most nano-silica particles are used for industrial purposes with one size (suspension) and two hydrophobic sizes (separation of oil and water) [42, 63, 69, 103].

Silicon dioxide nanoparticles, also known as silica nanoparticles or nanosilica, are the basis of biomedical research due to their molecular stability, low toxicity, and ability to bond with a wide range of molecules and polymers. In many experimental studies about colloidal suspensions of matter in the form of aqueous solution and aerosol solution, the suspended phase should contain homogeneous particles of the same shape and size. The rest of such suspended particles have many experimental and theoretical advantages. These particles not only facilitate the calibration process for analytical equipment but also facilitate data simplification, evaluation, and interpretation of experiments investigating the physiological properties or effects of colloids and aerosols. In this case, the particle size distribution and shape of particles will not affect the results [5]. Some suspended suspensions consisting of particles in the colloidal size range are spherical and natural in form with many polymers. In aerosol fields, they are primarily used as model materials for calibration purposes. Transducers that produce suspended particle molecules from dissolved or volatile materials are used in various aerosol studies [42, 63, 69, 103].

Silica is used in various forms in various products, including silica powder, silica gel, silica foam, silica aerogel, etc. Due to its high chemical resistance, good thermal stability, and compatibility with many materials, silica powder is used in many applications such as the bases of catalysts or absorbents. Also, it is a unique material in the industry. Silica aerogel with hydrophobic properties can be used as an adsorbent to remove organic compounds from aqueous solutions or catalysts with good dispersion in the fat (oil) phase. There are many techniques to produce these silica materials, including supercritical drying, sol–gel method, vapor phase reaction, and thermal decomposition technique. Among these methods, the sol–gel way is a suitable and economical process for synthesizing silica materials. Figure 6 shows poly amidoamine-grafted cellulose nanofibers at low pH were multifunctional. Pores have a positive effect on absorption. The absorber showed internal cavities and a high specific area [108].

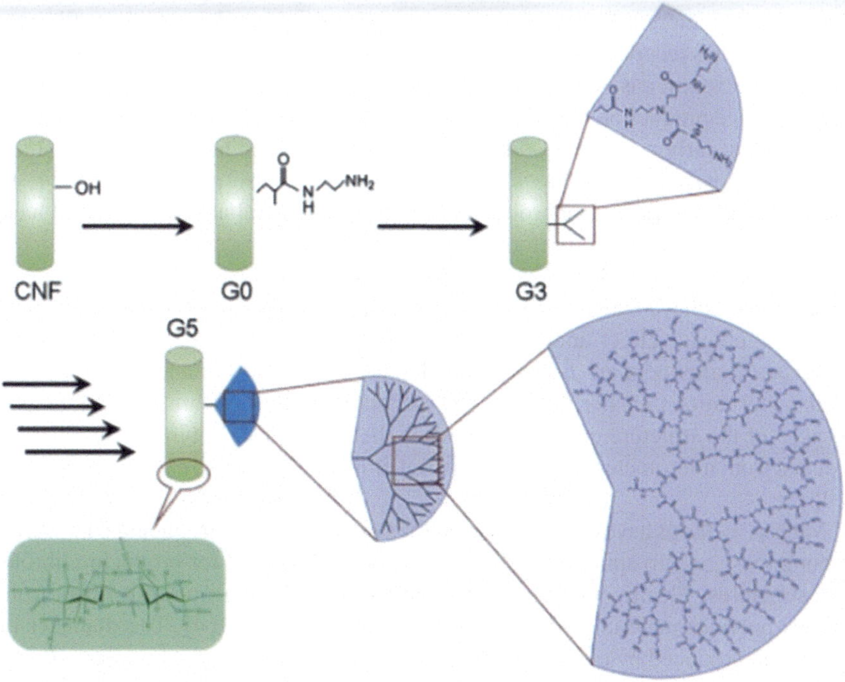

Fig. 6 Synthesis of cellulose nanofibers grafted with poly amidoamine [108]

2.14 Removal of Dyes from Textile Effluents with Natural Nano Biopolymers

Various chemical pollutants threaten the environment in the air, water, soil, and sediments. Today, industrial wastewater produces a large amount of organic and non-organic pollutants. Dyes are an essential range of contaminants used in many industries, such as textile, paper, printing, food, cosmetics, etc. Most industrial paints have undesirable properties such as toxicity and mutagenicity [80]. Unfortunately, most are stable and resistant to photon decomposition, biodegradation, and oxidizing [81]. Pollution caused by industrial wastewater is a common problem in many countries.

Removing color from water is very important because pigments strongly influence water quality. Deficient concentrations of dyes (less than 1 mg/l) can be seen in the water. In addition, many colors cause health problems such as allergic dermatitis, skin irritation, cancer, and human mutation.

Since dyes are aromatic organic compounds that absorb light in the visible region and are considered one of the most critical pollutants for natural ecosystems, it is necessary to treat colored wastewater before discharging them into the environment. Many methods, including absorption, chemical oxidation, electrochemical oxidation, and photocatalytic oxidation are used to remove the color from wastewater. Physical absorption is a more effective method to remove paint from the outgoing effluents. This method has received much attention due to its advantages such as high effectiveness and efficiency, easy operation and application, easy access to many adsorbents, and no sludge production [68].

Activated carbon is an adsorbent that is widely used in surface absorption. However, activated carbon has slow kinetics and low adsorption capacity due to large molecules in the mesoporous structure [50]. The ideal adsorbent should have a system with stable pores, uniform distribution, and a high contact surface.

In recent years, nanofibers have attracted the attention of many researchers because these materials have nanometer dimensions compared with the standard scale; they have a significant effect and improvement in physical, chemical, and biological properties. Nanofibers are fibers whose diameter is 1–100 nm and whose length-to-diameter ratio is greater than 50. The recent research of scientists in this field is because of the structure of nanofibers when the diameter of polymer nanofibers is reduced from micrometer scale to submicron or nanometer scale, exciting properties [20].

Various methods have been introduced for synthesizing and producing nanofibers. Among these, electrospinning is the most successful method for producing fibrous nanostructures. Electrospinning is a simple method of producing thin fibers from various polymer materials affected by an electrostatic field [64]. The diameter of nanofibers produced by electrospinning can be easily controlled at less than 10 nm, which is still a big challenge for other methods. Today, many fields of application are known for these nanofibers, among which hydrogen storage, fuel cell, environmental engineering, biotechnology, and filtration can be mentioned [70].

Direct yellow dye, an anionic dye belonging to the azo group, is widely used in various processes such as dyeing silk, wool, leather, cotton, textile printing, paper printing, and veterinary drugs due to its stability in acidic and alkaline solutions. Some harmful side effects of using direct yellow color12 were reported, such as eye damage in humans and animals [54]. Investigations showed that wastewater with an azo structure is dangerous for the environment. It is necessary to remove these types of colors in polluting compounds [36, 37].

2.15 Nanofiber Membranes

Carbon nanofiber is a type of carbon fiber on a nanoscale. Cylindrical nanostructures with graphene layers wrapped into perfect cylinders are called carbon tubes. This material can be classified into hollow and solid carbon nanofibers according to structural characteristics. Its diameter is generally in the range of 10 to 500 nm, and its length of 0.5–200 μm. Carbon nanofibers, which have a higher degree of crystal orientation, are preferred for electrical and thermal conductivity. Nanocarbon fiber has low density, high modulus, high strength, high conductivity, and thermal stability of carbon fiber made by the chemical vapor deposition growth method [71].

Carbon nanofibers (CNF) have been used in various applications, from aerospace materials as reinforcing materials or conductive fillers to improve the mechanical, electrical, and thermal properties of polymer matrix composites to the energy industry as anode materials for applications. Lithium-ion batteries improve energy storage and lifecycle characteristics.

The diameter of carbon nanofibers has a significant effect on their application. But controlling the nanofiber diameter is very time-consuming and expensive. Carbon nanofibers, like carbon nanotubes, are one of the promising options for strengthening different types of polymer fields. The diameter of these nanoparticles is about 10–500 nm.

The first report on carbon nanofibers, which dates back to 1889, is the synthesis of filamentous carbon by Hughes and his colleagues. They used a methane/hydrogen gas mixture and created carbon filaments through gas pyrolysis, carbon deposition, and filament growth. However, a proper understanding of these fibers came much later when their structure was analyzed by electron microscopy.

The dominant commercial methods for manufacturing: Catalytic chemical vapor deposition includes various types of thermal as well as with the help of plasma. Molecules are decomposed in the gas phase at high temperatures. Carbon is deposited on a substrate in the presence of an intermediate metal catalyst. The growth of fibers is done around the catalyst particles. The stimuli used in the synthesis are chromium, manganese, iron/nickel, molybdenum, copper, nickel, palladium, vanadium, and cobalt.

This work involved catalytic chemical vapor deposition synthesizing CNF at moderate temperatures of 700 and 800 °C. Non-ferromagnetic metal complexes of La, Nb, and Ti, dispersed on porous NaX-type zeolite support, were tested as new catalysts. CNF with a diameter of 30–200 nm was obtained. Transmission electron microscopy images show transition metal nanoparticles encapsulated by CNFs.

X-ray diffraction patterns showed the crystal structures of La (FCC), Nb (BCC), and Ti (HCP) formed on zeolite. Superconducting magnetic resonance rings of Nb encapsulated in CNF at 2 K. Raman spectra showed that all samples have graphitic and amorphous carbon structures. Based on SEM images and Raman spectra, these three metals all catalyze the synthesis of CNFs.

Recently, the preparation of metal oxide-supported carbon nanofiber composites through electrospinning has been widely studied. Carbon nanocomposites are used in various applications: 1—Energy conversion and storage, 2—Capacitive ionization, 3—Electrode materials, 4—Catalysis, 5—Adsorption/separation, 6—Oil spill repair, and 7—Biomedicine, such as gene transfer. The unique structure of these porous carbon nanofibers leads to good electrochemical performance, such as high reversible capacity and good cycle stability when used as anodes for rechargeable Li-ion batteries[7] [11–13, 46].

---

[7] https://bismoot.com/blog.

2.16 Porous Nanofibers

Polymer chains are connected through covalent bonds, and nanofibers' diameter will also differ according to the type of applied polymer and production method. Still, all nanofiber polymers have exceptional properties such as high surface-to-volume ratio and many holes. They have good mechanical strength and flexibility in performance, which are the characteristics that distinguish them, and this is when we compare them with their competitors, i.e., microfibers. Nanofibers are produced with the help of many different methods, including 1—electrospinning, 2—stretching, 3—self-organization, and 4—template synthesis. Among the features of the mentioned method, we can say the ability to produce fragile fibers with controllable diameters, compositions, and appropriate orientations. By using new melting processing methods, which are also suitable for industrial mass production, scientists and engineers have produced nanofibers with a diameter of 36 nm, which are fragile fibers. As mentioned earlier, nanofibers have very different commercial and technology-oriented uses for tissue engineering, drug delivery, cancer diagnosis, lithium-air batteries, optical sensors, and air filtration.

For the first time, nanofibers were created more than four centuries ago by electrospinning methods. After developing such a method, an English physicist named William Gilbert realized that there is a type of electrostatic attraction between liquids, and he did this through experiments. In 1887, the British physicist published a text on the development and production of nanofibers, and in 1901, the American inventor was able to obtain the first patent for the electrospinning method. Anton Form Hals was the first person who succeeded in producing nanofibers between 1934 and 1944, and he published his first invention and explained the experimental process of producing nanofibers. Harold Simmons also invented a device in 1966, with the help of which he had a variety of light and thin nanofibers with many different designs. At the end of the twentieth century, electrospinning and nanofibers gained more fame among scientists and researchers. The electrospinning method is still developing today[8] [4].

2.17 Application of Nanotechnology in the Textile Industry

Nanotechnology has attracted the attention of the world community because this technology can offer vast potential for a wide range of final needs. The unique and new properties of nanomaterials have attracted not only the attention of scientists and researchers but also the attention of business owners due to their high economic advantage. Nanotechnology has a high economic potential for the textile industry. Nanotechnology can provide high durability for materials because nanoparticles have a high surface area to volume ratio and high surface energy, offering better affinity

[8] https://asp-co.org/blog/314-nanofibre.

for fabrics and leading to increased durability performance. In addition, a coating of nanoparticles on materials will not affect breathability or body feel.

There are many examples of industries in that nanotechnology increases their textiles, which currently include some applications in the sports industry, skin care products, space technology, clothing, and material technologies for better prediction in vast environments. Textiles improved with nanotechnology material are considered a way to improve the properties of textiles, which make the fabric last longer and offer more diverse colors. Nanotechnology can also lead to new capabilities, such as energy storage and communication. Some attractive examples of nano-enhanced textiles on the market include stain-resistant fabrics and wrinkle-resistant fibers in body-warming textiles that use phase change materials (PCMs) in response to body temperature changes. Nano socks improved with silver nanoparticles. Silver works against infection and horrible smell [90].

2.18 The Application of Nanotechnology in the Properties of Textile Materials

The most widely used nanomaterials used in nanomedicine are polymer nanoparticles. These types of nanoparticles, having various properties, draw a step toward a bright future for improving the quality of treatment using modern nanomedicine methods. Polymer nanoparticles are used in multiple therapeutic techniques, such as targeted drug release, vaccines, tissue engineering, and various imaging methods. Scientific research at the frontier of knowledge regarding the use of polymer nanoparticles in cancer, neurogenic disorders, and cardiovascular diseases is being carried out at the international level.

The various properties of polymer nanoparticles have made it possible to provide understandable and reassuring solutions for therapeutic, diagnostic, preventive, and biological challenges in nanomedicine research. However, the challenges of the physiological system are very complex.

Cells exhibit diverse responses at the nanoscale level. Nanomaterials and biological compounds are essential in terms of toxicity.

The first category of polymer nanoparticles used in nanomedicine was non-biodegradable nanoparticles. Release systems based on non-biodegradable nanoparticles were designed to be quickly and effectively removed from the bloodstream through feces or urine. These polymer nanoparticles are not easily decomposed, and their accumulation in body tissues causes toxicity. Non-biodegradable polymer particles are used in various aspects of nanomedicine, including targeted drug release, wound healing dressings, and antimicrobial medical coatings. Chronic poisoning and inflammatory reactions are the side effects of using non-biodegradable materials. These side effects have led to research on alternative options. It is due to their lower toxicity and ability to create a specific drug release pattern and increase biocompatibility. Biodegradable polymers include; kinetic polymers and natural polymers.

We can mention poly(lactide) (PLA) copolymers (PLGA), PCL polycarbonate and poly(amino acids) and natural polymers such as chitosan, alginate, gelatin, and albumin.

In addition, nanotechnology provides the ability to combine multiple therapeutic agents and control and target drug release by summarizing the research results on the relationships between the physical and chemical properties of nanomaterials and the biological functions of the surrounding environment[9] [11–13, 27].

2.19 Self-Cleaning Cloths

The use of particular complements to gain more abilities in textiles has been noticed by humanity long ago.

In recent years, nanoparticles have been used in various industries, especially textile industries. New complementary materials are the main focus of researchers' research in the textile industry. Due to the increasing human need to produce textiles with greater efficiency, the need for more research in the field of textile product completion becomes more apparent. Among these, the removal of water and oil is considered one of the main goals of textile industries. The basis of the self-cleaning phenomenon is inspired by nature because their leaves can self-distinguish and repel water and dirt. Today, nano compounds are used to create self-cleaning clothes using natural laws. To have chemical particles in nanometer dimensions are one of the main things that have been tried to be used as much as possible in completing textile goods. Chemical particles in nanometer dimensions are considered in textile products. The use of nanoparticles in supplements reduces these problems to a great extent. At the same time, the uniquely high surface of these particles is considered another advantage of these particles. The basic strategy for creating the phenomenon of self-differentiation is shown in Fig. 7.

The production of fabrics with self-cleaning properties, especially in military clothing, is significant, so most American army uniforms have benefited from this addition. Using materials with self-cleaning properties will be beneficial to prevent the spread of pollution for car, train, and airplane seat covers or for patients' clothes and hospital bed sheets.

Clothing is considered one of the human needs. Therefore, it is used every day in all human societies. On the other hand, the quality and price of clothes used in each organization differ according to the community's welfare level and cultural status. Therefore, this product is mainly used in more prosperous societies. It seems that countries like America and some European countries are consumers of this clothing.

He talked about society's need for it. Because the way it is produced and demanded depends on certain conditions such as people's living conditions, cultural level, and other parameters, it seems that by reducing the production costs of this product, it

---

[9] https://paper.nano.ir/print/1/4687.

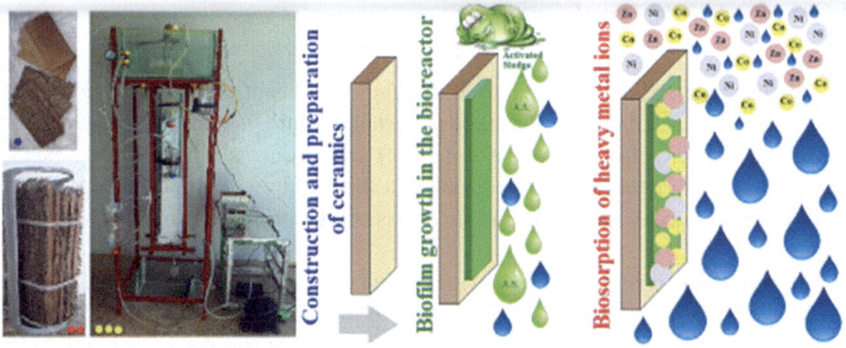

Fig. 7 Biological nanofiber textile application in the wastewater industry [11]

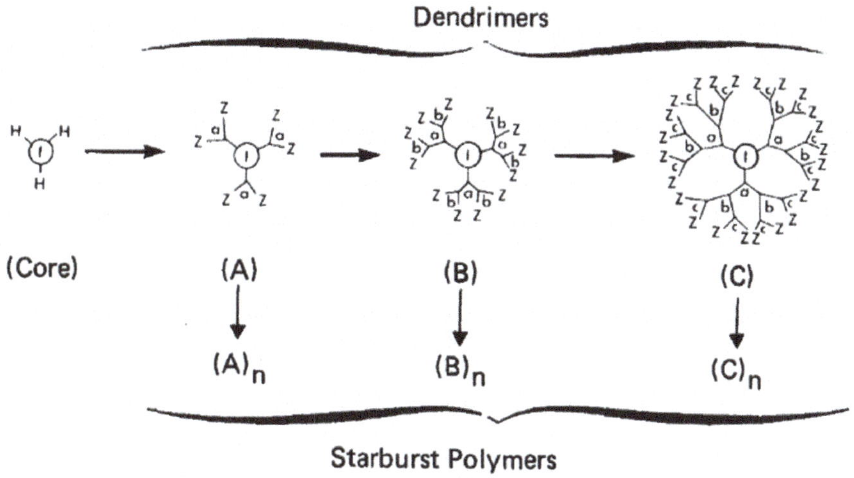

Fig. 8 Starburst branching for self-cleaning cloths [95]

can be introduced as one of the essential goods. But currently, it is impossible to have a proper estimate of the need for this product now and in the coming years.

Nanoparticles often have a positive charge. For example, Ag+ and Ti+ particles, as the most widely used nanoparticles, are positively charged cations. The main point regarding the additions made with these materials is how to place these particles on the textile fiber (Fig. 8). Since the compelling force in keeping these particles on the thread is through ionic bonds, it is necessary to create suitable and sufficient harmful sites on the fiber[10] [30].

[10] https://etojihi.com/self-cleaning-clothes-production-plan/.

2.20 Antistatic Final Coatings

An antistatic agent is a compound used to treat materials or their surfaces to reduce or eliminate the accumulation of static electricity. Static charge may be generated by the triboelectric effect or by a non-contact process using a high-voltage power source. Static charge may be introduced into the mold on a surface as part of the label printing process. One of the applications of this property is in epoxy flooring products.

The role of an antistatic agent is to make the surface of the material itself slightly conductive by its conductivity or by absorbing moisture from the air so that some humectants can be used. Molecules of an antistatic agent, similar to molecules of a surfactant, often have hydrophilic and hydrophobic regions. The hydrophobic side interacts with the material's surface, while the hydrophilic side interacts with the moisture in the air and binds water molecules together. Antistatic agents provide cost-effective protection for short-term applications, but other applications require long-term protection or lower necessary resistance to prevent sparking and protect electronics from electrostatic discharge.

An antistatic agent for treating coatings may also comprise an ionic liquid or a salt solution in an ionic liquid. Indium tin oxide can be used as an antistatic clear window coating. Conductive polymers such as conductive polymer nanofibers, especially polyaniline nanofibers, can also be used. In general, these systems are not very durable for layers; mainly, antimony tin oxide is used for durable systems, often in its nano form, then formulated into a final coating. Fabric antistatic is used as a moisture-absorbent or conductive material in textiles. This material creates a significant antistatic function in textiles by changing the conductivity of the surface and core of the fibers. Antistatic materials are chemical additives that prevent or reduce the accumulation of electric charges on the surface by creating some electrical conductivity to the surface. Antistatic agents are often added to the bulk or texture of the fibers.

Static electricity or electrostatic charge causes many problems processing textile materials, especially those made of hydrophobic synthetic fibers. In most dry textile processes, yarns and fabrics rush on different surfaces, creating an electrostatic charge due to frictional force. This electric charge can cause fibers and threads to fold.

During the spinning, weaving, and finishing yarn and fabric, friction is created, and hydrophobic fibers tend to generate a static charge. Still, since cellulose fibers are good conductors, they do not generate a static electric charge, while synthetic fibers are not good conductors, create a dormant electric account. Therefore, synthetic materials are coated with a chemical substance known as antistatic to prevent the production of static charge.

Antistatic is used to remove the unwanted effects of electrostatic charge produced during the production process and wear of woven and knitted fabrics in synthetic fibers. The electrostatic charge causes undesirable adhesion and, thus, textile damage. For this reason, to solve this problem, an antistatic or antistatic chemical substance is used.

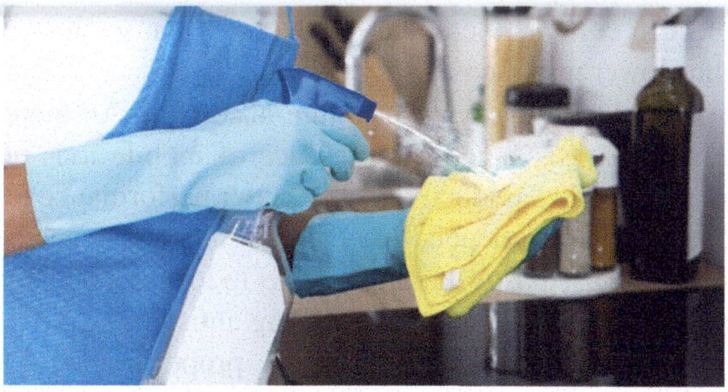

Fig. 9 The necessity of using antistatic in textiles

Fabrics made solely of water-repellent fibers, such as polyester, tend to create a static charge that results in clothes clinging to the body or annoying sparking noises when putting on or taking off clothes. Static electricity can cause problems preparing and producing textile materials, especially those made from hydrophobic synthetic fibers (Fig. 9).

Some polyamines may react with poly glycols to form a durable hydrophilic group in textiles. Carbon deposition or metal coatings (e.g., nanosilver) may also lead to increased fabric conductivity and reduced static charge accumulation[11] [33, 34, 100].

Static load occurs in many stages of textile production and during wear, that is, the use of textile products. This problem is more common in fibers that have very little moisture. Therefore, it often occurs in hydrophobic synthetic fibers, e.g., polyester, polyamide, and polyacrylic fibers.

Natural fibers also create static charge if they dry too much, for example, in an environment with low humidity or winter. During yarn production, static on the filament leads to yarn shedding and windage. Therefore, appropriate preparation steps are added during fiber production to reduce static or static electricity. Fixed charges on the fabric can make it difficult to transport and lead to clothes clinging, ultimately affecting the comfort and beauty of the clothes.

Even small electric shocks can occur as a result of a static charge. At the end of continuous fabric dryers, where the fabric is transported over a longer distance without direct contact with any metal parts, such as tent dryers, the dried material has a high static charge at the exit. Antistatic properties in clothing are often created using appropriate polishing techniques. A layer of electrically insulating material is deposited by adding chemicals to the fiber, which shows significant electrical conductivity for rapid neutralization of static charge or static electricity.

The critical part is the deposition of humectants that absorb sufficient water to form a conductive layer. As a result, this type of antistatic coating depends on water absorption and the weather conditions around the fiber sample. The antistatic material's primary mechanism of action is to increase the fiber surface's conductivity (equivalent to reducing surface resistance) and frictional forces through lubrication.

Surface resistance is a material characteristic, and its numerical value equals the voltage gradient ratio to current density. Resistance is the resistance of fibers to

[11] https://www.jahaneshimi.com.

electric current. Increased conductivity causes less charge, faster dissipation, and thus lower initial cost.

This layer is usually moisture-absorbent. An increase in moisture content leads to higher resistance. The presence of mobile ions on the surface is significant in increasing the conductivity. The better performance of moisture-absorbent antistatic coatings largely depends on the humidity of the surrounding environment when the fibers are used. Low humidity leads to lower conductivity (higher resistance) and, as a result, more problems when using antistatic.

Many fabrics acquire antistatic properties based on the first mechanism, i.e., increased load reduction and surface conductivity.

Silicone emulsions, polyethylene emulsions, polyethylene glycerol, poly quaternary salts, and acrylic polymers can be used for this purpose. Silicone emulsion becomes antistatic by reducing the friction between fibers, and it also benefits. Additionally, soil release property adds softness and softness[12] [39–41].

2.21 *Nanotechnology for Wrinkle-Free Processing*

Resin is commonly used to add wrinkle resistance to a piece of fabric. Using resin reduces fabric resistance, water permeability, color, and wear resistance. For this purpose, researchers have used nanosilica to restore the wrinkle resistance of silk and cotton. The result showed that when nanosilica is used, it can effectively improve the wrinkle resistance of silk. As a result, it makes clothes anti-wrinkle. Nanotechnology is improving with discovery. Nanotechnology used in consumer goods is increasing. At the same time, the industry of nanotechnology products is expected to grow at a high speed. The fabric sector has been affected by nanotechnology. Advances in nanoparticle performance have been rapid over the past few years, mainly in textiles. Nanofibers can improve properties such as water repellency, UV protection, antibacterial, antistatic, and anti-wrinkle. Research on nanotechnology to enhance the performance or create unique operations of textile fabrics is booming. There is a high probability that nanotechnology will penetrate all areas of material in the coming years, and we will see a bright future for nano fabric. If this content was helpful to you, share it with your friends and colleagues, and if you have any questions, share them with us in the comments section.

Gold and silver nanoparticles provide antibacterial, antifungal, and odor-resistant properties. When silver nanoparticles are placed on the fabric, they can kill the bacteria that cause resistance to the smell of the clothes. It also helps prevent the growth of bacteria and fungi. Nanomilizers are used for water-repellency properties, and titanium dioxide is used for UV protection. Nanosilica and nano-titanium dioxide improve silk's wrinkle resistance and make clothes anti-wrinkle. Carbon nanotubes are essential due to their high strength and excellent electrical conductivity. Carbon

[12] https://www.jahaneshimi.com.

nanotubes provide electrical conductivity, fire resistance, antistatic, waterproof, and high strength to fibers[13] [38].

2.22 *Antibacterial Final Coatings*

Antibacterial products produced with nanotechnology are one of the most widely used nanotechnology products known for having antimicrobial and antibacterial properties. As a catalyst, it can destroy more than 650 types of bacteria, viruses, and fungi, and at the same time, the presence of such features does not cause sensitivity in case of contact with human skin. By using this technology, we can give unique features to the surfaces, which in addition to improving efficiency, we educate the cost compared with the conventional antimicrobial methods. Compared with other antimicrobial methods (such as the use of chemicals in the completion of goods), nanotechnology has higher durability and efficiency, and its use in most of the standard processes in the industry without the need for special machinery and side processes is easy. It is possible.

Electroless nickel phosphoric (Ni–P) coatings have been widely used in chemical, mechanical, and electronic industries due to their high corrosion and wear resistance [1, 109]. In recent years, the incorporation of nanoparticles inside the Ni–P matrix coatings significantly improves their properties. It adds completely new features to the performance of the coating, which increases its application in various industries [109]. So far, Ni and Ni–P–TiO_2 nanocomposite coatings have been developed by adding PTFE nanoparticles or TiO_2 [1]. Liu and Zhao [62] found that Ni–P-PTFE nanocomposite coatings have better antibacterial properties than Ni–P or stainless-steel coatings. Al et al. [1] prepared Ni–P–TiO_2 nanocomposite coatings with different amounts of TiO_2 nanoparticles by the electroless method. Novakovic et al. [76] reported that the hardness and corrosion resistance improved by adding TiO_2 nanoparticles to Ni–P coatings.

Balaraju et al. [8] showed that incorporating TiO_2 nanoparticles in the Ni–P matrix did not affect electroless Ni–P coatings' structure and phase transformation behavior.

Chen et al. [23] showed that the microhardness of Ni–P–TiO_2 composite coatings increased significantly. In this article, Ni–P–TiO_2 nanocomposite coatings were coated on 316L stainless steel. The experimental results showed that Ni–P–TiO_2 coating reduces the adhesion of three bacterial species by 75% and 70%, respectively, compared with stainless and Ni–P coatings. The surface energy of Ni–P–TiO_2 electron cover increases significantly with increasing TiO_2 content after UV irradiation. The number of viable bacteria decreases with increasing energy levels caused by electron shells.

[13] https://kiantp.com.

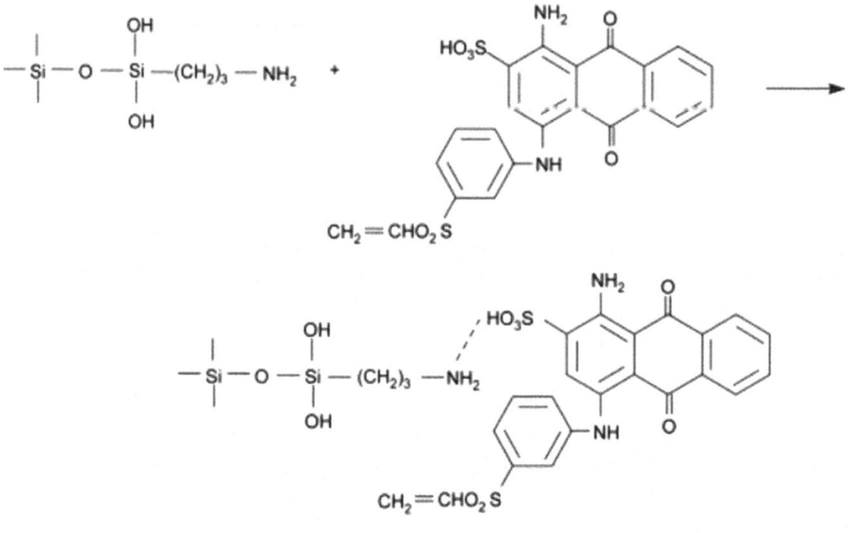

Fig. 10 Mechanism involves converting the β-sulfate-sulfonic group color to the vinyl sulfonic group. This figure shows the vinyl sulfonic group with a modified silica surface [5]

2.23 Modified Silica for Textile Dye Treatment

The increasing volume of industrial wastewater seriously threatens the natural environment. Industrial wastewater results are found in industrial and technological industries, i.e., in paint factories and textile, chemical, and petrochemical industries. Various techniques have been found to remove dyes from textile waste, including extraction [60], membrane processes, adsorption processes of great importance, and industrial adsorbents, including activated carbon, activated Al_2O_3, and zeolites [98].

Due to the presence of silanol groups in silica particles, they have a hydrophilic surface. Acidic groups are reactive.

Silanol groups in silica particles give a hydrophilic surface. Acidic groups allow chemical modification of the silica surface. The deformation of silica induces surface hydrophobicity and gives organic properties. The vinyl sulfonic group forms a covalent bond with the modified silica surface. [5]. The mechanism involves the conversion of the β-sulfate-sulfonic group of the dye to the vinyl sulfonic group. Under alkaline conditions, it cleaves the residual sulfate group as follows (Fig. 10).

3 Conclusion and Future Perspective

The use of conventional dyeing methods has partially met the needs of consumers. The need for high color constancy, increase in color strength with the help of less material, non-pollution, reduction of energy consumption, and dyes has become popular. Fibers like polypropylene use nanotechnology. Dendrimers, cyclodextrins, nanoclay, chitosan, and metal nanoparticles are nanomaterials introduced. They improved the dyeing process in the textile industry.

The results show that silica hybridized with dendrimers can be used as a selective adsorbent to purify waste dye solutions in the textile industry. Its application guarantees highly efficient color removal. The product containing silica can be used as a pigment in paint. Dendrimers are polymers with unique properties that are useful in biological systems. Solvent conditions can change predictably and are easily modified. Dendrimers are biocompatible compounds with low cytotoxicity and high permeability. Various types of dendrimers are commercially available and have already been used as candidates for receptor-ligand interaction, nanocarriers for survival and targeting, and in textile industries. Colors are also widely used. It is expected that new dendrimeric structures will continue to be developed. New methods were created for dendrimers' synthesis, modification, and derivatization that allows for binding, charge density, etc., and entirely new classes of textile industries and bioactive materials based on creating a macromolecular structure. The pigment obtained has a very low polydispersity with a medium diameter. The tiny particles do not tend to form secondary agglomerates.

References

1. Aal AA, Hassan HB, Rahim MA (2008) Nanostructured Ni–P–TiO2 composite coatings for electrocatalytic oxidation of small organic molecules. J Electroanal Chem 619:17–25
2. Aksu Z, Ertuğrul S, Dönmez G (2010) Methylene Blue biosorption by Rhizopus arrhizus: effect of SDS (sodium dodecylsulfate) surfactant on biosorption properties. Chem Eng J 158(3):474–481
3. Alam MA, Jahan A, Khan MW (2017) A review on dendrimers and metallodendrimers, the important compounds as catalyst in material chemistry. Adv Mater 6:52
4. Andrady AL (2008) Science and technology of polymer nanofibers. John Wiley & Sons
5. Andrzejewska A, Krysztafkiewicz A, Jesionowski T (2007) Treatment of textile dye wastewater using modified silica. Dyes Pigm 75(1):116–124
6. Arami M, Limaee NY, Mahmoodi NM (2008) Evaluation of the adsorption kinetics and equilibrium for the potential removal of acid dyes using a biosorbent. Chem Eng J 139(1):2–10
7. Balapure KH, Jain K, Chattaraj S, Bhatt NS, Madamwar D (2014) Co-metabolic degradation of diazo dye—Reactive blue 160 by enriched mixed cultures BDN. J Hazard Mater 279:85–95
8. Balaraju J, Narayanan TS, Seshadri S (2006) Structure and phase transformation behaviour of electroless Ni–P composite coatings. Mater Res Bull 41(4):847–860
9. Balla W, Essadki A, Gourich B, Dassaa A, Chenik H, Azzi M (2010) Electrocoagulation/electroflotation of reactive, disperse and mixture dyes in an external-loop airlift reactor. J Hazard Mater 184(1–3):710–716
10. Baruah JB (2019) Principles and advances in supramolecular catalysis. CRC Press
11. Beni AA, Esmaeili A (2019) Design and optimization of a new reactor based on biofilm-ceramic for industrial wastewater treatment. Environ Pollut 255:113298
12. Beni AA, Esmaeili A (2020) Biosorption, an efficient method for removing heavy metals from industrial effluents: a review. Environ Technol Innov 17:100503
13. Beni AA, Esmaeili A (2020) Fabrication of 3D hydrogel to the treatment of moist air by solar/wind energy in a simulated battery recycle plant salon. Chemosphere 246:125725

14. Benkhaya S, M'rabet S, El Harfi A (2020) A review on classifications, recent synthesis and applications of textile dyes. Inorg Chem Commun 115:107891
15. Bhattacharyya R, Ray SK (2015) Removal of congo red and methyl violet from water using nano clay filled composite hydrogels of poly acrylic acid and polyethylene glycol. Chem Eng J 260:269–283
16. Bo Z, Schäfer A, Franke P, Schlüter AD (2000) A facile synthetic route to a third-generation dendrimer with generation-specific functional aryl bromides. Org Lett 2(11):1645–1648
17. Boas U, Heegaard PM (2004) Dendrimers in drug research. Chem Soc Rev 33(1):43–63
18. Bogush G, Tracy M, Zukoski Iv C (1988) Preparation of monodisperse silica particles: control of size and mass fraction. J Non-Cryst Solids 104(1):95–106
19. Buhleier E (1978) "Cascade"-and "nonskid-chain-like" syntheses of molecular cavity topologies
20. Campos AFC, Michels-Brito PH, da Silva FG, Gomes RC, Gomide G, Depeyrot J (2019) Removal of direct yellow 12 from water using CTAB-coated core-shell bimagnetic nanoadsorbents. J Environ Chem Eng 7(2):103031
21. Carneiro PA, Umbuzeiro GA, Oliveira DP, Zanoni MVB (2010) Assessment of water contamination caused by a mutagenic textile effluent/dyehouse effluent bearing disperse dyes. J Hazard Mater 174(1–3):694–699
22. Chen K-C, Wu J-Y, Liou D-J, Hwang S-CJ (2003) Decolorization of the textile dyes by newly isolated bacterial strains. J Biotechnol 101(1):57–68
23. Chen W, Gao W, He Y (2010) A novel electroless plating of Ni–P–TiO_2 nano-composite coatings. Surf Coat Technol 204(15):2493–2498
24. Chis AA, Dobrea C, Morgovan C, Arseniu AM, Rus LL, Butuca A, Juncan AM, Totan M, Vonica-Tincu AL, Cormos G (2020) Applications and limitations of dendrimers in biomedicine. Molecules 25(17):3982
25. Choucair M, Thordarson P, Stride JA (2009) Gram-scale production of graphene based on solvothermal synthesis and sonication. Nat Nanotechnol 4(1):30–33
26. Crini G (2006) Non-conventional low-cost adsorbents for dye removal: a review. Biores Technol 97(9):1061–1085
27. Dadashi F, Esmaeili A (2021) Optimization, in-vitro release and in-vivo evaluation of bismuth-hyaluronic acid-melittin-chitosan modified with oleic acid nanoparticles computed imaging-guided radiotherapy of cancer tumor in eye cells. Mater Sci Eng, B 270:115197
28. Denkewalter RG, Kolc J, Lukasavage WJ (1981) Macromolecular highly branched homogeneous compound based on lysine units. Google Patents
29. Elmoubarki R, Mahjoubi F, Tounsadi H, Moustadraf J, Abdennouri M, Zouhri A, El Albani A, Barka N (2015) Adsorption of textile dyes on raw and decanted Moroccan clays: kinetics, equilibrium and thermodynamics. Water Resour Indus 9:16–29
30. Esmaeili A, Asgari A (2015) In vitro release and biological activities of Carum copticum essential oil (CEO) loaded chitosan nanoparticles. Int J Biol Macromol 81:283–290
31. Esmaeili A, Beni AA (2014) A novel fixed-bed reactor design incorporating an electrospun PVA/chitosan nanofiber membrane. J Hazard Mater 280:788–796
32. Esmaeili A, Beni AA (2015) Novel membrane reactor design for heavy-metal removal by alginate nanoparticles. J Ind Eng Chem 26:122–128
33. Esmaeili A, Haseli M (2017) Electrospinning of thermoplastic carboxymethyl cellulose/poly (ethylene oxide) nanofibers for use in drug-release systems. Mater Sci Eng, C 77:1117–1127
34. Esmaeili A, Haseli M (2017) Optimization, synthesis, and characterization of coaxial electrospun sodium carboxymethyl cellulose-graft-methyl acrylate/poly (ethylene oxide) nanofibers for potential drug-delivery applications. Carbohydrate Polymers
35. Esmaeili A, Kalantari M (2012) Bioremoval of an azo textile dye, Reactive Red 198, by Aspergillus flavus. World J Microbiol Biotechnol 28(3):1125–1131
36. Esmaeili A, Khodaei M (2018) Encapsulation of rifampin in a polymeric layer-by-layer structure for drug delivery. J Biomed Mater Res, Part A 106(4):905–913
37. Esmaeili A, Khoshnevisan N (2016) Optimization of process parameters for removal of heavy metals by biomass of Cu and Co-doped alginate-coated chitosan nanoparticles. Biores Technol 218:650–658

38. Esmaeili A, Loghmani K (2016) Removal of monoethylene glycol from gas field wastewater using Aspergillus tubingensis and a new bioreactor. Waste Biomass Valor 7(1):151–156
39. Esmaeili A, Rustaiyan A, Nadimi M, Masoudi S, Tadayon F, Sedaghat S, Ebrahimpur N, Hajyzadeh E (2005) Volatile constituents of Centaurea depressa MB and Carduus pycnocephalus L. two compositae herbs growing wild in Iran. J Essential Oil Res 17(5):539–541
40. Esmaeili A, Sadeghi E (2014) The efficiency of Penicillium commune for bioremoval of industrial oil. Int J Environ Sci Technol 11(5):1271–1276
41. Esmaeili A, Saremnia B (2012) Preparation of extract-loaded nanocapsules from Onopordon leptolepis DC. Ind Crops Prod 37(1):259–263
42. Gallis KW, Araujo JT, Duff KJ, Moore JG, Landry CC (1999) The use of mesoporous silica in liquid chromatography. Adv Mater 11(17):1452–1455
43. Ghaneian MT, Ghanizadeh G, Gholami M, Ghaderinasab F (2010) Application of eggshell as a natural sorbent for the removal of reactive red 123 dye from synthetic textile wastewater. Zahedan J Res Med Sci 11(4)
44. Gosavi VD, Sharma S (2014) A general review on various treatment methods for textile wastewater. J Environ Sci Comp Sci Eng Technol 3(1):29–39
45. Gupta V (2009) Application of low-cost adsorbents for dye removal–a review. J Environ Manage 90(8):2313–2342
46. Haghi A, Akbari M (2007) Trends in electrospinning of natural nanofibers. physica status solidi (a) 204(6):1830–1834
47. Hawker CJ, Wooley KL, Fréchet JM (1993) Unimolecular micelles and globular amphiphiles: dendritic macromolecules as novel recyclable solubilization agents. J Chem Soc, Perkin Trans 1(12):1287–1297
48. Hessien M, Rashad M, Zaky R, Abdel-Aal E, El-Barawy K (2009) Controlling the synthesis conditions for silica nanosphere from semi-burned rice straw. Mater Sci Eng, B 162(1):14–21
49. Hoffmann F, Cornelius M, Morell J, Fröba M (2006) Silica-based mesoporous organic–inorganic hybrid materials. Angew Chem Int Ed 45(20):3216–3251
50. Huang C-H, Chang K-P, Ou H-D, Chiang Y-C, Chang E-E, Wang C-F (2011) Characterization and application of Ti-containing mesoporous silica for dye removal with synergistic effect of coupled adsorption and photocatalytic oxidation. J Hazard Mater 186(2–3):1174–1182
51. Hunger K (2007) Industrial dyes: chemistry, properties, applications. John Wiley & Sons
52. Inoue K (2000) Functional dendrimers, hyperbranched and star polymers. Prog Polym Sci 25(4):453–571
53. Jadidi K, Esmaeili M, Kalantari M, Khalili M, Karakouzian M (2020) A review of different aspects of applying asphalt and bituminous mixes under a railway track. Materials 14(1):169
54. Janssen PA, Selwood BL, Dobson SR, Peacock D, Thiessen PN (2003) To dye or not to dye: a randomized, clinical trial of a triple dye/alcohol regime versus dry cord care. Pediatrics 111(1):15–20
55. Jauris IM, Fagan SB, Adebayo MA, Machado FM (2016) Adsorption of acridine orange and methylene blue synthetic dyes and anthracene on single wall carbon nanotubes: a first principle approach. Comput Theor Chem 1076:42–50
56. Khajeh M, Amin MM, Taheri E, Fatehizadeh A, McKay G (2020) Influence of co-existing cations and anions on removal of direct red 89 dye from synthetic wastewater by hydrodynamic cavitation process: an empirical modeling. Ultrason Sonochem 67:105133
57. Khan AJ, Song J, Ahmed K, Rahim A, Volpe PLO, Rehman F (2020) Mesoporous silica MCM-41, SBA-15 and derived bridged polysilsesquioxane SBA-PMDA for the selective removal of textile reactive dyes from wastewater. J Mol Liq 298:111957
58. Kheraldine H, Rachid O, Habib AM, Al Moustafa A-E, Benter IF, Akhtar S (2021) Emerging innate biological properties of nano-drug delivery systems: a focus on PAMAM dendrimers and their clinical potential. Adv Drug Deliv Rev 178:113908
59. Kresge CT, Leonowicz M, Roth WJ, Vartuli J, Beck J (1992) Ordered mesoporous molecular sieves synthesized by a liquid-crystal template mechanism. Nature 359(6397):710–712
60. Kunreuther H, Gowda MR (2012) Integrating insurance and risk management for hazardous wastes. Springer Science & Business Media

61. Li T, Wang T (2008) Preparation of silica aerogel from rice hull ash by drying at atmospheric pressure. Mater Chem Phys 112(2):398–401
62. Liu C, Zhao Q (2011) The CQ ratio of surface energy components influences adhesion and removal of fouling bacteria. Biofouling 27(3):275–285
63. Liu L, Wang Y, Feng K, Li Y, Li W, Zhao C, Zhao Y (2006) Preparation of boron carbon nitride thin films by radio frequency magnetron sputtering. Appl Surf Sci 252(12):4185–4189
64. Liu Z, Sun DD, Guo P, Leckie JO (2007) An efficient bicomponent TiO_2/SnO_2 nanofiber photocatalyst fabricated by electrospinning with a side-by-side dual spinneret method. Nano Lett 7(4):1081–1085
65. Mahmoodi NM, Hayati B, Arami M, Lan C (2011) Adsorption of textile dyes on pine cone from colored wastewater: kinetic, equilibrium and thermodynamic studies. Desalination 268(1–3):117–125
66. Mansur HS, Vasconcelos WL, Lenza RS, Oréfice RL, Reis EF, Lobato ZP (2000) Sol–gel silica based networks with controlled chemical properties. J Non-Cryst Solids 273(1–3):109–115
67. McNeil SE (2005) Nanotechnology for the biologist. J Leukoc Biol 78(3):585–594
68. Mousa KM, Taha AH (2015) Adsorption of Reactive Blue dye onto natural and modified wheat straw. J Chem Eng Process Technol 6(6):1
69. Nadrah P, Planinšek O, Gaberšček M (2014) Stimulus-responsive mesoporous silica particles. J Mater Sci 49(2):481–495
70. Nasouri K, Bahrambeygi H, Rabbi A, Shoushtari AM, Kaflou A (2012) Modeling and optimization of electrospun PAN nanofiber diameter using response surface methodology and artificial neural networks. J Appl Polym Sci 126(1):127–135
71. Nateri AS, Hasanzadeh M (2009) Using fuzzy-logic and neural network techniques to evaluating polyacrylonitrile nanofiber diameter. J Comput Theor Nanosci 6(7):1542–1545
72. Newkome GR, Shreiner C (2010) Dendrimers derived from 1→3 branching motifs. Chem Rev 110(10):6338–6442
73. Newkome GR, Yao Z, Baker GR, Gupta VK (1985) Micelles. Part 1. Cascade molecules: a new approach to micelles. A [27]-arborol. J Org Chem 50(11):2003–2004
74. Nigam P, Banat IM, Singh D, Marchant R (1996) Microbial process for the decolorization of textile effluent containing azo, diazo and reactive dyes. Process Biochem 31(5):435–442
75. Niu Y, Qu R, Chen H, Mu L, Liu X, Wang T, Zhang Y, Sun C (2014) Synthesis of silica gel supported salicylaldehyde modified PAMAM dendrimers for the effective removal of Hg (II) from aqueous solution. J Hazard Mater 278:267–278
76. Novakovic J, Vassiliou P, Samara K, Argyropoulos T (2006) Electroless NiP–TiO_2 composite coatings: their production and properties. Surf Coat Technol 201(3–4):895–901
77. Ostovar F, Hasanzadeh M (2020) A review of widely used industrial dyes and their removal methods from water and wastewater. J Environ Res Technol 7(7):29
78. Pal N, Bhaumik A (2013) Soft templating strategies for the synthesis of mesoporous materials: inorganic, organic–inorganic hybrid and purely organic solids. Adv Coll Interface Sci 189:21–41
79. Patil P, Shedbalkar U, Kalyani D, Jadhav J (2008) Biodegradation of Reactive Blue 59 by isolated bacterial consortium PMB11. J Ind Microbiol Biotechnol 35(10):1181–1190
80. Qadri S, Ganoe A, Haik Y (2009) Removal and recovery of acridine orange from solutions by use of magnetic nanoparticles. J Hazard Mater 169(1–3):318–323
81. Qu S, Huang F, Yu S, Chen G, Kong J (2008) Magnetic removal of dyes from aqueous solution using multi-walled carbon nanotubes filled with Fe_2O_3 particles. J Hazard Mater 160(2–3):643–647
82. Rani C, Jana A, Bansal A (2012) Potential of different white rot fungi to decolourize textile azo dyes in the absence of external carbon source. Environ Technol 33(8):887–896
83. Robinson T, McMullan G, Marchant R, Nigam P (2001) Remediation of dyes in textile effluent: a critical review on current treatment technologies with a proposed alternative. Biores Technol 77(3):247–255
84. Rouquerol J, Avnir D, Fairbridge C, Everett D, Haynes J, Pernicone N, Ramsay J, Sing K, Unger K (1994) Recommendations for the characterization of porous solids (technical report). Pure Appl Chem 66(8):1739–1758

85. Roy R, Kim JM (1999) Amphiphilic p-tert-butylcalix [4] arene scaffolds containing exposed carbohydrate dendrons. Angew Chem Int Ed 38(3):369–372
86. Royer B, Cardoso NF, Lima EC, Vaghetti JC, Simon NM, Calvete T, Veses RC (2009) Applications of Brazilian pine-fruit shell in natural and carbonized forms as adsorbents to removal of methylene blue from aqueous solutions—kinetic and equilibrium study. J Hazard Mater 164(2–3):1213–1222
87. Sahasrabudhe MM, Saratale RG, Saratale GD, Pathade GR (2014) Decolorization and detoxification of sulfonated toxic diazo dye CI Direct Red 81 by Enterococcus faecalis YZ 66. J Environ Health Sci Eng 12(1):1–13
88. Sajid M, Nazal MK, Baig N, Osman AM (2018) Removal of heavy metals and organic pollutants from water using dendritic polymers based adsorbents: a critical review. Sep Purif Technol 191:400–423
89. Schlomach J, Kind M (2004) Investigations on the semi-batch precipitation of silica. J Colloid Interface Sci 277(2):316–326
90. Selseleh Hassan-Kiadehi M, Amoozegar MA, Asad S (2017) Evaluation of biodecolorization of the textile azo dye by halophilic archaea. Biol J Microorganism 6(23):1–17
91. Sowinska M, Urbanczyk-Lipkowska Z (2014) Advances in the chemistry of dendrimers. New J Chem 38(6):2168–2203
92. Stöber W, Fink A, Bohn E (1968) Controlled growth of monodisperse silica spheres in the micron size range. J Colloid Interface Sci 26(1):62–69
93. Tahmasebi M, Esmaeili A, Bambai B (2022) New method of identifying morphine in urine samples using nanoparticle-dendrimer-enzyme hybrid system. Arab J Chem 15(2):103630
94. Tan KB, Vakili M, Horri BA, Poh PE, Abdullah AZ, Salamatinia B (2015) Adsorption of dyes by nanomaterials: recent developments and adsorption mechanisms. Sep Purif Technol 150:229–242
95. Tomalia DA, Baker H, Dewald J, Hall M, Kallos G, Martin S, Roeck J, Ryder J, Smith P (1985) A new class of polymers: starburst-dendritic macromolecules. Polym J 17(1):117–132
96. Tunç Ö, Tanacı H, Aksu Z (2009) Potential use of cotton plant wastes for the removal of Remazol Black B reactive dye. J Hazard Mater 163(1):187–198
97. Turnbull WB, Stoddart JF (2002) Design and synthesis of glycodendrimers. Rev Mol Biotechnol 90(3–4):231–255
98. Turner R (1991) Municipal solid waste management: an economic perspective
99. Valkenberg MH, Hölderich WF (2002) Preparation and use of hybrid organic–inorganic catalysts. Catalysis Rev 44(2):321–374
100. Van Langenhove L (2013) Smart textiles for protection: an overview. Smart Textiles for Protect:3–33
101. Virendrabhai RJ. Cloning and heterologous expression of genes involved in AZO dye decolorization and metabolic engineering for efficient bioremediation
102. Vreeland RH, Litchfield C, Martin E, Elliot E (1980) Halomonas elongata, a new genus and species of extremely salt-tolerant bacteria. Int J Syst Evol Microbiol 30(2):485–495
103. Wetzel ED, Lee Y, Egres R, Kirkwood K, Kirkwood J, Wagner N (2004) The effect of rheological parameters on the ballistic properties of shear thickening fluid (STF)-kevlar composites. In: AIP conference proceedings, 2004, vol 1. American Institute of Physics, pp 288–293
104. Witoon T, Chareonpanich M, Limtrakul J (2008) Synthesis of bimodal porous silica from rice husk ash via sol–gel process using chitosan as template. Mater Lett 62(10–11):1476–1479
105. Yilmaz AE, Boncukcuoğlu R, Kocakerim M, Karakaş İH (2011) Waste utilization: The removal of textile dye (Bomaplex Red CR-L) from aqueous solution on sludge waste from electrocoagulation as adsorbent. Desalination 277(1–3):156–163
106. Zaky R, Hessien M, El-Midany A, Khedr M, Abdel-Aal E, El-Barawy K (2008) Preparation of silica nanoparticles from semi-burned rice straw ash. Powder Technol 185(1):31–35
107. Zdravkov B, Čermák J, Šefara M, Janků J (2007) Pore classification in the characterization of porous materials: a perspective. Open Chem 5(2):385–395
108. Zhao J, Zhang X, He X, Xiao M, Zhang W, Lu C (2015) A super biosorbent from dendrimer poly (amidoamine)-grafted cellulose nanofibril aerogels for effective removal of Cr (VI). J Mater Chem A 3(28):14703–14711

109. Zhao Q, Liu Y (2005) Electroless Ni–Cu–P–PTFE composite coatings and their anticorrosion properties. Surf Coat Technol 200(7):2510–2514
110. Zohra B, Aicha K, Fatima S, Nourredine B, Zoubir D (2008) Adsorption of Direct Red 2 on bentonite modified by cetyltrimethylammonium bromide. Chem Eng J 136(2–3):295–305

Nanohybrid-Based Catalysts for Degradation of Dyes from Aqueous Solution

Burcu Palas

Abstract Water pollution has become a major issue for the environmental and human health due to the discharge of improperly treated recalcitrant pollutants into the receiving water bodies. Dye-containing industrial effluents cause attention due to their large discharge scales. Efficient removal of dyes from industrial effluents is a significant research topic since the presence of dyes reduces the penetration of light into water and the aquatic life is affected negatively even at low concentrations of dyes. Use of efficient, stable and environmentally friendly catalysts is substantial to operate under mild reaction conditions and decrease the cost of the treatment. With the developments in nanotechnology, the use of nanohybrid catalysts in wastewater treatment processes has provided many advantages for the degradation of persistent organic pollutants including dyes. Recently, various types of catalysts such as graphene, carbon nanotube, semiconductor, ferrite, and polymer based nanohybrids have been effectively used for the degradation of dyes in different wastewater treatment processes including photocatalytic oxidation, sonocatalytic degradation, catalytic wet air oxidation, and ozonation. Nanohybrid catalysts have benefits over conventional catalysts due to the synergistic effect of their components. The large surface areas, adjustable sizes, stability, recoverability and reusability of the nanohybrid catalysts make them suitable candidates for wastewater treatment applications.

Keywords Nanohybrid catalyst · Organic pollutant · Wastewater treatment · Dye degradation

---

B. Palas (✉)
Faculty of Engineering, Chemical Engineering Department, Ege University, 35100, Bornova İzmir, Turkey
e-mail: burcu.palas@ege.edu.tr

© The Author(s), under exclusive license to Springer Nature Singapore Pte Ltd. 2023
A. Ahmad et al. (eds.), *Nanohybrid Materials for Treatment of Textiles Dyes*, Smart Nanomaterials Technology, https://doi.org/10.1007/978-981-99-3901-5_15

1 Introduction

Extensive developments in nanoscience resulted in the development of numerous nanomaterials that have improved features when compared to their bulk materials. Nanomaterials have enhanced optical, mechanical, magnetic, electrical, and thermal characteristics due to their size depending physicochemical properties including high aspect ratio and surface energy, surface Plasmon resonance, and quantum confinement effect [57].

Nanocomposites are defined as the materials in which one of the components is within the size range of 1–100 nm. One of the components in the nanocomposite structure is generally a nanocluster or a nanoblock. Nanohybrid term is used when a covalent bond occurs between two different components or when the inorganic component of the hybrid material is formed in situ from individual building blocks [26]. Nanohybrids comprise of two dissimilar or similar components constituting a single entity with either improved or new properties. The advanced properties of the nanohybrid materials provide many benefits in various fields including biomaterials, optics, electronics, sensing, coating, energy storage, and catalysis [7].

Hybrid materials at nanometer scale are developed by the combination of multidisciplinary technologies including nanotechnology, biotechnology, information and cognitive technologies leading to creation of innovative multifunctional systems. Hybrid materials have been employed in diverse fields such as eco-environment, power generation, electronics, and production of fine chemicals and polymers. Recently, these materials are increasingly integrated into different areas such as biomimicry, medicine, healthcare, renewable energy generation, agriculture, food production, and wastewater treatment [79].

Among various types of wastewaters, textile industry effluents create an important environmental pollution problem due to the dyes they contained. Dyes are persistent organic compounds, which cause intense color in the textile wastewaters. Textile mills discharge highly colored effluents with a dye content varying between 10 and 200 mg/L leading to esthetic and environmental problem [93].

Development of effective treatment methods and catalysts for decolorization of textile industry effluents is significant for several reasons. Dyes are highly visible even at low concentrations. The presence of dyes in water bodies affects the penetration of light into water and reduces the photosynthetic activity of the aquatic flora. The environmental effect of heavy metals in wastewater is also a significant concern since some of the dyes are metalized [107]. The discharge of dyestuffs into the aquatic environment has attracted concern due to genotoxic, mutagenic, and carcinogenic impacts of some of the dyes. Even though the dyes have low toxicity toward mammals generally, the transformation products such as aromatic amines generated during biodegradation can be harmful [35].

Since dyes are designed to resist decomposition with time, sunlight exposure, oxidizing agents, diluted acids, detergents and soaps, they cannot be easily removed by traditional treatment processes due to their complex chemical structures and synthetic origins [72, 73]. Therefore, dyes have a potential risk for bioaccumulation in case of not applying effective treatment methods.

Advanced oxidation processes (AOPs) have shown great promise in water and wastewater treatment, including the decomposition of toxic materials, emerging contaminants, and persistent compounds. Advanced oxidation processes are based on generation of reactive oxygen species (ROS) including hydroxyl radicals ($\cdot OH$), superoxide radicals ($\cdot O_2^-$), and sulfate radicals ($SO_4^{\cdot -}$). Advanced oxidation methods achieve high degradation efficiencies through oxidation reactions with hydroxyl radicals, often assisted by catalytic and photocatalytic processes. Hydroxyl radicals are strong oxidants and react with a wide variety of pollutants rapidly and nonselectively [32, 34, 76, 111].

For practical applications, advanced oxidation catalysts should have certain characteristics such as high catalytic activity, nonselectivity, long-term thermal stability at high temperatures, mechanical stability, and chemical stability under a wide range of reaction conditions [82]. Nanohybrid materials has a great potential to be used as catalyst and provide high degradation efficiencies in advanced oxidation processes. In this chapter, the application of advanced oxidation methods including Fenton-like oxidation, photocatalytic oxidation, sonocatalytic oxidation, and catalytic wet air oxidation for dye removal in the presence of various nanohybrid catalysts are investigated. The reported performances of different types of nanohybrid catalysts containing carbon based materials (e.g. graphitic carbon nitride, graphene and its derivatives, etc.), perovskites, ferrites and layered double hydroxides are reviewed.

2 Classification of Dyes

Dyes are chromophore carrying complex structured substances, which gives them the ability to absorb light (a part of visible spectrum) so that the unabsorbed light is reflected as color. Dyes are generally composed of aromatic hydrocarbons and often contain conjugated double bonds and aromatic amines. The most common chromophores are nitro, azo, nitroso, carbonyl and alkenes. The chromogenic molecules have dyeing properties by the addition of auxochromes, which impart to the dye the ability of electrolytic dissociation, and modify the color. The auxochromes can be acidic or basic. Textile dyes contain functional groups (i.e. carboxylic, amine and azo) that cannot be degraded by conventional treatment methods [13, 86].

Dyes are classified according to several methods. For instance, they can be classified by the method of application to the substrate. Based on their industrial applications, dyes are classified as direct, reactive, vat, disperse, etc. In addition to the applications, dyes can also be classified depending on their chromophores and chemical structures [39]. Dyes are classified as acridine, azo, arylmethane, anthroqinone, nitro, xanthene and quinine–amine, etc. based on their structures. The dye classes are presented in Fig. 1 [3, 91] and the chemical structures of the commonly used dyes are shown in Table 1 [13, 87].

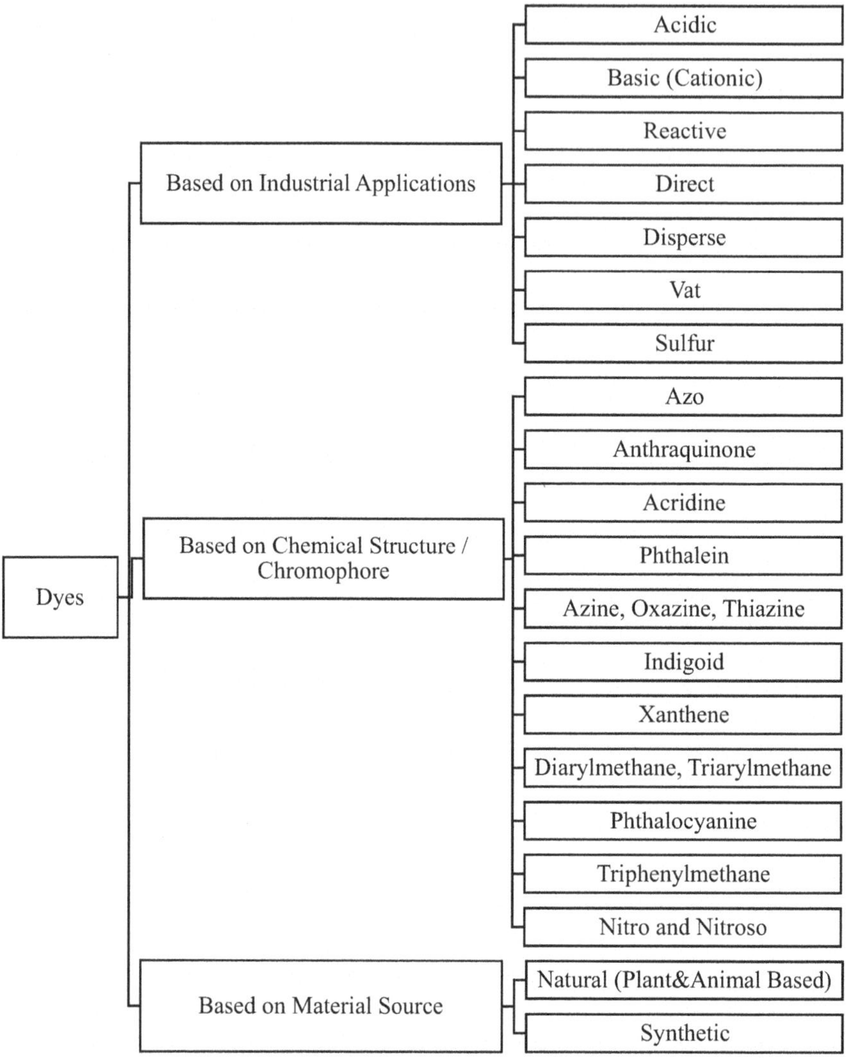

Fig. 1 Classification of dyes according to their applications, chemical structure, and the source of the material

The largest group of the synthetic colorants is the azo dyes. Azo dyes account for more than 60% of total dyes used in industry. They are used in a wide variety of industries including textile, paper manufacturing, leather, food, cosmetics, printing, etc. Azo dyes are characterized by nitrogen to nitrogen double bonds that are attached to two moieties of which at least one is aromatic group. Some examples for azo dyes are methyl orange, methyl red, Reactive Black 5, Reactive Red 2, Tartrazine, Direct Blue 160, Basic Yellow 15, Basic Blue 41. The color of these dyes is determined by

Table 1 Common dye structures and chromophores

Class	Structure/Chromophore	Class	Structure/Chromophore
Azo	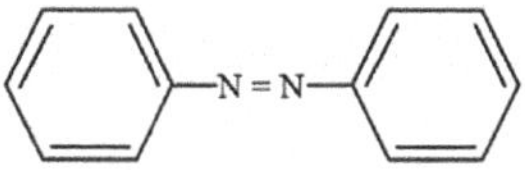	Xanthene	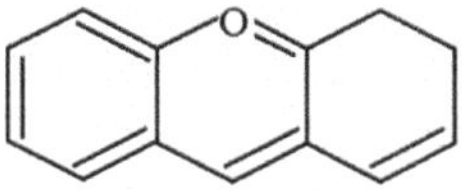
Anthraquinone	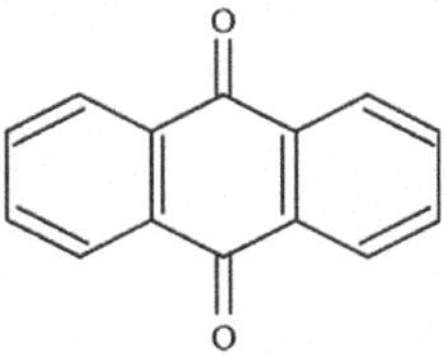	Oxazin	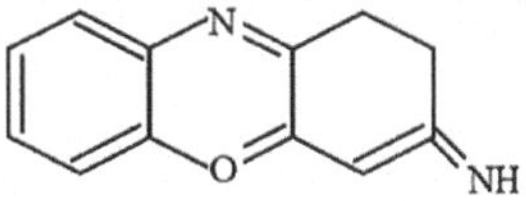
Acridine	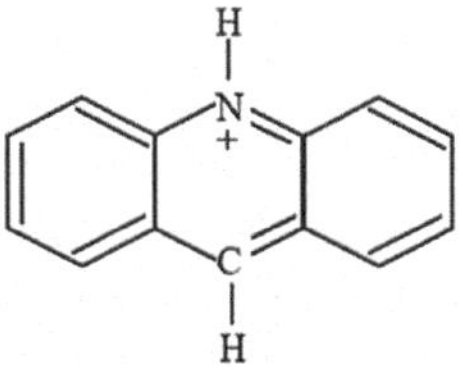	Quinoneimine	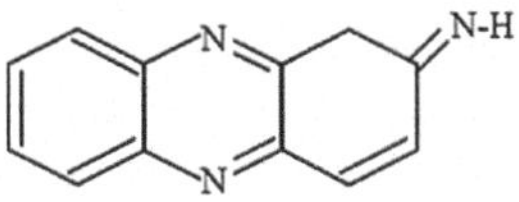
Diarylmethane	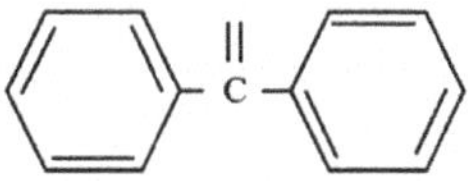	Thiazin	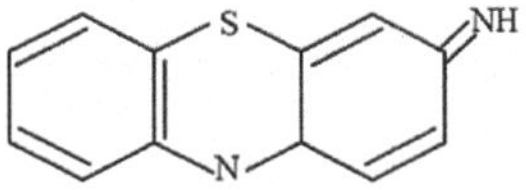
Nitro	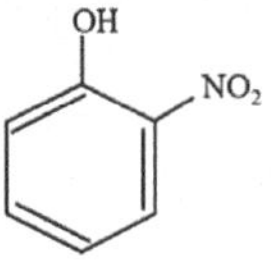	Triarylmethane	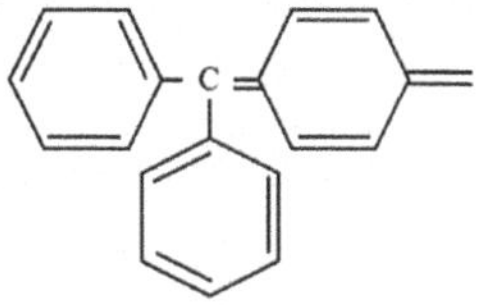

the azo bonds, their chromophores and auxochromes. The breakage of the azo bonds results in decolorization of dye [11, 14, 84].

Acidic, basic, direct, and reactive dyes are water soluble, whereas the disperse and vat dyes are insoluble. Textile dyes are applied to various types of fibers including wool, nylon, acrylic, polyesters, and cotton. The fixation rates of different type of dyes generally varied between 50 and 98%. Among various classes of dyes, reactive dyes are the most commonly used dyes due to their bright color, flexibility and ease of application. However, some environmental problems arise from the use of reactive dyes since reactive dyes have a low degree of fixation (50–90%). The remaining part is released as waste effluents and intensely contaminated wastewaters are produced [61, 88].

3 Nanohybrid-Based Catalysts Used in Advanced Oxidation Processes for the Degradation of Dyes

Since many types of dyes and their breakdown products are toxic, it is significant to use effective treatment methods for the removal of dyes from industrial wastewaters. Various chemical, physical, and biological methods have been applied for the removal dyes. Conventional wastewater treatment methods are generally insufficient and have several drawbacks such as high cots and formation of solid wastes. In most of the conventional process, the pollution is transferred from one phase to another and a secondary waste is generated which needs to be disposed. For instance, accumulation of concentrated sludge causes a disposal problem in coagulation process. In adsorption and ion exchange processes, regeneration of adsorbents and resins brings additional costs. Conventional biological treatment methods require a large land area and are limited by sensitivity of microorganisms toward toxicity of complex structured pollutants [23, 80].

The bottleneck of the conventional wastewater treatment methods can be overcome by the application of advanced oxidation methods. Advanced oxidation processes (AOPs) are considered to be one of the most suitable technologies for the treatment of organic wastewaters. High mineralization efficiencies, strong oxidation ability, complete decomposition of organic substances, fast oxidation rate and no secondary pollution are the advantages of advanced oxidation methods [63, 64].

Advanced oxidation processes comprise Fenton oxidation, ozonation, sonocatalysis, electrochemical oxidation, photocatalytic oxidation, super critical water oxidation (SCWO), catalytic wet oxidation (CWAO), and various combinations of advanced oxidation constituents such as hydrogen peroxide, oxygen, ozone, and light sources. Classification of advanced oxidation processes are shown in Fig. 2 [10, 46].

Some of the common advanced oxidation methods and the nanohybrid catalysts used in these catalytic treatment processes are introduced below.

3.1 *Fenton-Based Processes*

Among various advanced oxidation processes, Fenton oxidation is one of the most widely used methods for the decolorization and mineralization of persistent pollutants. Fenton oxidation has unique advantages such as strong oxidizing ability, low

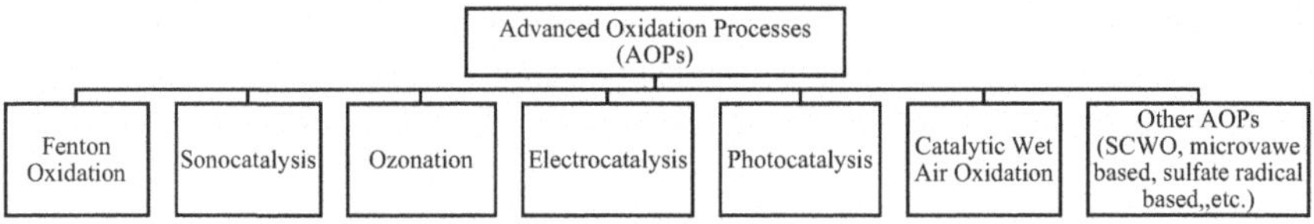

Fig. 2 Classification of advanced oxidation processes

equipment cost, adjustable operational conditions, and nonselective decomposition of contaminants. Fenton oxidation eliminates highly toxic organic compounds and refractory pollutants which most of the conventional wastewater treatment processes fail. Additionally, Fenton oxidation can be carried out under mild reaction conditions near to ambient temperature and pressure and it does not require complex reactor systems.

3.1.1 Fenton and Fenton-Like Oxidation

In the traditional homogeneous Fenton process, Fe^{2+} react with H_2O_2 in acidic medium to produce highly reactive $^{\cdot}OH$, oxidizing ferrous ion to ferric ion. Then, $^{\cdot}OH$ react with organic pollutants in a fast free radical chain reaction to mineralize pollutants nonselectively and to convert into harmless products such as CO_2, H_2O, and mineral salts [27, 51, 95].

$$Fe^{2+} + H_2O_2 \rightarrow Fe^{3+} + ^{\cdot}OH + OH^-$$
$$Fe^{2+} + ^{\cdot}OH \rightarrow Fe^{3+} + OH^-$$
$$^{\cdot}OH + H_2O_2 \rightarrow HO_2^{\cdot} + H_2O$$
$$Fe^{3+} + HO_2^{\cdot} \rightarrow Fe^{2+} + O_2 + H^+$$
$$\text{Organic pollutants} + ^{\cdot}OH \rightarrow H_2O + \text{Intermediates} \rightarrow CO_2 + H_2O$$

Since classical Fenton process demands acidic media (around pH 3) and high iron concentration, large amount of Fe^{2+}, H_2O_2 and acid are consumed for wastewater treatment. Neutralization of the effluents up to pH 6–9 requires alkali chemicals and generates sludge, which need to be disposed. In addition, the remaining iron ions may hinder the reuse of treated wastewater. When the iron concertation in water is above 2 mg/L, it cannot be reused in the dyeing processes. In order to overcome these limitations, heterogeneous Fenton-like oxidation is applied in the presence of various solid catalysts including metal oxides, metal sulfides, carbon-based materials, metal organic frameworks, clays, etc. [21]. Heterogeneous Fenton-like oxidation has proven prominent advantages over classic Fenton oxidation such as no sludge formation and applicability in a wide range of pH.

In the presence of various transition metals, the following Fenton-like oxidation reactions occur [59]:

$$M^{+n} + H_2O_2 \rightarrow M^{+n+1} + ^{-}OH + ^{\cdot}OH$$
$$^{\cdot}OH + H_2O_2 \rightarrow HO_2^{\cdot} + H_2O$$
$$HO_2^{\cdot} + M^{+n+1} \rightarrow M^{+n} + H^+ + O_2$$

The recent studies show that the nanohybrid and nanocomposite catalysts have been used in Fenton like oxidation processes for the degradation of dyes effectively. Kalita et al. [52] used Fe-Cu@HAp (HAp: hydroxyapatite) nanocomposite catalyst for Fenton-like oxidation of dyes. At 20 mg/L initial dye concentration, pH 7.5, and 8 mM H_2O_2 loading the degradation efficiencies were evaluated as 92% and 90% for methylene blue and malachite green, respectively [52]. Yang et al. [114] investigated the catalytic behavior of yolk-shell Fe_3O_4@MOF-5 nanocomposite in Fenton-like oxidation process. It is demonstrated that the nanocomposite exhibited high catalytic activity for the removal of methylene blue. Complete degradation was achieved in 60 min. Yolk-shell Fe_3O_4@MOF-5 has a good stability and recyclability as a heterogeneous Fenton-like catalyst [114]. Song et al. [98] tested the catalytic performance of Fe_3O_4-modified ultrasmall graphene oxide nanocomposites in heterogeneous Fenton-like oxidation process. It is reported that the catalyst effectively cleaned low concentration methylene blue wastewater. Complete degradation was achieved in 120 min at 60 °C, 2.92 mg/L of catalyst loading, and 10 mM H_2O_2 concertation. According to the kinetic study, Fenton-like oxidation follows the first-order reaction kinetic model [98].

3.1.2 Photo Fenton-Like Oxidation

Fenton-like oxidation is often assisted by UV–visible light irradiation of ultrasound irradiation. Photo-Fenton oxidation is the combination of Fenton reagents and UV–Visible light irradiation ($\lambda < 600$ nm) resulting in formation of extra hydroxyl radicals by the following additional reactions [45, 83].

$$H_2O_2 + hv \rightarrow 2^{\cdot}OH$$

$$Fe^{+2} + H_2O_2 \rightarrow Fe(HO)^{2+} + ^{\cdot}OH$$

$$Fe(HO)^{2+} + hv \rightarrow Fe^{+2} + ^{\cdot}OH$$

In the presence of dyestuffs, visible light irradiation may also reduce Fe^{3+} to Fe^{2+} due to the intermolecular electron transfer from excited dye state:

$$\text{Dye} + \text{Visible Light} \rightarrow \text{Dye}*$$

$$\text{Dye} * + Fe^{3+} \rightarrow Fe^{2+} + \text{Dye}*$$

Studies reporting photo Fenton-like oxidation performances of nanohybrid and nanocomposites catalysts exist in the literature. Gonçalves et al. [36] compared the Fenton and photo Fenton-like oxidation performances of CoCr/Fe_3O_4-LDH and CoCr-LDH (LDH: Layered double hydroxide) nanoparticles for the removal of methylene blue and Reactive Black 5 dyes. Magnetite containing composite material was proven more efficient than pure layered double hydroxide due to its larger surface area and larger pores. The dye degradation efficiencies reached over 90% in 90 min by Fenton-like oxidation. The presence of UV light irradiation increased the oxidation rate and the same degradation efficiencies were obtained in a much shorter reaction time. The degradation efficiencies increased over 90% in 30 min in photo Fenton-like oxidation process. The recyclability tests demonstrated that the decolorization efficiencies did not decreased significantly after five cycles for both Fenton and photo Fenton oxidation processes [36]. Arshad et al. [9] investigated the catalytic performance of graphene/Fe_3O_4 nanocomposites by using H_2O_2 under UV light irradiation. The degradation efficiencies were evaluated as 99.24% and 43% in the presence of nanocomposite catalyst and bare Fe_3O_4, respectively. Graphene/Fe_3O_4 showed high stability after five continuous cycles, whereas the degradation efficiency of Fe_3O_4 nanoparticles decreased remarkably [9].

3.1.3 Sono-Assisted Fenton-Like Oxidation

In sono-Fenton processes, ultrasonic dissociation of water and molecular oxygen improve the hydroxyl radical generation in addition to the in-situ generation of H_2O_2 [16]:

$$H_2O + Ultrasound \rightarrow {}^{\cdot}H + {}^{\cdot}OH$$

$$O_2 + Ultrasound \rightarrow 2{}^{\cdot}O$$

$$2H^{\cdot} + O_2 \rightarrow H_2O_2$$

According to the literature survey on sono-assisted Fenton-like oxidation of dyes, high removal efficiencies are obtained by using various types of nanohybrid catalysts. Saemian et al. [89] synthesized a nanocomposite comprising of $CoFe_2O_4$ ferrite nanoparticles, porous silica and Copper metal–organic framework for the degradation of methylene blue in sono-Fenton-like oxidation process. In the presence of $CoFe_2O_4$/SiO_2/Cu-MOF catalyst and hydrogen peroxide, 98% dye removal was accomplished in 30 min. The degradation reactions followed pseudo-second-order kinetic model, revealing that the dye removal occurred dominantly through the multi-site interactions due to the different active sites on nanocomposite surface [89]. Xiao et al. [109] investigated the removal of Rhodamine B in an ultrasonic-assisted heterogeneous Fenton process by using a copper magnetite nanohybrid catalyst. Complete dye degradation was achieved in 30 min. Reaction rate constant obtained over Cu–Fe_3O_4/Cu/C catalyst was much larger than that over Cu/C and over Fe_3O_4/C. Additionally, the nanohybrid delivered good reusability after four cycles. The dominant reactive species were hydroxyl and superoxide radicals [109].

3.1.4 Electro Fenton-Like Oxidation

Electro Fenton oxidation is another widely applied advanced oxidation processes used for the elimination of water pollutants. Nanohybrid materials can be used as electro Fenton catalyst or electrode material [118] for the degradation of dyestuffs. Electrochemical advanced oxidation methods based on Fenton's chemistry are economic and environmentally friendly techniques that have gained attention for wastewater remediation. These methods possess some important advantages such as superior versatility, high efficiency, flexibility, and environmental compatibility. In the electro Fenton processes, in situ generated or externally added H_2O_2 is utilized for hydroxyl radical production [33, 74]. The electro Fenton reactions that occur in the wastewater treatment systems [38]:

In situ production of H_2O_2 at the cathode:

$$O_2 + 2H^+ + 2e^- \rightarrow H_2O_2$$

Hydroxyl radical generation by Fenton reaction:

$$Fe^{2+} + H_2O_2 \rightarrow Fe^{3+} + OH^- + {}^{\cdot}OH$$

Regeneration of ferrous ions at the cathode:

$$Fe^{3+} + e^- \rightarrow Fe^{2+}$$

Additionally, the parallel water molecules in the reaction medium can be oxidized to oxygen at the anode generating more hydroxyl radicals [17, 66]:

$$H_2O \rightarrow {}^{\cdot}OH + H^+ + e^-$$

Pollutant removal potential of nanohybrid and nanocomposite catalysts in electro Fenton process is investigated in several research studies. Cruz et al. [24] investigated the mineralization of Acid Black 210 dye in synthetic and tannery wastewater in the presence of $CoFe_2O_4$/Natural organic matter catalyst. Dye mineralization performances of heterogeneous electro-Fenton and electrochemical oxidation processes were compared. Electro Fenton process showed higher efficiency in comparison to electrochemical oxidation. In the photo Fenton oxidation process, 95% Acid Black 210 dye mineralization was accomplished in 7 h. The nanohybrid catalyst maintained good reusability in three cycles with 10% decrease in mineralization efficiency [24]. Fayazi and Ghanei-Motlagh [31] used sepiolite/pyrite nanocomposite catalyst for the electro Fenton oxidation of methylene blue. Platinum sheet and graphite plate were used as anode and cathode, respectively, in an undivided electrochemical cell. Almost complete dye mineralization was achieved in 75 min at pH 3, 1 g/L of catalyst loading, 50 mg/L of initial dye concentration and 150 mA current density. Sep/FeS_2 nanocomposite was reported as a promising electro Fenton catalyst for industrial applications due to its catalytic capability and low cost [31].

Some of the recent studies on dye removal in Fenton based processes including Fenton-like oxidation, photo Fenton-like oxidation, ultrasound assisted Fenton-like oxidation, electro Fenton oxidation in the presence of nanohybrid and nanocomposite catalysts are summarized in Table 2.

Table 2 Fenton-like oxidation of dyes in the presence of nanohybrid and nanocomposite catalysts

Nanohybrid catalyst	Method	Dye	Operating conditions	Removal efficiency	References
Fe-Cu@HAp (HAp: hydroxyapatite)	Fenton-like oxidation	Methylene blue, Malachite green	Initial dye concentration: 20mg/L Catalyst loading: 0.08g/L Reaction time: 45min pH = 7.5, room temperature, $[H_2O_2]_o = 8mM$	92% methylene blue and 90% malachite green degradation	[52]
Yolk-shell Fe_3O_4@MOF-5 (MOF: Metal organic framework)	Fenton-like oxidation	Methylene blue	Initial dye concentration: 50mg/L Catalyst loading: 1g/L Reaction time: 60 min pH = 4, $T = 30$ °C, $[H_2O_2]_o = 30$ mM	100% degradation	[114]
Fe_3O_4/usGO (usGO: Ultrasmall graphene oxide)	Fenton-like oxidation	Methylene blue	Initial dye concentration: 5 mg/L Catalyst loading: 0.00292 g/L Reaction time: 120 min pH = 3, $T = 60$ °C, $[H_2O_2]_o = 10$ mM	100% degradation	[98]
CoCr/Fe_3O_4-LDH and CoCr-LDH (LDH: Layered double hydroxide)	Fenton and photo-Fenton oxidation	Methylene blue, Reactive Black 5	Initial dye concentration: 30–100 mg/L for Methylene Blue, 50–200 mg/L for Reactive Black 5 Catalyst loading: 1 g/L Reaction time: 40–180 min pH = 7, $T = 25$ °C, 0.10 M H_2O_2 (50%)/H_2O	Fenton-like oxidation: Above 90% degradation Photo Fenton-like oxidation: 92–100%	[36]
Graphene/Fe_3O_4	Photo-Fenton oxidation	Methyl orange	Initial dye concentration: 20 mg/L Catalyst loading: 0.2 g/L Reaction time: 30 min, pH = 3 1 mL H_2O_2 for 100 mL dye solution	99.24% degradation	[9]

(continued)

Table 2 (continued)

Nanohybrid catalyst	Method	Dye	Operating conditions	Removal efficiency	References
$CoFe_2O_4/SiO_2$/Cu-MOF (MOF: Metal organic framework)	Sono-Fenton-like oxidation	Methylene blue	Initial dye concentration: 5 mg/L Catalyst loading: 0.01 g/L Reaction time: 30 min $[H_2O_2]_o = 0.1$ mM	98% degradation	[89]
Cu–Fe_3O_4/Cu/C	Ultrasonic-assisted heterogeneous Fenton oxidation	Rhodamine B	Initial dye concentration: 10 mg/L Catalyst loading: 0.1 g/L Reaction time: 30 min $[H_2O_2]_o = 10$ mM	100% degradation	[109]
$CoFe_2O_4$/NOM (NOM: natural organic matter)	Electro Fenton oxidation	Acid Black 210	Initial dye concentration: 55 mg/L Catalyst loading: ~0.21 g/L Anode: Boron doped diamond Cathode: Gas diffusion electrode Current density: 28.2 mA/cm^2 Reaction time: 420 min, pH = 6	95% mineralization	[24]
Sepiolite/pyrite (Sep/FeS_2)	Electro Fenton oxidation	Methylene blue	Initial dye concentration: 50 mg/L Catalyst loading: 1 g/L Anode: Pt sheet Cathode: Graphite plate Current intensity: 150 mA Reaction time: 75 min, pH = 3	Almost total mineralization	[31]

3.2 Photocatalytic Oxidation

Photocatalysis is a highly effective method for the removal of organic contaminants due to its low cost and easy operation. Photocatalytic oxidation can be operated at ambient temperature and pressure effectively. The principle of photocatalytic degradation is established based on the photo-generation of charge carriers of semiconductor materials leading to oxidation and reduction reactions. During the photocatalytic degradation, absorption of light photons by the photocatalyst results in generation of holes (h^+) and electrons (e^-) in valence and conduction band edge, respectively. The photocatalytic mechanism includes the reaction of photoexcited electron with oxygen in the conduction band and the reaction between water and hole participating in the valence band. Consequently, superoxide, hydroxyl and hydroperoxyl radicals are formed. Reactive oxygen species effectively oxidize recalcitrant organic and inorganic pollutants present in industrial wastewaters [15, 43]. Additionally, the high oxidative potential of the holes in the photocatalyst permits lead to the direct oxidation of dyes to reactive intermediates [25, 84].

$$\text{Photocatalyst} + hv \rightarrow \mathrm{e}^-_{\mathrm{cb}} + \mathrm{h}^+_{\mathrm{vb}}$$
$$\mathrm{H_2O} + \mathrm{h}^+_{\mathrm{vb}} \rightarrow {}^{\cdot}\mathrm{OH} + \mathrm{H}^+$$
$$h^+_{vb} + \mathrm{Dye} \rightarrow \mathrm{Dye}^+$$
$$\mathrm{O_2} + \mathrm{e}^-_{\mathrm{cb}} \rightarrow + {}^{\cdot}\mathrm{O}_2^-$$
$${}^{\cdot}\mathrm{O}_2^- + \mathrm{H}^+ \rightarrow \mathrm{HO}_2^{\cdot}$$
$${}^{\cdot}\mathrm{OH} + \mathrm{Dye} \rightarrow \text{Degradation products}$$

3.2.1 Carbon-Based Nanohybrid Catalysts

Traditional catalysts used in photocatalytic oxidation processes such as titanium dioxide and zinc oxide present high bandgaps and electron/hole recombination rates that limit their use under visible light. In order to improve the adsorption capacities, catalytic activities and treatment performances of photocatalysts, porous hybrid catalysts with larger specific surface areas and highly porous structures were developed. Porous nanohybrid catalysts have important advantages comprising of abundant active sites for photocatalytic reactions and light-harvesting enhancement [37].

Therefore, nanohybrid materials have been developed by combining traditional catalysts and carbon-based materials to increase the wastewater treatment efficiencies in photocatalytic processes. Carbon based porous nanomaterials include graphene and graphene-like materials, graphitic carbon nitride, carbon nanofibers, carbon nanotubes, fullerenes, and carbon-based quantum dots, etc. Their unique dimensions and multifunctional properties have enabled them to be used in diverse areas including energy generation and storage, water and wastewater treatment, and biomedical applications. Carbon nanomaterials have been used to synthesize various nanohybrid catalysts employed in dye removal processes [81].

Graphene Containing Nanohybrids

Graphene is one of the most popular two-dimensional graphitic carbon materials. It has a flexible structure and a high surface area. Due to its unique thermal, mechanical, optical and electrical properties, graphene and graphene like materials have many applications including photocatalysis, photoelectrochemistry, electronic devices, and energy storage (e.g. lithium-ion batteries and solar cells) [40, 58, 116]. Photocatalytic applications of graphene-based materials comprise of hydrogen production, photocatalytic reduction of carbon dioxide to hydrocarbon fuels, and pollutant degradation. In comparison to pure materials, graphene based hybrid materials exhibit improved photocatalytic performances since the graphene content reduces the photoexcited electron–hole recombination [20].

According to the literature survey, high dye removal efficiencies can be achieved in the presence of graphene based nanohybrid photocatalysts. Alwan et al. [5] studied on photocatalytic degradation of Rhodamine B in the presence of TiO_2/rGO nanocomposites. Under the optimum conditions, dye removal efficiency was evaluated as ~94.55%. TiO_2/rGO nanocomposite showed good cycling stability for dye removal [5]. Bashir et al. [12] tested the photocatalytic activity of rGO/Ag/MoO_3 (rGO: Reduced graphene oxide) catalyst. The hybrid photocatalyst demonstrated remarkably higher catalytic activity for dye degradation than pure MoO_3 and Ag/MoO_3 photocatalysts [12].

Graphitic Carbon Nitride Containing Nanohybrids

As a carbon-based polymeric material, graphitic carbon nitride (g-C_3N_4) gains great attention due to the availability of its raw materials, high stability, hardness, tunable optical properties, chemical inertness, and nontoxicity. Fast recombination of photoinduced electron–hole pairs results in low photocatalytic activity to produce reactive oxygen species that can degrade the dyes and limits the photocatalytic applications of pure g-C_3N_4. To enhance the photocatalytic performance and reach maximum pollutant removal, g-C_3N_4 containing hybrid materials have been synthesized due to the wider visible light absorption range and improved photocatalytic ability of the hybrid materials [103, 110].

Zedan et al. [117] investigated the photocatalytic performance of plasmonic graphitic carbon nitride based nanohybrids. Methylene blue removal reached to 97–98% within 20 min whereas complete degradation of methyl orange was achieved in 100 min by using Au@gCN and Ag@gCN (gCN: graphitic carbon nitride) photocatalysts [117]. Mohammad et al. [68] used zinc oxide-graphitic carbon nitride nanohybrid as electrochemical sensor and photocatalyst. An efficient photocatalytic degradation of Chicago Sky Blue, Congo Red, and methylene blue was obtained [68].

In addition to g-C_3N_4-metal nanohybrids, metal free carbon based nanohybrids can also be effectively used for the photocatalytic degradation of dyes. For instance, graphene-g-C_3N_4 nanohybrids showed high photocatalytic activity for the dye removal from aqueous solutions. Zhu et al. [120] reported that 98.5% Rhodamine degradation was obtained in 90 min by using GO/g-C_3N_4/CC (GO: graphene oxide, CC: carbon cloth) nanohybrids [120].

3.2.2 Perovskite-Based Nanohybrid Catalysts

Perovskites with a general formula ABO_3 (A: rare earth metal and B: transition metal) have attracted special attention among various types of semiconductors due to their diverse applications. The key properties of perovskites comprise of easy fabrication, high solar adsorption, sensor applications and utilization in photocatalytic processes and photovoltaic cells [2, 30]. The catalytic activity of perovskite materials is mainly associated to transition metal ion whereas thermal resistance is ascribed to rare earth metal [6].

$LaFeO_3$ and $BiFeO_3$ based catalysts are extensively used in wastewater treatment applications. $BiFeO_3$ perovskite is considered as a suitable replacement for the conventional photocatalysts due to its narrow band-gap energy (2–2.6 eV), high chemical stability, and multiferroic behavior having ferroelectric and magnetic properties [71, 101]. By synthesizing composite materials, the charge separation efficiency and charge transfer can be improved which enhance the photocatalytic activity in turn.

As an emerging visible light active photocatalyst, $LaFeO_3$ has the advantages of non-toxicity, high stability, cost effectiveness, and narrow band gap energy (~2.0 eV). However, it has some shortcomings including low surface area and low separation efficiency of electron–hole pairs that can be overcome by several strategies such as constructing hybrid materials with other semiconductors [112].

According to the literature survey on photocatalytic degradation of dyes, perovskite-based nanohybrid and nanocomposite materials exhibit high catalytic activity. Koyyada et al. [56] studied on photocatalytic degradation of Rhodamine B in the presence of $BiFeO_3/Fe_2O_3$ nanocomposites. Under visible light irradiation, 82.1% and 95.7% degradation efficiencies were achieved with $BiFeO_3$ and $BiFeO_3/Fe_2O_3$, respectively [56]. Li et al. [60] investigated degradation of organic pollutants in the presence of $LaFeO_3/CeO_2$ nanocompostite. Removal efficiencies of 95.9% and 85.5% were achieved for methylene blue and Rhodamine B after 150 min and120 min, respectively [60]. Iqbal et al. [48] tested photocatalytic activity of Lanthanum and Manganese codoped Bismuth Ferrite/Ti_3C_2 MXene nanohybrids for the degradation of Congo Red. It is reported that almost complete degradation was achieved in 30 min [48]. Samran et al. [90] investigated the photocatalytic performance of $BiFeO_3/BiVO_4$ nanocomposites. Dye degradation efficiencies obtained with pure $BiFeO_3$ and pure $BiVO_4$ were 9.27% and 13.09%, respectively. Use of $BiFeO_3$/

$BiVO_4$ nanocomposite increased the removal efficiencies remarkably and the degradation efficiencies varied between 31.93% and 69.27% for different mole ratios of $BiFeO_3$:$BiVO_4$ [90].

The wastewater treatment performances of perovskite catalysts can be improved by constructing perovskite and carbon-based hybrid materials. For instance, reduced graphene oxide supplement provide a larger surface area and enhances the photocatalytic activity. The reduced graphene oxide nanolayers promote the photoexcited charge carriers to the interfaces and the charge separation process is prolonged [69].

3.2.3 Spinel Ferrite-Based Nanohybrid Catalysts

Magnetic nanoparticles gained attention due to their catalytic and superparamagnetic properties enabling them to be separated easily by using external magnets. Particularly, ferrites are one of the most significant magnetic nanoparticles. Ferrites are classified as spinel (MFe_2O_4, M = Co, Cu, Mn, Ni, Zn, etc.), garnet ($M_3Fe_5O_{12}$, M = rare earth metals), hexagonal, and orthoferrite ($MFeO_3$, M = rare earth metals) based on crystal structures. Among these, spinel ferrites come into prominence due to their excellent magnetic properties, chemical stability, tunable shape and size, and wide applications in various fields including water and wastewater treatment, biomedicine, and electronics [54, 85]. Ferrites are considered as potential alternatives to conventional photocatalysts since they can absorb lower energy photons ($h\nu \sim 2$ eV) [78]. Ferrite based nanohybrid catalysts have been used for dye removal in photocatalytic processes. For instance, Janani et al. [49] investigated the photocatalytic performance of $CdO/ZnFe_2O_4$ catalysts under white light. The dye degradation efficiencies obtained in the presence of $CdO/ZnFe_2O_4$, CdO, and $ZnFe_2O_4$ were 79.1%, 71.8%, and 66.4%, respectively [49]. Palanisamy et al. [77] studied on methylene blue removal in the presence of magnetically recoverable $ZnS–WO_3–CoFe_2O_4$ nanohybrid catalyst under visible light irradiation and 95.97% degradation was achieved in 3 h [77].

In addition to the use of metal oxide-ferrite nanohybrids, magnetically recoverable ferrite-carbon based nanohybrid materials have also been employed in photocatalytic processes [42, 50, 94]. Use of hybrid catalysts composed of ferrites and carbon-based materials improves the adsorption and photocatalytic performances and reduces the toxic metal leaching and agglomeration.

3.2.4 Layered Double Hydroxide-Based Nanohybrid Catalysts

Layered double hydroxides (LDH) are a type of anionic clay with lamellar structures made of octahedral sheets. Their general formula is $[M^{2+}_{1-x}M^{3+}_x(OH)_2](A^{n-})_{x/n}\cdot mH_2O]$ where M^{2+} and M^{3+} are divalent (e.g. Co^{2+}, Cu^{2+}, Mg^{2+}, Ni^{2+}, Zn^{2+}, etc.) and trivalent cations (e.g. Al^{3+}, Cr^{3+}, and Fe^{3+}) and A^{n-} is interlayer anion (e.g. $CO_3{}^{2-}$, NO^{3-}, Cl^-, and $SO_4{}^{2-}$). Large surface areas, high crystallinity, and structural stability make layered double hydroxides to be a promising candidate for environmental remediation applications [19, 44, 113]. Layered double hydroxide containing nanohybrid materials have been effectively used in photocatalytic processes for the removal of dyes from aqueous solutions.

Khamesan et al. [55] synthesized graphitic-C_3N_4/ZnCr-layered double hydroxide nanocomposites and investigated the azo dye removal performance of the photocatalyst. 99.8% Rhodamine B removal was accomplished in 90 min under UV/vis radiation [55]. Ao et al. [8] studied on photocatalytic degradation of methyl orange and Rhodamine B under UV light irradiation in the presence of BiOBr–Co–Ni–NO_3 layered double hydroxide nonocomposites. The photocatalytic degradation performance of the nanocomposite was significantly higher than the performances of BiOBr and layered double hydroxide [8].

3.2.5 Supramolecular Nanohybrid Catalysts

The use of novel supramolecular catalysts in advanced oxidation processes has attracted considerable attention recently. Organic semiconductor photocatalysts is of interest due to their tunable optical and electronic characteristics, versatility, adjustability, and low cost. High synthesis selectivity and broad range of light absorption makes the organic semiconductors suitable for photocatalytic applications. Perylene diimide based catalysts are known to be one of the best organic semiconductors with their high thermal and photochemical stability, and strong electron-accepting character [62]. The major advantage of supramolecular catalysts is the enhancement of electron transfer between the components, leading to improvement of the photocatalytic performance and increase in the reaction rates [100].

Supramolecular nanohybrid and nanocomposite catalysts have been used in photocatalytic oxidation of various organic pollutants including dyes. Cai et al. [18] tested the photocatalytic activity of Perylenediimide/silver nanohybrid materials. Both the antibacterial effect and dye removal performance of the nanohybrid materials were investigated. It is reported that Rhodamine B concentration decreased sharply in the presence of fiber-like nanohybrids and visible light irradiation improved antibacterial effect of Perylenediimide/silver nanohybrid via Ag^+ release during the photocatalysis. Radical quenching studies showed that superoxide radicals and hole play critical roles in the photocatalytic degradation [18]. Wang et al. [104] investigated the photocatalytic performance of supramolecular H_{12}SubPcB-OPhCOPh/TiO_2 Z-scheme nanohybrid catalyst. The degradation efficiencies of tetracycline, methyl orange, and bromophenol blue were evaluated as 96.2%, 98%, and 100%, respectively, which were remarkably higher than the efficiencies obtained with pristine TiO_2. According to the radical trapping experiments, the dominant reactive species were the hydroxyl and superoxide radicals [104].

Mardiroosi et al. [65] synthesized supramolecular nanocomposite catalysts by grafting of perylene diimide (PDI) onto graphitic carbon nitride nanosheets (PCN). The metal organic framework, UiO-66 nanoparticles were grown on the PCN to obtain PCN@UiO-66 hybrid catalyst. Rhodamine B removal performance of the catalyst was tested and 99% degradation efficiency was achieved. The catalyst exhibited good stability and the degradation efficiency was still 91% ± 1% after four consecutive runs. Superoxide radical was determined to be the main species responsible for Rhodamine B degradation [65]. Shen et al. [96] used graphene oxide–ZnCdS supramolecular nanocomposites for the degradation of organic dyes under visible light irradiation. The degradation efficiencies of methyl orange varied between 55 and 85%, while the degradation efficiencies of Rhodamine B changed between 55 and 65% in the presence of RGO–ZnCdS nanohybrids with different reduced graphene oxide contents [96].

The photocatalytic reaction conditions and dye removal performances reported in recent studies are summarized in Table 3 based on the type of nanohybrid and nanocomposite catalysts.

3.3 Sonocatalytic Oxidation

Sonocatalysis is a promising advanced oxidation technology for the removal of organic dyes from wastewaters due to its safe and easy operation, energy conservation, enhanced mass transfer, high removal efficiency, in situ formation of oxidizing species, and no need for additional oxidants [102].

Sono-assisted advanced oxidation methods are applicable for the degradation a variety of organic pollutants in wastewaters. The inclusion of ultrasound increases the efficiency of the treatment processes since the ultrasound passing through water facilitates breaking chemical bonds and results in generation of free radicals. Sonolysis mechanism comprises of nucleation, cavitation, bubble dynamics/interactions, thermodynamic and chemical processes [1, 105, 108].

Under ultrasound irradiation, the waves interact with dissolved gases in the water leading to acoustic cavitation. The compression and expansion cycles occur by ultrasonic waves. Waves in the expansion period form bubbles. The energy from compression and expansion cycles of the ultrasound waves is absorbed by bubbles which results in growth of bubbles up to the critical size and then collapse. Hot spots are generated during the bubble collapse. The hot spots trigger the reactions to split the H_2O molecules into a hydrogen atom and hydroxyl radical [70].

$$H_2O + Ultrasound \rightarrow {}^{\cdot}H + {}^{\cdot}OH$$
$$O_2 + Ultrasound \rightarrow 2\,{}^{\cdot}O$$
$$H_2O + {}^{\cdot}O \rightarrow 2\,{}^{\cdot}OH$$

The degradation of azo dyes by sonolysis can be expressed by the following reactions [47]:

$$\text{Dye} + {}^{\cdot}OH \rightarrow [\text{Dye - OH adduct}]^* \rightarrow \text{Oxidized dye} + CO_2 + H_2O$$
$$\text{Dye} + Ultrasound \rightarrow R^{\cdot} + \text{Dissociated dye fragments} + C_2H_4$$

Table 3 Photocatalytic degradation of dyes in the presence of magnetic nanohybrid and nanocomposite catalysts

Catalyst type	Nanohybrid/ Nonocomposite catalyst	Dye	Operating conditions	Removal efficiency	Dominant reactive species	References
Carbon material containing	TiO_2/rGO (rGO: reduced graphene oxide)	Rhodamine B	Initial dye concentration: 15 mg/L Catalyst loading: 1.2 g/L Reaction time: 120 min Visible light irradiation	~ 94.55% degradation	Hole	[5]
	rGO/Ag/MoO_3 (rGO: reduced graphene oxide)	Methylene blue	Initial dye concentration: 5 mg/L Catalyst amount: 0.04 g Reaction time: 70 min Visible light irradiation	98.2% degradation		[12]
	Au@Graphitic carbon nitride and Ag@Graphitic carbon nitride	Methylene blue and Methyl orange	Initial dye concentration: 10 mg/L Catalyst concentration: 1 g/L Reaction time: 20 min for methylene blue, 100 min for methyl orange Visible light irradiation	97–98% methylene blue degradation complete degradation of methyl orange	Hydroxyl	[117]

(continued)

Table 3 (continued)

Catalyst type	Nanohybrid/ Nonocomposite catalyst	Dye	Operating conditions	Removal efficiency	Dominant reactive species	References
	GO/g-C_3N_4/CC (GO: graphene oxide, CC: carbon cloth)	Rhodamine B	Initial dye concentration: 5 mg/L 3 × 3 cm^2 catalyst for 30 mL dye solution Reaction time: 90 min Visible light irradiation	98.5% degradation	Superoxide and Hydroxyl	[120]
	Zinc oxide-graphitic carbon nitride	Chicago Sky Blue, Congo Red and Methylene blue	Initial dye concentration: 10 mg/L Catalyst concentration: 0.2 g/L Reaction time: 45–150 min UV light irradiation	~ 85–99.6% degradation		[68]
Perovskite or ferrite containing	$BiFeO_3/Fe_2O_3$	Rhodamine B	Initial dye concentration: 2 × 10^{-5} M Catalyst loading: 1 g/L Reaction time: 80 min Visible light irradiation	95.7% degradation	Hole and hydroxyl	[56]

(continued)

Table 3 (continued)

Catalyst type	Nanohybrid/ Nonocomposite catalyst	Dye	Operating conditions	Removal efficiency	Dominant reactive species	References
	$LaFeO_3/CeO_2$	Methylene blue, Rhodamine B	Initial dye concentration: 10^{-5} M Catalyst loading: 1 g/L Reaction time: 150 min for methylene blue, 120 min for Rhodamine B Visible light irradiation	95.9% methylene blue, 85.5% Rhodamine B degradation	Hole and hydroxyl	[60]
	$CdO–ZnFe_2O_4$	Methylene blue	Initial dye concentration: 25 mg/L Catalyst loading: 0.1 g/L Reaction time: 160 min Visible light irradiation	79.1% degradation	Hole and hydroxyl	[49]

(continued)

Table 3 (continued)

Catalyst type	Nanohybrid/ Nonocomposite catalyst	Dye	Operating conditions	Removal efficiency	Dominant reactive species	References
	$ZnS–WO_3–CoFe_2O_4$	Methylene blue	Initial dye concentration: 50 mg/L Catalyst amount: 0.05 g Reaction time: 180 min Visible light irradiation	95.97% degradation	Superoxide and hydroxyl	[77]
	La- and Mn–Co doped Bismuth Ferrite/Ti_3C_2 MXene ($BLFO/Ti_3C_2$ and $BLFMO-5/Ti_3C_2$)	Congo Red	Initial dye concentration: 100 mg/L Catalyst loading: 1 g/L Reaction time: 30 min Visible light irradiation	Almost 100% degradation		[48]
	$BiFeO_3/BiVO_4$	Rhodamine B	Initial dye concentration: 2×10^{-5} M Catalyst loading: 1 g/L Reaction time: 120 min Visible light irradiation	69% degradation	Hydroxyl and hole	[90]

(continued)

Table 3 (continued)

Catalyst type	Nanohybrid/ Nonocomposite catalyst	Dye	Operating conditions	Removal efficiency	Dominant reactive species	References
Layered double hydroxide containing	Graphitic-C_3N_4/ ZnCr-layered double hydroxide	Rhodamine B	Initial dye concentration: 5 mg/L Catalyst loading: 1 g/L Reaction time: 90 min UV/vis light irradiation	99.8% degradation	Superoxide	[55]
	BiOBr/Co–Ni layered double hydroxide	Methyl orange, Rhodamine B	Initial dye concentration: 20 mg/L methyl orange, 10 mg/L Rhodamine B Catalyst loading: 0.2 g/L Reaction time: 30 min for methyl orange, 15 min for Rhodamine B UV light irradiation	Complete degradation	Hole and superoxide	[8]
Supramolecular catalysts	Perylene diimide/silver (BF@AgNPs)	Rhodamine B	Initial dye concentration: 10 μM Reaction time: 60 min Visible light irradiation	Higher than 95% degradation	Superoxide and hole	[18]

(continued)

Table 3 (continued)

Catalyst type	Nanohybrid/ Nonocomposite catalyst	Dye	Operating conditions	Removal efficiency	Dominant reactive species	References
	UiO-66 supported on Perylene diimide functionalized g-C_3N_4 (PCN@UiO-66)	Rhodamine B	Initial dye concentration: 30 mg/L Catalyst loading: 1 g/L Reaction time: 140 min Visible light irradiation	99% degradation	Superoxide	[65]
	H_{12}SubPcB-OPhCOPh/ TiO_2 (SubPc-4/TiO_2)	Methyl orange, Bromophenol blue	Initial dye concentration: 10 mg/L Catalyst loading: 1 g/L Reaction time: 180 min Visible light irradiation	98% methyl orange and 100% bromophenol blue degradation	Superoxide and hydroxyl	[104]
	Reduced graphene oxide–ZnCdS (RGO–ZnCdS)	Methyl orange, Rhodamine B	Initial dye concentration: 10 μM Catalyst loading: 0.02 g/L Reaction time: 30 min Visible light irradiation	55%-85% for Methyl orange 55%-65% for Rhodamine B		[96]

where [Dye-OH adduct]* is an excited intermediate product generated from the addition of hydroxyl ion onto –C–N– or –N=N– bonds, and R˙ is an organic radical decomposed from dye molecule.

The contaminant removal potential of ultrasound is based on the thermolytic decomposition of water to produce hydroxyl and hydrogen radicals, as well as dissolved gases and volatile solutes that favors the oxidation reactions and formation of reactive species. In addition, ultrasound improves the surface properties of solid particles and increases the rate of chemical reactions leading to the reduction of chemical consumption and waste sludge generation. The use of heterogeneous catalysts accelerates sonochemical reactions [29].

The sonocatalytic degradation of dyes has been performed effectively in the presence of various nanohybrid and nanocomposite catalysts. Ferrite based nanocatalysts have been widely used in sonocatalytic processes. For instance, Han et al. [41, 42] prepared a magnetic nanohybrid sonocatalyst composed of lanthanum dioxide carbonate and zinc ferrite loaded reduced graphene oxide (LZF-rGO) for ultrasound-assisted degradation of methyl orange. It is reported that 75.9% dye removal was accomplished at 0.2 g/L catalyst loading and 0.71 W/cm^2 and ultrasonic power intensity in 65 min [41]. Nirumand et al. [75] synthesized a ternary magnetic catalyst including metal organic framework, reduced graphene oxide, and ferrite. The sonocatalytic performance of MIL-101(Cr)/RGO/$ZnFe_2O_4$ was tested for the removal of Congo Red, methylene blue, Rhodamine B, and methyl orange. Complete Congo Red degradation was achieved whereas the methylene blue and Rhodamine B degradation efficiencies were over 96% and the methyl orange removal efficiency was 80% [75]. Siadatnasab et al. [97] studied on sonocatalytic degradation of methylene blue, methyl orange and Rhodamine B by using magnetic CuS/CFO (CFO: $CoFe_2O_4$) nanohybrids. Methylene blue degradation efficiency of sonolysis/H_2O_2, sonocatalysis in the presence of CuS/H_2O_2, CFO/H_2O_2 and CuS/CFO/H_2O_2 systems were evaluated as 6%, 62%, 23%, and 100%, respectively. Methyl orange and Rhodamine B removal efficiencies were calculated as 83% and 72%, respectively, in CuS/CFO/H_2O_2 treatment system. The most effective reactive species was found to be the hydroxyl radical [97].

Since the oxidation of organic compounds by using only sonication requires high energy and long reaction times to achieve complete degradation, ultrasound irradiation is often used in hybrid processes to improve the efficiency of sonodegradation [28]. For the removal of dyes, ultrasonication is mostly combined with photocatalysis. Al-Musawi et al. [4] prepared MWCNT–$CuNiFe_2O_4$ nanocomposites (MWCNT: multi-walled carbon nanotube) for the degradation of Acid Blue 113. Complete dye degradation, 93% total organic carbon removal, and 95% chemical oxygen demand removal were achieved in 30 min. Radical trapping studies demonstrated that the holes and hydroxyl radical were the main reactive species for dye removal [4]. Selim et al. [92] prepared a nanohybrid catalyst from a polyamide crosslinked carbon dot-polymer and subsequent deposition of Pd nanoparticles. The sonocatalytic activity of Pd@CD-CONH catalyst was tested for Rhodamine B and methylene blue degradation. In the absence of light, 99% Rhodamine B removal was achieved in 5 min. Visible light assistance did not improved the removal efficiency significantly. The same degradation efficiency was obtained in 20 min in case of methylene blue removal. Radical scavenging studies showed that superoxide radicals played a major role in dye degradation mechanism, while hydroxyl ions also contributed to the removal [92].

Mohamed et al. [67] used ZnO incorporated carbon nanotubes (CNT) or graphene oxide (GO) nanohybrid catalysts for improved sonophotocatalyis of methylene blue. ZnO(70)/CNT nanohybrid catalyst showed the highest catalytic performance and 99% dye degradation was achieved whereas 54% degradation efficiency was obtained in the presence of ZnO(90)/GO in 180 min. Radical scavenging experiments indicated the significance of hydroxyl radicals followed sequentially by electrons and holes [67]. Some of the recent studies on sonocatalytic and sonophotocatalytic degradation of dyes in the presence of nanohybrid and nanocomposite catalysts are summarized in Table 4.

3.4 Catalytic Wet Air Oxidation

Wet air oxidation is a liquid phase oxidation process in which oxygen molecules in air are used as oxidant to degrade the organic pollutants. It is an effective technology for the treatment of intensely polluted wastewaters. The toxic and hazardous organic pollutants are converted into harmless products without the emissions of NO_x, SO_2, hydrochloric acid, dioxins, furans, fly ash, etc. The oxidation products are generally carbon dioxide, water, inorganic salts, and simple biodegradable compounds. The water contaminants are oxidized under high temperatures (generally varying between 125 and 325 °C) and high pressures in a range of 5–200 atm. Though wet air oxidation is a useful method for treating industrial wastewaters that are too toxic for biological treatment and too dilute for incineration, severe operating conditions increase the cost of the treatment and limit its practical applications [22, 99, 115].

The limitations of the wet air oxidation can be overcome by using suitable catalysts, which shows high performance at mild reaction conditions. Catalytic wet air oxidation is an environmentally friendly method, which can be operated even at ambient temperature and atmospheric pressure without using any expensive oxidants. Homogenous catalysts, noble metal catalysts, and metal oxide catalysts have been extensively used in catalytic wet air oxidation processes. The homogenous catalysts, such as sulfates of iron, copper and manganese show quite high catalytic performances in mineralization of organic pollutants. However, the additional metal separation step of the dissolved metal ions increases the cost of the treatment. Noble metal catalysts and transition metal oxide catalysts used in catalytic wet air oxidation process suffer from deactivation occurring as a result of partial coverage of the active surface by carbonaceous materials. In addition, most of the traditional catalysts have some leaching problems [53, 119].

In this frame, development of innovative and ecofriendly catalysts for catalytic wet air oxidation of resistant organic compounds is a significant research area. Nanohybrid catalysts can be effectively used in catalytic wet air oxidation of various dyes.

Table 4 Sonocatalytic degradation of dyes in the presence of nanohybrid and nanocomposite catalysts

Nanohybrid catalyst type	Method	Dye	Operating conditions	Removal efficiency	Dominant reactive species	References
MWCNT–$CuNiFe_2O_4$ (MWCNT: multi-walled carbon nanotubes)	Sonophotocatalysis	Acid Blue 113	Initial dye concentration: 50 mg/L Catalyst loading: 0.6 g/L Reaction time: 30 min Ultrasonic frequency = 35 kHz UV light irradiation	100% dye degradation, 93% TOC removal, 95% COD removal	Hole and hydroxyl	[4]
Pd nanoparticles anchored onto carbon dot–polymer matrix (Pd@CD-CONH)	Sonocatalysis	Rhodamine B, Methylene blue	Initial dye concentration: 4.79 mg/L Rhodamine B, 10^{-5} M methylene blue Catalyst loading: 0.5 g/L Reaction time: 5 min for Rhodamine B, 20 min for methylene blue Ultrasonic power:100 W	99% dye degradation	Superoxide ion	[92]
$La_2O_2CO_3$ and $ZnFe_2O_4$ loaded reduced graphene oxide (LZF-rGO)	Sonocatalysis	Methyl orange	Initial dye concentration: 20 mg/L Catalyst loading: 0.2 g/L Reaction time: 65 min Ultrasonic power intensity: 0.71 W/cm^2	75.9% degradation		[41]
ZnO/CNTs, ZnO/GO (CNT: carbon nanotubes, GO: graphene oxide)	Sonophotocatalysis	Methylene blue	Initial dye concentration: 20 mg/L Catalyst loading: 0.5 g/L Reaction time: 180 min Ultrasonic power: 60 W Visible light irradiation	54–99% degradation	Hydroxyl	[67]

(continued)

Table 4 (continued)

Nanohybrid catalyst type	Method	Dye	Operating conditions	Removal efficiency	Dominant reactive species	References
MIL-101(Cr)/ RGO/ $ZnFe_2O_4$ (RGO: reduced graphene oxide)	Sonocatalysis	Congo Red, Methylene blue, Rhodamine B, Methyl orange	Initial dye concentration: 25 mg/L Catalyst loading: 0.5 g/L Reaction time: 2 min for Congo Red, 50 min for methylene blue and Rhodamine B, 70 min for methyl orange $[H_2O_2]_o = 40$ mM Ultrasonic frequency: 37 kHz	Removal of 100% Congo Red, over 96% methylene blue and Rhodamine B, 80% methyl orange	Hydroxyl	[75]
CuS/CFO ($CFO{:}CoFe_2O_4$)	Sonolysis/H_2O_2 and Sonocatalysis	Methylene blue, Methyl orange, Rhodamine B	Initial dye concentration: 25 mg/L Catalyst loading: 0.5 g/L Reaction time: 30 min $[H_2O_2]_o = 4$ mM Ultrasonic power:100 W	100% methylene blue, 83% methyl orange, 72% Rhodamine B degradation	Hydroxyl	[97]

Wang et al. [106] used nanohybrid bimetallic catalysts for the degradation of brilliant green, safranine O, and methylene blue in catalytic wet air oxidation process. The hybrid of $Na_2Mo_4O_{13}$ and MoO_3 (NM) was prepared by hydrothermal method, and Au–Pd nanoparticles were deposited on $Na_2Mo_4O_{13}/\alpha\text{-}MoO_3$ hybrid material to form Au–Pd(1:1)/NM-400 catalyst. The nanohybrid catalyst showed a good catalytic activity, reusability, and stability as well as being usable at ambient conditions. It is reported that dye removal efficiencies were still high after five times reuse. The mechanism studies revealed that the main active radical was superoxide during the catalytic wet air oxidation of dyes [106].

4 Future Aspects and Conclusions

With the growing concern of environment and water pollution, complete mineralization of organic compounds in wastewaters becomes more necessary than just transferring the wastes simply from one phase to another. Dyes have been used in many of the manufacturing industries including textile, leather, paint, food processing, pharmaceutical, cosmetics, plastic, etc. Discharge of dye containing wastewaters threaten the environment due to the toxicity and mutagenic effects of some of dyes and the reduction of the light penetration, which suppresses the photosynthetic activity and growth of the aquatic flora. Color removal from the textile industry effluents has also become more significant than ever because of the stricter regulations for discharge standards and increasing awareness of environment and water scarce.

Among available wastewater treatment methods, advanced oxidation processes have the greatest potential for ecofriendly degradation of industrial pollutants including persistent dyes. Advanced oxidation processes have significant advantages such as high removal efficiency, cost effectiveness, and secondary pollution control. With ongoing research, an increasing number of novel catalysts have been synthesized to be utilized in wastewater remediation applications. Development of nanotechnology gives rise to the production of multifunctional nanohybrid materials that can be used in treatment of industrial wastewaters. Application of nanohybrid and nanocomposite materials give excellent results in dye degradation studies. Carbon based (e.g. graphene and its derivatives, graphitic carbon nitride, etc.), perovskite and ferrite based, and clay based nanohybrids have been effectively used in various advanced oxidation processes, mostly in photocatalytic oxidation and photo assisted processes.

Many of dyes are completely degraded in short reaction times varying between a few minutes to several hours in the presence of nanohybrid catalysts. Even though the aqueous solutions of dyes are often completely decolorized, analyzing the toxicity of the treated solutions or identification of oxidation products is also significant since the partial oxidation of dyes may result in the generation of toxic intermediates and products.

The catalyst production and wastewater treatment costs are significant parameters determining the applicability of large-scale applications of nanohybrid materials in dye removal processes. Limited number of studies on cost analysis of dye degradation is available in literature and more attention should be devoted on cost analysis to assess the applicability of nanohybrid catalysts in wastewater treatment processes at industrial scale. Additionally, in a practical application environment, such as real textile wastewater, many factors can affect the efficiency of the treatment process. Interaction of various pollutants with each other and with reactive oxygen species influence the dye removal performance of the catalysts in real wastewaters. Therefore, particular consideration is required to express the catalytic performance of nanohybrid catalysts not only for the treatment aqueous solutions of dyes but also for the treatment of dye containing real wastewaters.

In conclusion, the use in nanohybrid catalysts in dye degradation processes gives promising results and it is recommended that more future studies focus on toxicity analysis, cost analysis, pilot scale applications and real wastewater applications in addition to enhancing the catalytic performance excellence.

References

1. Abdurahman MH, Abdullah AZ, Shoparwe NF (2021) A comprehensive review on sonocatalytic, photocatalytic, and sonophotocatalytic processes for the degradation of antibiotics in water: synergistic mechanism and degradation pathway. Chem Eng J 413:127412. https://doi.org/10.1016/j.cej.2020.127412
2. Acharya S, Kandi D, Parida K (2020) CdS QD decorated $LaFeO_3$ nanosheets for photocatalytic application under visible light irradiation. ChemistrySelect 5:6153–6161. https://doi.org/10.1002/slct.202000220
3. Ajmal A, Majeed I, Malik RN, Idriss H, Nadeem MA (2014) Principles and mechanisms of photocatalytic dye degradation on TiO_2 based photocatalysts: a comparative overview. RSC Adv 4:37003–37026. https://doi.org/10.1039/c4ra06658h
4. Al-Musawi TJ, Mengelizadeh N, Taghavi M, Shehu Z, Balarak D (2022) Capability of copper–nickel ferrite nanoparticles loaded onto multi-walled carbon nanotubes to degrade acid blue 113 dye in the sonophotocatalytic treatment process. Environ Sci Pollut Res 29:51703–51716. https://doi.org/10.1007/s11356-022-19460-z
5. Alwan SH, Salem KH, Alshamsi HA (2022) Visible light-driven photocatalytic degradation of Rhodamine B dye onto TiO_2/rGO nanocomposites. Mater Today Commun 33:104558. https://doi.org/10.1016/j.mtcomm.2022.104558
6. Aman D, Zaki T, Mikhail S, Selim SA (2011) Synthesis of a perovskite $LaNiO_3$ nanocatalyst at a low temperature using single reverse microemulsion. Catal Today 164:209–213. https://doi.org/10.1016/j.cattod.2010.11.034
7. Amir MNI, Halilu A, Julkapli NM, Ma'amor A (2020) Gold-graphene oxide nanohybrids: a review on their chemical catalysis. J Ind Eng Chem 83:1–13. https://doi.org/10.1016/j.jiec.2019.11.029
8. Ao Y, Wang D, Wang P, Wang C, Hou J, Qian J (2015) A BiOBr/Co-Ni layered double hydroxide nanocomposite with excellent adsorption and photocatalytic properties. RSC Adv 5:54613–54621. https://doi.org/10.1039/c5ra05473g
9. Arshad A, Iqbal J, Ahmad I, Israr M (2018) Graphene/Fe_3O_4 nanocomposite: interplay between photo-Fenton type reaction, and carbon purity for the removal of methyl orange. Ceram Int 44:2643–2648. https://doi.org/10.1016/j.ceramint.2017.08.157
10. Awofiranye OS, Modise SJ, Naidoo EB (2020) Overview of polymer–TiO_2 catalyst for aqueous degradation of pharmaceuticals in heterogeneous photocatalytic process. J Nanoparticle Res 22. https://doi.org/10.1007/s11051-020-04877-9

11. Ayalkie Gizaw B, Gabbiye Habtu N (2022) Catalytic wet air oxidation of azo dye (reactive red 2) over copper oxide loaded activated carbon catalyst. J Water Process Eng 48:102797. https://doi.org/10.1016/j.jwpe.2022.102797
12. Bashir S, Habib A, Jamil A, Alazmi A, Shahid M (2022) Fabrication of Ag-doped MoO_3 and its nanohybrid with a two-dimensional carbonaceous material to enhance photocatalytic activity. Adv Powder Technol 33:103482. https://doi.org/10.1016/j.apt.2022.103482
13. Benkhaya S, M'rabet S, El Harfi A (2020) A review on classifications, recent synthesis and applications of textile dyes. Inorg Chem Commun 115:107891. https://doi.org/10.1016/j.inoche.2020.107891
14. Benkhaya S, M'rabet S, El Harfi A (2020) Classifications, properties, recent synthesis and applications of azo dyes. Heliyon 6. https://doi.org/10.1016/j.heliyon.2020.e03271
15. Bhuvaneswari K, Palanisamy G, Bharathi G, Pazhanivel T, Upadhyaya IR, Kumari MLA, Rajesh RP, Govindasamy M, Ghfar A, Al-Shaalan NH (2021) Visible light driven reduced graphene oxide supported ZnMgAl LTH/ZnO/g-C_3N_4 nanohybrid photocatalyst with notable two-dimension formation for enhanced photocatalytic activity towards organic dye degradation. Environ Res 197:111079. https://doi.org/10.1016/j.envres.2021.111079
16. Bokare AD, Choi W (2014) Review of iron-free Fenton-like systems for activating H_2O_2 in advanced oxidation processes. J Hazard Mater 275:121–135. https://doi.org/10.1016/j.jhazmat.2014.04.054
17. Brillas E (2022) Fenton, photo-Fenton, electro-Fenton, and their combined treatments for the removal of insecticides from waters and soils. A review. Sep Purif Technol 284:120290. https://doi.org/10.1016/j.seppur.2021.120290
18. Cai Y, Cheng W, Ji C, Su Z, Yin M (2021) Perylenediimide/silver nanohybrids with visible-light photocatalysis enhanced antibacterial effect. Dye Pigment 195:109698. https://doi.org/10.1016/j.dyepig.2021.109698
19. Chaillot D, Bennici S, Brendlé J (2021) Layered double hydroxides and LDH-derived materials in chosen environmental applications: a review. Environ Sci Pollut Res 28:24375–24405. https://doi.org/10.1007/s11356-020-08498-6
20. Chen D, Zhang H, Liu Y, Li J (2013) Graphene and its derivatives for the development of solar cells, photoelectrochemical, and photocatalytic applications. Energy Environ Sci 6:1362–1387. https://doi.org/10.1039/c3ee23586f
21. Chen Q, Yao Y, Zhao Z, Zhou J, Chen Z (2021) Long term catalytic activity of pyrite in heterogeneous Fenton-like oxidation for the tertiary treatment of dyeing wastewater. J Environ Chem Eng 9:105730. https://doi.org/10.1016/j.jece.2021.105730
22. Chung J, Lee M, Ahn J, Bae W, Lee YW, Shim H (2009) Effects of operational conditions on sludge degradation and organic acids formation in low-critical wet air oxidation. J Hazard Mater 162:10–16. https://doi.org/10.1016/j.jhazmat.2008.05.038
23. Çinar Ö, Yaşar S, Kertmen M, Demiröz K, Yigit NÖ, Kitis M (2008) Effect of cycle time on biodegradation of azo dye in sequencing batch reactor. Process Saf Environ Prot 86:455–460. https://doi.org/10.1016/j.psep.2008.03.001
24. Cruz DRS, de Jesus GK, Santos CA, Silva WR, Wisniewski A, Cunha GC, Romão LPC (2021) Magnetic nanostructured material as heterogeneous catalyst for degradation of AB210 dye in tannery wastewater by electro-Fenton process. Chemosphere 280. https://doi.org/10.1016/j.chemosphere.2021.130675
25. Daneshvar N, Salari D, Khataee AR (2004) Photocatalytic degradation of azo dye acid red 14 in water on ZnO as an alternative catalyst to TiO2. J Photochem Photobiol A Chem 162:317–322. https://doi.org/10.1016/S1010-6030(03)00378-2
26. De Clippel F, Dusselier M, Van De Vyver S, Peng L, Jacobs PA, Sels BF (2013) Tailoring nanohybrids and nanocomposites for catalytic applications. Green Chem 15:1398–1430. https://doi.org/10.1039/c3gc37141g
27. Duarte F, Maldonado-Hódar FJ, Madeira LM (2011) Influence of the characteristics of carbon materials on their behaviour as heterogeneous Fenton catalysts for the elimination of the azo dye Orange II from aqueous solutions. Appl Catal B Environ 103:109–115. https://doi.org/10.1016/j.apcatb.2011.01.016

28. Dükkanci M (2019) Heterogeneous sonocatalytic degradation of Bisphenol-A and the influence of the reaction parameters and ultrasonic frequency. Water Sci Technol 79:386–397. https://doi.org/10.2166/wst.2019.065
29. Eren Z, Ince NH (2010) Sonolytic and sonocatalytic degradation of azo dyes by low and high frequency ultrasound. J Hazard Mater 177:1019–1024. https://doi.org/10.1016/j.jhazmat.2010.01.021
30. Fatima S, Ali SI, Iqbal MZ (2020) Congo Red dye degradation by graphene nanoplatelets/doped bismuth ferrite nanoparticle. Catalysts
31. Fayazi M, Ghanei-Motlagh M (2020) Electrochemical mineralization of methylene blue dye using electro-Fenton oxidation catalyzed by a novel sepiolite/pyrite nanocomposite. Int J Environ Sci Technol 17:4541–4548. https://doi.org/10.1007/s13762-020-02749-2
32. Gągol M, Przyjazny A, Boczkaj G (2018) Wastewater treatment by means of advanced oxidation processes based on cavitation—a review. Chem Eng J 338:599–627. https://doi.org/10.1016/j.cej.2018.01.049
33. Ganiyu SO, Zhou M, Martínez-Huitle CA (2018) Heterogeneous electro-Fenton and photoelectro-Fenton processes: a critical review of fundamental principles and application for water/wastewater treatment. Appl Catal B Environ 235:103–129. https://doi.org/10.1016/j.apcatb.2018.04.044
34. Gautam P, Kumar S, Lokhandwala S (2019) Advanced oxidation processes for treatment of leachate from hazardous waste landfill: a critical review. J Clean Prod 237:117639. https://doi.org/10.1016/j.jclepro.2019.117639
35. Gita S, Shukla SP, Saharan N, Prakash C, Deshmukhe G (2019) Toxic effects of selected textile dyes on elemental composition, photosynthetic pigments, protein content and growth of a freshwater Chlorophycean alga Chlorella vulgaris. Bull Environ Contam Toxicol 102:795–801. https://doi.org/10.1007/s00128-019-02599-w
36. Gonçalves RGL, Lopes PA, Resende JA, Pinto FG, Tronto J, Guerreiro MC, de Oliveira LCA, de Castro Nunes W, Neto JL (2019) Performance of magnetite/layered double hydroxide composite for dye removal via adsorption, Fenton and photo-Fenton processes. Appl Clay Sci 179:105152. https://doi.org/10.1016/j.clay.2019.105152
37. González-González RB, Parra-Saldívar R, Alsanie WF, Iqbal HMN (2022) Nanohybrid catalysts with porous structures for environmental remediation through photocatalytic degradation of emerging pollutants. Environ Res 214. https://doi.org/10.1016/j.envres.2022.113955
38. Gopinath A, Pisharody L, Popat A, Nidheesh PV (2022) Supported catalysts for heterogeneous electro-Fenton processes: recent trends and future directions. Curr Opin Solid State Mater Sci 26:100981. https://doi.org/10.1016/j.cossms.2022.100981
39. Gregory P (1990) Classification of dyes by chemical structure. In: The chemistry and application of dyes. Springer US, Boston, MA, pp 17–47. https://doi.org/10.1007/978-1-4684-7715-3_2
40. Han C, Chen Z, Zhang N, Colmenares JC, Xu Y-J (2015) Hierarchically CdS decorated 1D ZnO nanorods-2D graphene hybrids: low temperature synthesis and enhanced photocatalytic performance. Adv Funct Mater 25:221–229. https://doi.org/10.1002/adfm.201402443
41. Han J, Jun BM, Heo J, Kim S, Yoon Y, Park CM (2019) Heterogeneous sonocatalytic degradation of an anionic dye in aqueous solution using a magnetic lanthanum dioxide carbonate-doped zinc ferrite-reduced graphene oxide nanostructure. Ecotoxicol Environ Saf 182:109396. https://doi.org/10.1016/j.ecoenv.2019.109396
42. Han J, Jun BM, Heo J, Lee G, Yoon Y, Park CM (2019) Highly efficient organic dye removal from waters by magnetically recoverable $La_2O_2CO_3/ZnFe_2O_4$-reduced graphene oxide nanohybrid. Ceram Int 45:19247–19256. https://doi.org/10.1016/j.ceramint.2019.06.173
43. Hasija V, Raizada P, Sudhaik A, Sharma K, Kumar A, Singh P, Jonnalagadda SB, Thakur VK (2019) Recent advances in noble metal free doped graphitic carbon nitride based nanohybrids for photocatalysis of organic contaminants in water: a review. Appl Mater Today 15:494–524. https://doi.org/10.1016/j.apmt.2019.04.003
44. He X, Qiu X, Hu C, Liu Y (2018) Treatment of heavy metal ions in wastewater using layered double hydroxides: a review. J Dispers Sci Technol 39:792–801. https://doi.org/10.1080/01932691.2017.1392318

45. Herney-Ramirez J, Vicente MA, Madeira LM (2010) Heterogeneous photo-Fenton oxidation with pillared clay-based catalysts for wastewater treatment: a review. Appl Catal B Environ 98:10–26. https://doi.org/10.1016/j.apcatb.2010.05.004
46. Hisaindee S, Meetani MA, Rauf MA (2013) Application of LC-MS to the analysis of advanced oxidation process (AOP) degradation of dye products and reaction mechanisms. TrAC—Trends Anal Chem 49:31–44. https://doi.org/10.1016/j.trac.2013.03.011
47. Ince NH (2018) Ultrasound-assisted advanced oxidation processes for water decontamination. Ultrason Sonochem 40:97–103. https://doi.org/10.1016/j.ultsonch.2017.04.009
48. Iqbal MA, Ali SI, Amin F, Tariq A, Iqbal MZ, Rizwan S (2019) La-and Mn-codoped bismuth ferrite/Ti_3 C 2 MXene composites for efficient photocatalytic degradation of Congo red dye. ACS Omega 4:8661–8668. https://doi.org/10.1021/acsomega.9b00493
49. Janani B, Syed A, Sruthi L, Sivaranjani PR, Elgorban AM, Bahkali AH, Zaghloul NSS, Badawy MM, Das A, Khan SS (2021) Visible light driven photocatalytic activity and efficient antibacterial activity of $ZnFe_2O_4$ decorated CdO nanohybrid heterostructures synthesized by ultrasonic-assisted method. Colloids Surfaces A Physicochem Eng Asp 628:127307. https://doi.org/10.1016/j.colsurfa.2021.127307
50. Javed H, Rehman A, Mussadiq S, Shahid M, Khan MA, Shakir I, Agboola PO, Aboud MFA, Warsi MF (2019) Reduced graphene oxide-spinel ferrite nano-hybrids as magnetically separable and recyclable visible light driven photocatalyst. Synth Met 254:1–9. https://doi.org/10.1016/j.synthmet.2019.05.013
51. Jiang Y, Ran J, Mao K, Yang X, Zhong L, Yang C, Feng X, Zhang H (2022) Recent progress in Fenton/Fenton-like reactions for the removal of antibiotics in aqueous environments. Ecotoxicol Environ Saf 236:113464. https://doi.org/10.1016/j.ecoenv.2022.113464
52. Kalita J, Das B, Dhar SS (2022) Synergistic effect of iron and copper in hydroxyapatite nanorods for Fenton-like oxidation of organic dye. Colloids Surfaces A Physicochem Eng Asp 643:128750. https://doi.org/10.1016/j.colsurfa.2022.128750
53. Keav S, Martin A, Barbier J, Duprez D (2010) Deactivation and reactivation of noble metal catalysts tested in the catalytic wet air oxidation of phenol. Catal Today 151:143–147. https://doi.org/10.1016/j.cattod.2010.01.025
54. Kefeni KK, Mamba BB, Msagati TAM (2017) Application of spinel ferrite nanoparticles in water and wastewater treatment: a review. Sep Purif Technol 188:399–422. https://doi.org/10.1016/j.seppur.2017.07.015
55. Khamesan A, Esfahani MM, Ghasemi JB, Farzin F, Parsaei-Khomami A, Mousavi M (2022) Graphitic-C3N4/ZnCr-layered double hydroxide 2D/2D nanosheet heterojunction: mesoporous photocatalyst for advanced oxidation of azo dyes with in situ produced H_2O_2. Adv Powder Technol 33:103777. https://doi.org/10.1016/j.apt.2022.103777
56. Koyyada G, Goud BS, Devarayapalli KC, Shim J, Vattikuti SVP, Kim JH (2022) $BiFeO_3$/Fe_2O_3 electrode for photoelectrochemical water oxidation and photocatalytic dye degradation: a single step synthetic approach. Chemosphere 303:135071. https://doi.org/10.1016/j.chemosphere.2022.135071
57. Kumarage S, Munaweera I, Kottegoda N (2022) Contemporary, multidisciplinary roles of mesoporous silica nanohybrids/nanocomposites. ChemistrySelect 7. https://doi.org/10.1002/slct.202200574
58. Kusiak-Nejman E, Morawski AW (2019) TiO_2/graphene-based nanocomposites for water treatment: a brief overview of charge carrier transfer, antimicrobial and photocatalytic performance. Appl Catal B Environ 253:179–186. https://doi.org/10.1016/j.apcatb.2019.04.055
59. Lee YN, Lago RM, Fierro JLG, González J (2001) Hydrogen peroxide decomposition over $Ln_{1-x}A_xMnO_3$ (Ln = La or Nd and A = K or Sr) perovskites. Appl Catal A Gen 215:245–256. https://doi.org/10.1016/S0926-860X(01)00536-1
60. Li L, Pan L, Zhang D, Rong J (2022) Ultrasonic-assisted synthesis of $LaFeO_3$/CeO_2 heterojunction for enhancing the photocatalytic degradation of organic pollutants. Mater Sci Semicond Process 152. https://doi.org/10.1016/j.mssp.2022.107058

61. Li T, Guthrie JT (2010) Colour removal from aqueous solutions of the reactive azo dye Remazol Black B using the immobilised cells (Shewanella Strain J18 143)–cellulose-g.co-monomer system. J Water Resour Prot 02:77–84. https://doi.org/10.4236/jwarp.2010.21009
62. Li Y, Zhang XL, Liu D (2021) Recent developments of perylene diimide (PDI) supramolecular photocatalysts: a review. J Photochem Photobiol C Photochem Rev 48. https://doi.org/10.1016/j.jphotochemrev.2021.100436
63. Liu L, Chen Z, Zhang J, Shan D, Wu Y, Bai L, Wang B (2021) Treatment of industrial dye wastewater and pharmaceutical residue wastewater by advanced oxidation processes and its combination with nanocatalysts: a review. J. Water Process Eng. 42:102122. https://doi.org/10.1016/j.jwpe.2021.102122
64. Ma D, Yi H, Lai C, Liu X, Huo X, An Z, Li L, Fu Y, Li B, Zhang M, Qin L, Liu S, Yang L (2021) Critical review of advanced oxidation processes in organic wastewater treatment. Chemosphere 275:130104. https://doi.org/10.1016/j.chemosphere.2021.130104
65. Mardiroosi A, Mahjoub AR, Fakhri H, Boukherroub R (2021) Design and fabrication of a perylene dimiide functionalized g-C_3N_4@UiO-66 supramolecular photocatalyst: insight into enhancing the photocatalytic performance. J Mol Struct 1246:131244. https://doi.org/10.1016/j.molstruc.2021.131244
66. Meijide J, Dunlop PSM, Pazos M, Sanromán MA (2021) Heterogeneous electro-Fenton as "Green" technology for pharmaceutical removal: a review. Catalysts 11:1–22. https://doi.org/10.3390/catal11010085
67. Mohamed MM, Ghanem MA, Khairy M, Naguib E, Alotaibi NH (2019) Zinc oxide incorporated carbon nanotubes or graphene oxide nanohybrids for enhanced sonophotocatalytic degradation of methylene blue dye. Appl Surf Sci 487:539–549. https://doi.org/10.1016/j.apsusc.2019.05.135
68. Mohammad A, Ahmad K, Qureshi A, Tauqeer M, Mobin SM (2018) Zinc oxide-graphitic carbon nitride nanohybrid as an efficient electrochemical sensor and photocatalyst. Sens Actuators, B Chem 277:467–476. https://doi.org/10.1016/j.snb.2018.07.086
69. Mohan S, Sivakumar B, Kulangara RV, Subramanian B (2016) Visible light driven photocatalytic efficiency of rGO-Ag-$BiFeO_3$ ternary nanohybrids on the decontamination of dye-polluted water: an amalgamation of 1D, 2D and 3D systems. ChemistrySelect 1:6961–6971. https://doi.org/10.1002/slct.201601275
70. Moradi M, Elahinia A, Vasseghian Y, Dragoi EN, Omidi F, Mousavi Khaneghah A (2020) A review on pollutants removal by Sono-photo-Fenton processes. J Environ Chem Eng 8:104330. https://doi.org/10.1016/j.jece.2020.104330
71. Mostafaloo R, Mahmoudian MH, Asadi-Ghalhari M (2019) $BiFeO_3$/Magnetic nanocomposites for the photocatalytic degradation of cefixime from aqueous solutions under visible light. J Photochem Photobiol A Chem 382:111926. https://doi.org/10.1016/j.jphotochem.2019.111926
72. Motola M, Dworniczek E, Satrapinskyy L, Chodaczek G, Grzesiak J, Gregor M, Plecenik T, Nowicka J, Plesch G (2019) UV light-induced photocatalytic, antimicrobial, and antibiofilm performance of anodic TiO_2 nanotube layers prepared on titanium mesh and Ti sputtered on silicon. Chem Pap 73:1163–1172. https://doi.org/10.1007/s11696-018-0667-4
73. Natarajan S, Bajaj HC, Tayade RJ (2018) Recent advances based on the synergetic effect of adsorption for removal of dyes from waste water using photocatalytic process. J Environ Sci (China) 65:201–222. https://doi.org/10.1016/j.jes.2017.03.011
74. Nidheesh PV, Gandhimathi R (2012) Trends in electro-Fenton process for water and wastewater treatment: an overview. Desalination 299:1–15. https://doi.org/10.1016/j.desal.2012.05.011
75. Nirumand L, Farhadi S, Zabardasti A, Khataee A (2018) Synthesis and sonocatalytic performance of a ternary magnetic MIL-101(Cr)/RGO/$ZnFe_2O_4$ nanocomposite for degradation of dye pollutants. Ultrason Sonochem 42:647–658. https://doi.org/10.1016/j.ultsonch.2017.12.033
76. O'Shea KE, Dionysiou DD (2012) Advanced oxidation processes for water treatment. J Phys Chem Lett 3:2112–2113. https://doi.org/10.1021/jz300929x

77. Palanisamy G, Bhuvaneswari K, Bharathi G, Pazhanivel T, Grace AN, Pasha SKK (2021) Construction of magnetically recoverable ZnS–WO_3–$CoFe_2O_4$ nanohybrid enriched photocatalyst for the degradation of MB dye under visible light irradiation. Chemosphere 273:129687. https://doi.org/10.1016/j.chemosphere.2021.129687
78. Park CM, Kim YM, Kim KH, Wang D, Su C, Yoon Y (2019) Potential utility of graphene-based nano spinel ferrites as adsorbent and photocatalyst for removing organic/inorganic contaminants from aqueous solutions: a mini review. Chemosphere 221:392–402. https://doi.org/10.1016/j.chemosphere.2019.01.063
79. Park DH, Hwang SJ, Oh JM, Yang JH, Choy JH (2013) Polymer-inorganic supramolecular nanohybrids for red, white, green, and blue applications. Prog Polym Sci 38:1442–1486. https://doi.org/10.1016/j.progpolymsci.2013.05.007
80. Patel H (2018) Charcoal as an adsorbent for textile wastewater treatment. Sep Sci Technol 53:2797–2812. https://doi.org/10.1080/01496395.2018.1473880
81. Patel KD, Singh RK, Kim HW (2019) Carbon-based nanomaterials as an emerging platform for theranostics. Mater Horizons 6:434–469. https://doi.org/10.1039/c8mh00966j
82. Poyatos JM, Muñio MM, Almecija MC, Torres JC, Hontoria E, Osorio F (2010) Advanced oxidation processes for wastewater treatment: state of the art. Water Air Soil Pollut 205:187–204. https://doi.org/10.1007/s11270-009-0065-1
83. Rahim Pouran S, Abdul Aziz AR, Wan Daud WMA (2015) Review on the main advances in photo-Fenton oxidation system for recalcitrant wastewaters. J Ind Eng Chem 21:53–69. https://doi.org/10.1016/j.jiec.2014.05.005
84. Rauf MA, Meetani MA, Hisaindee S (2011) An overview on the photocatalytic degradation of azo dyes in the presence of TiO_2 doped with selective transition metals. Desalination 276:13–27. https://doi.org/10.1016/j.desal.2011.03.071
85. Reddy DHK, Yun YS (2016) Spinel ferrite magnetic adsorbents: alternative future materials for water purification? Coord Chem Rev 315:90–111. https://doi.org/10.1016/j.ccr.2016.01.012
86. Rehman R, Raza A, Yasmeen F, Dar A, Al-thagafi ZT, Meraf Z (2022) Recent literature review of significance of polypyrrole and its biocomposites in adsorption of dyes from aqueous solution. Adsorpt Sci Technol 2022:1–18. https://doi.org/10.1155/2022/7047832
87. Rodríguez Couto S (2009) Dye removal by immobilised fungi. Biotechnol Adv 27:227–235. https://doi.org/10.1016/j.biotechadv.2008.12.001
88. Rudakiya DM, Tripathi A, Gupte S, Gupte A (2019) Fungal bioremediation: a step towards cleaner environment. In: Advancing frontiers in mycology & mycotechnology. Springer Singapore, Singapore, pp 229–249. https://doi.org/10.1007/978-981-13-9349-5_9
89. Saemian T, Hossaini Sadr M, Tavakkoli Yaraki M, Gharagozlou M, Soltani B (2022) Synthesis and characterization of $CoFe_2O_4/SiO_2$/Cu-MOF for degradation of methylene blue through catalytic sono-Fenton-like reaction. Inorg Chem Commun 138:109305. https://doi.org/10.1016/j.inoche.2022.109305
90. Samran B, lunput S, Tonnonchiang S, Chaiwichian S (2019) $BiFeO_3/BiVO_4$ nanocomposite photocatalysts with highly enhanced photocatalytic activity for rhodamine B degradation under visible light irradiation. Phys B Condens Matter 561:23–28. https://doi.org/10.1016/j.physb.2019.02.049
91. Sardar M, Manna M, Maharana M, Sen S (2021) Remediation of dyes from industrial wastewater using low-cost adsorbents, pp 377–403. https://doi.org/10.1007/978-3-030-47400-3_15
92. Selim A, Kaur S, Dar AH, Sartaliya S, Jayamurugan G (2020) Synergistic effects of carbon dots and palladium nanoparticles enhance the sonocatalytic performance for rhodamine B degradation in the absence of light. ACS Omega 5:22603–22613. https://doi.org/10.1021/acsomega.0c03312
93. Selvaraj V, Swarna Karthika T, Mansiya C, Alagar M (2021) An over review on recently developed techniques, mechanisms and intermediate involved in the advanced azo dye degradation for industrial applications. J Mol Struct 1224:129195. https://doi.org/10.1016/j.molstruc.2020.129195

94. Shakir I, Agboola PO, Haider S (2021) Manganese spinel ferrite-reduced graphene oxides nanocomposites for enhanced solar irradiated catalytic studies. Ceram Int 47:28367–28376. https://doi.org/10.1016/j.ceramint.2021.06.254
95. Shangguan Z, Yuan X, Jiang L, Zhao Y, Qin L, Zhou X, Wu Y, Chew JW, Wang H (2022) Zeolite-based Fenton-like catalysis for pollutant removal and reclamation from wastewater. Chinese Chem Lett 33:4719–4731. https://doi.org/10.1016/j.cclet.2022.01.001
96. Shen J, Huang W, Li N, Ye M (2015) Highly efficient degradation of dyes by reduced graphene oxide-ZnCdS supramolecular photocatalyst under visible light. Ceram Int 41:761–767. https://doi.org/10.1016/j.ceramint.2014.08.135
97. Siadatnasab F, Farhadi S, Khataee A (2018) Sonocatalytic performance of magnetically separable CuS/$CoFe_2O_4$ nanohybrid for efficient degradation of organic dyes. Ultrason Sonochem 44:359–367. https://doi.org/10.1016/j.ultsonch.2018.02.051
98. Song S, Wang Y, Shen H, Zhang J, Mo H, Xie J, Zhou N, Shen J (2019) Ultrasmall graphene oxide modified with Fe_3O_4 nanoparticles as a Fenton-like agent for methylene blue degradation. ACS Appl Nano Mater 2:7074–7084. https://doi.org/10.1021/acsanm.9b01608
99. Sun M, Liu HH, Zhang Y, Zhai LF (2020) Degradation of bisphenol A by electrocatalytic wet air oxidation process: kinetic modeling, degradation pathway and performance assessment. Chem Eng J 387:124124. https://doi.org/10.1016/j.cej.2020.124124
100. Tamaki Y, Ishitani O (2017) Supramolecular photocatalysts for the reduction of CO_2. ACS Catal 7:3394–3409. https://doi.org/10.1021/acscatal.7b00440
101. Tariq A, Ali SI, Akinwande D, Rizwan S (2018) Efficient visible-light photocatalysis of 2D-MXene nanohybrids with Gd3+- and Sn4+-codoped bismuth ferrite. ACS Omega 3:13828–13836. https://doi.org/10.1021/acsomega.8b01951
102. Vasseghian Y, Dragoi EN, Almomani F, Le VT (2022) A comprehensive review on MXenes as new nanomaterials for degradation of hazardous pollutants: deployment as heterogeneous sonocatalysis. Chemosphere 287:132387. https://doi.org/10.1016/j.chemosphere.2021.132387
103. Vaya D, Kaushik B, Surolia PK (2022) Recent advances in graphitic carbon nitride semiconductor: structure, synthesis and applications. Mater Sci Semicond Process 137:106181. https://doi.org/10.1016/j.mssp.2021.106181
104. Wang B, Bozal-Ginesta C, Zhang RQ, Zhou B, Ma HX, Jiao LY, Xu L, Liu EZ, Wang C, Li Z (2021) A supramolecular H12SubPcB-OPhCOPh/TiO_2 Z-scheme hybrid assembled via dimeric concave-ligand π-interaction for visible photocatalytic oxidation of tetracycline. Appl Catal B Environ 298:120550. https://doi.org/10.1016/j.apcatb.2021.120550
105. Wang J, Lv Y, Zhang L, Liu B, Jiang R, Han G, Xu R, Zhang X (2010) Sonocatalytic degradation of organic dyes and comparison of catalytic activities of CeO_2/TiO_2, SnO_2/TiO_2 and ZrO_2/TiO_2 composites under ultrasonic irradiation. Ultrason Sonochem 17:642–648. https://doi.org/10.1016/j.ultsonch.2009.12.016
106. Wang P, Liang YN, Zhong Z, Hu X (2020) Nano-hybrid bimetallic Au-Pd catalysts for ambient condition-catalytic wet air oxidation (AC-CWAO) of organic dyes. Sep Purif Technol 233:115960. https://doi.org/10.1016/j.seppur.2019.115960
107. Waranusantigul P, Pokethitiyook P, Kruatrachue M, Upatham ES (2003) Kinetics of basic dye (methylene blue) biosorption by giant duckweed (Spirodela polyrrhiza). Environ Pollut 125:385–392. https://doi.org/10.1016/S0269-7491(03)00107-6
108. Wood RJ, Lee J, Bussemaker MJ (2017) A parametric review of sonochemistry: control and augmentation of sonochemical activity in aqueous solutions. Ultrason Sonochem 38:351–370. https://doi.org/10.1016/j.ultsonch.2017.03.030
109. Xiao J, Lai J, Li R, Fang X, Zhang D, Tsiakaras P, Wang Y (2020) Enhanced ultrasonic-assisted heterogeneous Fenton degradation of organic pollutants over a new copper magnetite (Cu-Fe_3O_4/Cu/C) nanohybrid catalyst. Ind Eng Chem Res 59:12431–12440. https://doi.org/10.1021/acs.iecr.0c01613
110. Xie K, Fang J, Li L, Deng J, Chen F (2022) Progress of graphite carbon nitride with different dimensions in the photocatalytic degradation of dyes: a review. J Alloys Compd 901. https://doi.org/10.1016/j.jallcom.2021.163589

111. Xin L, Hu J, Wu X, Huang K, Huang X (2022) Fenton-like degradation of carmine dyes based on artificial intelligence modeling and optimization of reduced graphene oxide loaded iron-cobalt-nickel trimetallic nanocomposites. Mater Today Commun 31:103463. https://doi.org/10.1016/j.mtcomm.2022.103463
112. Xu J, Liu C, Niu J, Zhu Y, Zang B, Xie M, Chen M (2020) Synthesis of $LaFeO_3/Bi_3NbO_7$ p-n heterojunction photocatalysts with enhanced visible-light-responsive activity for photocatalytic reduction of Cr(VI). J Alloys Compd 815:152492. https://doi.org/10.1016/j.jallcom.2019.152492
113. Yan J, Chen Y, Qian L, Gao W, Ouyang D, Chen M (2017) Heterogeneously catalyzed persulfate with a CuMgFe layered double hydroxide for the degradation of ethylbenzene. J Hazard Mater 338:372–380. https://doi.org/10.1016/j.jhazmat.2017.05.007
114. Yang R, Peng Q, Yu B, Shen Y, Cong H (2021) Yolk-shell Fe_3O_4@MOF-5 nanocomposites as a heterogeneous Fenton-like catalyst for organic dye removal. Sep Purif Technol 267:118620. https://doi.org/10.1016/j.seppur.2021.118620
115. Yang S, Liu Z, Huang X, Zhang B (2010) Wet air oxidation of epoxy acrylate monomer industrial wastewater. J Hazard Mater 178:786–791. https://doi.org/10.1016/j.jhazmat.2010.02.009
116. Yu X, Lin D, Li P, Su Z (2017) Recent advances in the synthesis and energy applications of TiO_2-graphene nanohybrids. Sol Energy Mater Sol Cells 172:252–269. https://doi.org/10.1016/j.solmat.2017.07.045
117. Zedan M, Zedan AF, Amin RM, Li X (2022) Visible-light active metal nanoparticles@carbon nitride for enhanced removal of water organic pollutants. J Environ Chem Eng 10. https://doi.org/10.1016/j.jece.2022.107780
118. Zhang G, Zhou Y, Yang F (2015) FeOOH-catalyzed heterogeneous electro-fenton system upon Anthraquinone@Graphene nanohybrid cathode in a divided electrolytic cell: catholyte-regulated catalytic oxidation performance and mechanism. J Electrochem Soc 162:H357–H365. https://doi.org/10.1149/2.0691506jes
119. Zhang W, Ye B, Zhong Z, Jiang Y, Zhou R, Liu Z, Hou Z (2022) Catalytic wet air oxidation of toxic containments over highly dispersed Cu(II)/Cu(I)-N species in the framework of g-C3N4. J Hazard Mater 424. https://doi.org/10.1016/j.jhazmat.2021.127679
120. Zhu B, Li X, Lang J, Wang Y, Yang J (2022) Construction of novel metal-free graphene oxide/graphitic carbon nitride nanohybrids on carbon cloth and efficient degradation of RhB under visible light irradiation. Appl Surf Sci 599:154019. https://doi.org/10.1016/j.apsusc.2022.154019

Dye Removal Using Magnetized Nanohybrid Adsorbent

Akansha Mehta

Abstract Water is one of our prized resources, which is essential for all living beings. Water is tied to so many things: food, energy, economy, politics, and so many other things. However, more than one third of renewable freshwater is utilized in industrial, agricultural, and domestic purposes which leads the water to get contaminated with several geogenic and synthetic contaminants like, dyes, pesticides, fertilizers, heavy metals, radionuclides, etc. The contamination through dye sector and the sectors related to dye application (tannery, paper, textile) are the most recognizable among the polluting industries based on composition of effluents. The dye removal through adsorption is the most attractive technique predominantly because of cheaper adsorbent, easy regeneration and the technique doesn't require any pretreatment step before its application. Currently, magnetic nanomaterials are proven to be right expedient for the removal of organic and inorganic dyes from the wastewater, restricted with few drawbacks when used alone. Consequently, developing magnetic nanohybrids can recuperate the efficiency and feasibility of magnetic nanomaterials. In this regard, magnetic nanohybrid will greatly possess both the properties as nanomaterial as well as magnetism. The content of this book chapter is to provide a comprehensive overview of application of magnetic nanohybrid in the adsorption of dyes.

Keywords Magnetic nanohybrids · Adsorbent · Dye removal · Recovery

---

A. Mehta (✉)
Centre for Functional and Surface Functionalized Glass, Alexander Dubček University of Trenčín, Trenčín, Slovakia
e-mail: akansha.akansha@tnuni.sk

© The Author(s), under exclusive license to Springer Nature Singapore Pte Ltd. 2023

A. Ahmad et al. (eds.), *Nanohybrid Materials for Treatment of Textiles Dyes*,
Smart Nanomaterials Technology, https://doi.org/10.1007/978-981-99-3901-5_16

1 General Introduction

1.1 Water Crisis

Clean water is mandatory for life. When water gets polluted, it becomes unhealthy for everyone. Water pollution has become an increasing problem on our earth, which is affecting human and animal life [1, 2]. Water pollution occurs when harmful chemicals, toxic waste, and other particles enters water bodies such as rivers, ponds, sea, oceans, and so on gets dissolved gets dissolved in them, lies suspended in the water, or get deposited on the bed. This results in the degradation of the quality of the water. Not only these harms aquatic ecosystems but the pollutants also reach ground water, which ends up in the public as contaminated water, which we include in our daily activity including drinking [3, 4]. But the most crucial question is, what causes water pollution? Well sometime water pollution can occur due to natural activity such as volcano eruptions, animal waste, algae blooms, and residue from storms and floods. Everyday sewage and sometimes even garbage from cities are dumped into oceans resulting in polluting the water terrifically. Other factor which leads to water pollutions are from oil spills, fossil fuel combustion, chemical fertilizers, industrial waste, and pesticides. Once this water gets consumed by human beings, it can lead to diseases such as hepatitis that can be fatal. Also, in many poor nations there is always an outbreak of cholera and other infections due to contaminated water [5, 6].

1.2 Dye Types and Toxicity

Dyes are usually synthetic fabrics that don't breakdown in aqueous media, but the ones that breaks create harmful substances like acids and other compounds. There are various types of dyes, which can be classified in terms of their method of application to the substrate [7]. Chromophore and auxochromes are two key modules of dye molecule, the chromophore is responsible for the colour of the dye and auxochrome is the supplement of chromophore which renders the molecule solubility in the water and provides the affinity towards the fibre. In few cases the dyes are classified according to their solubility in aqueous media, like acid, azo, basic, and direct dyes are soluble in water, moreover disperse, mordant and sulphur dyes are insoluble in water. The dyes show very intense colour in water even at minute concentration, which can decrease the oxygen solubility and transparency of the water [8, 9].

Predictably dyes contain toxic functional groups like, amines, aromatic groups, toxic metal ions like cadmium, lead, copper, zinc, cobalt, etc. Due to the presence of toxic metal ions and organic groups, dyes are recognized as very harmful towards human and aquatic life. Dyes containing water has severe effects on human health especially on human brain, reproductive system, DNA, liver, kidney, nerve system,

Table 1 Different types of synthetic dyes and their toxicity

Dye type	Toxicity	Solubility in water	Application	Example
Reactive dye	Allergic conjunctivitis, rhinitis, occupational asthma or other allergic reactions	Soluble	Wool, silk, nylon	Reactive red
Acid dye	Acute oral toxicity and neurotoxicity	Water soluble	Wool, cosmetics, leather	Acid blue 78
Azo dye	Clastogenicity, hypersensitivity, environmental toxicity, estrogenicity, genotoxicity, photoinduced toxicity, mutagenicity, and acute oral toxicity	Water soluble	Silk, cotton, nylon	Aniline yellow
Basic dye	Microbial toxicity, mutagenicity, hematotoxicity, nucleic acid damage, teratogenicity, photodynamic toxicity, reproductive toxicity, and effluent toxicity	Water soluble	Ink, medicines	Methylene blue
Disperse dye	Bacterial toxicity, yeast toxicity, algal toxicity, protozoan toxicity, genotoxicity, microbial toxicity, carcinogenicity, cytotoxicity, mutagenicity	Insoluble in water	Polyamide, nylon, plastics	Disperse red 60
Direct dye	Bacterial toxicity, yeast toxicity, algal toxicity, protozoan toxicity, genotoxicity, microbial toxicity, carcinogenicity, cytotoxicity, mutagenicity	Water soluble	Leather, paper, cotton	Direct yellow 11
Mordant dye	Mutagenicity, carcinogenicity, and genotoxicity	Insoluble	Anodised aluminium, wool	Mordant red 11
Sulphur dye	DNA damage, mutagenicity, cytotoxicity, carcinogenicity, and genotoxicity	Insoluble	Cotton, paper, silk	Thiazine

and skin allergies. Table 1 illustrates the usual classification of dyes along with their toxicity, application, and solubility in water [10, 11].

1.3 Dye Removal Techniques

Noticing the adverse toxic effects of dyes in water (In Sect. 1.2), its essential to remove dyes from wastewater to protect human and aquatic life. There are several techniques to remove dye containing effluents from wastewater. The technologies can be divided into three method categories: physical, chemical, biological as depicted in Fig. 1a and every method has their advantages and disadvantages in terms of environmental affability, design, and separation efficiency [12, 13]. As seen from Fig. 1b, there is rapid progress in the number of articles published on dye removal, more than 3689 articles were published from 2016 to 2020 [10]. However, it is difficult to retain on single technique that prevail the dye effluent issue.

The research has been focused on the methods that reduce the cost of treatment and remove the maximum percentage of pollutants. Among all the treatments depicted in the physical processes are more frequent [14]. The chemical methods are restricted because of their adverse side effects on environment. The biological methods are very simple and ecological in compare with physical methods and usually various microorganisms are used for the decolouration of dyes. Conversely, the biological methods require strict environment conditions like; pH, temperature, and nutrition, also the biological reactor needs space requirement and time, which increases the complexity of the treatment technique [15]. Now, coming back to the physical methods, it has two common sub-techniques; one adsorption and other is membrane separation. The membrane separation is very efficient and includes, nano, micro, and ultrafiltration,

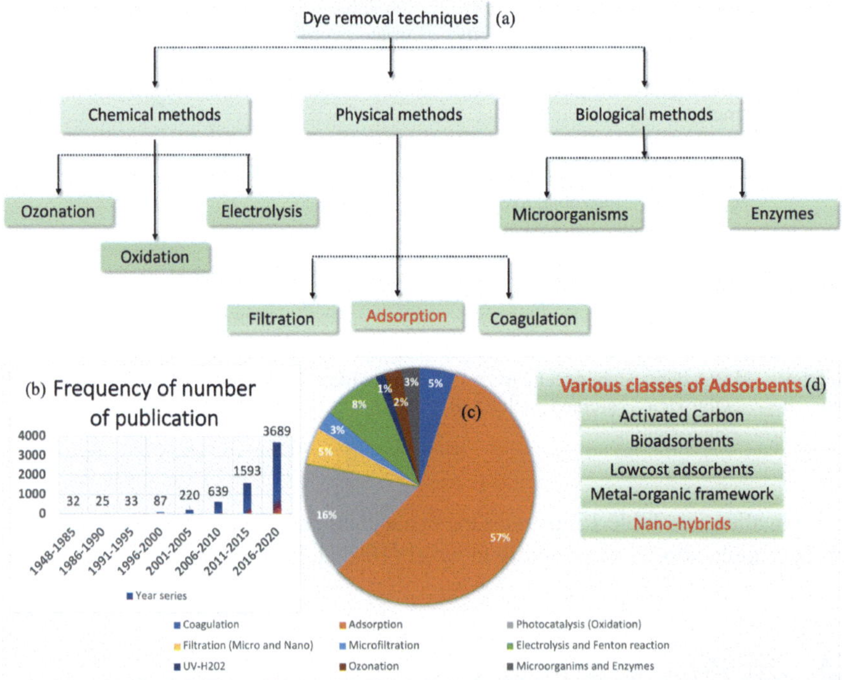

Fig. 1 **a** Dye removal treatment techniques. **b** Recent record on frequency of publication year wise. **c** Pie chart of literature available of different dye removal techniques. **d** Various classes of adsorbents

but its usage in dye removal is not so common. As, the services life of membranes is very limited, and it is easy to get polluted. Predicting through the literature the adsorption is the most idealist method of removing dyes due to cost/time effectiveness as highlighted in Fig. 1c through pie chart [16]. In this book chapter the adsorption technique will be discussed further.

1.4 Adsorption of Dyes

Adsorption is a surface phenomenon and can be classified as chemisorption or physisorption, depending on the adsorbed species adsorbed on the adsorbent surface. Probably, in the fifteenth century we gather the enhance understanding of water adsorption in the book entitled 'L'urchitetturu' from one Italian architect and painter Leon Battistata Alberti. Then, 300 years later, Scheele in around 1773 started to investigate profoundly about adsorption phenomenon. The first article tried to separate dye molecules with adsorption method was published by Chapman and Siebold in 1912, however the study was more analytical with limited technology. Furthermore, for around 10 years, the study was continued to remove dyes from aqueous solution with the sodium nitrate, barium nitrate, and lead nitrate as material. Later, grounded by the previous study, Gibby and Argument investigated the adsorption of methylene blue, Bordeaux Extra Congo Red, Solway Ultra Blue Indigo and Carmine using as adsorbent materials with mercury interfaces, also the adsorption percentage was calculated with Gibbs equation. Later, in half nineteenth century, Ewing and Liu investigated the adsorption of orange II and crystal violet by TiO_2 (anatase and rutile) and zinc oxide. Through the investigation, almost after nine days the adsorption equilibrium was found, although high temperature can hasten the reaction equilibrium. Following, Haldeman and Emmett studied the adsorption of alkyl orange dyes on silica gel fabricated during the adsorption of dyes. Later, Prasad and Dey prepared an additional thermodynamic analysis for dye adsorption, estimating the heats of adsorption of Fuchsine and Congo Red with several trials of hydrous thorium oxide. During the period of 1971–2000, many changes were released regarding the adsorption of dyes from aqueous solution in regard with the development of new materials and investigation with several parameters like; temperature and pH, adsorption capacity, and kinetics pf the reaction. Sethuraman and Raymahashay investigated the kinetics adsorption study of catonic and anionic dyes with montmorillonite and kaolinite clays [17, 18].

Recently, in twenty-first century, almost all the adsorption materials were introduced and studied theoretically by the researchers. In recent years, many adsorbents such as activated carbon, natural silica, synthetic silica, zeolites, biomass, cellulose, biochar, nanomaterials, clays, chitosan, xerogel, and polymers have been used for dye removal as depicted in Fig. 1d. Generally, few shortcomings like; low selectivity, low adsorption capacity, and poor reusability are typically associated with the adsorbents [19, 20]. Thus, exploring additional adsorbents with high reusability and higher adsorption capacity is extremely desired for the dye removal from wastewater for environmental protection.

2 Magnetic Nanohybrids

2.1 Emerging Synthesis Techniques

In general, nanoparticles own a series of exceptional chemical and physical properties. Metal oxide, metals, ceramics, carbon derivatives, and polymers at nanoscale have relatively higher ratio of surface area to particle size and demonstrate different electrical, magnetic, and optical properties in compared to that of macroscopic particles. Subsequently, the combination of two or more nanomaterials, which is denoted as hybrid nanomaterials, displays multi-functionalities found to be very effectual to remove dyes from wastewater [21].

Several synthesis methods have been reported for preparation of magnetic sorbents in which the magnetization process of the sorbents including physical attachment (blending and ultrasonic), hydrothermal/solvothermal procedure, co-precipitation, microemulsions, and hydrothermal method, self-assembly combustion reaction, emulsion evaporation, and sonochemical route. These synthetic approaches are categorized into three different methods, i.e. physical methods, chemical methods, and biological methods and are explained well in Table 2 [22–24].

2.2 Functionalized Magnetic Nanohybrids

The adsorption capacity of the any source of sample is highly dependent of the physical properties like surface area, porosity, and surface charge. As explained in above section the synthesis methods greatly help in improving all these physical properties of the material. Merely, the desired characteristics can be improved up to certain limit. Henceforth to further improve the efficiency of the conventional materials functionalization step is needed [25, 26]. Functionalization can bring about significant morphological and chemical improvements in an adsorbent especially MNHs. Morphological enhancements include more porosity, greater surfaces, more profuse functional sites 'ready' for housing interactions, also extra favourable porosity distributions. Also, bare MNPs shows a lot of agglomeration during the adsorption reaction which might decrease the reusable efficiency of the material. Grafting of functional groups like hydroxyl, carboxyl, or amide groups onto the virgin material are exemplars of such chemical improvements [21]. Numerous materials like surfactants, carbon nanotubes (CNTs), graphene/graphene oxide (G/GO), layered double hydroxide (LDHs), metal organic frameworks (MOFs), molecularly imprinted polymers (MIPs), restricted access materials (RAMs), and covalent organic frameworks

Table 2 Main synthesis method of magnetic nanohybrids with advantages and limitations

Synthesis technique	Magsorbents examples	Advantages	Limitations
Physical methods			
Mechanical milling	Nanocrystalline magnesium ferrite (α-Fe_2O_3 and MgO)	Simple and inexpensive	Contamination of the product and particles have wide size distribution
Electron beam lithography	Nanosized iron oxide NPs	Low cost-effective and it do not require any expensive chemical or produce any hazardous waste as in wet chemistry methods	The NPs produced through this method are not monodispersed
Electrical explosion of wires	1D iron oxide nanostructure	This method is a green technique considering that no waste is generated	The metal particles produced through this method are very pure restricted to the initial purity of metal wire
Chemical methods			
Co-precipitation synthesis	Manganese ferrite ($MnFe_2O_4$) were formed by using ferric chloride ($FeCl_3$) and manganese (II) chloride ($MnCl_2$)	Includes the use of less harmful materials and procedures	Difficult to control the shape of nanohybrids via co-precipitation
Hydrothermal synthesis/ Solvothermal reaction	Chitosan-coated Fe_3O_4 NPs	Crystal formation of nanoparticles	High pressure and temperature
Thermal decomposition	Monodispersed iron oxide magnetic nanoparticles by polymer-catalysed decomposition of $Fe(CO)_5$	Magnetic nanohybrids prepared by this method have high crystallinity, controlled size, and well-defined shape	Production of toxic organic-soluble solvents, which limits its application in the biomedical field
Sol–gel method	Silica-coated MNPs	Controlling the composition, shape, and size of magnetic nanohybrids	Require prolong reaction time and involve toxic organic solvents
Biological methods			
Microbes	Magnetotactic bacteria	Efficiency, eco-friendly, and clean process	Mechanism of formation of NPs by using microorganisms and plants is not well understood and still under investigation

(continued)

Table 2 (continued)

Plants	Fe_3O_4 nanoparticles by plantain peel extract	Efficiency, eco-friendly, and clean process	Less yield and NPs dispersion still need to be investigated
Templates (membranes, DNA, etc.)	Gold coated magnetic nanoparticles	Comparatively biocompatible	

(COFs) are reported mainly for the functionalization of MNPs to fabricate distinctive MNHs and some examples are mentioned in Table 3.

In addition to the synthesis of MNPs. Huge development has also been made in the fabrication of quantum dots, nanosilicas and gold NPs making it conceivable to integrate diverse functional NPs into one single nanoentity [25]. Additionally, the schematic structure of these magnetic sorbents is shown in Fig. 2.

Though, functionalization of Magsorbents progresses the separation efficiency of the sorbents and also upturns their advantages, still more elaborations are required to improve their dispersion and hydrophobicity.

2.3 Sustainable Magnetic Nanohybrid Adsorbents

Green adsorbents would indeed better meet the objective of the future circular economy where there will be no 'waste'. This book presents advanced methods and adsorbents for the removal of metals and dyes. Adsorbents include carbon nanostructures, biomass, cellulose, polymers, clay, composites, and chelating materials.

Research has recently focused on the development of greener adsorbents, which are both cheap and renewable, to replace classical, fossil fuel-derived adsorbents [45, 46]. Several types of biomaterials, such as guava leaf, tea waste, ground coffee, orange peel, peanut shell, and saw dust have been investigated for the removal of various pollutants including pharmaceuticals, dyes from wastewater. The agricultural waste products are great option and have attracted many researchers because of their large abundance and low cost. Natural biopolymers are important class of adsorbents that are also highly efficient and budget-friendly but displays an disadvantage of their separation from aqueous medium [47]. Some of the important MNHs fabricated from greener ways are explained in Table 4 with examples.

Table 3 Functionalized magnetic nanohybrids as adsorbents for dye removal

Adsorbent	Dye type	Adsorptive capacity/ Efficiency	Adsorption kinetics and mechanism	References
Iron–carbon hybrid magnetic nanosheets	MB and CR	185.2 and 531.9 mg g^{-1}, respectively	Pseudo-second-order kinetics; Langmuir isotherm; chemisorption	[28]
Magnetic multi-wall carbon nanotube	MB, neutral red, and brilliant cresyl blue	15.74 mg g^{-1}, 20.33 mg g^{-1}, and 23.55 mg g^{-1}, respectively	Pseudo-second-order kinetics; Freundlich isotherm; van der Waals interactions occurring	[29]
Magnetite/MWCNTs	MB	55 mg g^{-1}	Pseudo-second-order kinetics; Langmuir isotherm; electrostatic interactions, π–π dispersion interaction, hydrogen bonding, and electron donor–acceptor complex formation	[30]
Magnetite-loaded multi-walled carbon nanotubes	MB	48.06 mg g^{-1}	Pseudo-second-order kinetics; Langmuir isotherm; electrostatic attraction and π–π stacking interactions	[31]
Magnetic GO/ poly(vinyl alcohol) composite gels	MB and methyl violet (MV)	270.94 and 221.23 mg g^{-1}, respectively	Pseudo-second-order kinetics, Langmuir isotherm; strong electrostatic attraction and complexation	[32]
GO/Fe_3O_4 nanohybrids	MB	~100%	Adsorption mechanism followed by π–π interaction	[33]
$Ca(OH)_2$/Na_2FeO_4 modified fly ash	Orange II	24.8 mg g^{-1}	Pseudo-second-order kinetics; Langmuir isotherm; electrostatic interaction, hydrogen bonding	[34]

(continued)

Table 3 (continued)

Pure magnetic materials Fe_3O_4	Rhodamine 6G	150–600 mg g^{-1}	Fluorescence resonance energy transfer between the dye molecules and Fe_3O_4 magnetic nanoparticle	[35]
Humic acid–Fe_3O_4	Malachite green (MG)	79.3 mg g^{-1}	Pseudo-second-order kinetics; Freundlich isotherm; electrostatic attraction, π–π interaction, and hydrogen bonding	[36]
Magnetic graphene oxide modified by chloride imidazole ionic liquid	Anionic: Glenn Black R (GR) and Orange IV (OIV) cationic: acridine orange (AO) and crystal violet (CV)	GR, OIV, AO, and CV were 588.24, 57.37, 132.80, and 69.44 mg g^{-1}, respectively	Pseudo-second-order kinetics; Langmuir isotherm; electrostatic interaction	[37]
Magnetic graphene oxide	MB and orange G	64.23 mg g^{-1} and 20.85 mg g^{-1}, respectively	Pseudo-second-order kinetics; Langmuir isotherm; electrostatic interaction	[38]
Fe_3O_4/CeO_2	Azo dye [Acid Black 210 (AB210)]	93.08 mg g^{-1}	Pseudo-second-order kinetics; Langmuir isotherm; chemisorption	[39]
Fe_3O_4/GO	MB and MO	666.7 and 714.3 mg g^{-1}, respectively	Langmuir isotherm; electrostatic interaction	[40]
Fe–Mn–Zr metal oxide	MO and eosin yellow (EY)	196.07 and 175.43 mg g^{-1}, respectively	Pseudo-second-order kinetics; Langmuir isotherm; surface adsorption and pore diffusions	[41]
Core@double-shell structured magnetic halloysite nanotube	MB	714.29 mg g^{-1} at 318.15 K	Pseudo-second-order kinetics; Langmuir isotherm; complexes formation, electrostatic interaction and π–π stacking interaction	[42]

(continued)

Table 3 (continued)

Fe_3O_4@C@polyaniline trilaminar core–shell	MO	120.2 mg g^{-1}	Pseudo-second-order kinetics; Langmuir isotherm; chemisorption followed by electrostatic attraction and π–π interaction	[43]
Fe_3O_4/graphite oxide nanosheet/citric acid-crosslinked β-cyclodextrin polymer	MB	173 mg g^{-1}	Pseudo-first-order or the intraparticle diffusion model; Sips isotherm; electrostatic attraction, the Lewis acid–Lewis base interaction, the host–guest interaction, and the π–π interaction	[44]

Recent advances on the removal of dyes from wastewater using various adsorbents: a critical review

3 Applications of Magnetic Nanohybrids as Dye Adsorbents

3.1 Critical Factors Influencing Dye Adsorption

Dye adsorption is a result of two mechanisms (adsorption and ion exchange) and is influenced by many factors, such as dye/adsorbent interaction, initial dye concentration, adsorbent's surface area, temperature, pH, and reaction time. The main advantage of adsorption recently became the use of low-cost materials, which reduces the procedure cost. Some of the extremely important factors are described in Table 5. In addition to the critical factors below, some external factors like material characteristics, source of food waste adsorbent, genetic evolution, growth location, radiation time, climatic conditions, microwave power, with by-the product utilization must be assessed [53].

3.2 Mechanism of Dye Adsorption

The different adsorbates adsorb on various adsorbents in distinctive ways. Usually, in literature four stage mechanism for dye adsorption is proposed. In stage1: the diffusion of dye from bulk of the solution to the adsorbent, in stage 2: Pore diffusion occurs, in stage 3: multilayer adsorption of dye molecules, in stage 4: monolayer

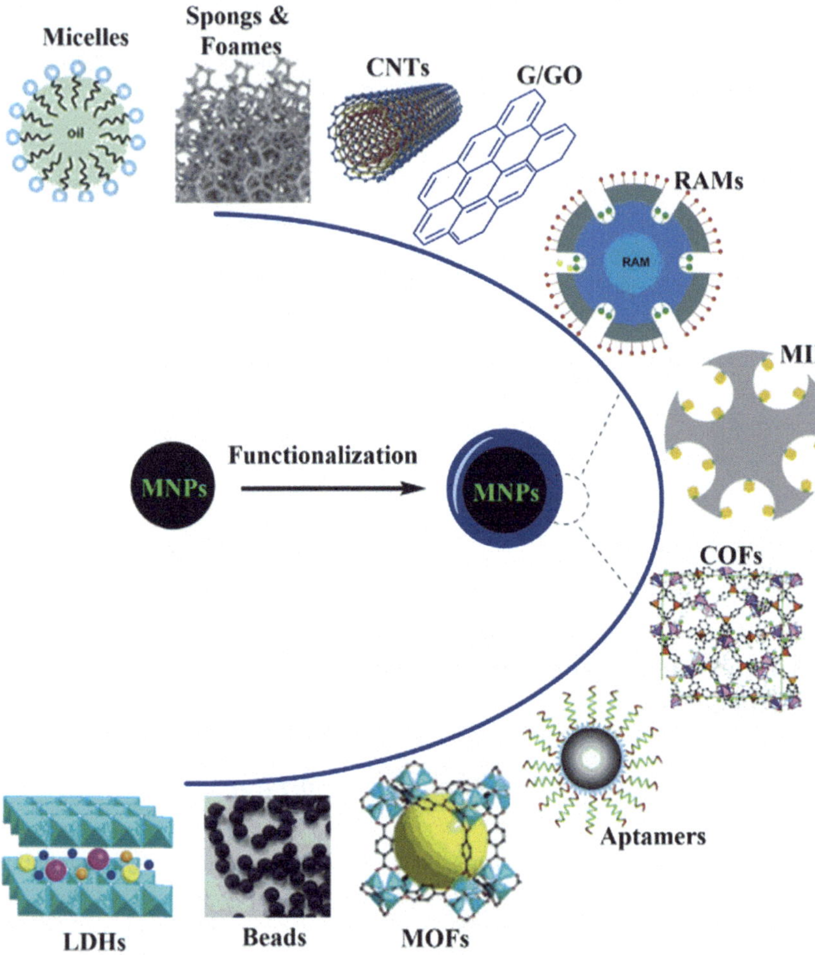

Fig. 2 Schematic illustration of different magnetic nanohybrids; MNPs; referred as magnetic nanoparticles (reprinted from the recent advances in magnetic sorbents and their applications [27])

adsorption of dye molecules [54]. The mechanism by which magnetic nanohybrid adsorb dyes from aqueous media is highly dependent on the nature of dye, pH, structure, and functional groups. In case of magsorbents various possible interactions that might occur are hydrogen bonding, π–π stacking, van der Waals forces and electrostatic interactions as shown in Fig. 3. According to the literature, anionic and cationic dyes are negatively and positively charged in the aqueous solution and depending on the surface charge of magsorbents the interactions might occurs [10]. In context to Zheng et al., the adsorption of anionic dyes, like MO and CR is achieved by ion exchange phenomena and electrostatic attraction. Moreover, the ion exchange mechanism involves the exchange of ions between a solid phase (adsorbent) and liquid (dye solution). In context with report of Fe_3O_4/graphene/polypyrrole nanocomposite for

Table 4 Type of adsorbents for various dyes with adsorption capacity from sustainable sources

Magnetic nanohybrids (MNHs)	Remarks and examples	References
Polymeric-based MNHs	Utilization of bio-wastes as a precursors/chemical source of reagents and the use of biopolymers like cellulose, chitosan, alginated, and agarose, prepared out of natural sources making them to be deliberated as renewable polymers	[48]
Carbon-based MNHs	Magnetic carbon nanodots/graphene oxide nanocomposite (Fe_3O_4@C-nanodot@GO) following two easy steps: the synthesis of carbon nanodots using a hydrothermal method with pasteurized cow milk; and the magnetization and incorporation of GO to the nanodots by a one-step hydrothermal approach	[49]
MOFs and COFs-based MNHs	Microwave-enhanced high temperature ionothermal method for rapid synthesis of magnetic nanoporous carbon materials on the basis of Fe-based MOF-type materials. The enhanced synthesis is considered as a safe, controllable, and rapid technology with a great potential for industry applications	[50]
ILs-based MNHs	Preparation of a sorbent made of magnetic cellulose NPs with 1-butyl-3-methylimidazolium hexafluoro phosphate ([C4MIm][PF6]) IL as coating. In this case, the IL was attached to the activated MNPs through a simple and direct mixture in methanol, followed by a vortex stirring for 5 min at room temperature	[51]
Magnetic biochar	Pyrolysis process to reduce the pinewood sawdust which is found abudantly from the wood processing factory by developing a Fe_3O_4-loaded magnetic biochar in the presence of $FeCl_3$ as the metallic solution to provide the magnetic effect on the magnetic biochar produced	[52]

the adsorption of methylene blue (MB), the electrostatic interaction was responsible for the effective adsorption of dye. Likewise, the MB dye molecules have C=C bonds plus contain π electrons which may interrelate with the benzene ring π electrons, existing in the nanohybrids via π–π electron coupling.

The pH of the solution also plays the critical role in the electrostatic interaction of adsorbate and adsorbent. From, the literature it is predicted that the higher pH of the solution makes the adsorbent surface more negative that helps to increase the electrostatic interaction with the positively charged dye. Also, the functionalization with different metal ions can influence the mechanistic study of the dyes [55]. At the

Table 5 Critical factors affecting the adsorption process

Critical factors influencing dye adsorption			
Initial dye concentration	Depends on active binding sites	On surface as well as on adsorbed chosen	Extremely helpful to understand the kinetic models
Adsorbent dosage	Higher adsorbent more dye removal efficiency	After reaching an equilibrium, more dosage doesn't affect much	Also depends on particle size
pH of the solution	Varies for anionic and cationic dye	Lower pH offers less adsorption rate	Optimum range:4–8
Temperature	Increase in temperature leads to decrease in adsorption efficiency	Temperature can be used as an indicator to identify if the reaction is exothermic or endothermic	Thermodynamic studies are important to understand the relationship between adsorption capacity and temperature
Reaction time	Greater contact time with adsorbent means more dye removal	Interactions taking place at their peak in the initial 10–30 min	Reaction time can show a considerable variation in adsorption capacity when it reaches equilibrium

end note, it can be concluded that the electrostatic interactions is the supreme means of removing cationic and anionic dyes from aqueous media.

3.3 Magnetic Recovery

The magnetic recovery is very important factor to improve the economic feasibility of the magnetic nanohybrids. The prime purpose of the recovery process is to reestablish the adsorption capacity of used adsorbent and to make the technology cost-effective [56]. Using magnetic nanoparticles as catalysts or support combines the advantage of magnetic recovery as well as advantage of nano-size particles which gives excellent catalytic activity even after several cycles. Magnetic nanohybrids shows supermagnetism; which is a form of magnetism applied when the size of ferrimagnetic or ferromagnetic nanoparticles is as low as 10–20 nm. Unfortunately, magnetic nanoparticles show agglomeration due to their small interparticle distance and presence of van der Waals forces. In solution to it, they are functionalized to form nanohybrids which consists of three steps; coating of outer shell of magnetic nanoparticle, formation of core–shell structure and formation of magnetic core.

Typically, the most popular techniques used for recycling of the catalysts are thermal, organic, solvent, supercritical fluid, centrifugation, microwave irradiation and filtration. Choosing, the regeneration technique is highly dependent upon the

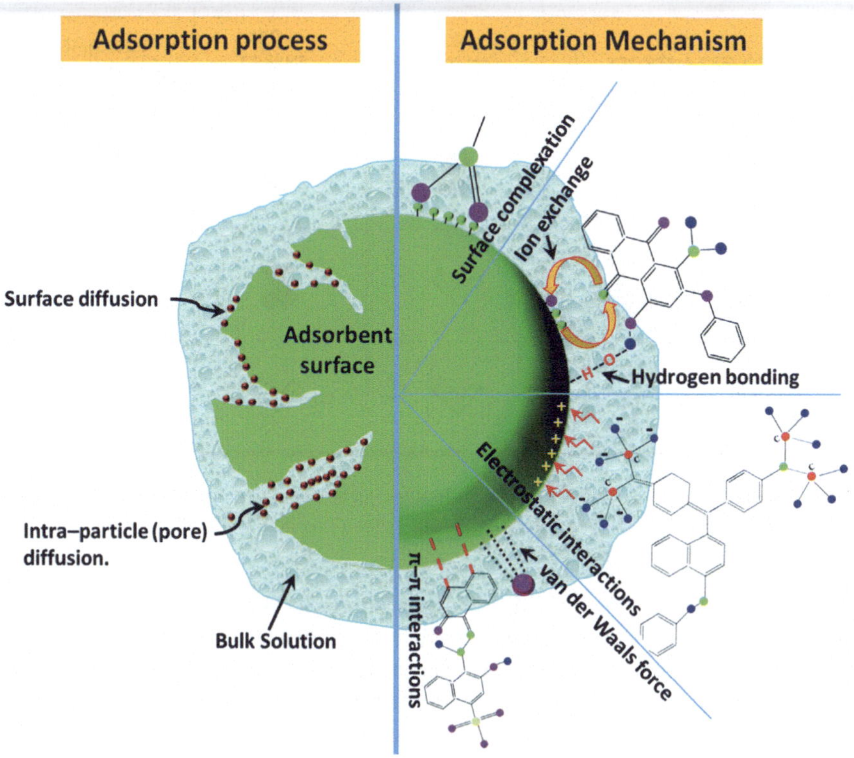

Fig. 3 Possible adsorption process and mechanism pathways (reprinted Recent advances on the removal of dyes from wastewater using various adsorbents: a critical review [10])

adsorbent chosen and the contaminate which needs to be removed. In accordance with green chemistry rules, using external magnetic for recovery reduces the derived waste, needs lower consumption, and shows higher efficiency [57]. Lastly, when recycling of the magnetic catalysts is no longer possible; it is possible to recover the functionalized metals or separation of various valuable components before final disposal. Specifically, bio metallurgy, hydrometallurgy, and pyrometallurgy combined with DC plasma arc technology is highly considerable in this work [57] (Fig. 4).

4 Conclusion, Critiques, and Future Remarks

This book chapter researched the issues related to dye sector and their remediation techniques by magnetic nanohybrids adsorbents. The studies showed that the expansion of magnetic nanohybrid adsorbents is an assuring solution to the problem of dye effluents in wastewater and blocking them from being discharged into the environment. The main highlighting parts of the chapter are the prime sources of dyes

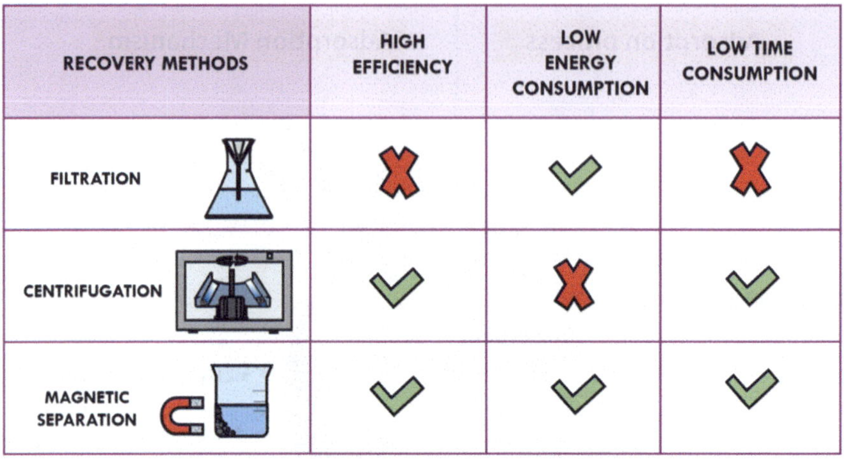

Fig. 4 Comparison of recovery methods: filtration, centrifugation, and magnetic separation (Reprinted from Recovery/Reuse of Heterogeneous Supported Spent Catalysts [57])

effluents, types of existing dyes, different procedures for the fabrication of magnetic nanohybrids and different techniques for removal of dyes from wastewater.

The main critique identified during literature review is; do these magnetic nanohybrids at large-scale production still retain their quality and can replace the present commercial adsorbents? The answer relies on two most important factors; one is cost-effective/value for money and other is environmental risk. The first factor is achieved as the researchers has proved to synthesize the magnetic nanohybrids with very cheap approaches. The second factor related to the toxicity of nanohybrids is still a question of debate. This can be solved by underlying careful attention on the transformation routes of the adsorption catalysts and fully recovery of nanohybrids after the reaction.

The concluding future remarks is to adopt realistic approaches even at the laboratory experiments. Though few levels of control are required to define and characterize a dye adsorption system. Farther accurate methodology towards system development needs to be adopted by integrating research on mixtures of dyes and real effluents, also very limited literature reports examining the mixtures of dyes are available. Second remark is regarding very limited reports on degradation pathways, although few examples proposing a degradation pathway, which shows identified fragments by analytical techniques are reported, but the evidence of further reaction steps are missing. This can be achieved by calculating the existence of free radicals and careful control on the kinetics of the reaction. Thus, extensive research is needed to be done in near future.

References

1. Qadri R, Faiq MA (2020) Freshwater pollution: effects on aquatic life and human health. In: Fresh water pollution dynamics and remediation. Springer, Singapore, pp 15–26
2. Dongre RS, Sadasivuni KK, Deshmukh K, Mehta A, Basu S, Meshram JS, Al-Maadeed MA, Karim A (2019) Natural polymer based composite membranes for water purification: a review. Polymer-Plastics Technology and Materials 58(12):1295–1310
3. Singh J, Yadav P, Pal AK, Mishra V (2020) Water pollutants: origin and status. In: Sensors in water pollutants monitoring: role of material. Springer, Singapore, pp 5–20
4. Mehta A, Mishra A, Basu S, Shetti NP, Reddy KR, Saleh TA, Aminabhavi TM (2019) Band gap tuning and surface modification of carbon dots for sustainable environmental remediation and photocatalytic hydrogen production–a review. J Environ Manage 250:109486
5. Jayaswal K, Sahu V, Gurjar BR (2018) Water pollution, human health and remediation. In: Water remediation. Springer, Singapore, pp 11–27
6. Clarke R (2013) Water: the international crisis. Routledge
7. Clark M (ed) (2011) Handbook of textile and industrial dyeing: principles, processes and types of dyes. Elsevier
8. Kumar A, Dixit U, Singh K, Gupta SP, Beg MS (2021) Structure and properties of dyes and pigments. In: Dyes and pigments-novel applications and waste treatment. IntechOpen
9. Khattab TA, Abdelrahman MS, Rehan M (2020) Textile dyeing industry: environmental impacts and remediation. Environ Sci Pollut Res 27(4):3803–3818
10. Dutta S, Gupta B, Srivastava SK, Gupta AK (2021) Recent advances on the removal of dyes from wastewater using various adsorbents: a critical review. Mater Adv 2:4497–4531
11. Moattari RM, Mohammadi T (2021) Hybrid adsorbents for dye removal from wastewater. In: Green adsorbents to remove metals, dyes and boron from polluted water. Springer, Cham, pp 405–451
12. Saravanan A, Kumar PS, Jeevanantham S, Karishma S, Tajsabreen B, Yaashikaa PR, Reshma B (2021) Effective water/wastewater treatment methodologies for toxic pollutants removal: processes and applications towards sustainable development. Chemosphere 280:130595
13. Shindhal T, Rakholiya P, Varjani S, Pandey A, Ngo HH, Guo W, Ng HY, Taherzadeh MJ (2021) A critical review on advances in the practices and perspectives for the treatment of dye industry wastewater. Bioengineered 12(1):70–87
14. Katheresan V, Kansedo J, Lau SY (2018) Efficiency of various recent wastewater dye removal methods: a review. J Environ Chem Eng 6(4):4676–4697
15. Pavithra KG, Jaikumar V (2019) Removal of colorants from wastewater: a review on sources and treatment strategies. J Ind Eng Chem 75:1–9
16. Donkadokula NY, Kola AK, Naz I, Saroj D (2020) A review on advanced physico-chemical and biological textile dye wastewater treatment techniques. Rev Environ Sci Biotechnol 19(3):543–560
17. Králik M (2014) Adsorption, chemisorption, and catalysis. Chem Pap 68(12):1625–1638
18. Dąbrowski A (2001) Adsorption—from theory to practice. Adv Coll Interface Sci 93(1–3):135–224
19. Ardila-Leal LD, Poutou-Piñales RA, Pedroza-Rodríguez AM, Quevedo-Hidalgo BE (2021) A brief history of colour, the environmental impact of synthetic dyes and removal by using laccases. Molecules 26(13):3813
20. Bonilla-Petriciolet A, Mendoza-Castillo DI, Reynel-Ávila HE (eds) (2017) Adsorption processes for water treatment and purification. Springer, Berlin
21. Ul-Islam M, Ullah MW, Khan S, Manan S, Khattak WA, Ahmad W, Shah N, Park JK (2017) Current advancements of magnetic nanoparticles in adsorption and degradation of organic pollutants. Environ Sci Pollut Res 24(14):12713–12722
22. Mohammed L, Gomaa HG, Ragab D, Zhu J (2017) Magnetic nanoparticles for environmental and biomedical applications: a review. Particuology. 30:1–4
23. Mashkoor F, Nasar A (2020) Magsorbents: potential candidates in wastewater treatment technology–a review on the removal of methylene blue dye. J Magn Magn Mater 500:166408

24. Shukla S, Khan R, Daverey A (2021) Synthesis and characterization of magnetic nanoparticles, and their applications in wastewater treatment: a review. Environ Technol Innov 24:101924
25. Hao R, Xing R, Xu Z, Hou Y, Gao S, Sun S (2010) Synthesis, functionalization, and biomedical applications of multifunctional magnetic nanoparticles. Adv Mater 22(25):2729–2742
26. Zhang X, Qian J, Pan B (2016) Fabrication of novel magnetic nanoparticles of multifunctionality for water decontamination. Environ Sci Technol 50(2):881–889
27. Faraji M, Shirani M, Rashidi-Nodeh H (2021) The recent advances in magnetic sorbents and their applications. TrAC, Trends Anal Chem 141:116302
28. Manippady SR, Singh A, Basavaraja BM, Samal AK, Srivastava S, Saxena M (2020) Iron–carbon hybrid magnetic nanosheets for adsorption-removal of organic dyes and 4-nitrophenol from aqueous solution. ACS Appl Nano Mater 3(2):1571–1582
29. Gong JL, Wang B, Zeng GM, Yang CP, Niu CG, Niu QY, Zhou WJ, Liang Y (2009) Removal of cationic dyes from aqueous solution using magnetic multi-wall carbon nanotube nanocomposite as adsorbent. J Hazard Mater 164(2–3):1517–1522
30. Ghaedi M, Khajehsharifi H, Yadkuri AH, Roosta M, Asghari A (2012) Oxidized multiwalled carbon nanotubes as efficient adsorbent for bromothymol blue. Toxicol Environ Chem 94(5):873–883
31. Ai L, Zhang C, Liao F, Wang Y, Li M, Meng L, Jiang J (2011) Removal of methylene blue from aqueous solution with magnetite loaded multi-wall carbon nanotube: kinetic, isotherm and mechanism analysis. J Hazard Mater 198:282–290
32. Cheng Z, Liao J, He B, Zhang F, Zhang F, Huang X, Zhou L (2015) One-step fabrication of graphene oxide enhanced magnetic composite gel for highly efficient dye adsorption and catalysis. ACS Sustainable Chem Eng 3(7):1677–1685
33. Jiao T, Liu Y, Wu Y, Zhang Q, Yan X, Gao F, Bauer AJ, Liu J, Zeng T, Li B (2015) Facile and scalable preparation of graphene oxide-based magnetic hybrids for fast and highly efficient removal of organic dyes. Sci Rep 5(1):1
34. Gao M, Ma Q, Lin Q, Chang J, Ma H (2017) Fabrication, and adsorption properties of hybrid fly ash composites. Appl Surf Sci 396:400–411
35. Phoemphoonthanyakit S, Seeharaj P, Damrongsak P, Locharoenrat K (2019) Effect of adsorption characteristics of rhodamine 6G dye solution in Fe_3O_4 magnetic nanoparticles on fluorescence quantum yield. Journal of Spectroscopy 2019:1–5
36. Abate GY, Alene AN, Habte AT, Addis YA (2021) Adsorptive removal of basic green dye from aqueous solution using humic acid modified magnetite nanoparticles: kinetics, equilibrium and thermodynamic studies. J Polym Environ 29(3):967–984
37. Wang H, Wei Y (2017) Magnetic graphene oxide modified by chloride imidazole ionic liquid for the high-efficiency adsorption of anionic dyes. RSC Adv 7(15):9079–9089
38. Deng JH, Zhang XR, Zeng GM, Gong JL, Niu QY, Liang J (2013) Simultaneous removal of Cd (II) and ionic dyes from aqueous solution using magnetic graphene oxide nanocomposite as an adsorbent. Chem Eng J 226:189–200
39. Gao S, Zhang W, Zhou H, Chen D (2018) Magnetic composite Fe_3O_4/CeO_2 for adsorption of azo dye. J Rare Earths 36(9):986–993
40. Khajeh M, Barkhordar A (2020) Fe_3O_4/Graphene oxide composite for adsorption of methylene blue and methyl orange in water treatment. J Appl Spectrosc 87(4):701–707
41. Singh NH, Kezo K, Debnath A, Saha B (2018) Enhanced adsorption performance of a novel Fe-Mn-Zr metal oxide nanocomposite adsorbent for anionic dyes from binary dye mix: response surface optimization and neural network modeling. Appl Organomet Chem 32(3):e4165
42. Zhan Y, Wan X, He S, Yang Q, He Y (2018) Design of durable and efficient poly (arylene ether nitrile)/bioinspired polydopamine coated graphene oxide nanofibrous composite membrane for anionic dyes separation. Chem Eng J 333:132–145
43. Yao W, Shen C, Lu Y (2013) Fe3O4@ C@ polyaniline trilaminar core–shell composite microspheres as separable adsorbent for organic dye. Compos Sci Technol 87:8–13
44. Jiang LW, Zeng FT, Zhang Y, Xu MY, Xie ZW, Wang HY, Wu YX, He FA, Jiang HL (2021) Preparation of a novel Fe_3O_4/graphite oxide nanosheet/citric acid-crosslinked β-cyclodextrin polymer composite to remove methylene blue from water. Adv Powder Technol 32(2):492–503

45. Gutiérrez-Serpa A, González-Martín R, Sajid M, Pino V (2021) Greenness of magnetic nanomaterials in miniaturized extraction techniques: a review. Talanta 225:122053
46. Mehta A, Karbouche K, Kraxner J, Elsayed H, Galusek D, Bernardo E (2022) Upcycling of pharmaceutical glass into highly porous ceramics: from foams to membranes. Materials 15(11):3784
47. Sharma A, Mangla D, Chaudhry SA (2022) Recent advances in magnetic composites as adsorbents for wastewater remediation. J Environ Manage 306:114483
48. Pandey N, Shukla SK, Singh NB (2017) Water purification by polymer nanocomposites: an overview. Nanocomposites. 3(2):47–66
49. Nasrollahzadeh M, Sajjadi M, Iravani S, Varma RS (2021) Carbon-based sustainable nanomaterials for water treatment: state-of-art and future perspectives. Chemosphere 263:128005
50. Jabbari V, Veleta JM, Zarei-Chaleshtori M, Gardea-Torresdey J, Villagrán D (2016) Green synthesis of magnetic MOF@ GO and MOF@ CNT hybrid nanocomposites with high adsorption capacity towards organic pollutants. Chem Eng J 304:774–783
51. Isosaari P, Srivastava V, Sillanpää M (2019) Ionic liquid-based water treatment technologies for organic pollutants: current status and future prospects of ionic liquid mediated technologies. Sci Total Environ 690:604–619
52. Thines KR, Abdullah EC, Mubarak NM, Ruthiraan M (2017) Synthesis of magnetic biochar from agricultural waste biomass to enhancing route for waste water and polymer application: a review. Renew Sustain Energy Rev 67:257–276
53. Rápó E, Tonk S (2021) Factors affecting synthetic dye adsorption; desorption studies: a review of results from the last five years (2017–2021). Molecules 26(17):5419
54. Singh NB, Nagpal G, Agrawal S (2018) Water purification by using adsorbents: a review. Environ Technol Innov 11:187–240
55. Faria PC, Orfao JJ, Pereira MF (2004) Adsorption of anionic and cationic dyes on activated carbons with different surface chemistries. Water Res 38(8):2043–2052
56. Mehta A, Colusso E, Kraxner J, Galusek D, Bernardo E (2022) Waste-derived glass as a precursor for inorganic polymers: from foams to photocatalytic destructors for dye removal. Ceramics International 48:27631–27636
57. Miceli M, Frontera P, Macario A, Malara A (2021) Recovery/reuse of heterogeneous supported spent catalysts. Catalysts 11(5):591

The Impact of Textile Dyes on the Environment

Tanzina Akter, Anica Tasnim Protity, Modhusudon Shaha, Mohammad Al Mamun, and Abu Hashem

Abstract Textile industries use large amounts of natural and synthetic dyes for fabric processing. These dyes have been used to brighten clothes for more than 4000 years, from ancient Egypt to the present day. The application of dyes in the textile industry is expanding day by day. These colored effluents used in the textile industry are highly salted and heavily contaminated and are released into the environment in massive quantities. Due to the inefficiency of the dyeing process, up to 200,000 tons of these colors are wasted as effluents every year in the textile industry during dyeing and finishing operations. Significant volumes of dyes are released into waterbodies by the textile industry, causing substantial environmental pollution. According to estimates, 12–15% of these dyes are discharged into manufacturing processes' effluents, contaminating the environment. So, there is a clear connection between the textile industry and environmental pollution. In this chapter, we have focused on the toxicities of textile dyes, including their impact on water quality parameters such as pH, biochemical oxygen demand (BOD), chemical oxygen demand (COD), total organic carbon (TOC), and suspended solids (TSS), followed by their impact on animal health (both land and aquatic animals), fisheries, plant (terrestrial and aquatic flora) growth and development, as well as on microbiota. To mitigate the harmful effects of textile dyes, it is highly promising to develop techniques for the treatment

T. Akter · A. T. Protity · M. Shaha · A. Hashem (✉)
Microbial Biotechnology Division, National Institute of Biotechnology, Ganakbari, Ashulia, Savar, Dhaka 1349, Bangladesh
e-mail: hashemnib04@yahoo.com

T. Akter
School of Optometry and Vision Science, University of New South Wales, Sydney NSW 2052, Australia

A. T. Protity
Department of Biological Sciences, Northern Illinois University, Dekalb, IL, USA

M. Al Mamun
Department of Chemistry, Jagannath University, Dhaka 1100, Bangladesh

M. Al Mamun · A. Hashem
Nanotechnology and Catalysis Research Centre, Institute for Advanced Studies, University of Malaya, 50603 Kuala Lumpur, Malaysia

© The Author(s), under exclusive license to Springer Nature Singapore Pte Ltd. 2023
A. Ahmad et al. (eds.), *Nanohybrid Materials for Treatment of Textiles Dyes*,
Smart Nanomaterials Technology, https://doi.org/10.1007/978-981-99-3901-5_17

of textile dyes. The application of nanohybrid materials may play an important role in that case.

Keywords Environment · Impact · Pollution · Textile dyes · Water quality

1 Introduction

Since the beginning of the industrial revolution, pollution has become one of the world's most important problems [58]. Textile industries are key contributors not only to the worldwide market but also to environmental pollution. Recently, water pollution due to the failure of textile industries to appropriately discard their wastewater has been one of the most important challenges affecting the entire world. Wastewater with dyes is a massive pollutant for the natural environment, as the textile industries generate huge amounts of intensely colored wastewater, including a wide range of persistent contaminants. More than 10,000 different artificial pigments and dyes are extensively used in different paper and textile industries, having a substantial adverse effect on both health and the ecosystem [5]. Synthetic and natural fibers, including cellulosic ingredients (for example, rayon, cotton, and linen derived from plants), protein fabrics such as wool, silk, and mohair, as well as manmade fabrics like acrylic, polyester, and nylon, can all be generated in the textile industry using both the dry and wet approaches. However, those factories also use a wide range of poisonous chemicals at various stages of the procedure, including softening, sizing, desizing, lightening, and finishing reagents [99]. The laundry phase is the last step in the fabric coloring process, which causes the removal of extra color plus pigments (containing many potentially carcinogenic and highly toxic chromophoric species including mercury, cadmium, chromium, arsenic, and lead). Those additional dyes do not attach firmly to the cloth but are eliminated as waste into the aquatic atmosphere, such as rivers, lakes, ponds, streams, seas, and oceans, with no previous treatment. Consequently, the constant discharge of raw wastewater from numerous fabric industries creates alarming ecotoxicological risks for both human health, and the aquatic environment. Even the toxic materials are transported along long routes with the effluent. After that, they stay in the soil and water for a long time, posing serious health risks to living things, reducing the soil's ability to hold water, and stopping marine plants from making oxygen through photosynthesis. This creates anoxic environments for both aquatic plants and animals [41].

There are many dyes that are classified into several groups in accordance with their structure, origin, and application. Of these synthetic colors, azo, reactive, direct, mordant, acidic, basic, sulfide, and disperse dyes are extensively applied by the textile industry [20, 161]. Among them, the most commonly used dyes are azo dyes, including Methylene Blue, Crystal Violet, Congo Red, Gentian Violet, Disperse Red 1, Reactive Green 19, Direct Blue 14, Sudan I, Disperse Orange, Disperse Yellow 7, and Reactive Blue 2, which have various toxic and genotoxic effects on both aquatic plants and animals [3].

Those dyes destroy the aesthetic value of aquatic bodies by raising chemical and biochemical O_2 demand, in that way damaging the photosynthesis process, hindering herbal growth, entering into the food chain, delivering resistance and bio-accretion, and actively endorsing mutagenicity, carcinogenicity, and toxicity [125]. The more important ecological concern in conjunction with dye is the transmission of sunlight by preventing it from inserting into the aquatic photic region, impacting on the photosynthesis process. The increase of solids in contaminated wastewater exhausts the O_2 content that is urgent for plantlets and disrupts the osmotic equilibrium of seeds [61]. Moreover, it reduces the chlorophyll levels of plants, which affects the natural photosynthesis, depletes the growth of plants, and reduces the plant's dry matter accumulation [148]. Furthermore, textile dye effluents can trigger an accumulation of proline in plants, implying that they are producing stress [42]. Even the industrial effluents containing azo dyes cause a negative impact on germination and plant flowering, including genotoxic effects such as micronucleus and chromosomal aberrations, affecting the seeds of *Allium* cepa [3].

Potentially carcinogenic and highly toxic textile dyes are also linked to a range of animal and human illnesses [78, 163]. It can cause dermatitis, difficulties in the central nervous system, skin rash, eye irritation, and allergic reactions such as allergic conjunctivitis, contact dermatitis, occupational asthma, and rhinitis. Furthermore, long-term exposure to these waste effluents can cause cancers, such as colorectal cancer, bladder cancer, and colon cancer.

Aquatic organisms, including fish, which provide protein, can initiate symptoms like fever, hypertension, and cramps if they are eaten by humans while containing dyes [6, 149]. The occurrence of textile pigments and dyes in effluent causes highly colored liquids with a fluctuating pH and a high degree of BOD, COD, TOC, and TSS. These TSS impede the water flow through the gills of fish, blocking gas exchange and causing a reduction in growth rate or death [18]. However, the textile effluents contaminated fish consumption as a food, which has a potential impact on human health [174]. Textile dyes and pigments containing stagnant water environments cause a lack of O_2, where bacteria decay about 7–8 mg of organic entities, which is enough to reduce the amount of O_2 in 1 L of water [105]. Organic matter, including textile dye wastage accumulation in waterways, causes an obnoxious taste, bacterial growth, erroneous pigmentation, and pestilent odors [98].

Trihalomethanes, which are chlorination byproducts applied to get rid of hazardous microbes, are formed when chlorine combines with organic colors during water treatment [173]. Azure-B is a cationic dye that can intercalate with the helical DNA and duplex RNA structures (the main metabolite of methylene blue) [55]. Mitotic poisoning from the color crystal violet (triarylmethane) results in chromosomal destruction and aberrant metaphase growth [109]. The reticular cell sarcoma of the vagina and bladder has been associated with this highly carcinogenic chemical [104]. Additionally, crystal violet may result in chemical cystitis, irritation of the digestive system and skin, as well as renal and respiratory failure in people [109].

Therefore, dye-containing wastewater should be effectively treated using cost-effective, eco-friendly technology to avoid negative effects on the environment, fish and animal health, soil and plant microbiota, as well as natural water resources.

Additionally, in order to circumvent the contamination of the environment and to promote sustainability, adequate and effective management approaches are also required due to the increasing number of textile businesses and the enormous quantities of effluent containing dyes. There are numerous approaches and methods utilized for the treatment of wastewater containing textile dyes, each with advantages and disadvantages. It can be treated chemically, physically, biologically, or by an amalgamation of these treatments, depending on the effluent quality, technical feasibility, and cost. Numerous remediation technologies have already been developed, but they don't seem to be able to entirely remove dyes from the environment. In the era of nanoscience and nanotechnology, nanomaterials can significantly improve the effectiveness of dye wastewater treatment. In recent days, there has been a quick growth in the use of nanoparticles, which has led to the development of novel, creative, extremely effective, and affordable dye remediation technologies. Large-surface area nanomaterials can alter surface properties and have unique electron-conducting properties, which makes them excellent candidates for the management of effluent containing dyes. More research is necessary in this field to build a cutting-edge, waste-free method and to lower threats to the environment and public health while scaling up from the laboratory to the pilot scale.

In this chapter, we will go over the impact of dye-containing textile sewer water on water quality as measured by pH, BOD, COD, TOC, and TSS, as well as their impact on animal health (both land and aquatic animals), fisheries, plant growth and development, and microbiota, followed by bioremediation.

2 Textile Dye Characteristics and Impacts

The majority of synthetic colors come from petrochemical chemicals, and they can be bought in powder, liquid, paste, or granule form [49]. They have a lot of possible uses, such as quick and stable pigmentation with different types of fabrics, a wide range of color pigments and shades, easy manipulation, stability against different outside factors, and low energy use [156].

Visible light can be absorbed by the organic substances used in textile colors. They dye a substance sustainably while simultaneously reflecting or dispersing complementary radiation [105]. Auxochromes, which are utilized to fix dyes, the matrix, which contains the remaining atoms in the molecules, and chromophoric groups, which provide color, make up these molecules [17]. When dyeing and/or printing, these three groups can each have unique independent characteristics such as color and material adhesion ability [134]. The single-bond and double-bond arrangements of these groups match the unstable mesomeric state, which lets light pass through [105].

The common characteristics of dyes are given below:

- Dyes should be temporarily or permanently soluble in water. Different dye classes possess different solubilizing groups and hence exhibit a different extent of solubility.
- They should be able to impart color, and the colors should be intense, so that they require very little quantity for dyeing.
- They should have the ability to stand up to drying, washing, cleaning, or light exposure and have chemical stability.
- Dyes should be absorbed and retained by the fabric material to be dyed. The dye should be substantial or have an affinity for one or more types of textile fibers. The substantivity of a dye can be imparted by the introduction of various chemical groups into it. For example, cationic, anionic, polar, phenolic, and a large number of benzene groups. They might have reactive groups, like chlorotriazine, that can form covalent bonds with the fiber.

When released into the environment in untreated or partially treated forms, the majority of synthetic dyes have negative effects [73, 153]. The processes involved in textile dyeing (dying, fixing, washing, etc.) use a significant quantity of water, and up to 15% of applied colors can escape from the textile fibers and end up in wastewater. As a result, a significant amount of inappropriate discharge is continually rejected [169, 170]. High biological and chemical oxygen demands (BOD and COD) are present in dye effluents, and they are also very high in organic and inorganic pollutants like formaldehyde, benzidine, heavy metals, nitrates, naphthol, soaps, chromium compounds, chlorinated compounds, and dyes and pigments [39, 52]. Even after some treatment operations, a number of hazardous substances are still present in the wastewater [95, 160]. As a result, they cause severe pollution of the air, soil, vegetation, and water resources, as well as human diseases [4]. Figure 1 depicts the overall effects of textile dyes on environmental substrates.

3 Textile Dyes and Water Quality

The dyeing of textiles has the potential to seriously harm the environment, including the water supply. Approximately 10,000 distinct types of dyes and pigments are reportedly utilized in the industry, and over 7×10^5 metric tons of synthetic dyes are created globally each year [119, 175]. An important process in the textile manufacturing sector is dying. The batch continuous, or semi-continuous, process can be used to dye textile fabrics. The batch procedure is the most commonly used to dye textile fabrics of the three [128]. The type of material (cellulosic components derived from plant sources, such as rayon, linen, and cotton, protein fabrics derived from animal resources, such as mohair, wool, and silk; and synthetic fabrics made artificially, such as polyester, nylon, and acrylic; quality requirements in the dyed fabric; and size of dye lots are just a few of the factors that determine which process must be used. The factories' dyeing and finishing processes consume a significant amount of water. In

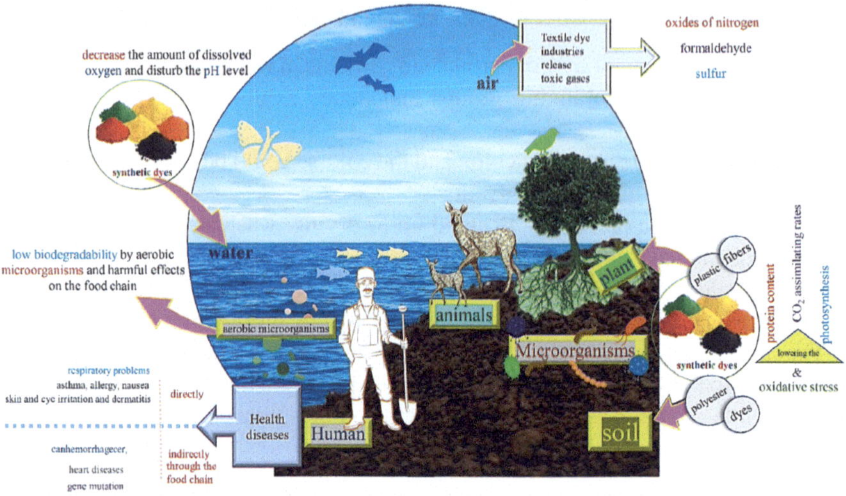

Fig. 1 Overall impacts of textile dyes on environmental substrates. Reprinted from Ref. [156] with permission under Creative Commons Attribution (CC BY) license

terms of both the volume produced and the effluent's composition, the effluents from textile factories are the most adulterating [16, 40, 145] elements. Textile effluents contain both organic and inorganic chemicals [44]. A percentage of the dyes that are applied to the fabrics during the dyeing process remain unfixed, and the washout of the unfixed colors causes high concentrations of these dyes in textile effluents. These colors have harmful side effects in addition to an undesirable appearance. Their breakdown byproducts could contaminate the soil and sediment in the area. Additionally, in recent years, there have been substantial sources of severe water contamination issues due to the increased demand for textile yields, their increased manufacturing, and the use of synthetic colors [40, 119]. Figure 2 shows the water pollution caused by the release of dyeing and printing liquids.

Every day, the textile industry dumps a large amount of dye into the water. Small aquatic creatures consume these textile microfibers and dye. They are then consumed by tiny fish, which are then swallowed by larger fish, introducing dye and plastic textile fiber into the human food chain.

3.1 Effluent Characteristics of Textile Industry Processes

Dry and wet procedures are used in the manufacturing of fiber in textile mills. Sizing, deizing, sourcing, bleaching, mercerizing, dyeing, printing, and finishing methods make up the wet process [11, 106]. Strong acids, strong alkalis, starch, surface-active chemicals, sodium hypochlorite, inorganic chlorinated compounds, bleaching agents, organic compounds like dye, finishing chemicals, thickening agents, metal

Fig. 2 Water pollution caused on by the discharge of textile dyes and printing ink. Reprinted from Ref. [122] with permission under Creative Commons Attribution (CC BY) license

salts, wetting, and dispensing agents, etc. are employed at each stage. To enhance the quality of the items, various dyes, and multicolors are employed. All these chemicals and elements are detrimental to the environment, especially the water bodies where they are discharged. According to Holkar et al. [63], Table 1 lists the processing processes that are involved along with the principal contaminants in the wastewater discharge from each phase.

Table 1 Processes used to treat fibers and related contaminants in textile effluent

Process	Effluent composition	Nature
Sizing	Starch, carboxymethyl cellulose (CMC), waxes, polyvinyl alcohol (PVA), and wetting agents	High in BOD and COD
Desizing	Starch, fats, CMC, PVA, pectins, and waxes	High in BOD, COD, SS, and dissolved solids (DS)
Bleaching	Cl_2, sodium hypochlorite, NaOH, H_2O_2, acids, $NaSiO_3$, surfactants, sodium phosphate, and short cotton fiber	High alkalinity and high TSS
Mercerizing	Cotton wax and sodium hydroxide	Low BOD, high pH, and high DS
Dyeing	Reducing agents, dyestuffs urea, acetic acid, oxidizing agents, detergents, and wetting agents	High BOD, DS, low SS, strongly colored, and heavy metals
Printing	Starches, pastes, urea, gums, oils, cross-linkers, binders, acids, thickeners, reducing agents, and alkali	Highly colored, high BOD, oily appearance, TSS slightly alkaline, and low BOD

3.2 Impact of Textile Dyes on Water Quality

The effluents from textile dyeing are what pollute the water bodies in the surrounding environment. According to reports, textile effluents themselves have no negative effects on groundwater [71]. Textile dyeing effluent, which contains a lot of color and organic compounds produced by dyeing and finishing salts, is released by textile mills in enormous quantities as harmful toxic waste. The combination of sulfur, naphthol, heavy metals, vat dyes, and a few auxiliary chemicals results in an exceedingly toxic effluent. Additionally, it is possible to find other dangerous compounds, including formaldehyde-based color fixing elements, non-biodegradable dyeing substances, and hydrocarbon-based softeners. Due to the presence of colorful and oily scum, colloidal elements that increase turbidity generate an aggressive appearance of water as well as a foul smell [84].

The mix of chemicals found in wastewater from the textile industry differs from factory to factory and country to country. The technique, the kind of fabric produced, the factory equipment, the chemicals employed, the weight of the fabric, current fashion trends, and the season all affect this variety [23]. Real wastewater composition varies greatly, according to Kehinde et al. findings [91]. The cause is thought to be the fact that these effluents were not passed from a particular treatment stage or that they were not taken from the same exact step in the textile production operations. Metal contamination is caused by the presence of colors and chemicals used in the textile production process, such as salts, caustic soda, sodium carbonate. According to Hussein [67], lead, mercury, chromium, zinc, and iron are the key elements that damage the environment. Cobalt, copper, and chromium are the three primary metals, according to Adinew [1], and they are present in the dye chromophores in textile effluents [1, 67].

Dye concentrations in textile effluent are listed in a wide range of values, starting at 10–7000 mg/l. According to a study by Laing [102], there are 10–50 mg/l of dye in the textile effluent, 60 mg/l of reactive dyes are used in cotton factories, according to Shelley [152], and Gahr et al. recorded a concentration of 100–200 mg/l in the same year [48, 102, 152]. Koprivanac et al. [100] calculated a concentration of 7000 mg/l, which was astronomically high compared to other references [100].

Dye is poisonous, and it can lower the amount of dissolved oxygen in receiving bodies of water. Aquatic life is threatened by dissolved oxygen reduction, which also lowers the overall quality of the water. It is estimated that dyeing losses only make up 10–30% of the total BOD5 and 2–5% of the total COD5, while dye bath chemicals make up as much as 25–35% [71]. Acetic acid has a high BOD5 and can contribute 50–90% of the dye house BOD5 when used in disperse dyes, cationic dyes, and acid dyes on polyester, acrylic fibers, and protein-derived fibers (wool, silk, and nylon) [71].

Numerous physicochemical properties of textile dyeing effluents from various national resources have been measured in different studies. According to a study done in Bangladesh, the excessive use of bleaching powder in the Bangladeshi industry is the cause of the alkaline nature of the textile dyeing effluents (Table 2) [141].

The electrical conductivity of the effluent was recorded to be greater than above the Department of environment's (Bangladesh's) regulation limit of 1200 μS/cm, ranging from 250 to 7950 μS/cm (Department of Environment). The higher electrical conductivity indicates that the textile effluent from Bangladesh contains a significant number of ionic compounds (Table 2) [141]. BOD and COD readings were discovered to almost always exceed the limits for discharge. Higher COD levels are a sign of a toxic situation due to the presence of physiologically resistant organic elements, whereas higher BOD levels reflect the effluents' pollution intensity [143]. The significant rise in COD levels relative to BOD also draws attention to the high amounts of heavy metals present in the effluents [141]. Because the TDS and TSS were larger, more oxygen was required for the formation of the organic and inorganic particles in the textile dyeing effluents [164].

In a different study, effluent from two states in Pakistan, Faisalabad, and Karachi (Table 4), as well as from the key four textile-oriented industrial districts of India's Punjab, Tamil Nadu, Maharashtra, and South Gujarat (Table 3), was characterized [71]. According to reports from that study, the pH, COD, BOD, TSS, and TDS of textile dyeing effluents exceeded both the National Environmental Quality Standard (NEQS) limit and the Board of Indian Standards (BIS) standard allowed limits (Tables 3 and 4, respectively) [71].

Table 2 Physical and chemical characteristics of textile dyeing effluents taken from various parts of Bangladesh

Area	Temp. (°C)	pH	TDS (mg/l)	COD (mg/l)	BOD5 (mg/l)	EC (μs/cm)	References
Dhaka (DEPZ)	37–65	8.7–10	460–5981	508	90–460	250–7950	[82]
Gazipur	34.7–48.8	8.9–10	531–1006	–	560–965	0.88–1701	[54, 159]
Narshindi	–	5–14	127–2676	–	–	–	[70]
Narayanganj	50	6.8–11	152–1011	268–1275	60–450	592–1696	[159]
Chittagong	25–55	8.9–11	685–1338	487–1120	140–420	1108–1907	[159]
Pabna	–	6.8–11	152–1011	208–1275	60–450	592–169	[159]
Standard (*DoE, BD)	50	6.5–9	2100	200	50	1200	[49]

Table 3 Physical and chemical characteristics of textile dyeing effluents collected in four different Indian regions

Area	pH	COD (mg/l)	BOD5 (mg/l)	TS/TSS (mg/l)	TDS (mg/l)	References
Punjab	4.3–11.9	120–3000	108–790	450–51,450	430–49,440	[89]
Kongu, Tamil Nadu	8.66	3080	970	7116	242,220	[51]
Maharashtra	7.9–9.5	548–816	158–226	128–192		[7]
South Gujrat	4–13	32–58	72–93	40–160 (TS)	8–20	[123]
Standard (BIS)	5.5–9	250	30	2100	150	[49]

Table 4 Physical and chemical characteristics of textile dyeing effluents collected from two Pakistani regions

Area	Temp. (°C)	pH	TDS (mg/L)	TSS (mg/L)	COD (mg/L)	BOD5 (mg/L)	References
Faisalabad		8–14	2700–4200	66	590–880	211–487	[118]
Karachi	36–49.2	7.5–11.55	2469–7295	934–1875	115–705	85–653	[69]
Standard (NEQS)	Up to 40	6–9	3500	150	200	80	[69]

4 Impact of Textile Dyes on Animals

4.1 Impact on Land Animals' Health

Textile dyes are very toxic and potentially carcinogenic [157]. Textile pollutants, including dyes, are potential determinants of environmental degradation and multiple ailments in animals, for instance, damage to organs and the central nervous system [94]. Lacasse and Baumann investigated the carcinogenic properties of 4000 dyes used in industries and observed that more than 100 of them produce carcinogenic amines [101]. The dyes are accumulated in sediments and soil due to their recalcitrance in aerobic conditions, particularly in typical treatment plants, and are transported to public water supply systems [166]. Some are partially degraded to azo-type compounds, generating toxic aromatic amines, followed by the accumulation of potentially mutagenic and carcinogenic substances [73, 166].

The mutagenic and recalcitrant nature of the dye components affects the structure and functioning of the ecosystems [138]. These dyes release heavy metals such as nickel, copper, cobalt, and, above all, chromium [24, 33]. Long-term exposure to these complex metal dye components brings profound changes to the animal's health [73].

Textile dye intermediate compounds, for instance, benzidine and 2-naphthylamine, are associated with dermatitis, tumorigenesis in the exposure sites, and damage to the central nervous system in animals [32]. Once inside the body, some dye components substitute for the enzymatic cofactors, resulting in the deactivation of the enzymatic functionalities, followed by metabolic disorders [36]. Furthermore, the severe toxicity of textile dyes in animals is primarily caused by oral consumption and inhalation, triggering severe irritations to the eyes and skin [32, 35]. The animals exposed to the reactive dyes may suffer contact dermatitis, rhinitis, allergic conjunctivitis, and other allergic responses [65]. Azo-lipophilic dyes are widely used in a variety of industrial sectors, including textiles. When animals eat these dyes, their intestinal flora uses enzymes to change them into cancer-causing aromatic amines [130].

4.2 *Impact on Fish and Other Aquatic Animals*

Textile dyes have a significant negative impact on aquatic animals and fish. Once the dye compounds are released into the aquatic ecosystem, fish and other aquatic life can absorb the heavy metal ions and build up these heavy metals in their bodies [86, 165].

A study with textile dye wastewater on freshwater fish, *Mastacembelusarmatus*, a proteinous edible freshwater fish, showed that the dye-exposed fish has a differential level of ionic regulation in its liver, kidney, and muscle tissues [77]. *Mastacembelusarmatus* had a decrease in the concentration of sodium and chloride ions and an increase in the concentration of calcium, potassium, and magnesium ions [77, 120]. Exposure to the textile industry effluent resulted in unusual behavior, including erratic swimming, rapid opercular movement, hyper-excitation, and a thick mucus layer in the teleost fish *Poeciliarecticulata* [144]. Furthermore, these fish showed an infiltration of hemocytes into the lumen and showed symptoms of disintegration of the intestinal villi [144, 151]. Likewise, a freshwater fish, *Oreochromis mossambicus,* showed a hematological change after exposure to the textile effluent [87]. They also demonstrated liver histological changes such as hyperaemia, necrosis, and degeneration [87, 110]. The nutritional value of fish is also affected by textile effluents. After exposure to the textile dyes, the overall amount of protein was significantly reduced in a freshwater fish, *Catlacatla*, and in a freshwater crab, *Spiralothelphusahydrodroma* [66, 126]. The details of the effect of toxic dye-containing wastewater on aquatic fauna such as fish and the diseases associated with it are depicted in Fig. 3.

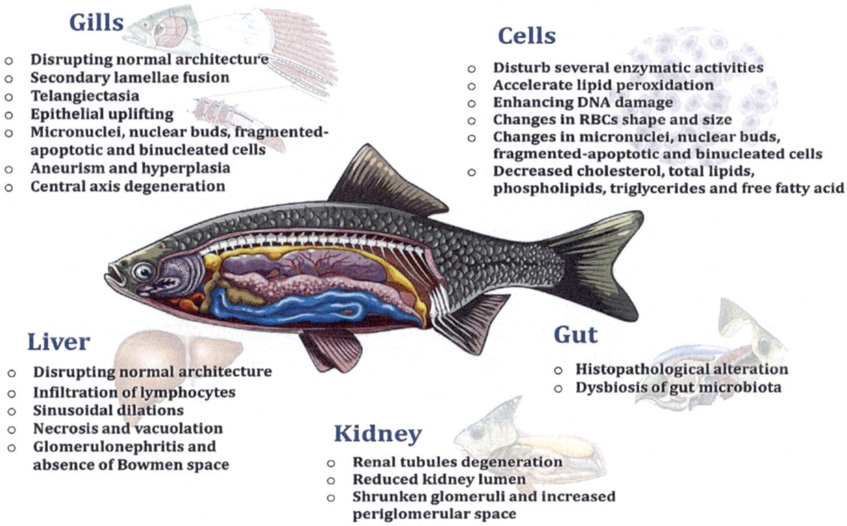

Fig. 3 The effect of toxic dye-containing wastewater on aquatic Fauna such as fish and the diseases associated with it. Reprinted from Ref. [3] with permission. Copyright (2022) Elsevier

5 Impacts of Textile Dye on Soil and Plants

5.1 Harmful Impacts of Textile Dye on Soil and Plants

The two industries that use the most water and are responsible for soil and water pollution are dyeing and printing. The groundwater reservoir in the industrial zone is constantly filled with harmful compounds, which are then used for irrigation. These dangerous chemicals can migrate from the soil to plant tissues and build up there. They can harm plants directly and also pose a risk to people's and animals' health. Textile effluent has a high biological oxygen demand (BOD) and chemical oxygen demand (COD), is brightly colored, and contains non-biodegradable chemicals. When these effluents are dumped into bodies of water untreated, they color the water and seriously harm aquatic life. Excessive color build up in effluent pollutes water bodies and blocks light from penetrating, which in turn hinders aquatic flora's ability to photosynthesize. The detrimental effects of many industrial effluents on plant growth have been well-documented by scientists, and dye wastewater has also been found to be poisonous to a number of crop plants [107]. The effect of textile effluent on the structure of cells and plant growth indicates that longer exposure to textile effluent may cause distortion of the cell [28].

It was discovered that the phytoremediation process used to remediate textile effluent affected plant development. When compared to control plants, there were decreases in plant height (8, 12, 30, and 9%), wet weight (20, 21, 42, and 20%), dry weight (27, 20, 51, and 20%), leaf size (9, 6, and 13%), and flower production (9, 8, 12, and 6%), respectively [28]. To evaluate the usefulness of textile effluent, numerous researchers irrigated a variety of crops using dye used in textile industries. According to Hayyat et al. [62], Sorghum vulgare Pers CV-5000 showed a similar decline in vegetative development indices, including plant height, number of leaves, number of senescent leaves, rate of photosynthesis, and rate of transpiration. When exposed to textile effluent, a wheat cultivar's dry weight was observed to decrease [90]. The overall production of peanut varieties, according to Saravanamoorthy and Kumari [142], was similarly impacted. According to Begum et al. [15], rice yield and growth were similarly impacted. According to Faryal et al. [45], the degradation of soil structure caused by various contaminants and textile effluent finally results in a reduction in output.

Malaviya et al. [108] showed that they also have a favorable effect on the germination and growth of *Pisum sativum* at lower concentrations of dying and industrial effluent. The growth of root and shoot length was delayed because nutrients were depleted to dangerous levels at 100% dye effluent concentration. At least for proper agricultural productivity, the germination of seed is an important stage that assures reproduction and, consequently, the control of a dynamic population in it. This soil pollution provokes plant growth inhibition by triggering oxidative pressure, reducing the protein content, photosynthesis, and CO_2 assimilation rates [19, 166].

5.2 Phytoremediation of Textile Dye

Utilizing plants to remove textile dye pollutants from the environment is known as "phytoremediation". To remove dye contaminants from the soil, it employs wild or transgenic plants [129]. Extraction, degradation, stabilization, and volatilization are all parts of phytoremediation. The term "phytoextraction" describes the process of removing metals or organics from soils by encapsulating them in plant biomass. Rhizofiltration is the process by which chemicals are removed from aqueous sources by plant roots. Phytodegradation or phytotransformation is the use of plants for the intake, storage, and breakdown of organic contaminants. By immobilizing or binding pollutants to the soil matrix, phytostabilization reduces their bioavailability, whereas phytovolatilization makes use of plants to absorb toxins from the growth matrix, change them, and release them back into the atmosphere. Uptake, transfer, and sequestration into the vacuole are all ways that the detoxification process might be carried out. The process of metabolization is completed by oxidation, reduction, hydrolysis, and conjugation with amino acids, glucose, or glutathione. Since the biomass made by bioremediation can be used to make the economy better without making more CO_2, phytoremediation is also seen as a method that could be profitable [116].

Natural industrial effluent can be used as a source of nutrients for the soil as well as an organic fertilizer. Using wastewater for irrigation has a number of benefits that reduce both short-term and long-term environmental problems. From crop to crop, industrial wastewater has different effects [72]. Textile effluents can be utilized for irrigation safely once they have undergone enough treatment and been diluted to 25% [132]. Textile dye can be phytoremediated with plants to reduce environmental pollution. The capacity for textile dye removal at laboratory sizes by nursery grown plants of *A. amellus*, *T. patula*, *P. grandiflora*, and *G. grandiflora* has already been documented [96, 124, 167]. After in vitro (plant tissue culture) experiments, it was discovered that these plants directly contribute to dye clearance. The true indicator of the color of water is the dye indicator value, which doesn't depend on the type of color of the dye and gives an exact value for the color of water and textile wastewater [85]. It was discovered that planting these garden decorative plants significantly decreased the color indicator value of accumulated dye in impacted soil caused by genuine effluent. Grown for a month on the ridges of a wetland and watered with textile effluent, *T. patula*, *A. amellus*, *P. grandiflora*, and *G. grandiflora* could lower the color indicator values in soil by 8, 20, 27, 39, 47, and 59%; 11, 16, 27, 32, 44, and 50%; 09, 15, 23, 29, 41, and 46%; 18, 25, 40, 47, 66, and 73%, as seen. When compared to *T. patula*, *A. amellus*, and *P. grandiflora*, *G. grandiflora* was shown to exhibit superior decolorization, as seen by the comparably larger clearance of accumulated dye from the soil [28]. In other studies, it was discovered that an artificial lagoon with a 60,000 L capacity that was planted with *Ipomoea aquatica* showed a color clearance of 70% within 8 days [135]. The color indicator value of textile industry effluent was reported to be reduced by up to 59, 62, and 76%, respectively, after 4 days in plantations of *Paspalum scrobiculatum*, *Typha angustifolia*, and their consortium along wetland ridges (91.4 m × 1.2 m × 0.6 m) [29]. While soil control (no plant) showed only a 22% reduction in the color indicator values of dyes in textile effluent, cultivation of Asparagus densiflorus along ridge beds of a high-rate transpiration system showed a reduction of up to 67% [168]. Green HE4B Remazol Red, Direct Red 5B, and Brilliant Blue R decolorization has also been shown to be significantly influenced by *A. amellus*, *B. malcolmii*, *Z. angustifolia*, *G. pulchella*, and *P. grandiflora* lignin peroxidase, Veratryl alcohol oxidase, tyrosinase, dichlorophenolindophenol reductases, and laccase [79, 80, 96, 97]. Even though phytoremediation might make the dye less harmful, it could hurt plant growth, development, and survival in some plants, as shown in Fig. 4.

The comparatively slow speed of phytoremediation is a significant disadvantage. Enzymatic activity may be disrupted, or significant amounts of reactive oxygen species may be produced by cumulative toxicants [116]. Recently, nanoparticles have found great application in various fields [56–59, 64, 92, 115] for their advanced features, including the nanoremediation of textile dye effluents. Applying nanoparticles to reduce toxins through oxidation or reduction processes is known as "nanoremediation" [116].

Fig. 4 Phytoremediation and its effect on plants: **a**, **f** normal plants before remediation; **b–e** *E. crassipes* after adsorption of dye; **g** control *S. natans* following remediation; **h** *S. natans* when remediated with a 500 mg/L dye solution; **i** *S. natans'* leaves after remediation. Reprinted from Ref. [137] with permission under Creative Commons Attribution (CC BY) license

6 Impact of Textile Dyes on Microbiota

Microorganisms, which comprise 15% of total living bodies [14], are usually unicellular and microscopic. Despite being small, their contribution to the ecosystem is not insignificant. Microbes set their "feet" on earth 4 billion years ago [38]. General populations often consider microbes as pathogenic bugs. The truth is that only 5% of them are the causative agents of infectious diseases in humans, plants, and animals. Most of them are beneficial to us. Some of them have an essential role in nutrient cycling. Some contribute to the production of cheese, yogurt, and other fermented foods. Microorganisms found in our intestines help us with digestion and vitamin synthesis. Surprisingly, a 70 kg human body contains more than 30 trillion bacterial cells [147].

Microbes play a crucial role in the environment by decomposing complex organic matter into simple carbon. This whole process serves two purposes. (1) It helps nature facilitate degradation, and (2) It keeps the environmental cycles running. A general example of bioremediation is the degradation of plant litter by cellulose-degrading fungi. Fungi are the front-line soldiers fighting this biodegradation battle. Cellulose is a complex natural polymer that is difficult to decompose. Cellulosic carbohydrates are degraded by the extracellular cellulolytic enzymes released by the fungi, resulting in glucose oligomers, which are further catabolized by several bacterial species while releasing carbon into the environment [155]. In a nutshell, nature has all the machinery for biodegradation and chemical recycling. However, this is not the case for artificial "difficult-to-degrade" compounds, which we call xenobiotics. Nature is still changing to deal with the complete breakdown of xenobiotics. As a result, these xenobiotics keep piling up in our environment every day, which has serious effects on our ecosystem.

Textile-based dyes are an excellent example of compounds with a xenobiotic nature. Commercial textile dyes were introduced into the world in the late nineteenth century during the Industrial Revolution. Humans have since transitioned from wearing handwoven handicraft natural clothes made of cotton, silk, or wool to various synthetic fibered apparel such as nylon, rayon, polyester, and so on. Due to their ease of chemical preparation, cost-effectiveness, durability, and broad color spectrum compared to natural dyes, different artificial dyes eventually overtook the natural dye market. Acid dyes, azo dyes, basic dyes, and reactive dyes are some of the common names of lab-based artificial dyes. The combination of industrially improved manufactured fiber and the discovery of high-quality synthetic dye resulted in the boom of the multibillion-dollar textile industry. Because of the cheap workforce and weak law enforcement practices, China, India, and Bangladesh are becoming central hotspots for the textile industries, where they manufacture, process, and export textiles, clothing, and ready-made garments. Although this industry makes a lot of money, the lack of green manufacturing practices creates a huge pollution problem. The industry uses a lot of water and dumps heavy metals, dyes made from chemicals, and wastewater into rivers. This pollutes the major rivers a lot.

6.1 *Impact of Textile Dyes on Environmental Microorganisms*

When toxic wastewater containing textile dyes is released into the water, it not only increases the biological and chemical oxygen demands but also alters the pH and salinity and impairs the transparency of the water [50, 93, 104]. All these consequences pose a significant threat to the water microbiota. Sunlight cannot penetrate the colored layer of topwater, which causes the slow death of the aqueous diatoms, phytoplankton, and microalgae. Due to the inhibition of photosynthesis, the water sphere does not get enough oxygen, which in turn causes the death of aquatic life [104]. A high concentration of textile dyes inhibits the growth of a cyanobacterium named *Spirulina platensis* [50]. Even a low concentration of textile dye wastewater effluent containing indigo dye and heavy metal mixtures can significantly reduce the growth and alter the morphology of *Scenedesmus quadricauda*, a freshwater microalga [140]. One study showed that the total heterotrophic bacterial and fungal count and diversity were increased in a river site contaminated with textile discharge compared to the non-industrial site of the river [76].

Moreover, the dyes can adsorb from the polluted water into microbial cells, thereby introducing themselves into the food chain [113]. 5 mg/L Basic Violet 3 dye toxicity reduces the mean survival rate of freshwater microbes to 20.7% [113]. Water-soluble dyes can be taken up by microbes and degraded as carcinogenic compounds [34]. Some dyes, such as Disperse Yellow 7, can be taken up by sediment in water and cause genotoxicity [13].

People living around textile processing industries often use textile effluent water for soil irrigation, which results in the bioaccumulation of textile dyes in the soil [139]. One study showed that textile dye-contaminated wastewater significantly lowers the microbial population [93]. One gram of red S3B dye per kg of soil led to a 33% decrease in viable microbial count [8]. Another study showed that the azo dyes, e.g., Reactive Black 5, can be retained in the soil for up to several weeks, alter the microbial community's structure, and reduce the growth of *Pseudomonas fluorescence* [68].

6.2 *Dye-Based Toxicity to Microorganisms*

Textile dyes are toxic to microbes in a variety of ways. Some are considered less toxic than others, and some are more toxic. Due to their recalcitrant nature, toxic textile dyes are difficult to decompose; therefore, they gradually bioaccumulate in the soil and nutrient cycles. In humans and large animals, this bioaccumulation and biomagnification of dyes causes cancer, organ failure, a neural defect, hypersensitivity, etc. [104]. For microbes, the overall impact of textile dyes is also significantly harmful. According to Sharma et al., Sulfur Red Brown 360 (SRB), Jade Green 2G (JG), Reactofix Turquoise Blue 5GFL (RTB), and Direct Scarlet 4BS dye induce DNA damage, mutation, and cytotoxicity in *Bacillus subtilis*, a common soil *bacterium*

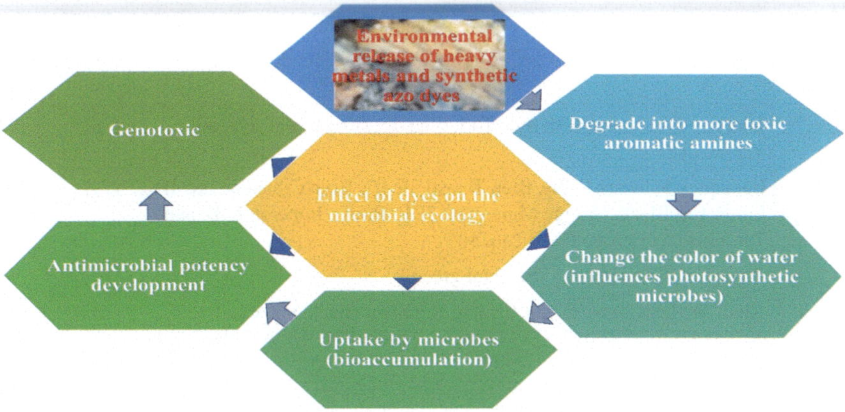

Fig. 5 Effect of textile dyes on environmental microorganisms

[150]. They also proved that the dye toxicity is dose-dependent, meaning the higher the dose, the more toxic the dye's effect on the bacterium [150]. Figure 5 summarizes the probable consequences of textile dyes on environmental microorganisms.

Textile dyes and their degradative byproducts can cause genetic mutations in the bacterial DNA [25, 74, 88, 104, 133]. A common approach is using the Ames test, which is performed to determine if a compound has the potential to cause genetic mutations [111, 112]. This test involves an auxotrophic mutant of *Salmonella typhimurium*, which cannot grow without histidine. The bacterial culture is treated with a potential mutagen. If the compound can induce a mutation in *Salmonella* DNA, it reverses the mutation, thereby allowing the organism to grow without histidine. One study concluded that out of 53 dyes, 15 were positive for the Ames test [74]. Moreover, with time, the parent azo dyes are chemically degraded into aromatic amine residues, which are also mutagenic [25]. Some of the azo dyes are degraded by a bacterial enzyme named "azoreductase". Upon degradation, the Direct Black 28 azo dye is converted to another mutagenic compound called benzidine [27]. This bacterial biotransformation of azo dyes occurs under an anaerobic condition by human intestinal microflora [27, 30, 34], which implies a potential reason for textile workers' having a high rate of colorectal cancer [154]. Khan et al. reported that the textile wastewater is positive for the Ames test [93]. Table 5 shows a representative list of some dyes that are mutagenic using the Ames test.

The mechanism by which textile dyes cause genotoxicity varies. For example, Disperse Orange 1 dye causes frameshift mutation, base pair substitution, and DNA damage in the *Salmonella* spp. assay [104], whereas Azure-B dye can intercalate between the DNA and duplex RNA [32]. Upon release into the environment or upon ingestion, these toxic azo dyes are further transformed by microbial enzymes into aromatic amines, which are even more harmful. The amines released from Direct Blue 14 and Sudan I dyes and Disperse Yellow 7 dyes are more carcinogenic than the parent molecules [13, 32].

Table 5 The list of dyes that show positive results in the Ames test

Serial no.	Dye name	References
1	Jaka red dye, Brown 3BS dye, Tork blue dye, Royal blue dye, and Yellow dye	[88]
2	Bemaplex Schwarz C-2B, Bleu Terasil 3R-02, Brun Cibanone 2RMP, Lumacron Black, Orange Minerprint 3RL, Rouge Imperon K-B, and Turquoise Cibacrone P-GR Liq. 50%	[74]
3	Trypan red, Basic blue 7, Brilliant green	[127]
4	All of the benzidine (Direct Brown 95, Direct Black 38, Acid Red 85, etc.) and o-tolidine dyes (Direct Red 2, Direct Blue 53, Acid Red 114)	[133]
5	Methyl orange	[34]
6	Disperse Red 13 (DR13) and its degradation products	[31]

6.3 The Antimicrobial Role of Dyes

Though their inhibitory mechanisms vary, dyes are often considered "antimicrobial" [2, 9, 21, 103]. Sometimes, the dye molecules are sufficient to inhibit bacterial growth, such as when azo dyes are inhibitory to *Pseudomonas aeruginosa* [81, 131]. Sometimes, the heavy metals present in the dyes are responsible for the inhibition, such as the copper molecules present in the benzidine dyes and C.I. direct blue 218 dye, which drastically reduce the survival rate of crustacean plankton *Daphnia Magna* [12].

Paul Ehrlich announced methylene blue as the panacea for malaria in 1891 [21]. Later on, the concept of using dyes to inhibit bacterial infection gained popularity with the use of rosaniline dye to treat African sleeping diseases (caused by the microscopic parasite *Trypanosoma brucei*) and syphilis (caused by the spirochete, *Treponema pallidum*) [21]. Before the discovery of penicillin antibiotics, scientists used two chemically modified synthetic dyes named: (1) arsenicals to treat protozoan and spirochete infections, and (2) sulfonamide to treat bacterial, urinal, and vaginal infections. After World War II, more than a dozen sulfonamide-based drugs were made available in the drugstore [103]. These sulfa drugs selectively kill bacteria by inhibiting the folic acid biosynthesis pathway while not affecting the human host much, except for mild side effects, e.g., nausea and vomiting [171]. Triphenylmethane (TPM), crystal violet, malachite green, and brilliant green are other dyes used as antimicrobials and antiseptics [21]. Oral consumption of crystal violet is used for pinworm treatment in humans and livestock [10]. However, these early studies on antimicrobial dyes were forgotten after penicillin was found, developed, and made in large quantities in the 1940s.

Natural dyes can be antimicrobial, too. Curcumin, the active ingredient in turmeric, is inhibitory to 70% of *Escherichia coli* at a concentration as low as 0.05% [53]. The henna plant is another source of natural dye. Alcoholic and oil extracts of henna exhibited antimicrobial activity against beta-hemolytic streptococci, *Staphylococcus aureus*, *S. epidermidis*, and *P. aeruginosa* [2]. A natural blue dye obtained from Cinchona tree bark extract has been used to treat malarial infection caused by *Plasmodium* sp. since at least the early seventeenth century [162]. The name of the dye is known to all, and it is none other than quinine. Table 6 summarizes some dyes with antimicrobial properties.

6.4 Bioremediation of Textile Dyes Using Microorganisms

The microorganism can be used as a catalyst to turn the chemical energy in organic waste into useful energy and byproducts that are not harmful [60]. According to most reports, bacteria can metabolize toxic dyes and use them as a source of carbon and nitrogen [104, 136, 158]. One study showed that the dye-containing wastewater significantly lowered the microbial population [93]. Soil bacteria can degrade the azo dyes into non-toxic forms using several enzymes, mainly hydroxylases, azoreductases, laccases, peroxidases, and oxygenases [139, 158]. Although dyes can be degraded and decolored using various physicochemical methods, i.e., adsorption, ion exchange, photocatalysis, ozonation, etc., they are not eco-friendly [104, 136, 158]. On the contrary, biodegradation using microorganisms will provide a promising era of green industry where the microbes will actively participate in detoxifying, decolorizing, and degrading the dye molecules in the environment. For textile-contaminated wastewater treatment, a cell-free system can be used by applying microbial enzymes to the residual dyes. As the enzymes are unstable outside cells, bacterial consortiums, single species or multiple species bacterial cultures, and activated sludge can also be added for improved microorganism-mediated bioremediation of textile effluent.

One study showed that 20 bacterial strains and 11 bacterial strains obtained from activated sludge and river water, respectively, can decolorize TPM dyes, Bengal rose dyes, and azo dyes. The efficiency rate, on the other hand, varies [172]. *Klebsiella*, *Pseudomonas*, *Bacillus*, *Rhodococcus*, *Arthrobacter soli* BS5, and *Shigella* sp. are some familiar names of bacteria involved in the decolorization and degradation of recalcitrant dyes [3, 75, 93]. Not only heterotrophic soil bacteria, but algae, fungi, and single-cell yeast can also play an important role in degrading the toxic dye molecules into non-toxic byproducts. *Aspergillus niger* and *Phanerochaetechrysosporium*, *Rhizopus oryzae*, *Trametes versicolor*, and other fungi secrete extracellular enzymes that help decolorize direct textile dyes [136, 146]. Phototrophic cyanobacteria Bioremediation of textile dye components is done by *Phormidiumautumnale*, *Lyngbyalagerlerimi*, *Chlorella vulgaris*, *Nostoc lincki*, *Elkatothrixviridis*, *Oscillatoria rubescens*, and *Volvox aureus* [37, 43].

Bacteria are considered the new soldiers against environmental pollution. Whenever there is selection pressure, whether natural or artificial, they will evolve to alleviate the pressure. Surprisingly, minimal studies report bacterial impacts due to the textile dye in soil and water. Instead, most studies focus on the success stories of

Table 6 Some common dyes, their applications, and microorganisms they are inhibitory to

Serial no.	Dye name	Dye type	Application	Antimicrobial activity against microbiota (if any), mechanism (if any)	References
1	Bis-mono-azo a1,b1 and diazo a2, b2	Acid	Silk and wool	*Escherichia coli*, *Staphylococcus aureus*	[114]
2	Thiazolidin-4-one derivatives	Reactive	Silk, wool, cotton, and polyester fabrics	*E. coli, S. aureus, Aspergillus niger*	[9]
3	Chrysodine	Azo	Silk, wool, cotton	*Bacillus*sp, *Streptococcus*sp, *Staphylococcus* sp.	[47]
4	Rosalinine	Azo	Fungicide, fabric dye, ink	Antibacterial against gram positive bacteria, antifungal, antiviral	[21, 83]
5	Sudan dyes	Azo	Textile dye, food coloring agent	Affect growth rate of intestinal microflora, e.g., *Clostridium perfringens, Lactobacillus rhamnosus, Enterococcus faecalis, E. coli*	[121]
6	Triphenylme Fung & Miller thane (TPM) dye, anionic TPM dye, aluminon	Triaryl methane	Nylon, wool, silk, cotton	Inhibition of the glutamine synthesis, and block protein synthesis. Effective against *E. coli, S. aureus*	[21]
7	Malachite green	Triphenyl methane	Leather, textile, paper	Antifungal, antiparasitic, inhibiting the growth of zygomycetes or spreading molds	[22, 46]

(continued)

Table 6 (continued)

Serial no.	Dye name	Dye type	Application	Antimicrobial activity against microbiota (if any), mechanism (if any)	References
8	Crystal violet	Triphenyl methane	Leather, textile, paper ink	*Bacillus cereus, B. polymyxa, B. subtilis, Micrococcus sp, Sarcina sp, S. aureus, Streptococcus, bovis, S. faecalis S. lactis, E. coli, Proteus vulgaris, Pseudomonas aeruginosa, Salmonella typhimurium, Shigella flexneri, Serratia marcescens*	[47]
9	Acrilflavin chloride, 3-Aminoacridine and Aminoacridine, Proflavine	Basic	Textiles, topical antiseptic	Antiviral, Antifungal, Antimicrobial, Induce DNA damage and mutation	[127]
10	Brilliant green	Cationic	Pulp and paper, dying industries	*Enterobacter aerogenes, P. aeruginosa, S. marcescens.* Mutagenic in bacterial DNA	[47, 127]
11	Methylene blue	Cationic	Dying paper, hair, cotton, wool	Oxidative DNA damage in gram positive bacteria	[47, 127]
12	197Rhodamine 6G	Water-soluble fluorescent dye	Silk, cotton, leather	Inhibitory against gram negative bacteria, induce mutation, inhibit aerobic growth of *Saccharomyces cerevisiae*	[26, 47, 117]

bacteria actively taking part in degrading textile dyes, thereby cleaning up the environment. To sum up what has been said about how textile dyes affect microorganisms, they are not as harmful to humans and other animals as they are to microorganisms. In fact, microorganisms use them well.

7 Conclusion and Recommendation

Textile dyes have a complex nature due to the presence of various dyes and other organic agents that, under various operating conditions, can accumulate in natural sources such as soil and water after untreated disposal and may be toxic to ecosystems and human life. In this chapter, we tried to illustrate the detrimental consequences that textile dyes have on the water quality, health of animals, plants, soil, and microbiota. As a result of these detrimental consequences, it is critical to remove dye contaminants from the biosphere. All living beings depend on water to survive, but according to estimates, approximately 800 million people worldwide still lack access to clean drinking water that is suitable for domestic use. In recent years, the widespread contamination of natural water supplies by organic and inorganic pollutants from textile dye has become a concern for many nations. Textile wastewater contains a variety of harmful xenobiotics that are found in high concentrations and are dangerous to the environment and the general public's health. Because textile factories all over the world routinely release millions of gallons of severely polluted effluent, toxic dye-containing wastewater is a significant issue. On the other hand, the treatment of wastewater that contains dye is a considerable barrier because no specialized and practical method exists to effectively address this issue. There are numerous established and new methods of treating wastewater that contains dyes. Physical and chemical techniques seem to be efficient at removing or degrading dye from wastewater that contains dye. These methods, however, entail expensive operating expenses and undesired consequences.

The microbial and phytoremediation approach to dye-containing wastewater clean-up is more economical, environmentally benign, and socially and politically acceptable when compared to physical and chemical techniques. However, one disadvantage of biological methods is that they are slow, require a lengthy duration of administration, and are less effective. As a result, additional research is required to find a high-tech, waste-free method, and to lower the dangers to the environment and public health associated with the transition from the lab to the pilot scale. Chemical and nano-based treatments, on the other hand, are new technologies that speed up textile dye remediation from the environment. Because of the versatile properties of nanoparticles in reducing the costs and time of wastewater clean-up operations, nanotechnology application in wastewater purification has a promising future. Applying nanoparticles to reduce toxins through oxidation or reduction processes is known as "nanoremediation". Additionally, it serves as a catalyst to quicken the velocity of the reaction. Materials with a size of 100 nm or less, a sizable surface area, and a high surface reactivity are known as nanomaterials. Nanoparticles and their composites have been found to decompose organic contaminants, according to numerous studies. There is still much to learn about the science and engineering behind bioremediation. Researchers hope that a better understanding of the bioremediation process will help them improve their method and solve the dye contamination problem.

References

1. Adinew B (2012) Textile effluent treatment and decolorization techniques—a review. Bulgarian J Sci Educ 21(3):434–456
2. Al-Rubiay KK, Jaber NN, Al-Mhaawe B, Alrubaiy LK (2008) Antimicrobial efficacy of henna extracts. Oman Med J 23(4):253
3. Al-Tohamy R, Ali SS, Li F, Okasha KM, Mahmoud YA-G, Elsamahy T, Jiao H, Fu Y, Sun J (2022) A critical review on the treatment of dye-containing wastewater: ecotoxicological and health concerns of textile dyes and possible remediation approaches for environmental safety. Ecotox Environ Safe 231:113160
4. Aldalbahi A, El-Naggar ME, El-Newehy MH, Rahaman M, Hatshan MR, Khattab TA (2021) Effects of technical textiles and synthetic nanofibers on environmental pollution. Polymers 13(1):155
5. Ali SS, Al-Tohamy R, Sun J (2022) Performance of Meyerozyma caribbica as a novel manganese peroxidase-producing yeast inhabiting wood-feeding termite gut symbionts for azo dye decolorization and detoxification. Sci Total Environ 806:150665
6. Amer OA, Ali SS, Azab M, El-Shouny WA, Sun J, Mahmoud YA-G (2022) Exploring new marine bacterial species, Alcaligenes faecalis Alca F2018 valued for bioconversion of shrimp chitin to chitosan for concomitant biotechnological applications. Int J Biol Macromol 196:35–45
7. Amte G, Mhaskar TV (2012) Studies on textile-dyeing effluent from Bhiwandi city, Dist: Thane, Maharashtra India. I Control Pollution 28(2):197–199
8. Arshad H, Imran M, Ashraf M (2020) Toxic effects of Red-S3B dye on soil microbial activities, wheat yield, and their alleviation by pressmud application. Ecotox Environ Safe 204:111030
9. Ayalew H, Reda G, Gashaw T, Babu N, Upadhyay RK (2014) Antimicrobial and dyeing properties of reactive dyes with thiazolidinon-4-one nucleus. Int Scholarly Res. Notices 2014:894250
10. Azmi W, Sani RK, Banerjee UC (1998) Biodegradation of triphenylmethane dyes. Enzyme Microb Tech 22(3):185–191
11. Babu BR, Parande A, Raghu S, Kumar TP (2007) Cotton textile processing: waste generation and effluent treatment. J Cotton Sci 11(3):141–153
12. Bae JS, Freeman HS (2007) Aquatic toxicity evaluation of copper-complexed direct dyes to the Daphnia magna. Dyes Pigments 73(1):126–132
13. Balakrishnan VK, Shirin S, Aman AM, de Solla SR, Mathieu-Denoncourt J, Langlois VS (2016) Genotoxic and carcinogenic products arising from reductive transformations of the azo dye, Disperse Yellow 7. Chemosphere 146:206–215
14. Bar-On YM, Phillips R, Milo R (2018) The biomass distribution on Earth. Proc Natl Acad Sci 115(25):6506–6511
15. Begum R, Zaman M, Mondol A, Islam M, Hossain M (2011) Effects of textile industrial waste water and uptake of nutrients on the yield of rice. Bangladesh J Agric Res 36(2):319–331
16. Ben Mansour H, Houas I, Montassar F, Ghedira K, Barillier D, Mosrati R, Chekir-Ghedira L (2012) Alteration of in vitro and acute in vivo toxicity of textile dyeing wastewater after chemical and biological remediation. Environ Sci Pollut R 19(7):2634–2643

17. Benkhaya S, M'rabet S, El Harfi A (2020) A review on classifications, recent synthesis and applications of textile dyes. Inorg Chem Commun 115:107891
18. Berradi M, Hsissou R, Khudhair M, Assouag M, Cherkaoui O, El Bachiri A, El Harfi A (2019) Textile finishing dyes and their impact on aquatic environs. Heliyon 5(11):e02711
19. Bhatia D, Sharma NR, Singh J, Kanwar RS (2017) Biological methods for textile dye removal from wastewater: a review. Crit Rev Env Sci Tec 47(19):1836–1876
20. Bhattacharya S, Gupta AB, Gupta A, Pandey A (2018) Introduction to water remediation: importance and methods. In: Water remediation. Springer, pp 3–8
21. Boulos RA (2013) Antimicrobial dyes and mechanosensitive channels. Antonie Van Leeuwenhoek 104(2):155–167
22. Bragulat M, Abarca M, Bruguera M, Caba es F (1991) Dyes as fungal inhibitors: effect on colony diameter. Appl Environ Microbiol 57(9):2777–2780
23. Brik M, Schoeberl P, Chamam B, Braun R, Fuchs W (2006) Advanced treatment of textile wastewater towards reuse using a membrane bioreactor. Process Biochem 41(8):1751–1757
24. Brock T, Groteklaes M, Mischke P (2000) European coatings handbook. Vincentz Network GmbH & Co KG
25. Brüschweiler BJ, Merlot C (2017) Azo dyes in clothing textiles can be cleaved into a series of mutagenic aromatic amines which are not regulated yet. Regul Toxicol Pharm 88:214–226
26. Carignani G, Lancashire WE, Griffiths DE (1977) Extra-chromosomal inheritance of rhodamine 6G resistance in Saccharomyces cerevisiae. Mol Gen Genet 151(1):49–56
27. Cerniglia CE, Zhuo Z, Manning BW, Federle TW, Heflich RH (1986) Mutagenic activation of the benzidine-based dye direct black 38 by human intestinal microflora. Mutat Res 175(1):11–16
28. Chandanshive VV, Kadam SK, Khandare RV, Kurade MB, Jeon B-H, Jadhav JP, Govindwar SP (2018) In situ phytoremediation of dyes from textile wastewater using garden ornamental plants, effect on soil quality and plant growth. Chemosphere 210:968–976
29. Chandanshive VV, Rane NR, Tamboli AS, Gholave AR, Khandare RV, Govindwar SP (2017) Co-plantation of aquatic macrophytes Typha angustifolia and Paspalum scrobiculatum for effective treatment of textile industry effluent. J Hazard Mater 338:47–56
30. Chen H (2006) Recent advances in azo dye degrading enzyme research. Curr Protein Pept Sci 7(2):101–111
31. Chequer FMD, Lizier TM, de Felício R, Zanoni MVB, Debonsi HM, Lopes NP, de Oliveira DP (2015) The azo dye Disperse Red 13 and its oxidation and reduction products showed mutagenic potential. Toxicol In Vitro 29(7):1906–1915
32. Christie R (2007) Environmental aspects of textile dyeing. Elsevier
33. Christie R (2014) Colour chemistry. Royal Society of Chemistry
34. Chung K, Fulk G, Andrews A (1978) The mutagenicity of methyl orange and metabolites produced by intestinal anaerobes. Mutat Res 58:375–379
35. Clark M (2011) Handbook of textile and industrial dyeing: principles, processes and types of dyes. Elsevier
36. Copaciu F, Opriş O, Coman V, Ristoiu D, Niinemets Ü, Copolovici L (2013) Diffuse water pollution by anthraquinone and azo dyes in environment importantly alters foliage volatiles, carotenoids and physiology in wheat (*Triticum aestivum*). Water Air Soil Pollut 224(3):1–11
37. Dellamatrice PM, Silva-Stenico ME, Moraes LAB, Fiore MF, Monteiro RTR (2017) Degradation of textile dyes by cyanobacteria. Braz J Microbiol 48, 25–31
38. Dodd MS, Papineau D, Grenne T, Slack JF, Rittner M, Pirajno F, O'Neil J, Little CT (2017) Evidence for early life in Earth's oldest hydrothermal vent precipitates. Nature 543(7643):60–64
39. Donkadokula NY, Kola AK, Naz I, Saroj D (2020) A review on advanced physico-chemical and biological textile dye wastewater treatment techniques. Rev Environ Sci Biotechnol 19(3):543–560
40. Dos Santos AB, Cervantes FJ, Van Lier JB (2007) Review paper on current technologies for decolourisation of textile wastewaters: perspectives for anaerobic biotechnology. Bioresource Technol 98(12):2369–2385

41. Dutta S, Bhattacharjee J (2022) A comparative study between physicochemical and biological methods for effective removal of textile dye from wastewater. In: Development in wastewater treatment research and processes. Elsevier, pp 1–21
42. Ebency CIL, Rajan S, Murugesan A, Rajesh R, Elayarajah B (2013) Biodegradation of textile azo dyes and its bioremediation potential using seed germination efficiency. Int J Curr Microbiol App Sci 2(10):496–505
43. El-Sheekh MM, Gharieb M, Abou-El-Souod G (2009) Biodegradation of dyes by some green algae and cyanobacteria. Int Biodeter Biodegr 63(6):699–704
44. Elliott A, Hanby W, Malcolm B (1954) The near infra-red absorption spectra of natural and synthetic fibres. Brit J Appl Phys 5(11):377
45. Faryal R, Tahir F, Hameed A (2007) Effect of wastewater irrigation on soil along with its micro and macro flora. Pak J Bot 39(1):193
46. Foster FJ, Woodbury L (1936) The use of malachite green as a fish fungicide and antiseptic. The Progressive Fish-Culturist 3(18):7–9
47. Fung DY, Miller RD (1973) Effect of dyes on bacterial growth. Appl Microbiol 25(5):793–799
48. Gähr F, Hermanutz F, Oppermann W (1994) Ozonation-an important technique to comply with new German laws for textile wastewater treatment. Water Sci Technol 30(3):255
49. Ghaly A, Ananthashankar R, Alhattab M, Ramakrishnan V (2014) Production, characterization and treatment of textile effluents: a critical review. J Chem Eng Process Technol 5(1):1–19
50. Gita S, Hussan A, Choudhury T (2017) Impact of textile dyes waste on aquatic environments and its treatment. Environ Ecol 35(3C):2349–2353
51. Gomathi E, Rathika G, Santhini E (2017) Physico-chemical parameters of textile dyeing effluent and its impacts with casestudy. IJRCE 7(1):17–24
52. Gürses, A., Açıkyıldız, M., Güneş, K., & Gürses, M. S. (2016). Colorants in health and environmental aspects. In: Dyes and pigments. Springer, pp 69–83
53. Han S, Yang Y (2005) Antimicrobial activity of wool fabric treated with curcumin. Dyes Pigments 64(2):157–161
54. Hannan M, Rahman M, Haque M (2011) An investigation on the effects of bleaching powder with dyeing industries' effluents. J Civil Eng Des 39(1):77–89
55. Haq I, Raj A (2018) Biodegradation of Azure-B dye by Serratia liquefaciens and its validation by phytotoxicity, genotoxicity and cytotoxicity studies. Chemosphere 196:58–68
56. Hashem A, Marlinda AR, Hossain M, Al Mamun M, Shalauddin M, Simarani K, Johan MR (2022) A unique oligonucleotide probe hybrid on graphene decorated gold nanoparticles modified screen-printed carbon electrode for pork meat adulteration. Electrocatalysis 1–16
57. Hashem A, Hossain MAM, Marlinda AR, Mamun MA, Sagadevan S, Shahnavaz Z, Simarani K, Johan MR (2022) Nucleic acid-based electrochemical biosensors for rapid clinical diagnosis: advances, challenges, and opportunities. Crit Rev Cl Lab Sci 59(3):156–177. https://doi.org/10.1080/10408363.2021.1997898
58. Hashem A, Hossain MM, Rahman MA, Al Mamun M, Simarani K, Johan MR (2021) Nanomaterials based electrochemical nucleic acid biosensors for environmental monitoring: a review. Appl Surf Sci Adv 4:100064
59. Hashem A, Khalil I, Al Mamun M, Shalauddin M, Hossain MM, Ab Rahman M, Simarani K, Johan MR (2022) Applications of nanocellulose as biosensing platforms for the detection of functional biomacromolecules: a Review. Malay NANO Int J 2(1):15–45
60. Hashem A, Simarani K, Marlinda AR, Hossain M, Al Mamun M, Johan MR (2022) Application of microbial fuel cells as biosensors. In: Microbial fuel cells for environmental remediation. Springer, pp 349–387
61. Hassan MM, Carr CM (2018) A critical review on recent advancements of the removal of reactive dyes from dyehouse effluent by ion-exchange adsorbents. Chemosphere 209:201–219
62. Hayyat MU, Mahmood R, Hassan S, Rizwan S (2013) Effects of textile effluent on growth performance of Sorghum vulgare Pers CV. SSG-5000. Biologia 59:15–22
63. Holkar CR, Jadhav AJ, Pinjari DV, Mahamuni NM, Pandit AB (2016) A critical review on textile wastewater treatments: possible approaches. J Environ Manage 182:351–366

64. Hossain MAM, Uddin SMK, Sultana S, Hashem A, Rizou M, Aldawoud TMS, Galanakis CM, Johan MR (2021) DNA-based methods for species identification in food forensic science. In: Galanakis CM (ed) Food toxicology and forensics. Academic Press, pp 181–211. https://doi.org/10.1016/b978-0-12-822360-4.00007-8
65. Hunger K (2007) Industrial dyes: chemistry, properties, applications. Wiley
66. Hussain B, Yaseen H, Khalid Al G, Al-Misned F, Qasim M, Al-Mulhm N, Mahboob S (2021) A study on risk assessment of effect of hematoxylin dye on cytotoxicity and nephrotoxicity in freshwater fish: food and water security prospective research. Saudi J Biol Sci 28(4):2267–2271. https://doi.org/10.1016/j.sjbs.2021.01.019
67. Hussein FH (2013) Chemical properties of treated textile dyeing wastewater. Asian J Chem 25(16):9393–9400
68. Imran M, Shaharoona B, Crowley DE, Khalid A, Hussain S, Arshad M (2015) The stability of textile azo dyes in soil and their impact on microbial phospholipid fatty acid profiles. Ecotox Environ Safe 120:163–168
69. Imtiazuddin S, Mumtaz M, Mallick KA (2012) Pollutants of wastewater characteristics in textile industries. J Basic App Sci 8:554–556
70. Islam A, Guha AK (2013) Removal of pH, TDS and color from textile effluent by using coagulants and aquatic/non aquatic plants as adsorbents. Resources and Environment 3(5):101–114
71. Islam M, Mostafa M (2018) Textile dyeing effluents and environment concerns-a review. Journal of Environmental Science and Natural Resources 11(1–2):131–144
72. Islam MT, Islam T, Islam T, Repon MR (2022) Synthetic dyes for textile colouration: process, factors and environmental impact. Textile and leather review 5:327–373
73. Ito T, Adachi Y, Yamanashi Y, Shimada Y (2016) Long–term natural remediation process in textile dye–polluted river sediment driven by bacterial community changes. Water Res 100:458–465
74. Jäger I, Hafner C, Schneider K (2004) Mutagenicity of different textile dye products in Salmonella typhimurium and mouse lymphoma cells. Mutation Research/Genetic Toxicology and Environmental Mutagenesis 561(1–2):35–44
75. Jamee R, Siddique R (2019) Biodegradation of synthetic dyes of textile effluent by microorganisms: an environmentally and economically sustainable approach. European Journal of Microbiology and Immunology 9(4):114–118
76. Jamialahmadi K, Soltani F, Nabavi fard M, Behravan J, Mosaffa F (2014) Assessment of protective effects of glucosamine and N-acetyl glucosamine against DNA damage induced by hydrogen peroxide in human lymphocytes. Drug Chem Toxicol 37(4):427–432
77. Javed M, Usmani N (2016). Accumulation of heavy metals and human health risk assessment via the consumption of freshwater fish *Mastacembelus armatus* inhabiting, thermal power plant effluent loaded canal, vol 5(1). Springerplus, p 776. https://doi.org/10.1186/s40064-016-2471-3
78. Jin X, Wu C, Tian X, Wang P, Zhou Y, Zuo J (2021) A magnetic-void-porous MnFe2O4/carbon microspheres nano-catalyst for catalytic ozonation: preparation, performance and mechanism. Environmental Science and Ecotechnology 7:100110
79. Kabra AN, Khandare RV, Kurade MB, Govindwar SP (2011) Phytoremediation of a sulphonated azo dye Green HE4B by Glandularia pulchella (Sweet) Tronc.(Moss Verbena). Environ Sci Pollut R 18(8):1360–1373
80. Kagalkar AN, Jagtap UB, Jadhav JP, Bapat VA, Govindwar SP (2009) Biotechnological strategies for phytoremediation of the sulfonated azo dye Direct Red 5B using Blumea malcolmii Hook. Bioresource Technol 100(18):4104–4110
81. Kahraman S, Yalcin P (2005) Antibacterial effect of untreated and treated (Decolourized) dyes by agricultural wastes. Am J Biochem Biotechnol 1(1):50–53
82. Kamal AKI, Ahmed F, Hassan M, Uddin M, Hossain SM (2016) Characterization of textile effluents from Dhaka export processing zone (DEPZ) Area in Dhaka Bangladesh. Pollution 2(2):153–161

83. Kamyab A, McGill E, Firman J (2009) The use of hexamethyl-p-rosaniline chloride as an ingredient in poultry feed. Worlds Poult Sci J 65(3):475–480
84. Kant R (2011) Textile dyeing industry an environmental hazard
85. Kao C, Chou M, Fang W, Liu B, Huang B (2001) Regulating colored textile wastewater by 3/31 wavelength ADMI methods in Taiwan. Chemosphere 44(5):1055–1063
86. Karim ME, Sanjee SA, Mahmud S, Shaha M, Moniruzzaman M, Das KC (2020) Microplastics pollution in Bangladesh: current scenario and future research perspective. Chem Ecol 36(1):83–99. https://doi.org/10.1080/02757540.2019.1688309
87. Karthikeyeni S, Siva Vijayakumar T, Vasanth S, Ganesh A, Vignesh V, Akalya J, Thirumurugan R, Subramanian P (2015) Decolourisation of Direct Orange S dye by ultra sonication using iron oxide nanoparticles. J Exp Nanosci 10(3):199–208. https://doi.org/10.1080/17458080.2013.822107
88. Kaur H, Kainth GS (2020) Mutagenicity assessment of textile dyes using AMES test. Int. J. Adv. Study Res. Work 3(1):1–6
89. Kaur H, Sharma G (2015) Removal of dyes from textile industry effluent: a review. Int J Humanit Soc Sci 2(4):59–63
90. Kaushik P, Garg V, Singh B (2005) Effect of textile effluents on growth performance of wheat cultivars. Bioresource Technol 96(10):1189–1193
91. Kehinde F, Aziz HA (2014) Textile waste water and the advanced oxidative treatment process, an overview. International Journal of Innovative Research in Science, Engineering and Technology 3(8):15310–15317
92. Khalil I, Hashem A, Nath AR, Muhd Julkapli N, Yehye WA, Basirun WJ (2021) DNA/Nano based advanced genetic detection tools for authentication of species: strategies, prospects and limitations. Mol Cell Probes 59:101758. https://doi.org/10.1016/j.mcp.2021.101758
93. Khan S, Anas M, Malik A (2019) Mutagenicity and genotoxicity evaluation of textile industry wastewater using bacterial and plant bioassays. Toxicol Rep 6:193–201
94. Khan S, Malik A (2018) Toxicity evaluation of textile effluents and role of native soil bacterium in biodegradation of a textile dye. Environ Sci Pollut R 25(5):4446–4458. https://doi.org/10.1007/s11356-017-0783-7
95. Khandare RV, Govindwar SP (2015) Phytoremediation of textile dyes and effluents: current scenario and future prospects. Biotechnol Adv 33(8):1697–1714
96. Khandare RV, Kabra AN, Kurade MB, Govindwar SP (2011) Phytoremediation potential of Portulaca grandiflora Hook.(Moss-Rose) in degrading a sulfonated diazo reactive dye Navy Blue HE2R (Reactive Blue 172). Bioresource Technol 102(12):6774–6777
97. Khandare RV, Rane NR, Waghmode TR, Govindwar SP (2012) Bacterial assisted phytoremediation for enhanced degradation of highly sulfonated diazo reactive dye. Environ Sci Pollut R 19(5):1709–1718
98. Khattab TA, Rehan M, Hamouda T (2018) Smart textile framework: photochromic and fluorescent cellulosic fabric printed by strontium aluminate pigment. Carbohyd Polym 195:143–152
99. Kishor R, Purchase D, Saratale GD, Saratale RG, Ferreira LFR, Bilal M, Chandra R, Bharagava RN (2021) Ecotoxicological and health concerns of persistent coloring pollutants of textile industry wastewater and treatment approaches for environmental safety. J Environ Chem Eng 9(2):105012
100. Koprivanac N, Bosanac G, Grabaric Z, Papic S (1993) Treatment of wastewaters from dye industry. Environ Technol 14(4):385–390
101. Lacasse K, Baumann W (2012) Textile chemicals: environmental data and facts. Springer Science & Business Media
102. Laing I (1991) The impact of effluent regulations on the dyeing industry. Rev Prog Color Relat Top 21(1):56–71
103. Landecker H (2019) Antimicrobials before antibiotics: war, peace, and disinfectants. Palgrave Communications 5(1):1–11
104. Lellis B, Fávaro-Polonio CZ, Pamphile JA, Polonio JC (2019) Effects of textile dyes on health and the environment and bioremediation potential of living organisms. Biotechnology Research and Innovation 3(2):275–290

105. Lim S-L, Chu W-L, Phang S-M (2010) Use of Chlorella vulgaris for bioremediation of textile wastewater. Bioresource Technol 101(19):7314–7322
106. Liu R, Tian Q, Yang B, Chen J (2010) Hybrid anaerobic baffled reactor for treatment of desizing wastewater. Int J Environ Sci Technol 7(1):111–118
107. Mahawar P, Akhtar A (2016) Impact of dye effluent on growth and chlorophyll content of alfalfa (Medicago sativa L.). Annals of Plant Sciences 5(10):1432–1435
108. Malaviya P, Hali R, Sharma N (2012) Impact of dyeing industry effluent on germination and growth of pea (*Pisum sativum*). J Environ Biol 33(6):1075
109. Mani S, Bharagava RN (2016) Exposure to crystal violet, its toxic, genotoxic and carcinogenic effects on environment and its degradation and detoxification for environmental safety. Rev Environ Contam T 237:71–104
110. Manna A, Amutha C (2018) Early maturation and liver necrosis in the fingerling stage of Oreochromis mossambicus due to BPA can cause an ecological imbalance. RSC Adv 8(23):12894–12899. https://doi.org/10.1039/c7ra11432j
111. Maron DM, Ames BN (1983) Revised methods for the Salmonella mutagenicity test. Mutation Research/Environmental Mutagenesis and Related Subjects 113(3–4):173–215
112. McCann J, Choi E, Yamasaki E, Ames BN (1975) Detection of carcinogens as mutagens in the Salmonella/microsome test: assay of 300 chemicals. Proc Natl Acad Sci 72(12):5135–5139
113. Michaels GB, Lewis DL (1985) Sorption and toxicity of azo and triphenylmethane dyes to aquatic microbial populations. Environ Toxicol Chem 4(1):45–50
114. Mohamed FA, Elkhabiry S, Ismail IA, Attia AO (2022) Synthesis, application and antimicrobial activity of new acid dyes based on 3-amino-2-thioxo-4-thiazolidinone nucleus on wool and silk fabrics. Curr Org Synth 19(1):166–176
115. Mohammad Al Mamun YAW, Hossain MAM, Hashem A, Johan MR (2021) Electrochemical biosensors with Aptamer recognition layer for the diagnosis of pathogenic bacteria: barriers to commercialization and remediation. Trac-Trend Anal Chem145:116458
116. Mohan D, Shukla SP (2022) Hazardous consequences of textile mill effluents on soil and their remediation approaches. Cleaner Engineering and Technology 7:100434
117. Nestmann ER, Douglas GR, Matula TI, Grant CE, Kowbel DJ (1979) Mutagenic activity of rhodamine dyes and their impurities as detected by mutation induction in Salmonella and DNA damage in Chinese hamster ovary cells. Cancer Res 39(11):4412–4417
118. Nosheen S, Nawaz H, Rehman K-U (2000) Physicochemical characterization of effluents of local textile industries of Faisalabad-Pakistan. Int J Agric Biol 2(3):232–233
119. Ogugbue CJ, Sawidis T (2011) Bioremediation and detoxification of synthetic wastewater containing triarylmethane dyes by Aeromonas hydrophila isolated from industrial effluent. Biotechnol Res Int 2011:967925
120. Owolabi OD, Abdulkareem SI, Ajibare AO (2021) Haemato-biochemical and ionic regulatory responses of the hybrid catfish, Heteroclarias, to sublethal concentrations of palm oil mill effluents. Bulletin of the National Research Centre 45(1):220. https://doi.org/10.1186/s42269-021-00679-8
121. Pan H, Feng J, He G-X, Cerniglia CE, Chen H (2012) Evaluation of impact of exposure of Sudan azo dyes and their metabolites on human intestinal bacteria. Anaerobe 18(4):445–453
122. Parvin F, Islam S, Akm SI, Urmy Z, Ahmed S (2020) A study on the solutions of environment pollutions and worker's health problems caused by textile manufacturing operations. Biomedical Journal of Scientific & Technical Research 28(4):21831–21844
123. Patel R, Tajddin K, Patel A, Patel B (2015) Physico-chemical analysis of textile effluent. IJRSI 5(2):33–37
124. Patil AV, Jadhav JP (2013) Evaluation of phytoremediation potential of *Tagetes patula* L. for the degradation of textile dye Reactive Blue 160 and assessment of the toxicity of degraded metabolites by cytogenotoxicity. Chemosphere 92(2):225–232
125. Patil R, Zahid M, Govindwar S, Khandare R, Vyavahare G, Gurav R, Desai N, Pandit S, Jadhav J (2022) Constructed wetland: a promising technology for the treatment of hazardous textile dyes and effluent. In: Development in wastewater treatment research and processes. Elsevier, pp 173–198

126. Peláez-Cid AA, Velázquez-Ugalde I, Herrera-González AM, García-Serrano J (2013) Textile dyes removal from aqueous solution using *Opuntia ficus-indica* fruit waste as adsorbent and its characterization. J Environ Manage 130:90–97. https://doi.org/10.1016/j.jenvman.2013.08.059
127. Penninks A, Baert K, Levorato S, Binaglia M (2017) Dyes in aquaculture and reference points for action. EFSA J 15(7):e04920
128. Perkins WS (1991) A review of textile dyeing processes. Text Chem Color 23(8):23–27
129. Peuke AD, Rennenberg H (2005) Phytoremediation: molecular biology, requirements for application, environmental protection, public attention and feasibility. EMBO Rep 6(6):497–501
130. Piątkowska M, Jedziniak P, Olejnik M, Żmudzki J, Posyniak A (2018) Absence of evidence or evidence of absence? A transfer and depletion study of Sudan I in eggs. Food Chem 239:598–602
131. Pokharia A, Ahluwalia SS (2015) Toxicological effect of textile dyes and their metabolites: a review. Current Trends in Biotechnology and Chemical Researchl January-June 5(1):11–17
132. Pokhriya P, Rajput R, Nautiyal P, Panwar P, Pandey D, Daverey A, Arunachalam A, Shridhar V, Arunachalam K (2020) Impact assessment of textile effluent on health and microbiota of agricultural soil in Bhagwanpur (Uttarakhand) India. SN Applied Sciences 2(9):1–10
133. Prival MJ, Bell SJ, Mitchell VD, Peiperl MD, Vaughan VL (1984) Mutagenicity of benzidine and benzidine-congener dyes and selected monoazo dyes in a modified Salmonella assay. Mutation Research/Genetic Toxicology 136(1):33–47
134. Rajasimman M, Babu SV, Rajamohan N (2017) Biodegradation of textile dyeing industry wastewater using modified anaerobic sequential batch reactor–start-up, parameter optimization and performance analysis. J Taiwan Inst Chem E 72:171–181
135. Rane NR, Patil SM, Chandanshive VV, Kadam SK, Khandare RV, Jadhav JP, Govindwar SP (2016) Ipomoea hederifolia rooted soil bed and Ipomoea aquatica rhizofiltration coupled phytoreactors for efficient treatment of textile wastewater. Water Res 96:1–11
136. Rani B, Kumar V, Singh J, Bisht S, Teotia P, Sharma S, Kela R (2014) Bioremediation of dyes by fungi isolated from contaminated dye effluent sites for bio-usability. Braz J Microbiol 45:1055–1063
137. Rápó E, Posta K, Csavdári A, Vincze BÉ, Mara G, Kovács G, Haddidi I, Tonk S (2020) Performance comparison of *Eichhornia crassipes* and *Salvinia natans* on azo-dye (Eriochrome Black T) phytoremediation. Crystals 10(7):565
138. Rawat D, Mishra V, Sharma RS (2016) Detoxification of azo dyes in the context of environmental processes. Chemosphere 155:591–605
139. Rehman K, Shahzad T, Sahar A, Hussain S, Mahmood F, Siddique MH, Siddique MA, Rashid MI (2018) Effect of reactive Black 5 azo dye on soil processes related to C and N cycling. PeerJ 6:e4802
140. Rodrigues ILA, Motta MCS, Ferreira MA (2016) Representações sociais da tuberculose por enfermeiros. Revista Brasileira de Enfermagem 69:532–537
141. Roy R, Fakhruddin A, Khatun R, Islam M, Ahsan M, Neger A (2010) Characterization of textile industrial effluents and its effects on aquatic macrophytes and algae. Bangladesh Journal of Scientific and Industrial Research 45(1):79–84
142. Saravanamoorthy M, Kumari BR (2007) Effect of textile waste water on morphophysiology and yield on two varieties of peanut (*Arachis hypogaea* L.). Journal of Agricultural Technology 3(2):335–343
143. Sawyer CN, McCarty PL (1978) Chemistry for environmental engineering. McGraw-Hill
144. Selvaraj D, Leena R, Kamal D (2015) Toxicological and histopathological impacts of textile dyeing industry effluent on a selected teleost fish Poecilia reticulata. Asian Journal of Pharmacology and Toxicology 3(10):26–30
145. Şen S, Demirer G (2003) Anaerobic treatment of real textile wastewater with a fluidized bed reactor. Water Res 37(8):1868–1878
146. Sen SK, Raut S, Bandyopadhyay P, Raut S (2016) Fungal decolouration and degradation of azo dyes: a review. Fungal Biol Rev 30(3):112–133

147. Sender R, Fuchs S, Milo R (2016) Revised estimates for the number of human and bacteria cells in the body. PLoS Biol 14(8):e1002533
148. Sharma A, Kumar V, Shahzad B, Ramakrishnan M, Singh Sidhu GP, Bali AS, Handa N, Kapoor D, Yadav P, Khanna K (2020) Photosynthetic response of plants under different abiotic stresses: a review. J Plant Growth Regul 39(2):509–531
149. Sharma K, Sharma P, Dhiman SK, Chadha P, Saini HS (2022) Biochemical, genotoxic, histological and ultrastructural effects on liver and gills of fresh water fish Channa punctatus exposed to textile industry intermediate 2 ABS. Chemosphere 287:132103
150. Sharma MK, Sobti R (2000) Rec effect of certain textile dyes in Bacillus subtilis. Mutation Research/Genetic Toxicology and Environmental Mutagenesis 465(1–2):27–38
151. Sharma S, Sharma S, Pathak S, Sharma KP (2003) Toxicity of the azo dye methyl red to the organisms in microcosms, with special reference to the guppy (*Poecilia reticulata* Peters). Bull Environ Contam Toxicol 70(4):753–760. https://doi.org/10.1007/s00128-003-0047-8
152. Shelley T (1994) Dye pollution clean-up by synthetic mineral. Int Dyer 79:26–31
153. Singh L (2017) Biodegradation of synthetic dyes: a mycoremediation approach for degradation/decolourization of textile dyes and effluents. J Appl Biotechnol Bioeng 3(5):430–435
154. Singh Z, Chadha P (2016) Textile industry and occupational cancer. J Occup Med Toxicol 11(1):1–6
155. Sinsabaugh RL, Antibus RK, Linkins AE (1991) An enzymic approach to the analysis of microbial activity during plant litter decomposition. Agr Ecosyst Environ 34(1–4):43–54
156. Slama HB, Chenari Bouket A, Pourhassan Z, Alenezi FN, Silini A, Cherif-Silini H, Oszako T, Luptakova L, Golińska P, Belbahri L (2021) Diversity of synthetic dyes from textile industries, discharge impacts and treatment methods. Appl Sci 11(14):6255
157. Soo JAL, Makhtar MMZ, Shoparwe NF, Otitoju TA, Mohamad M, Tan LS, Li S (2021) Characterization and kinetic studies of poly(vinylidene fluoride-co-hexafluoropropylene) polymer inclusion membrane for the malachite green extraction. Membranes (Basel) 11(9):676. https://doi.org/10.3390/membranes11090676
158. Souza AGV, Maria TC, Saran LM, Alves LMC (2022) Enzymatic bioremediation of dyes from textile industry effluents. Intechopen
159. Sultana Z, Ali ME, Uddin MS, Haque MM (2013) Implementation of effluent treatment plants for waste water treatment. J Environ Prot 4(3):28740
160. Tabish TA, Memon FA, Gomez DE, Horsell DW, Zhang S (2018) A facile synthesis of porous graphene for efficient water and wastewater treatment. Sci Rep-Uk 8(1):1–14
161. Thangaraj S, Bankole PO, Sadasivam SK (2021) Microbial degradation of azo dyes by textile effluent adapted, *Enterobacter hormaechei* under microaerophilic condition. Microbiol Res 250:126805
162. Tolkushin A, Luchinin E, Kholovnya-Voloskova M, Zavyalov A (2020) History of aminoquinoline preparations: from cinchona bark to chloroquine and hydroxychloroquinon. Problemy Sotsial'noi Gigieny, Zdravookhraneniia i Istorii Meditsiny 28(Special Issue): 1118–1122
163. Tounsadi H, Metarfi Y, Taleb M, El Rhazi K, Rais Z (2020) Impact of chemical substances used in textile industry on the employee's health: epidemiological study. Ecotox Environ Safe 197:110594
164. Uwidia I (2013) Characterisation of textile wastewater discharges in Nigeria and its pollution implications. Global Journals of Research in Engineering 13(4):1–4
165. Vargas A, Paulino A, Nozaki J (2009) Effects of daily nickel intake on the bio-accumulation, body weight and length in tilapia (*Oreochromis niloticus*). Toxicol Environ Chem 91(4):751–759
166. Vikrant K, Giri BS, Raza N, Roy K, Kim K-H, Rai BN, Singh RS (2018) Recent advancements in bioremediation of dye: current status and challenges. Bioresource Technol 253:355–367
167. Watharkar AD, Jadhav JP (2014) Detoxification and decolorization of a simulated textile dye mixture by phytoremediation using *Petunia grandiflora* and *Gailardia grandiflora*: a plant–plant consortial strategy. Ecotox Environ Safe 103:1–8

168. Watharkar AD, Kadam SK, Khandare RV, Kolekar PD, Jeon B-H, Jadhav JP, Govindwar SP (2018) Asparagus densiflorus in a vertical subsurface flow phytoreactor for treatment of real textile effluent: a lab to land approach for in situ soil remediation. Ecotox Environ Safe 161:70–77
169. Yang C, Li L, Shi J, Long C, Li A (2015) Advanced treatment of textile dyeing secondary effluent using magnetic anion exchange resin and its effect on organic fouling in subsequent RO membrane. J Hazard Mater 284:50–57
170. Yaseen DA, Scholz M (2016) Shallow pond systems planted with Lemna minor treating azo dyes. Ecol Eng 94:295–305
171. Yun M-K, Wu Y, Li Z, Zhao Y, Waddell MB, Ferreira AM, Lee RE, Bashford D, White SW (2012) Catalysis and sulfa drug resistance in dihydropteroate synthase. Science 335(6072):1110–1114
172. Zabłocka-Godlewska E, Przystaś W, Grabińska-Sota E (2018) Possibilities of obtaining from highly polluted environments: new bacterial strains with a significant decolorization potential of different synthetic dyes. Water Air Soil Pollut 229(6):1–13
173. Zhao B, Xiao W, Shang Y, Zhu H, Han R (2017) Adsorption of light green anionic dye using cationic surfactant-modified peanut husk in batch mode. Arab J Chem 10:S3595–S3602
174. Zheng X, Fu D, Cheng J, Tang R, Chu M, Chu P, Wang T, Yin S (2021) Effects of hypoxic stress and recovery on oxidative stress, apoptosis, and intestinal microorganisms in *Pelteobagrus vachelli*. Aquaculture 543:736945
175. Zollinger H (1987) Azo dyes and pigments. Colour chemistry-synthesis, properties and applications of organic dyes and pigments, pp 92–100

Photoactive Titanium Dioxide Nanoparticles Hybrid for Dye Removal Under Light Irradiation

Mustaffa Ali Azhar Taib, Mohd Azam Mohd Adnan, Mohd Fadhil Majnis, and Nurhidayatullaili Muhd Julkapli

Abstract This chapter of book summarizes on the recent application of titanium dioxide (TiO_2) in nano-sized with hybridization of other materials for photoreduction of dye compounds under photon irradiation. The use of native TiO_2 as a photocatalyst is highlighted in the first section of the chapter due to its special properties, which include strong optical activity, excellent chemical stability, and nontoxicity. As a result, TiO_2 is frequently used to photodegrade a variety of dye contaminants found in wastewater, as it facilitates the breakdown of dye molecules when exposed to photons. However, pure TiO_2 is more prone to excitation when exposed to UV light because of its broad bandgap, and the rate of recombination of photogenerated species electron–hole (e^-/h^+) pairs is considerable. Despite this, the photocatalytic efficiency of pure TiO_2 is still relatively low. In order to exhibit a bandgap suitable for excitation by visible light, good charge separation efficiency, charge carrier and transport, and suitable positions of the valence band (VB) and conduction band (CB) for redox reactions, hybridization with other materials, such as nano-metal oxide, ceramic-based materials, and polymeric materials, comes into the picture. Given that the UV region makes up a relatively minor portion of the solar spectrum, visible light active nano-photocatalysts are favoured in this situation. It is important to note that the combination of TiO_2 with those materials causes visible light absorption in TiO_2 by altering its inner band structure by substituting oxygen in the TiO_2 lattice to decrease the optical energy bandgap, which subsequently overcomes the bandgap

M. A. A. Taib
Pusat Latihan Teknologi Tinggi (ADTEC) Taiping, PT15643, Kamunting Raya, Mukim Asam Kumbang, 34600 Kamunting, Perak, Malaysia

M. A. Mohd Adnan
Advanced Materials & Manufacturing Research Group (AMMRG), Faculty of Engineering and Life Sciences, University of Selangor, 45600 Bestari Jaya, Selangor, Malaysia

M. A. A. Taib · M. A. Mohd Adnan · N. Muhd Julkapli (✉)
Nanotechnology and Catalysis Research Center (NANOCAT), Level 3, Block A, Institute for Advanced Studies (IAS), Universiti Malaya, 50603 Kuala Lumpur, Malaysia
e-mail: nurhidayatullaili@um.edu.my

M. F. Majnis
School of Chemical Engineering, College of Engineering, Universiti Teknologi MARA (UiTM), 40450 Shah Alam, Selangor, Malaysia

© The Author(s), under exclusive license to Springer Nature Singapore Pte Ltd. 2023
A. Ahmad et al. (eds.), *Nanohybrid Materials for Treatment of Textiles Dyes*,
Smart Nanomaterials Technology, https://doi.org/10.1007/978-981-99-3901-5_18

energy, stability, dispersion, and recombination rate limitations for more effective dye molecule photoreduction. Finally, this book chapter brought new insight into applying photoactive nanotechnology as an alternative method for sustainable and economic dyes treatment. **Highlight:** (1) Treatment methods of dye removal. (2) Mechanism of TiO_2 photocatalysis process. (3) Hybrid TiO_2 nanoparticles photocatalysts for dye removal under solar-driven AOPs. (4) The recent development of hybrid materials of TiO_2 nanoparticles for dye removal is discussed. (5) Prospects and limitations of photocatalytic dye removal reactions.

Keywords Semiconductor · Photoactive materials · TiO_2 photocatalysts · Hybrid nanoparticles · Dye removal

1 Introduction: Conventional Treatment Methods of Dyes Removal

Since dyes are used widely and are found in many manufacturing and processing facilities, their discharge negatively influences the environment. With so many industries producing pollutants that seriously pollute the environment, protecting the environment becomes a challenging responsibility. Removing colour from dye house effluents has become more crucial for the environment because many organic dye chemicals are hazardous to humans. The paint and textile industries are the main producers of the vast majority of dyes released as effluent. Textile industry wastewater is highly coloured, complex, and variable, depending on the fibre processing and the associated dyestuffs and auxiliary chemicals used. There are various methods to remove dyes from wastewater. The conventional dye removal treatment process can be divided into three categories [1], as given in Table 1.

Various technologies have been developed in the last few decades to remove various toxic dyes from industrial effluents. There are three classifications of treatments for conventional dye removal, including physical treatment, chemical

Table 1 Conventional dye removal treatment process

Treatment process	Details
Primary	The primary treatment process consists primarily of removing suspended materials (e.g. pieces of fabrics, yarns, rags, and lint), grits, and oil using bar and fine screens
Secondary	The secondary treatment process uses microorganisms in aerobic or anaerobic conditions to remove colour, oil, and reduce biological oxygen demand
Tertiary	The tertiary treatment process is the final stage of the conventional treatment process that prepares the effluent for reuse. All pollutants, such as suspended solids, organic matter, heavy metals, and pathogens, are difficult to remove in the secondary treatment process; however, it can be eliminated via this process

treatment, and biological treatment. Table 2 shows the different classifications of treatments for traditional dye removal that are normally applied in industries.

2 TiO_2 Photocatalyst

There is intensive research on the photocatalytic activity of TiO_2. The mechanism of its degradation of dissolved organic compounds has been proposed by various researchers, especially those who pioneered the process decades ago. The mechanism of photocatalytic degradation of organic pollutants is explained as follows. Firstly, it is absorbed when the incident photon illuminates on TiO_2 photocatalyst. Only photons with an energy hv > E_g are absorbed. The absorption of photons leads to the excitation of electrons from the valence band into the conduction band, leaving holes in the valence band. This is called electron–hole (e–h) pairs formation. The electrons in the conduction band are diffused and transferred to the surface of TiO_2. These electrons (at the surface of TiO_2) reduced the absorbed oxygen into superoxide anions ($^{\cdot}O_2{}^{-}$) by the oxidation process [7].

Therefore, the holes in the valence band also are diffused and transferred to the surface of TiO_2. Water is absorbed and oxidized at the TiO_2 surface into H^+ and OH^- via water decomposition. The energy required to decompose water has to straddle between redox potential, which is 1.23 eV. When holes at the surface of TiO_2 react with OH-, hydroxyl radicals ($^{\cdot}OH$) are produced. The production of $^{\cdot}OH$ and $^{\cdot}O_2{}^{-}$ are the key to organic dye decomposition. Besides, dye decomposition can also happen by holes via a direct oxidation process. The decomposition of organic dye pollutants will turn into harmless substances such as water and carbon dioxide (CO_2). Throughout the whole process, the recombination of charges may happen. It may occur at the bulk or surface (defect) of TiO_2.

Photoexcitation: Light absorption (hv > E_g = 3.2 eV) on TiO_2 anatase.

$$TiO_2 + hv \rightarrow TiO_2\left(e_{CB}^{-} + h_{VB}^{+}\right) \tag{1}$$

Oxidation of oxygen (the first step of oxygen reduction; oxygen's oxidation degree passes from (0 to − 1/2)

$$TiO_2\left(e_{CB}^{-}\right) + O_2 \rightarrow TiO_2 + {}^{\cdot}O_2^{-} \tag{2}$$

Neutralization of OH^- groups by photoholes which produces $^{\cdot}OH$ radicals

$$\left(H_2O \Leftrightarrow H^+ + OH^-\right)_{ads} + TiO_2\left(h_{VB}^{+}\right) \rightarrow H^+ + {}^{\cdot}OH \tag{3}$$

Neutralization of $^{\cdot}O_2{}^{-}$ by protons

$${}^{\cdot}O_2^{-} + H^+ \rightarrow HO_2^{\cdot} \tag{4}$$

Table 2 Conventional dye removal treatment process

Classification	Treatment	Description	Scope of Application	Advantages	Limitation	References
Physical	Adsorption	Adsorbents made of high adsorption capacity materials that are used to adsorb dye molecules	Water soluble dyes, cationic dyes	Waste utilization, low costs, simple design, and operation, higher efficiency, no effect by toxic substances	Poor regeneration of adsorbent and small adsorption capacity	[1–4]
	Membrane technology	Dye wastewater containing dye particles is separated from clean water by passing it through a membrane	Disperse dyes	The decolourization effect is obvious, water recovery and reuse are effective, process can be easily scaled-up	High cost of cleaning for membrane clogging, easy membrane fouling, generate concentrated sludge	
Chemical	Electrochemical oxidation and electrochemical reduction	The application of electric current or a potential difference between two electrodes (anode and cathode) connected to a power source and producing strong oxidizing species to remove the organic pollutants	Dyes other than cationic dyes	The processing colour, COD, BOD, and TSS are effective, the process is flexible and adaptable	High unit power consumption, high iron consumption and high cost	[1–3]
	Coagulation and flocculation	Process of destabilization of charged particles dispersed in water by reducing the charge of the surface and the formation of large particle by accumulation	Disperse dyes, vat dyes, sulphur dyes	Cost-effective, large handling capacity, hydrophobic dye, high decolourization efficiency	Low decolourization ability for hydrophilic dyes, low COD removal efficiency, chemical sludge production, secondary pollution, non-ionic impurities in the effluent	

(continued)

Table 2 (continued)

Classification	Treatment	Description	Scope of Application	Advantages	Limitation	References
	Photocatalytic oxidation	Process combines UV irradiation with a substance (catalyst) which results in a reaction that changes malignant contaminants into water, carbon dioxide, and detritus	NA	The catalyst input is small, the treatment effect is good, and the reaction time is short	High catalyst price and reusability issues, low utilization of light energy, and complex reactor	
	Fenton reaction	The usage of Fenton's reagent (a mixture of catalyst and hydrogen peroxide)	NA	Fenton reagents have both oxidation and coagulation effects	The process condition is harsh, and the running cost is high, excessive sludge production	
	Ozonation	Ozone is produced from oxygen to eliminate dye particles	Direct dyes, acid dyes, basic dyes, reactive dyes	Small area, easy automatic control. Easy to adjust, zero sludge, no secondary pollution	High cost, unsatisfactory, has an extremely short half-life for only 20 min, produces toxic by-products, unstable method	
Biological	Aerobic process	The usage of microorganisms (bacteria and fungi) in removal of dyes	NA	Efficient for azo dyes removal, low operational cost	Slow process, suitable environment for microorganisms' growth is required, unstable effect, impact resistance	[1–3, 5, 6]

(continued)

Table 2 (continued)

Classification	Treatment	Description	Scope of Application	Advantages	Limitation	References
	Anaerobic process	The decomposition of complex organic compounds in the absence of oxygen	Azo dyes	Decolourization effect is good, and residence time is short, by-product can be used as energy sources	Produce a large amount of aniline, the effluent biological toxicity increased	
	Aerobic/Anaerobic process	A prepared sludge breaks down complex dye molecules	NA	High COD removal rate, good economy, no foam formation	Poor impact resistance, poor adaptability, longer reaction time, large land area is required, formation of methane and hydrogen sulphide as by-product	

Note
COD: Chemical Oxygen Demand
TSS: Total Suspended Solid
BOD: Biological Oxygen Demand

Transient hydrogen peroxide formation and dismutation of oxygen

$$2HO_2^{\cdot} \rightarrow H_2O_2 + O_2 \tag{5}$$

Decomposition of H_2O_2 and second reduction of oxygen

$$H_2O_2 + TiO_2\left(e_{CB}^{-}\right) \rightarrow {}^{\cdot}OH + OH^{-} \tag{6}$$

Oxidation of the organic reactant via successive attacks by $^{\cdot}OH$ radicals

$$R + {}^{\cdot}OH \rightarrow {}^{\cdot}R + H_2O \tag{7}$$

Direct oxidation by reaction with holes

$$R + h^{+} \rightarrow {}^{\cdot}R^{+} \rightarrow \text{degradation products} \tag{8}$$

Figure 1 shows several steps occurring during photocatalytic activity on spherical TiO_2 anatase when illuminated with photons (UV irradiation) whose energy is equal to or greater than their band gap energy, Eg (hv > Eg):

Based on the above-mentioned reaction, it is $^{\cdot}OH$ and $^{\cdot}O_2^{-}$ species that play a role in decomposing the organic (and sometimes inorganic) pollutant. The decomposition of organic pollutants into a less harmful by oxidation through reactions with these species is known as an advanced oxidation process (AOP). AOP involves two stages of oxidation: firstly, the formation of strong oxidants such as $^{\cdot}OH$, and secondly, the reaction of these oxidants with organic contaminants in water. Therefore, an ideal photocatalyst that can perform AOP must have the ability to

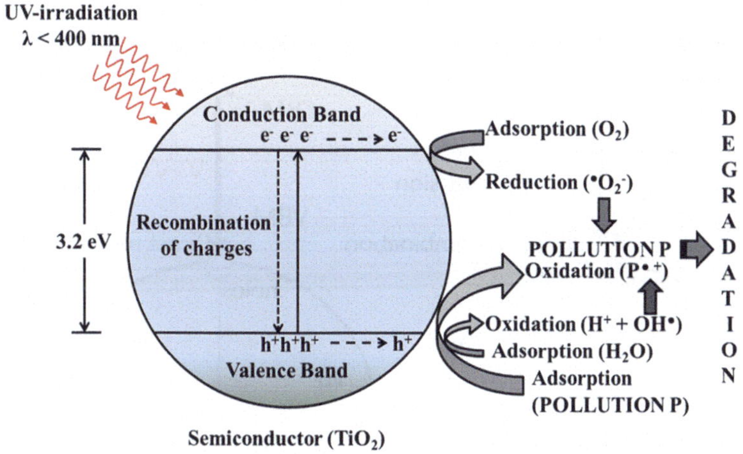

Fig. 1 Photocatalytic process over spherical TiO_2 anatase (adapted from [8])

- Absorb light and is activated under light (good optical properties). As seen from reaction 1, a photocatalyst must have the ability to absorb light (photon) to allow for the excitation of electrons from the valence band to the conduction band. Therefore, semiconductors with electrons in the valence and empty conduction band, as well as with a direct band gap are required.
- From electron–hole pairs efficiently but with slow recombination, the diffusion of electron and hole must be slow before they recombine.
- Long electron–hole lifetime, the travel path of electrons and holes must be long.
- Allow the electron–hole to diffuse to the surface of the catalysts without recombination so that the Reactions (3) and (4) are possible to occur.
- Adsorb water and dissociate it to OH^- and H^+ (Fig. 2)
- Ability to transfer electrons or holes to the OH^- and H^+.
- Adsorb the organic substances to be oxidized.
- Allow $^{\cdot}OH$ formation (Reaction 3)
- To perform the oxidation process on the adsorbed organic molecules.
- High aspect ratio for larger catalytic sites.

In terms of general material properties, the photocatalyst must be chemically and biologically inert to the environment, including against photocorrosion, easy to produce in a large scale, consist of the desired phases, be cost-effective, and should not have risk to humans and the environment (non-toxic). There are many semiconductors, such as TiO_2, zinc oxide (ZnO), zirconia dioxide (ZrO_2), iron oxide (Fe_2O_3), tungsten trioxide (WO_3), and cadmium sulphide (CdS) that have been examined and can fulfil the requirements mentioned above and properties. However, when ranking them among each other, TiO_2 appears to be the most excellent, except it does not

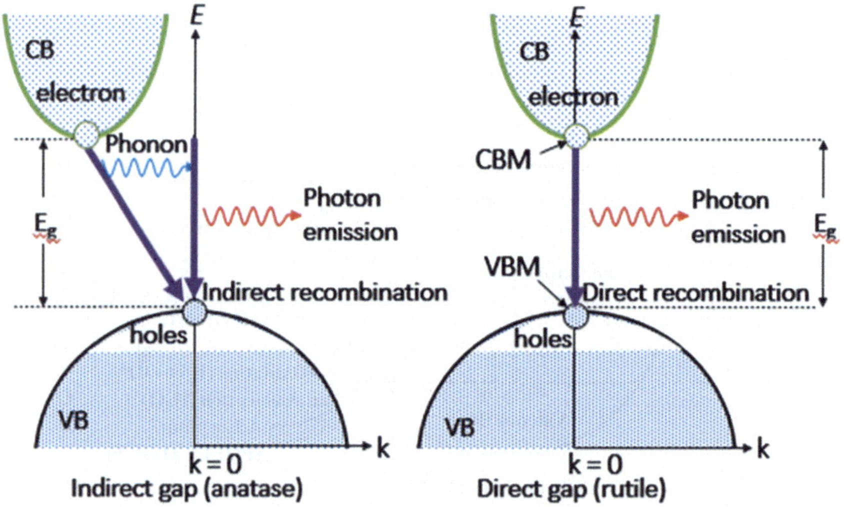

Fig. 2 Comparison of photogenerated electron and hole recombination processes within **a** indirect bandgap of anatase and **b** direct bandgap of rutile

efficiently activate by the whole spectrum of sunlight, and electron–hole pairs recombination is also too fast in this oxide. The minimum band gap energy required for photons to cause photogeneration of charge carriers over TiO_2 semiconductor such as anatase phase is 3.2 eV, corresponding to the wavelength of 388 nm, which is the blue region.

2.1 TiO_2: Synthesis Method

TiO_2 could be synthesized using various methods, including chemical, physical, and green chemistry. Table 3 provides a detailed comparison of the synthesis routes of TiO_2 nanoparticles-based catalysts for environmental applications. A variety of different techniques, including sol–gel method, hydrothermal method, vapour deposition, spray pyrolysis, electrophoretic deposition, solvothermal method, microwave method, and sonochemical method, has been extensively used in the past few years to produce nano-TiO_2. Though chemical synthesis of TiO_2 nanoparticles has widely been used owing to the ease of synthesis and effective control of the size and shape of nanoparticles, there still exist certain limitations, including high cost, the requirement of extreme temperatures and pressures, and eco-toxicity.

2.2 TiO_2: Characteristics and Properties

TiO_2 belongs to the family of transition metal oxides. Transition metal oxides are compounds composed of oxygen atoms bound to transition metal. This n-type metal oxide has been subjected to extensive academic and technological research for decades. This is due to its unique properties such as

- Electronic properties of TiO_2: Semiconductor with energy gap, E_g, of 3.2 eV.
- Optical = UV activation
- Highly photostable and chemically stable, robust, and environmentally friendly.
- Highly photoactive materials and easily synthesized and handled.
- Not classified as hazardous (according to the United Nations' (UN) Globally Harmonized System of Classification and Labelling of Chemicals (GHS).
- Inexpensive, which is low cost, non-toxic, and long-term stable.
- Intimate contact or proximity between the semiconductor and the electrolyte.
- With high dielectric constant, hardness, and transparency, TiO_2 films are applicable for storage capacitors in integrated electronics, protective coatings, and optical components.

The properties of TiO_2 in the form of nanoparticles are known to be affected by particle size, morphology, and crystal structure. There are three commonly known crystalline polymorphs of TiO_2: anatase, brookite, and rutile [20]. Brookite has an orthorhombic crystal structure, whereas anatase and rutile have tetragonal crystal

Table 3 Synthesis route of TiO_2 nanoparticles

No.	Method	Description	Advantages	Disadvantages	References
1	Sol–gel	This wet chemical method involves hydrolysis and condensation of metallo-organic alkoxide precursors for gel formation, followed by dip/ spin/ spray coating or screen printing	• Simplest, homogeneity, low cost, reliability, reproducibility, controllability • Produce very fine powders • Synthesis under low temperature i.e. < 100 °C • Easy monitoring and control of nanoparticle shape and size • Produce a variety of products like spherical, fine, and uniform-sized nanopowders	• High cost of precursors • Longer processing times • Residual fine pores and hydroxyls • High temperature (~500 °C) is required to form anatase nanocrystal	[9, 10]
2	Hydrothermal	Includes either a single or heterogeneous phase reaction in an aqueous solution at elevated temperatures and pressures to crystallize materials directly from the solution	• It provides a simple mode of operation • It has the ability to grow large high-quality crystals while maintaining a reasonable control of their chemical composition • Simplest, homogeneity, low cost, reliability, reproducibility, controllability	• Need of expensive autoclaves for pressurized equipment • The inability to monitor the crystal as it grows • Long synthesis duration	[10–13]

(continued)

Table 3 (continued)

No.	Method	Description	Advantages	Disadvantages	References
3	Chemical vapour deposition (CVD)	Materials or material precursors are condensed from a vapour state to form a solid-phase material. A thin film of metal oxide is formed on a heated substrate from a gaseous phase in a closed chamber at a relatively higher temperature	• Produce uniform films at low or high rates with good reproducibility • Flexible with regards to the shape of the substrate • Compatibility with good adhesion • Simultaneously coat multiple components • Control structure of crystal and generate uniform films with pure materials and high density • Flexibility of using a wide range of chemical precursors	• High cost • High reaction temperature • Mostly involve safety and contamination due to toxicity, flammable and/or explosive precursor gases • Presence of corrosive gases • Low deposition rates • Cannot control the stoichiometry of films using more than one material	[10, 13]
4	Physical vapour deposition (PVD)	It entails the atomic-level transfer of material onto a solid substrate. Rather than a chemical reaction between precursors, this is a physical process such as high temperature vacuum evaporation followed by condensation	• Suitable for any type of inorganic materials • Safer than other methods	• High cost	[12]

(continued)

Table 3 (continued)

No.	Method	Description	Advantages	Disadvantages	References
5	Spray pyrolysis	A nanoporous nebulizer sprays a solution containing a precursor onto the hot substrate in the furnace	• Operates at atmospheric pressure, well-controlled stoichiometry, homogeneity • Variety of precursors could be used • Low processing temperature • Cost-effective and can easily perform	• High cost of precursors • Longer processing times • Coatings are not uniform in thickness	[10, 12]
6	Electrophoretic deposition	Formation of coating on the charged surface takes place by the movement of charge particles in suspension under an appropriate electric field	• Simple and cheap • Uniform coating • Size and shape of nanoparticles can be controlled • High-quality coatings • Safety issues during the reaction process	• Volatile, toxic • Flammability • Costly • High electric field strengths are required	[10, 12]
7	Solvothermal	Heating process of both precursors and a solvent in a closed system at a temperature higher than the boiling point of the solvent used to generate desired material	• The procedure is well controlled through liquid-phase or multiphase chemical reactions	• Expensive equipment • High temperatures	[14]

(continued)

Table 3 (continued)

No.	Method	Description	Advantages	Disadvantages	References
8	Microwave	Interactions of the microwaves with the materials and the reaction medium under the action of the electromagnetic field	• Ease of use, energy saving, and high yield • Rapid heating rate and time-saving by several orders of magnitude • Suppressing side reactions and good reproducibility • Shorten the polymerization time • Uniform morphology and super selectivity	• Relatively expensive equipment • Unfeasible for reaction monitoring	[15–18]
9	Sonochemical	Molecules undergo a chemical reaction under powerful ultrasound radiation	• It involves high energies and pressure in a short time • Improves reaction rate • No additives needed • Reduced number of reaction steps	• Low efficiency and low yield • Inefficient energy	[19]

structures. TiO_2 is a wide bandgap semiconductor with limited activation in the ultraviolet (UV) region. Anatase appears to be an indirect bandgap semiconductor, whereas brookite and rutile belong to the direct bandgap semiconductors category. Due to direct transitions of photogenerated electrons from the conduction band (CB) to the valence band (VB) of anatase TiO_2 are impossible, indirect band gap anatase has a longer lifetime of photoexcited electrons and holes than direct band gap rutile and brookite [21, 22]. The bandgap for anatase is 3.2, 3.1 eV for brookite, and 3.0 eV for rutile. It is known that each of the phases has its own properties and characteristics suitable for a specific application. The general properties of TiO_2 polymorphs are outlined in Table 4.

For rutile and brookite, an electron only emits a photon for photogenerated charge carrier recombination indirect band gap semiconductors. However, for indirect band gap semiconductors, the recombination of photoexcited electrons and holes in anatase is aided by phonon [22]. As a result, because the excited electrons cannot recombine directly with holes, the photogenerated electron–hole lifetime in anatase is longer

Table 4 Properties of TiO_2 polymorph according to its different phases

Property	Anatase	Brookite	Rutile	References
Crystal structure	Tetragonal	Orthorhombic	Tetragonal	[11, 23–25]
Atom per unit cell, Z	4	8	2	
Crystal size (nm)	< 11	11–35	> 35	
Point group	4/mmm	2/mmm	4/mmm	
Space group	$I4_1/amd$	*Pbca*	$P4_2/mnm$	
Unit cell volume (nm)	0.1363	*NA*	0.0624	
Lattice parameter (nm)	$a = b = 0.3785$ $c = 0.9514$	$a = c = 0.5436$ $b = 0.9166$	$a = b = 0.4594$ $c = 0.2959$	
Density ($g\ cm^{-3}$)	3.84	4.12	4.26	
Calculated indirect bandgap (eV) (nm)	3.23–3.59 345.4–383.9	3.14 394.8	3.02–3.24 382.7–410.1	
Experimental band gap (eV) (nm)	~ 3.2 ~ 387	~ 3.1 ~ 399	~ 3.0 ~ 413	
Light absorption (nm)	< 390		< 415	
Refractive index (n_D)	2.54, 2.49	2.58	2.79, 2.903	
Solubility in H_2O	Insoluble	Insoluble	Insoluble	
Hardness (Mohs)	5.5–6	5.5–6	6–6.5	

than in rutile and brookite. Furthermore, anatase has the lowest average effective mass of photogenerated electrons and holes compared to rutile and brookite. The lowest effective mass implies the fastest migration of photogenerated electrons and holes from the interior to the surface of an anatase TiO_2 particle, resulting in the lowest photogenerated charge carrier recombination rate within an anatase TiO_2. As a result, it is not surprising that anatase has greater photocatalytic activity than rutile and brookite. Figure 2 depicts anatase's indirect bandgap, which prevents direct recombination of photogenerated electrons from the conduction band minimum (CBM) with holes from the valence band maximum (VBM), thus increasing the electron–hole lifetime relative to the direct bandgap of rutile. As a result, the diffusion length and reaction time of the electron and hole excited in the anatase increase. This is one of the reasons why, according to most reports, indirect semiconductors have better photocatalytic performance than direct semiconductors [22].

Both anatase and rutile are tetragonal in structure but differ slightly in crystal habit and more distinctly in cleavage. The Ti–Ti distances in anatase are larger, whereas the Ti–O distances are shorter than those in rutile. The octahedrons in anatase shared four edges forming the fourfold axis. It forms the eight-faced tetragonal dipyramids that come to sharp elongated points. The elongated is pronounced enough to distinguish this crystal form from octahedral crystals, but there is a similarity. The anatase form is thought to be more photoactive than rutile and has exhibited excellent catalytic activity performance under UV illumination. It is because anatase has a higher hydroxyl group surface adsorption capacity and a lower charge carrier recombination rate.

So, it is usually used as a catalyst for photocatalytic activity. It also has better electrical properties than other phases. Typically, chemically produced TiO_2 particles are amorphous. An amorphous structure implies the presence of a high number of defects (localized state) that act as traps and recombination centres for the electrons and holes. Since anatase is the favoured phase for photocatalytic or photocurrent generation, particles must be subjected to thermal treatment.

Brookite is the rarest mineral form and is difficult to obtain synthetically [22]. Brookite and anatase are metastable polymorphs that transform exothermically into rutile, the transformation is irreversible and occurs over a temperature range (400–1200 °C) depending on several factors such as the presence of impurities, particle size, and whether the oxide is supported over another material. Brookite's properties (density, band gap, and refractive index) are intermediate between rutile and anatase.

Rutile is the most common natural form of TiO_2. It is the most thermodynamically stable phase of TiO_2 form [23]. Rutile comes from the Latin word "rutilus" which means red, about the deep red colour observed when viewed by transmitted light. When viewed by transmitted light, the rutile exhibits a deep red colour. In rutile, the structure is based on octahedrons of TiO_2, which share two (out of twelve) edges of the octahedron with other octahedrons and form linear chains. It is the chains themselves that are arranged into a fourfold symmetry. It has a high refractive index which is preferably used in interference applications. Figure 3 depicts the different crystal structures of TiO_2 that comprise anatase, brookite, and rutile, respectively.

As shown in Fig. 3, all three crystal structures are made up of distorted octahedra, each representing a TiO_6 unit, with each Ti^{4+} at the centre of the unit and coordinating six O^{2-} ions. The way the octahedra assemble to form a TiO_6-based chain differs and is unique to each polymorph. However, only anatase and rutile are important and considered for application purposes. Both anatase and rutile phases have been extensively studied for photocatalytic applications, with the former being found to be the most suitable for photocatalytic reactions compared to brookite.

The specific electronic and ionic properties of TiO_2 strongly depend on their crystallographic form (amorphous, anatase, or rutile). The desired crystallographic form can be achieved after thermal annealing. The most desired crystal structure due to the highest electron mobility is anatase, which is applicable for many electron-conducting applications such as photoinduced reaction (photocatalytic electrodes or photoelectrochemical cells).

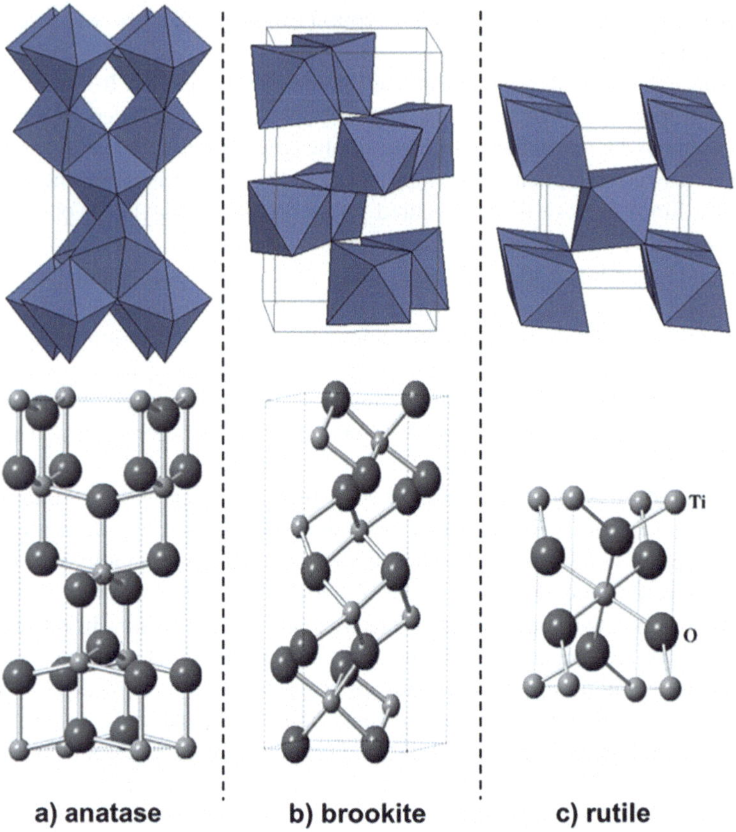

Fig. 3 Different crystal structure of TiO_2: **a** anatase, **b** brookite, and **c** rutile (adapted from [26]

2.3 TiO_2: Limitation and Prospect in Photocatalytic Reaction in Dyes Removal

Despite its increased use and development over the last few decades, TiO_2-based photocatalytic technology faces technical challenges that impede commercialization. More research and development in this area are required to overcome or mitigate these flaws and to expand the use of pure TiO_2-based particles for photocatalytic degradation. The wide band gap of pure TiO_2-based photocatalysis is one of its main limitations [26, 27]. As a result, the photoactivation of this material is limited to the UV radiation range. Sunlight contains 50% infrared radiation, 40% visible radiation, and only 10% ultraviolet radiation. As a result, TiO_2's photocatalytic efficiency in solar energy-based applications remains low and unable to make full use of sunlight. Several strategies have been used to overcome these disadvantages. The recent works to address the raised issue included doping [28], decoration, and structural modification of TiO_2 particles for applications in photocatalysis. At the same time, narrowing

TiO_2's band gap below 3.0 eV by the above modification strategies could improve photo energy capture.

The rapid charge carrier's recombination of photogenerated electrons and holes is another limitation in photocatalytic reaction [29]. According to time-resolved spectroscopic studies, nearly 90% of electrons and holes recombine immediately after photoexcitation, leaving only 10% of charge carriers for the photocatalytic process to continue. This is true for the vast majority of semiconductor materials, including pure TiO_2. Aside from direct charge carrier recombination, it is hypothesized that there are $Ti^{4+}OH$ groups on the TiO_2 surface that act as electron trapping centres and are converted to $Ti^{3+}OH$ species. Following that, $Ti^{3+}OH$ attracts holes and acts as recombination centres. Compared to crystalline TiO_2, amorphous TiO_2 has higher electron–hole recombination, reducing photocatalytic activity.

Furthermore, the modification by dye sensitization can also be used to improve the performance of TiO_2 as a photocatalyst [30]. Extending TiO_2's usable radiation range is a highly advantageous factor in dye sensitization. In general, dye-sensitized semiconductors are used in a variety of applications, including solar energy generation, photocatalytic water splitting, and environmental decontamination. Dye sensitization aims to reduce transparency in the visible range and achieve a longer electron lifetime through efficient charge separation. Through this process, the dye molecules are initially adsorbed onto the photocatalytic surface, which is capable of absorbing visible light [31].

When the semiconductor (TiO_2) is illuminated, electrons are injected into the CB, followed by photoexcitation of the molecules. This charge separation will suppress the excited electrons from recombining with oxidized dye molecules. To successfully sensitize TiO_2 with a dye, the dye must have certain characteristics. It is anticipated that it will have a strong anchoring group for strong interaction with the TiO_2 surface. Furthermore, the dye's energy levels and ground state redox potentials must be compatible with those of TiO_2. Anchoring groups include phosphonic acid groups, carboxylic acid groups, and carboxyl derivatives such as esters, acid chlorides, and salts [32–34]. Furthermore, ethers, silanes, salicylates, and acetylacetonate have been reported [29, 35].

3 TiO_2 Hybrid Photocatalysis in Dyes Removal

3.1 TiO_2: Metal Oxide Hybrid Photocatalysis

The combination of photocatalysis and solar technology is valuable for reducing water pollution from dyeing composites. TiO_2 is said to be the most studied and promising semiconductor photocatalyst due to its numerous advantageous characteristics, such as its nontoxicity, water insolubility, hydrophilicity, affordable availability, and durability against photocorrosion [36–38].

TiO_2 photocatalyst has received a lot of interest due to its promising excitation state and ability to cause the needed redox reactions without biased potential. The main mechanism underlying an extensive range of photochemistry and photo energy is the excitation of TiO_2 by photons with light energy more prominent than the band gap [39–41]. However, specific changes of TiO_2 have been made to improve its light sensitivity, dispersion, and recycling properties due to its tiny fraction of the solar spectrum in the UV region adsorption. TiO_2 is modified with support materials that not only significantly impact changing the electronic band structure, increasing the adsorption capacity and reaction rate as well as the photocatalyst's dispensability and reusability [39, 42].

Significant attention has been paid to the use of inorganic materials like metal and metal oxide as a support for TiO_2 photocatalysts in light energy conversion systems to improve charge separation. The photogenerated electrons are dispersed between TiO_2 and metal oxide while the metal oxide and TiO_2 are in contact. Until the system reaches equilibrium, electrons flow from TiO_2 to metal oxide. According to Fig. 4, metals captured the electrons produced by the excitation of TiO_2, and it is difficult for the electrons inside TiO_2 to return to their initial state. Metal ions can function as a carrier-trapping centre after metal doping, where metal ions greater than tetravalent are more likely to receive electrons than titanium ions and metal ions lower than tetravalent trap holes. The recombination of electron–hole pairs can then be stopped by ion doping, allowing TiO_2 to produce additional electrons and holes. As a result, photocatalytic efficiency of TiO_2 is increased.

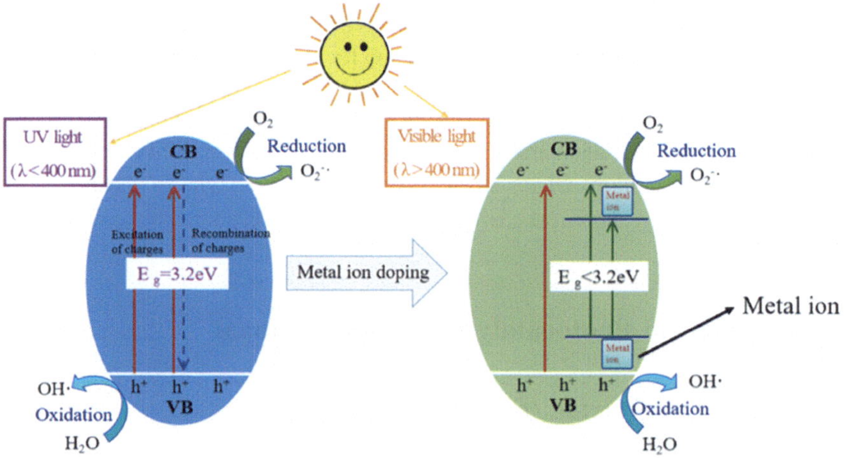

Fig. 4 Mechanism of metal ion doping TiO_2 [43]

3.1.1 TiO_2/ZnO Hybrid Photocatalysis

ZnO is yet another photocatalyst that has been extensively studied. Although the photocatalytic effectiveness of ZnO is not as high as TiO_2, it is nevertheless recognized as a competitive and viable replacement for TiO_2, owing to its nontoxicity, biocompatibility, low manufacturing cost, ease of crystallization, and high electron mobility [44–46]. According to Alshamsi, [47, 48], ZnO has good mechanical, electrical, optical, antibacterial, and photocatalytic properties that enable it to be used for the effective photocatalytic degradation of dyes.

ZnO exhibits strong UV absorbance at ambient temperature, high excitation binding energy (60 meV), and a wide direct bandgap width (3.37 eV) [49]. Beyond its stability, this exhibits photo-corrosive properties that lower photoactivity and are also characterized by a sluggish response to visible light [50]. TiO_2/ZnO nanocomposites were studied by Mohammadi et al. for their photocatalytic and antibacterial characteristics under visible light [51]. Compared to ZnO alone, the photocatalytic activities of the ZnO matrix with TiO_2 addition are improved.

Most recent studies have demonstrated the creation of TiO_2/ZnO composite heterojunction using the electrospinning technique. As a result, there was an effective photocatalytic degradation of methylene blue (MB) dye due to a decreased band gap. It also greatly stimulated an antibacterial effect [52–55]. Studies by Gnanasekaran et al. [56] and Balakrishnan and John [57] showed that the coupled TiO_2/ZnO system was developed by basic sol–gel and thermal decomposition synthesis techniques. The work's application entails eliminating a dye solution containing azo dyes such as methylene orange (MO) and MB. Under UV light, the catalyst showed 90% of its breakdown after 40 min. The addition of ZnO to TiO_2 advantages in decreasing particle build-up and aids in increasing active sites. When compared to bare TiO_2, it was found that the linked system had greater photocatalytic efficiency [52].

In another work by Tekin et al. [58], the thin film photocatalyst's ZnO/TiO_2 solution was made using the sol–gel technique, and the dip-coating method was used to coat the prepared solution on a quartz plate. The degradation of orange G dye on ZnO/TiO_2 photocatalyst was discovered to be in accordance with the Network Kinetic Model through kinetic studies that looked at the impacts of various initial dye concentrations, temperatures, and light intensities. Meanwhile, immobilized TiO_2 on termite hill (TiO_2/TH) and ZnO-promoted TiO_2/TH (ZnO–TiO_2/TH) were synthesized by Yusuff et al. [59] as a solar light responsive photocatalyst. With a band gap energy, specific surface area, pore volume, and pore diameter of 2.85 eV, 28.1 m^2/g, 0.15 cm^3/g, and 23.6 nm, respectively, the ZnO–TiO_2 composite supported by siliceous termite hills had been successfully developed and used as a photocatalyst to degrade MB dye in aqueous solution under solar irradiation. As a visible light responsive and inexpensive photocatalyst for the degradation and decolourization of wastewater, ZnO–TiO_2 hybrid photocatalyst had shown remarkable potential.

3.1.2 TiO_2/Fe_3O_4 Hybrid Photocatalysis

Zero-valence iron (Fe) is a viable method for discolouring industrial dye wastewater and quickly degrading dye molecules [60]. Fe is a powerful reducer that frequently forms Fe^{2+} (usually at pH values of 2–3), a powerful catalyst for the colouring, breakdown, and mineralization of organic contaminants in wastewater. When irradiated with light energy greater than their band gaps, active photocatalysts like Fe_3O_4 with a band gap of 2.2 eV can absorb visible light to produce conduction band electrons (e) and valence band holes (h^+) for the destruction of dye molecules [56, 61, 62], studied the magnetic separation of green synthesized Fe_3O_4 nanoparticles on photocatalytic activity of MO dye removal. The absence of the methyl orange tone was confirmed by the brilliant (UV) maintenance top's disappearance at 565 nm. The remarkable shading removal of the Fe_3O_4 nanorod was 100% after 110 min. This removal occurred due to the photochemical redox process and the usage of Fe_3O_4 nanorods with a large energy gap. The fact that the Fe_3O_4 photocatalyst could be recycled up to five times shows how stable it is. It might be used as a potential promising photocatalyst for dye-polluted water.

TiO_2 has significant drawbacks due to its large energy bandgap (~ 3.2 eV) and low visible light absorption. Another is the quick recombination of electron and hole pairs produced by photosynthesis [56, 63]. In this regard, it has been suggested that attaching TiO_2 to magnetic nanoparticles to create magnetic photocatalysts is a practical method for the beneficial separation and recycling of the photocatalyst [64]. In research work by Kermani et al. [64], a hydrothermal process was used to successfully create magnetic Fe_3O_4/TiO_2 nanocomposites as a useful and recyclable photocatalyst with varying Fe_3O_4 loading percentages. Under simulated solar light irradiation, the synthesized photocatalysts were employed to photo catalytically degrade rhodamine B as the model pollutant. Utilizing 20% Fe_3O_4/TiO_2 nanocomposite, the maximum degrading efficiency of 91% was attained after 120 min.

Kubo et al. [65] demonstrated the facile fabrication of HKUST-1 by a spray-assisted synthetic process, which allowed a controlled amount of nanoparticles to be incorporated into the Metal–Organic Framework (MOF). The same HKUST-1 particle may include both Fe_3O_4 and TiO_2 nanoparticles. Due to the Fe_3O_4 nanoparticles added, HKUST-1 and (Fe_3O_4, TiO_2)/HKUST-1 exhibit superparamagnetic, making it simple to separate them from an aqueous solution by using an external magnetic field. In addition, the nanocomposites have exceptionally high MB adsorption capabilities (> 700 mg/g) due to their huge surface area (> 1200 m^2/g). For reuse as an adsorbent, (Fe_3O_4, TiO_2)/HKUST-1 composites demonstrated good stability, promising for dye removal in wastewater.

Nadimi et al. [66] reported that coprecipitation was used to create Fe_3O_4-bentonite nanoparticles in a nitrogen environment. First, $FeCl_2$ and $FeCl_3$ were added to an aqueous suspension of bentonite. Then, using a sol–gel technique, TiO_2 was added to the Fe_3O_4-bentonite's surface. The azo dye MB was photodegraded using the synthesized photocatalyst. It was discovered that under UV irradiation, the removal effectiveness of MB approached 90% and that after six cycles, only a 20% mass loss occurred. As a result, the composite material demonstrated robust photocatalytic functionality and recycling capabilities.

3.1.3 TiO_2/CuO Hybrid Photocatalysis

TiO_2 is a semiconductor that can only be stimulated at shorter spectral wavelengths, typically in the UV region [67]. Creating semiconductor photocatalysts with increased light absorption capacity and charge separation efficiency has taken a lot of work. Due to a decrease in recombination rate and an increase in photostability, synthesizing semiconductor heterostructures is one of the developed strategies of particular interest [68]. Metal-free heterojunctions have demonstrated promising promise as photocatalyst components. By using p–n junctions such as CuO/TiO_2, Cu_2O/$BiVO_4$, and $CuBi_2O_4$/TiO_2, we can obtain several benefits: a longer lifetime of the charge carriers, improved charge separation, quick charge transfer to the catalyst, and separation of locally incompatible reduction and oxidation reactions in nanospace.

Çinar et al. [69] reported plate-like CuO particles produced hydrothermally and electrospun TiO_2 fibres were used to create CuO–TiO_2 p–n heterostructures, which were then used to high-efficiency photocatalytic degradation of MB. The effect of varying amounts of CuO decoration on TiO_2 fibres on microstructure, phase constitution, and optical and photocatalytic properties was studied. The findings of the photocatalytic testing showed that the number of CuO particles in the samples had an impact on the degrading behaviour. 1.25 wt% CuO–TiO_2 heterostructure had the highest efficiency for MB decomposition under UV light with a rate constant of 0.135 min^{-1}. The most significant rate constant for visible irradiation was measured to be 0.015 min^{-1} for 0.50 wt% CuO–TiO_2. It concluded that particle-fibre architecture, higher and extended light harvesting ability, more effective charge separation due to staggered band structure, and p–n junctions of the heterostructure samples were all credited with the better photocatalytic performance in comparison with pure TiO_2 fibres.

Singh et al. [70] presented a simple, one-step hydrothermal method for making CuO–Cu_2O nanorod/TiO_2 nanoparticle heterostructures (CTHS), which may be less expensive photocatalytic substitute. According to transmission electron microscopy (TEM) analysis, the average size of the manufactured heterojunction components was 10 nm for the functionalizing TiO_2 nanoparticles and 13 nm on average for the nanorod length and breadth. CTHS demonstrated more sunlight-induced photodegradation activity compared to TiO_2 nanoparticles. Additionally, Cu_XO/TiO_2 heterojunction was able to remove an organic pollutant (methylene blue, 10 mM) from toxic wastewater in 60 min. Band gap engineering, an active decrease in charge recombination, and better responsiveness to visible light are three leading causes of the superior photocatalytic efficacy of the heterojunction.

3.1.4 TiO_2/WO_3 Hybrid Photocatalysis

One effective way to overcome the inherent limitations and improve photocatalytic degradation is to combine TiO_2 with a suitable metal oxide. Metal oxide combinations, including ZnO, PdO, CuO, ZrO_2, Fe_2O_3, Bi_2WO_6, SnO_2, and WO_3 have all been studied so far [71]. WO_3 is a promising candidate to combine with TiO_2 out of these due to its small band gap (E_g: 2.4–2.8 eV) and the presence of Bronsted and Lewis acidic sites (W6 + species), which adsorb OH or H_2O to eliminate organic contaminants. Regarding the photocatalytic properties of WO_3/TiO_2 composites, multiple studies found that combining TiO_2 and WO_3 reduces interfacial photoinduced charge separation and electron–hole recombination, resulting in increased catalytic activity [72, 73].

Khan et al. [72] worked on sol–gel and crash precipitation techniques to generate titania tungstate photocatalysts (TiO_2/WO_3), which were then spray-dried to create a micro-sized hybrid material. The tetrahedral and monoclinic crystalline structures of TiO_2 and WO_3 in the hybrid material calcined at 600 °C were confirmed by X-ray diffraction. In the UV degradation of the model pollutant MB, $TW_{0.075}$ (TiO_2/WO_3 having 0.075 molar ratio of tungsten precursor) outperforms the other samples, converting 90% of MB in 100 min (30 min dark + 40 min light + 30 min dark), exhibiting energy storage capacity without UV irradiation. The species primarily used to break down pollutants are $^{\cdot}OH$ and $^{\cdot}O_2^-$.

Other work by Zhao et al. [74] successfully synthesized WO_3/TiO_2 mesoporous spheres (450 nm in diameter) made of multiple self-assembled anatase TiO_2 layers covered with WO_3. A simple wet-chemistry method is successfully used to create nano crystallites. The addition of monolayered WO_3 drastically alters mesoporous TiO_2 spheres' surface, photocatalytic, and optical characteristics. Mesoporous WO_3/TiO_2 spheres with these advantageous textural characteristics, coupled with the improved interfacial charge carrier separation, exhibit superior photocatalytic activity in the MB degradation process under UV, visible, and solar light irradiations. In addition to delivering more significant fluxes during membrane filtration in dead-end and cross-flow modes, the synthesized mesoporous WO_3/TiO_2 spheres also show promise as a photocatalyst for future uses in concurrent solar photocatalysis and membrane filtration processes.

3.2 TiO_2: Ceramic Hybrid Photocatalysis in Dyes Removal

The semiconductor material must stay intact once the dispersion has been totally dye-decolourized for ceramic-based support materials to function properly. This is a crucial indication that continued exposure to light will not cause the degraded particles to breakdown [75, 76]. Additionally, the ceramic support material must guarantee that after numerous cycles of usage, the efficiency of recycled semiconductor photocatalyst does not significantly decline [77, 78].

However, some issues have limited the development of ceramic materials with TiO_2 support for use as photocatalysts in dye removal on a broad scale. In order to gather knowledge for future large-scale applications, particularly for the massive volume of actual textile wastewater purification, the development of pilot-scale treatment systems must be thoroughly investigated [79]. It is important to note that no most ceramic hybrid photocatalytic system supported by TiO_2 is currently available for a comprehensive treatment that includes the necessary features, such as high mineralization efficiency for organic/synthetic dyes, simple retrieval from treated solution, and visible light photoactive [80, 81]. The development of such technology is still very recent, and much progress is required especially for the development of photocatalytic reactors, determination on optimize dyes removal mechanism, low cost with less energy consumption of synthesis process, improvement on selectivity of TiO_2 photocatalyst towards co-existence of dyes compounds and possibility in application of catalyst under green sun light irradiation [82, 83].

3.2.1 TiO_2/SiC Hybrid Photocatalysis

The three-dimensional substance silicon carbide (SiC), which has a high level of chemical and thermal stability, is often regarded consequently. The supported TiO_2 and the dye effluents have a substantial contact surface due to the SiC's structure. Furthermore, because of their macro-porosity, three-dimensional SiC foams are great options for support materials for TiO_2 photocatalysts because they produce hybrid materials with high internal surfaces and superior catalyst immobilization, particularly when applied using dip-coating techniques [84]. This hybridization also enhances the inadequate exchange surface and the potential for recovering and reusing the catalyst made of TiO_2 powder. In addition, it offers a solid anchor for TiO_2 on the surface of SiC. Other research claimed that the heterostructures of TiO_2 and SiC had the ability to significantly lower the rate of electron–hole pair recombination, which ultimately increased the effectiveness of the photocatalytic process in the elimination of dyes [85].

A slightly elevated, integrated Fe–TiO_2–SiC ternary system has been created using a low-cost sonochemical technique in the presence of citric acid. This integration system showed that an anatase crystalline TiO_2 phase with iron incorporation had produced a composite of Fe–TiO_2 over the whole surface of the SiC [86]. In contrast to commercial anatase TiO_2, the integrated catalyst displays good improved photocatalytic activity for the elimination of synthetic rhodamine B dyes under solar simulator irradiation. Additionally, the doping of TiO_2–SiC makes a promising photoreduction observation during the removal of dyes [87]. Under blue light LED irradiation, the produced N-doped TiO_2–SiC foam shown good photocatalytic activity with a high elimination of 96.3% MB synthetic dyes. N-doped TiO_2–SiC foam still exhibited little changes in its photocatalytic activity even after five cycles of MB photodegradation [88].

3.2.2 TiO_2/SiO_2 Hybrid Photocatalysis

TiO_2 particles can generally be easily separated at the nanoscale, although this is restricted by their extremely small surface areas and challenging dispersion. As a result, support materials like silicon dioxide (SiO_2) are synthesized in a precursor solution before being immobilized on a substrate made of titanium dioxide (TiO_2) [89, 90]. This may be done using the ex situ impregnation and coprecipitation methods. Both of these approaches provide various TiO_2 supported SiO_2 photocatalyst capabilities, as well as various catalytic and adsorption results for the reduction of dye pollutants. In other words, the ex situ process of impregnating TiO_2 onto a substrate of SiO_2 maintains and sustains both TiO_2 and SiO_2 properties [91, 92]. The development of a self-assembling thin film of TiO_2 on SiO_2 support materials, on the other hand, would result in the introduction of new hybrid materials between TiO_2 and SiO_2, which improves catalytic activities for the degradation of dye pollutants [93, 94]. While SiO_2 provides greater adsorption sites close to the TiO_2, the TiO_2 acts as the photoactive core to produce $^{\cdot}OH$ when exposed to light. There are several such domains on a single particle in genuine particles [95]. Whether the particles behave as straightforward mixes of the bulk materials, mixtures of much smaller entities like quantum particle domains, or entirely arbitrary intermixed phases is an interesting question.

The self-assembling TiO_2/SiO_2 photocatalyst at a 20:80 ratio showed increased photocatalytic decolourization of MB up to 81% due to SiO_2's adsorbability and the presence of a $^{\cdot}OH$ on the catalyst's surface to support photoreduction. This hybridization has also been found to increase the stability, activity, and light harvesting in the visible light spectrum [96, 97]. By creating novel hybrid materials between SiO_2 and TiO_2, research on the creation of SiO_2 precursor from rice husk silica dissolved in TiO_2 solution has claimed to improve the photo decolourization of MB [98]. This enhancement is the result of creating a chemical connection and expanding the specific surface area. In this instance, the rising specific surface area and chemical makeup of the photocatalyst determine the thin film's photocatalytic reactivity for the decolourization of MB dyes. The TiO_2/SiO_2 materials are said to be composed of matrix-isolated TiO_2 quantum particles, according to the catalysts' characterization.

In addition, the manner of azo dye removal on the surface of TiO_2/SiO_2 photocatalyst is linked to particle size and NH^{4+} production when using a TiO_2/SiO_2 hybrid photocatalyst to decolourized [99, 100]. The penta-heterocycle and benzene ring substitution N groups changed mostly to NH^{4+} ions, whereas the nitrogen of the azo group was altered to NH^{4+} ions and N_2 gas. The triazine ring and its substituted groups are substantially more stable over the 8-h photodegradation period when exposed to UV light. TiO_2 is reinforced by SiO_2 made using the sol–gel process was a more effective photocatalyst than TiO_2 by itself for the photocatalytic breakdown of rhodamine-6G [101, 102]. The presence of an adsorbent on the surface of SiO_2 was blamed for the improvement in efficiency. Rhodamine-6G was more concentrated close to the TiO_2 sites in the adsorbent phase than in the solution. Rhodamine-6G adsorption is helpful for the photocatalytic activity of the catalysts; however, the dye molecules do not adsorb on TiO_2 alone; instead, a TiO_2 slurry photodegrades the dyes' dissolved form [103].

With the aim of enhancing light sensitivity, reaction kinetics, pollutants compound selectivity, and recycling/separation at the conclusion of the treatment process, an improved ternary photocatalysis system comprising TiO_2-metal oxide supported SiO_2 has been discovered to be of interest [104, 105]. A magnetically separable TiO_2–Fe_3O_4/SiO_2 photocatalyst is created by mixing magnetic nanoparticles with TiO_2/SiO_2 photocatalyst. Organic dyes were completely decoloured by TiO_2–Fe_3O4/SiO_2 under ideal circumstances in less than 45 min. Photocatalyst load, dye concentrations, and photoreactor transmittance all play significant roles in this process. With roughly the same efficiency, the magnetic TiO–Fe_3O_4/SiO_2 has shown good cycling up to five cycles. The ultimate cost of the treatment procedure is anticipated to be decreased by the reuse of the photocatalyst [106].

In previous studies, the impact of TiO_2 supported SiO_2 covered cement-based materials has shown potential photocatalytic activities towards reduction of malachite green oxalate, MB, and MO under ultraviolet light [107]. The outcome demonstrated an improvement in the photocatalytic effects, which are dependent on the catalyst's structure and the pH of the used cement. Additionally, it has been shown that TiO_2 supported doped cerium-SiO_2 nanostructured, which were created using electrospinning, generate a superior photocatalyst for the breakdown of MB when exposed to artificial sunlight [108]. When RuO_2 is included in a TiO_2/SiO_2 photocatalyst, the effective surface area and the spacing of electron–hole pairs improve, increasing the photocatalyst's effectiveness [100]. Hydroxyl has shown to be the major photooxidant in this instance throughout the decolourization process [109].

3.2.3 TiO_2/Al_2O_3 Hybrid Photocatalysis

Due to its low quantum and restricted capacity to absorb visible light, TiO_2 still faces significant hurdles when used in actual waste water technology in addition to adsorption and homogeneity [110]. One of the most popular support materials for TiO_2 photocatalysts in dye degradation is alumina (Al_2O_3) because it may enhance the textural qualities of TiO_2, such as its high specific surface area and to some extent, the bulk of the material [111]. Due to the synergistic effects of Al_2O_3 and TiO_2, the synthetically created TiO_2 supported Al_2O_3 produced by decorating TiO_2 film on Al_2O_3 surface by wet chemical approach has produced high adsorption and photocatalytic capabilities even under weak light irradiation [112]. A lowering band gap energy of the hybrid TiO_2/Al_2O_3 down to $E_g = 2.94$ eV was reported in another work [113]. This indicates that Al_2O_3 in TiO_2 can improve photocatalytic performance by preventing the recombination of electron–hole. Al_2O_3 has been used as an adsorbent with a mesoporous structure for preconcentrating and separating amounts of coloured water molecules. Its substantial specific surface area and sturdy structure are advantageous for the adsorption of dyes and subsequent interactions with active sites. Instead, the mesoporous structure of Al_2O_3 helps sustain the hybrid system, improving its efficiency and stability during the breakdown of the dye molecules [114]. The quantity of dye molecules increased on the mesoporous surface of Al_2O_3, and a strong contact between the hybrid components was created to encourage enhanced photocatalytic activities [115].

Particle preparation by mechanical grinding has been used to enhance the photocatalytic activity of TiO_2/Al_2O_3 towards the degradation of dye pollutants. With the help of this technique, you may adjust the size of the particles for the desired use in water treatment by changing the particle size during the sol–gel synthesis. The increase of local TiO_2 sites on TiO_2/Al_2O_3 materials in comparison with other materials was proven by the photocatalytic activity of TiO_2/Al_2O_3 and TiO_2 towards the photocatalytic breakdown of dye molecules. The immobilization of size-selective TiO_2 nanoparticles, devoid of aggregation, into ultra-porous Al_2O_3 monoliths has improved the selectivity of TiO_2/Al_2O_3 as a photocatalyst [114]. The outcome demonstrated that the developed photocatalyst has greater dye degradation efficiency as a result of porosity and density structure. The TiO_2/Al_2O_3 photocatalyst is attractive as a stable yet large surface area of catalyst in dyes adsorption and reduction due to its low density and high porosity, tiny size of the structural unit, and evanescent intrinsic absorption in the UVA spectral region [116].

3.3 *TiO_2: Polymer Hybrid Photocatalysis in Dyes Removal*

In the last few decades, numerous photocatalytic processes have been investigated using metal oxide semiconductors in dyes removal and degradation, which typically possess excellent stability and a high capacity for sunshine collecting. TiO_2 has been the prototypical semiconductor photocatalyst due to its low cost, chemical stability, and eco-friendliness. Nonetheless, TiO_2 has the characteristic shortcoming of a wide band gap (3.2 eV for anatase TiO_2), resulting in a low visible light absorption efficiency. As a result, it is now preventing the widespread use of photocatalytic materials based on TiO_2 [117]. The nanoparticles can be modified on the surface of the photocatalyst to facilitate the ability for h^+ and e^- to separate and then significantly improve the photodegradation of water dyes molecules in much more efficient and effective ways [118]. With the help of the right co-catalyst, the range of light absorption can be increased, holes and photogenerated electrons can be separated more easily, and the photocatalysis overpotential of the catalyst can be lowered. Several strategies to achieve it include TiO_2 doping and co-doping with non-metals and metals and incorporation of TiO_2 into organometallic frames and polymer composites [119].

3.3.1 TiO_2/Chitosan Hybrid Photocatalysis

Chitosan (CS) is appropriate for photocatalytic activity with TiO_2 due to its excellent stability and metal ion and dye adsorption capacity [120]. The biopolymer matrix would absorb the dye molecules and make them readily accessible for continuous degradation by TiO_2, hence eliminating the recombination effects that impede the dye degradation process. Besides, the amine (–NH2) and hydroxyl (OH) groups present in chitosan make it an excellent sorbent for various organic contaminants. The addition of CS to TiO_2 photocatalysis improves adsorption, dispersion, and stability [121]. Balakrishnan et al. [122] stated the reasons that use TiO_2/chitosan are not limited to minimize catalyst loss, improvement in stability of catalyst, minimizing the aggregation of nano-sized particles, facilitating the recovery of photocatalyst, boosting the reusability of catalyst and assistance in continuous photocatalytic degradation at its site.

Sugashini et al. [123] used chitosan biopolymer in the form of nano chitosan (NCS) and carboxymethyl cellulose (CMC) as the support material for TiO_2 to overcome the drawbacks. These components are essential for increasing the active surface area, boosting the dye degradation performance, and enhancing the chemical stability of the composite catalyst. The dye degradation process is aided by the adsorption capacities of NCS and CMC, which bring the dye molecules close to the active sites of TiO_2 where $^{\cdot}OH$ are formed. Besides, holes and electrons readily react with dye species adsorbed on the surface of the TiO_2 catalyst to ensure effective dye degradation. Feng et al. [124] reported that the as-prepared TiO_2/CS-biochar composites exhibited excellent and durable UV–Vis light photocatalytic activity for the degradation of rhodamine B (RhB) aqueous solution, and the degradation rate constant was more than thirty times higher than that of pristine TiO_2.

3.3.2 TiO_2/Alumina–Silicate Polymer Hybrid Photocatalysis

To harness the photocatalytic ability of TiO_2 nanoparticles in organic polymeric materials, researchers have prepared TiO_2 particles coated with alumina (Al_2O_3) or silica (SiO_2) in the shape of islands. The Al_2O_3 or SiO_2 protected the organic polymeric compounds from undergoing decomposition by photocatalysis because the TiO_2 nanoparticles covered with Al_2O_3 or SiO_2 would not be directly in contact with the organic polymeric compounds [125]. Li et al. [126] reported that the loaded TiO_2 on such SiO_2 clay shows a higher photodegradation rate of methyl orange (MO) in aqueous by comparing it with pure TiO_2 particles. Yuan et al. [127] synthesized TiO_2 photocatalysts coated on foam nickel by the sol–gel method. In the formation of the photocatalyst, Al_2O_3–SiO_2 films were introduced as transition interlayers between TiO_2 and nickel metal foam to promote the photocatalytic activity of TiO_2. They found that TiO_2/Al_2O_3–SiO_2 composite films have much higher photocatalytic activity and stability for the degradation of gaseous acetaldehyde than TiO_2 films. The large specific surface area of Al_2O_3–SiO_2, which provides more active sites for future TiO_2 loading and photocatalytic reaction, is responsible for the considerable enhancement impact. The Al_2O_3–SiO_2 interlayer also prevents the transport of a photogenerated electron from TiO_2 to Ni metal, hence enhancing the photocatalytic activity of photocatalysts.

3.3.3 TiO_2/Polypyrrole (PPy) Hybrid Photocatalysis

As one of the conductive polymers, polypyrrole (PPy) with broadband absorption and high absorption rate for visible light has been widely used in biomaterials, supercapacitors, and other fields due to its simple preparation and good environmental stability. Indeed, as a conducting polymer PPy possessing a p-conjugated structure, has excellent electron mobility and high stability, and is readily available as a high mobility charge carrier, PPy can inhibit the recombination of electron–hole pairs and improve the charge separation efficiency of photocatalysts. As reported, PPy with unique structures has been widely used for the modification of semiconductors, such as BiOI/PPy and TiO_2/PPy, due to the convenient polymerization and modification on the surface of nanoparticles. Photocatalyst modified with PPy obtains stronger adsorption capacity, faster electron–hole separation rate, and higher light collection performance. Therefore, the photocatalytic activity of PPy-modified photocatalysts can be improved significantly. As reported, TiO_2/PPy had higher photocatalytic activity than that of the original TiO_2. At the same time, photocatalyst modified with PPy shows excellent stability after repeated use, and PPy is not degraded in the process of photocatalysis. In addition, the abundant functional groups on PPy can also provide enough active sites for NiCoP loading [118].

The improved photocatalytic performance of PPyNS–TiO_2 is due to the highest surface area and fewer defects in the nanostructure PPy and to the wide light absorption and increased charge separation efficiency owing to the heterojunction formation [128]. The PPyNS–TiO_2 composite photocatalyst shows high stability and recyclability. For example [129] reported that an organic/inorganic core–shell nanobelt, TiO_2@V_2O_5-(PPy) exhibited high photocatalytic performance and water stability. Besides that, the heterojunction between the TiO_2 nanobelt and V_2O_5 nanosheets improves the ability to absorb visible light and the separation efficiency of the photogenerated carrier. When covered with the conjugated polymer PPy, the organic/inorganic nanocomposite absorbs throughout the visible light region and exhibits enhanced photocatalytic performance. Due to the outstanding transfer and separation efficiency of the electrons and holes, the new catalytic system is applied to remove water-soluble organic pollutants such as tetracycline (TC), doxycycline, and oxytetracycline with high efficiency under visible light irradiation. In addition, the PPy nanolayer also prevents leakage of V_2O_5, which is beneficial to the construction of a stable heterostructure, and thus, the prepared photocatalyst can be recycled.

Conducting polymers (CPs) function as conventional organic semiconductors. The nanostructure of a conjugated polymer plays an essential role in its photocatalytic activity. The morphology and size of the photocatalytic nanomaterials have an important effect on their photocatalytic performance. The difference in photocatalytic properties between nano and bulk CPs may be the consequence of larger diameters, higher surface areas, and more flaws in bulk CPs, which are accountable for electron–hole recombination [128]. In UV irradiation, the photoinduced electron (e^-) in the CB of TiO_2 can be recombined with holes (h^+) in the highest occupied orbital (HOMO) of PPyNS via the intimate interface resulting in the separation of photogenerated charge carriers (Fig. 5a), which explains the decreased intensity

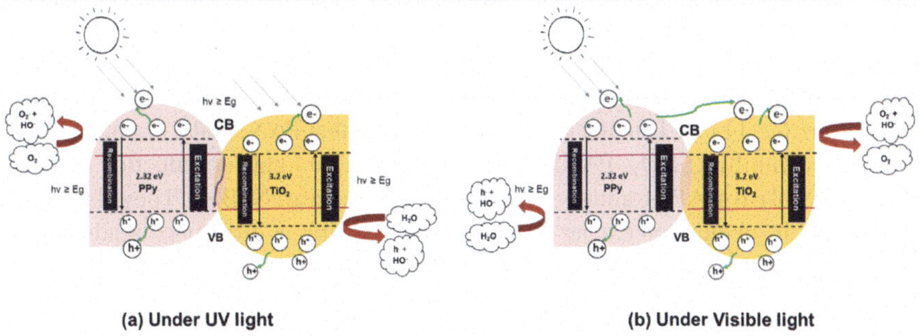

Fig. 5 Proposed photocatalytic mechanism for the PPyNS–TiO_2 system and charge transfer mechanisms under **a** UV light and **b** visible light irradiation

signal of PPyNS–TiO_2 in time-resolved microwave conductivity (TRMC) measurement at the excitation of 365 and 400 nm wavelength compared with bare TiO_2. In this case, the oxidative reaction and reductive species are produced in the VB band of TiO_2 ($h^+ + OH^- \rightarrow {}^{\cdot}OH$) and the lowest unoccupied orbital (LUMO) of PPyNS ($e^- + O_2 \rightarrow {}^{\cdot}O_2{}^-$), respectively.

This proposed mechanism is highly in agreement with the results of TRMC and photocatalytic properties under UV irradiation. However, the higher yield and longer lifetime of charge carriers from the TRMC measurements and higher photocatalytic activity in the presence of PPyNS–TiO_2 composite compared to pristine TiO_2 under visible light excitation, another possible photocatalytic mechanism can be proposed for PPyNS–TiO_2 (Fig. 5b). While for visible light irradiation, PPyNS can be photoexcited and act as a photosensitizer. Owing to the internal electric field function, the e^- in the LUMO of PPyNS could be transferred to that of CB of TiO_2. Photogenerated h^+ located on the HOMO of PPyNS can participate in the oxidative reaction. Hence, the redox reaction happens in the CB of TiO_2 (reductive reaction: $A \rightarrow A^-$) and HOMO of PPyNS (oxidative reaction: $D \rightarrow D^+$), and the effective charge separation can be extended the lifetime of generated e^- and h^+ and then enhanced photocatalytic activity.

3.3.4 TiO_2/Polyaniline Hybrid Photocatalysis

Polyaniline (PANI) as a conducting polymer has received positive attention in various practical applications, especially in electrical or optoelectronic devices and sensors, due to its ease of synthesis, low cost, excellent environmental stability, and unique physicochemical behaviour. Besides many other interesting properties, PANI possesses substantial surface area and porosity and has demonstrated favourable ability as an adsorbent that could remove azo dyes from wastewater [130, 131]. In addition, PANI is expected to interact with specific metal ions which have a strong affinity to the nitrogen of amine and imine functional groups of the polymer. Although noble metal coupling could be efficient in prolonging the surface charge separation, their cost-effectiveness for an industrial application is usually replaced by a more economical transition or non-metal doping metal [132]. Specifically, PANI was shown to be effective as a photosensitizer capable of modifying TiO_2 photocatalytic properties [119]. Thakare et al. [133] reported that PANI had a dramatic effect on TiO_2–polyaniline (TiO_2–PANI) composite in visible light for reduction of 4-NP (4-nitrophenol) to 4-AP (4-amino-phenol) in the presence of $NaBH_4$. They also found that NaBH4 is an electron carrier in addition to assisting in nitro-phenol adsorption on the catalyst. This has a significant impact on the product's performance.

4 Conclusion

Depending on the processed fibre and the auxiliary chemicals and dyestuffs employed, textile industry effluent is highly coloured, complicated, and variable. Physical, chemical, and biological treatments are the three categories of treatments for traditional dye removal. Due to its high efficiency in breaking down the dye molecules, photocatalysis using TiO_2 nanoparticles appears to be one of the most promising methods for dye reduction. TiO_2 nanoparticles have been produced using a wide range of diverse approaches during the past few years, including the sol–gel method, hydrothermal method, vapour deposition, spray pyrolysis, electrodeposition, solvothermal method, microwave method, and sonochemical method. However, specific changes of TiO_2 have been made to improve its light sensitivity, dispersion, and recycling properties due to its tiny fraction of the solar spectrum in the UV region adsorption.

To that point, TiO_2 is modified with support materials that not only significantly impact changing the electronic band structure, but increasing the adsorption capacity and reaction rate as well as the photocatalyst's dispensability and reusability. With regard to this, the summarized on the application of various metal oxide nanoparticles, including ZnO_2, Fe_3O_4, CuO, and WO_3 to reduce electron–hole recombination by allowing TiO_2 to produce additional electrons and holes for the efficiency of TiO_2 is highlighted in the chapter. Following that, combination of TiO_2 with ceramic-based materials, including SiC, SiO_2, and Al_2O_3 has produced an attractive photocatalyst with a stable yet large surface area of catalyst in dyes adsorption and reduction due to its low density and high porosity, tiny size of the structural unit, and evanescent intrinsic absorption in the wide light spectral region. In the meantime, hybridization of TiO_2 nanoparticles with polymeric-based materials such as chitosan, aluminium silicate polymer, polypyrrole, and polyaniline possesses substantial surface area and porosity and has demonstrated favourable ability as an adsorbent that treats the dyes molecules from wastewater much efficient.

Acknowledgements The first author of the manuscript is under scholar of the Postdoctoral Research Fellowship 2022 (PDRF2022), Public Service Department of Malaysia.

References

1. Sardar M, Manna M, Maharana M, Sen S (2021) Remediation of dyes from industrial wastewater using low-cost adsorbents. In: Green adsorbents to remove metals, dyes and boron from polluted water. Springer, Cham, pp 377–403
2. Katheresan V, Kansedo J, Lau SY (2018) Efficiency of various recent wastewater dye removal methods: a review. J Environ Chem Eng 6(4):4676–4697
3. Liu Q (2020) Pollution and treatment of dye waste-water. In: IOP conference series: earth and environmental science (vol 514, No. 5). IOP Publishing, p 052001
4. Srivatsav P, Bhargav BS, Shanmugasundaram V, Arun J, Gopinath KP, Bhatnagar A (2020) Biochar as an eco-friendly and economical adsorbent for the removal of colorants (dyes) from aqueous environment: a review. Water 12(12):3561
5. Dawood S, Sen T (2014) Review on dye removal from its aqueous solution into alternative cost effective and non-conventional adsorbents. Journal of Chemical and Process Engineering 1(104):1–11
6. Saratale RG, Saratale GD, Chang JS, Govindwar SP (2011) Bacterial decolorization and degradation of azo dyes: a review. J Taiwan Inst Chem Eng 42(1):138–157
7. Taib MAA, Razak KA, Jaafar M, Lockman Z (2017) Initial growth study of TiO_2 nanotube arrays anodised in KOH/fluoride/ethylene glycol electrolyte. Mater Des 128:195–205
8. Herrmann J-M (2005) Heterogeneous photocatalysis: state of the art and present applications In honor of Pr. RL Burwell Jr. (1912–2003), Former Head of Ipatieff Laboratories, Northwestern University, Evanston (Ill). Topics in Catalysis 34(1–4):49–65
9. Arun J, Nachiappan S, Rangarajan G, Alagappan RP, Gopinath KP, Lichtfouse E (2022) Synthesis and application of titanium dioxide photocatalysis for energy, decontamination and viral disinfection: a review. Environmental Chemistry Letters 1–24
10. Varshney G, Kanel SR, Kempisty DM, Varshney V, Agrawal A, Sahle-Demessie E, Varma RS, Nadagouda MN (2016) Nanoscale TiO_2 films and their application in remediation of organic pollutants. Coord Chem Rev 306:43–64
11. Nyamukamba P, Okoh O, Mungondori H, Taziwa R, Zinya S (2018) Synthetic methods for titanium dioxide nanoparticles: a review. In: Titanium dioxide-material for a sustainable environment, pp 151–1755
12. Pant B, Park M, Park SJ (2019) Recent advances in TiO_2 films prepared by sol-gel methods for photocatalytic degradation of organic pollutants and antibacterial activities. Coatings 9(10):613
13. Shahid M, McDonagh A, Kim JH, Shon HK (2015) Magnetised titanium dioxide (TiO_2) for water purification: preparation, characterisation and application. Desalin Water Treat 54(4–5):979–1002
14. Asim N, Ahmadi S, Alghoul MA, Hammadi FY, Saeedfar K, Sopian K (2014) Research and development aspects on chemical preparation techniques of photoanodes for dye sensitized solar cells. International Journal of Photoenergy 2014
15. Ambrožič G, Orel ZC, Zigon M (2011) Microwave-assisted non-aqueous synthesis of ZnO nanoparticles. Materiali in Tehnologije 45(3):173–177
16. Imoisili PE, Jen TC, Safaei B (2021) Microwave-assisted sol–gel synthesis of TiO_2-mixed metal oxide nanocatalyst for degradation of organic pollutant. Nanotechnol Rev 10(1):126–136
17. Lu Z, Chen F, He M, Song M, Ma Z, Shi W, Yan Y, Lan J, Li F, Xiao P (2014) Microwave synthesis of a novel magnetic imprinted TiO_2 photocatalyst with excellent transparency for selective photodegradation of enrofloxacin hydrochloride residues solution. Chem Eng J 249:15–26
18. Wang Z, Zhu J, Xu W, Sui J, Peng H, Tang X (2012) Microwave hydrothermal synthesis of perovskite $BiFeO_3$ nanoparticles: an insight into the phase purity during the microwave heating process. Mater Chem Phys 135(2–3):330–333

19. Modan EM, Plaiasu AG (2020) Advantages and disadvantages of chemical methods in the elaboration of nanomaterials. The Annals of "Dunarea de Jos" University of Galati. Fascicle IX, Metallurgy and Materials Science 43(1):53–60
20. Berahim MSA, Majnis MF, Taib MAA (2018) Formation of titanium dioxide nanoparticles by anodization of valve metals. Progress in Energy and Environment 7:11–19
21. Taib MAA, Alias N, Jaafar M, Razak KA, Tan WK, Shahbudin IP, Kawamura G, Matsuda A, Lockman Z (2020) Formation of grassy TiO_2 nanotube thin film by anodisation in peroxide electrolyte for Cr (VI) removal under ultraviolet radiation. Nanotechnology 31(43):435605
22. Zhang J, Zhou P, Liu J, Yu J (2014) New understanding of the difference of photocatalytic activity among anatase, rutile and brookite TiO_2. Phys Chem Chem Phys 16(38):20382–20386
23. Hanaor DA, Sorrell CC (2011) Review of the anatase to rutile phase transformation. J Mater Sci 46(4):855–874
24. Oi LE, Choo MY, Lee HV, Ong HC, Abd Hamid SB, Juan JC (2016) Recent advances of titanium dioxide (TiO_2) for green organic synthesis. RSC Adv 6(110):108741–108754
25. Stephen L (2020) Titanium dioxide versatile solid crystalline: an overview. In: Assorted dimensional reconfigurable materials
26. Etacheri V, Di Valentin C, Schneider J, Bahnemann D, Pillai SC (2015) Visible-light activation of TiO_2 photocatalysts: advances in theory and experiments. J Photochem Photobiol, C 25:1–29
27. Dong H, Zeng G, Tang L, Fan C, Zhang C, He X, He Y (2015) An overview on limitations of TiO_2-based particles for photocatalytic degradation of organic pollutants and the corresponding countermeasures. Water Res 79:128–146
28. Ghosh S, Das AP (2015) Modified titanium oxide (TiO_2) nanocomposites and its array of applications: a review. Toxicol Environ Chem 97(5):491–514
29. Peiris S, de Silva HB, Ranasinghe KN, Bandara SV, Perera IR (2021) Recent development and future prospects of TiO_2 photocatalysis. J Chin Chem Soc 68(5):738–769
30. Goulart S, Nieves LJJ, Dal Bó AG, Bernardin AM (2020) Sensitization of TiO_2 nanoparticles with natural dyes extracts for photocatalytic activity under visible light. Dyes Pigm 182:108654
31. Banerjee S, Pillai SC, Falaras P, O'shea KE, Byrne JA, Dionysiou DD (2014) New insights into the mechanism of visible light photocatalysis. J Phys Chem Lett 5(15):2543–2554
32. Reginato G, Calamante M, Zani L, Mordini A, Franchi D (2018) Design and synthesis of organic sensitizers with enhanced anchoring stability in dye-sensitized solar cells. Pure Appl Chem 90(2):363–376
33. Sharma K, Sharma V, Sharma SS (2018) Dye-sensitized solar cells: fundamentals and current status. Nanoscale Res Lett 13(1):1–46
34. Zhang L, Cole JM (2015) Anchoring groups for dye-sensitized solar cells. ACS Appl Mater Interfaces 7(6):3427–3455
35. Mozaffari S, Nateghi MR (2014) Effects of multi anchoring groups of catecholamine polymer dyes on the electrical characteristics of metal free dye-sensitized solar cells: a comparison study. Sol Energy 106:63–71
36. Krishnan T, Salida W, Mansor W (2020) Photocatalytic degradation of dyes by TiO_2 process in batch photoreactor. Letters in Applied NanoBioScience 9:1502–1512
37. Kumar S, Sharma R, Gupta A, Dubey KK, Khan AM, Singhal R, Kumar R, Bharti A, Singh P, Kant R, Kumar V (2022) TiO_2 based photocatalysis membranes: an efficient strategy for pharmaceutical mineralization. Science of The Total Environment 845:157221
38. Mohd Adnan MA, Muhd JN, Amir MNI, Maamor A (2019) Effect on different TiO_2 photocatalyst supports on photodecolorization of synthetic dyes: a review. Int J Environ Sci Technol 16:547–566
39. Chen D, Cheng Y, Zhou N, Chen P, Wang Y, Li K, Huo S, Cheng P, Peng P, Zhang R, Wang L (2020) Photocatalytic degradation of organic pollutants using TiO_2-based photocatalysts: a review. J Clean Prod 268:121725
40. Fonseca-Cervantes OR, Pérez-Larios A, Romero Arellano VH, Sulbaran-Rangel B, Guzmán González CA (2020) Effects in band gap for photocatalysis in TiO_2 support by adding gold and ruthenium. Processes 8(9):1032

41. Zhao D, Tang X, Qin X, Tang Z, Yuan D, Ju L (2022) Applying hydrogenation to stabilize N-TiO_2 and enhance its visible light photocatalytic activity. Catalysts 12(2):178
42. Balassa L, Ágoston Á, Kása Z, Hornok V, Janovák L (2022) Surface sulfate modified TiO_2 visible light active photocatalyst for complex wastewater purification: preparation, characterization and photocatalytic activity. J Mol Struct 1260:132860
43. Li R, Li T, Zhou Q (2020) Impact of titanium dioxide (TiO_2) modification on its application to pollution treatment—a review. Catalysts 10(7):804
44. Bai Y, Yu H, Li Z, Amal R, Lu GQ, Wang L (2012) In situ growth of a ZnO nanowire network within a TiO_2 nanoparticle film for enhanced dye-sensitized solar cell performance. Adv Mater 24(43):5850–5856
45. Chang X, Li Z, Zhai X, Sun S, Gu D, Dong L, Yin Y, Zhu Y (2016) Efficient synthesis of sunlight-driven ZnO-based heterogeneous photocatalysts. Mater Des 98:324–332
46. Han Z, Liao L, Wu Y, Pan H, Shen S, Chen J (2012) Synthesis and photocatalytic application of oriented hierarchical ZnO flower-rod architectures. J Hazard Mater 217:100–106
47. Intarasuwan K, Amornpitoksuk P, Suwanboon S, Graidist P (2017) Photocatalytic dye degradation by ZnO nanoparticles prepared from $X_2C_2O_4$ (X=H, Na and NH_4) and the cytotoxicity of the treated dye solutions. Sep Purif Technol 177:304–312
48. Simović B, Poleti D, Golubović AV, Matković A, Šćepanović M, Babić BM, Branković GO (2017) Enhanced photocatalytic degradation of RO16 dye using Ag modified ZnO nanopowders prepared by the solvothermal method. Processing and Application of Ceramics 11(1):27–38
49. Mitra M, Ghosh A, Mondal A, Kargupta K, Ganguly S, Banerjee D (2017) Facile synthesis of aluminium doped zinc oxide-polyaniline hybrids for photoluminescence and enhanced visible-light assisted photo-degradation of organic contaminants. Appl Surf Sci 402:418–428
50. Xiao FX (2012) Construction of highly ordered ZnO–TiO_2 nanotube arrays (ZnO/TNTs) heterostructure for photocatalytic application. ACS Appl Mater Interfaces 4(12):7055–7063
51. Mohammadi H, Ghorbani M (2018) Synthesis photocatalytic TiO_2/ZnO nanocomposite and investigation through anatase, Wurtzite and $ZnTiO_3$ phases antibacterial behaviors. Journal of Nano Research 51:69–77
52. Kanjwal MA, Barakat NA, Sheikh FA, Park SJ, Kim HY (2010) Photocatalytic activity of ZnO-TiO_2 hierarchical nanostructure prepared by combined electrospinning and hydrothermal techniques. Macromol Res 18(3):233–240
53. Pant B, Ojha GP, Kuk YS, Kwon OH, Park YW, Park M (2020) Synthesis and characterization of ZnO-TiO_2/carbon fiber composite with enhanced photocatalytic properties. Nanomaterials 10(10):1960
54. Petrovičová B, Dahrouch Z, Triolo C, Pantò F, Malara A, Patanè S, Allegrini M, Santangelo S (2021) Photocatalytic degradation of methylene blue dye by electrospun binary and ternary zinc and titanium oxide nanofibers. Appl Sci 11(20):9720
55. Someswararao MV, Dubey RS, Subbarao PSV (2021) Electrospun composite nanofibers prepared by varying concentrations of TiO_2/ZnO solutions for photocatalytic applications. Journal of Photochemistry and Photobiology 6:100016
56. Gnanasekaran L, Priya AK, Ghfar AA, Sekar K, Santhamoorthy M, Arthi M, Soto-Moscoso M (2022) The influence of heterostructured TiO_2/ZnO nanomaterials for the removal of azo dye pollutant. Chemosphere 308:136161
57. Balakrishnan M, John R (2020) Properties of sol-gel synthesized multiphase TiO_2 (AB)-ZnO (ZW) semiconductor nanostructure: an effective catalyst for methylene blue dye degradation. Iranian Journal of Catalysis 10:1–16
58. Tekin D, Kiziltas H, Ungan H (2020) Kinetic evaluation of ZnO/TiO_2 thin film photocatalyst in photocatalytic degradation of Orange G. J Mol Liq 306:112905
59. Yusuff AS, Thompson-Yusuff KA, Adeniyi OD, Olutoye MA (2022) Siliceous termite hill supported ZnO-TiO_2 as a solar light responsive photocatalyst: synthesis, characterization and performance in degradation of methylene blue dye. Surfaces and Interfaces 34:102360
60. AlSalka Y, Granone LI, Ramadan W, Hakki A, Dillert R, Bahnemann DW (2019) Iron-based photocatalytic and photoelectrocatalytic nano-structures: facts, perspectives, and expectations. Appl Catal B 244:1065–1095

61. Afzal S, Julkapli NM, Mun LK (2019) Response surface approach for visible light assisted photocatalytic degradation of ortho nitrophenol by magnetically separable TiO_2/CS nanocomposite. Mater Sci Semicond Process 99:34–43
62. Afzal S, Julkapli NM, Mun LK (2021) Visible light active TiO_2/CS/Fe_3O_4 for nitrophenol degradation: studying impact of TiO_2, CS and Fe_3O_4 loading on the optical and photocatalytic performance of nanocomposite. Mater Sci Semicond Process 131:105891
63. Dubey M, Kumar R, Srivastava SK, Joshi M (2021) Visible light induced photodegradation of chlorinated organic pollutants using highly efficient magnetic Fe_3O_4/TiO_2 nanocomposite. Optik 243:167309
64. Kermani M, Kakavandi B, Farzadkia M, Esra A (2018) Catalytic ozonation of high concentrations of catechol over TiO_2@Fe_3O_4 magnetic core-shell nanocatalyst: Optimization, toxicity and degradation pathway studies. Journal of Cleaner Production 192:597–607
65. Kubo M, Moriyama R, Shimada M (2019) Microporous and mesoporous materials facile fabrication of HKUST-1 nanocomposites incorporating Fe_3O_4 and TiO_2 nanoparticles by a spray-assisted synthetic process and their dye adsorption performances. Microporous Mesoporous Mater 280:227–235
66. Nadimi M, Ziarati SA, Aroon MA, Ebrahimian PA (2019) Photodegradation of methylene blue by a ternary magnetic TiO_2/Fe_3O_4/graphene oxide nanocomposite under visible light. Mater Chem Phys 225:464–474
67. Pelaez M, Nolan NT, Pillai SC, Seery MK, Falaras P, Kontos AG, Dunlop PS, Hamilton JW, Byrne JA, O'shea K, Entezari MH (2012) A review on the visible light active titanium dioxide photocatalysts for environmental applications. Appl Catal B 125:331–349
68. Xiong D, Chang H, Zhang Q, Tian S, Liu B, Zhao X (2015) Preparation and characterization of $CuCrO_2$/TiO_2 heterostructure photocatalyst with enhanced photocatalytic activity. Appl Surf Sci 347:747–754
69. Çinar B, Kerimoğlu I, Tönbül B, Demirbüken A, Dursun S, Kaya IC, Kalem V, Akyildiz H (2020) Hydrothermal/electrospinning synthesis of CuO plate-like particles/TiO_2 fibers heterostructures for high-efficiency photocatalytic degradation of organic dyes and phenolic pollutants. Materials Science in Semiconductor Processing 109:104919
70. Singh J, Juneja S, Soni RK, Bhattacharya J (2021) Sunlight mediated enhanced photocatalytic activity of TiO_2 nanoparticles functionalized CuO-Cu_2O nanorods for removal of methylene blue and oxytetracycline hydrochloride. J Colloid Interface Sci 590:60–71
71. Ren D, Huang H, Qi J, Zheng P (2020) One-pot template-free cross-linking synthesis of SiOx–SnO_2@C hollow spheres as a high volumetric capacity anode for lithium-ion batteries. Energ Technol 8(7):2000314
72. Khan H, Rigamonti MG, Patience GS, Boffito DC (2018) Spray dried TiO_2/WO_3 heterostructure for photocatalytic applications with residual activity in the dark. Appl Catal B 226:311–323
73. Li YF, Xu D, Oh JI, Shen W, Li X, Yu Y (2012) Mechanistic study of codoped titania with nonmetal and metal ions: a case of C^+ Mo codoped TiO_2. ACS Catal 2(3):391–398
74. Zhao T, Qian R, Zhou G, Wang Y, Lee WI, Pan JH (2021) Mesoporous WO_3/TiO_2 spheres with tailored surface properties for concurrent solar photocatalysis and membrane filtration. Chemosphere 263:128344
75. Negishi N, Sugasawa M, Miyazaki Y, Hirami Y, Koura S (2019) Effect of dissolved silica on photocatalytic water purification with a TiO_2 ceramic catalyst. Water Res 150:40–46
76. Vu TK, To HT, Negishi N (2022) Application of TiO_2-ceramic/UVA photocatalyst for the photodegradation of sulfamethoxazole. Sustainable Chemistry and Pharmacy 26:100617
77. Xing Z, Li J, Wang Q, Zhou W, Tian G, Pan K, Tian C, Zou J, Fu H (2013) A floating porous crystalline TiO_2 ceramic with enhanced photocatalytic performance for wastewater decontamination. Eur J Inorg Chem 2013(13):2411–2417
78. Xing Z, Zhou W, Du F, Zhang L, Li Z, Zhang H, Li W (2014) Facile synthesis of hierarchical porous TiO_2 ceramics with enhanced photocatalytic performance for micropolluted pesticide degradation. ACS Appl Mater Interfaces 6(19):16653–16660

79. Aguado MA, Anderson MA, Hill CG Jr (1994) Influence of light intensity and membrane properties on the photocatalytic degradation of formic acid over TiO_2 ceramic membranes. J Mol Catal 89(1–2):165–178
80. Kim H, Lee S, Han Y, Park J (2006) Preparation of dip-coated TiO_2 photocatalyst on ceramic foam pellets. J Mater Sci 41(18):6150–6153
81. Xie TH, Lin J (2007) Origin of photocatalytic deactivation of TiO_2 film coated on ceramic substrate. J Phys Chem C 111(27):9968–9974
82. Gong H, Wang L, Zhou K, Zhang D, Zhang Y, Adamaki V, Sergejevs A, Bowen C (2022) Improved photocatalytic performance of gradient reduced TiO_2 ceramics with aligned pore channels. Advanced Powder Materials 1(3):100025
83. Navalon S, Alvaro M, Garcia H, Escrig D, Costa V (2009) Photocatalytic water disinfection of cryptosporidium parvum and Giardia lamblia using a fibrous ceramic TiO_2 photocatalyst. Water Sci Technol 59(4):639–645
84. Marien CB, Le Pivert M, Azaïs A, M'bra IC, Drogui P, Dirany A, Robert D (2019) Kinetics and mechanism of Paraquat's degradation: UV-C photolysis vs UV-C photocatalysis with TiO_2/SiC foams. Journal of Hazardous Materials 370:164–171
85. Limpachanangkul P, Jedsukontorn T, Zhang G, Liu L, Hunsom M, Chalermsinsuwan B (2019) Comparative photocatalytic behavior of photocatalysts (TiO_2, SiC, Bi_2O_3, ZnO) for transformation of glycerol to value added compounds. Korean J Chem Eng 36(9):1527–1535
86. Sun W, Li X, Zou J, Guo H (2022) N-TiO_2-coated SiC foam for the treatment of dyeing wastewater under blue light LED irradiation. Coatings 12(5):585
87. M'bra IC, García-Muñoz P, Drogui P, Keller N, Trokourey A, Robert D (2019) Heterogeneous photodegradation of Pyrimethanil and its commercial formulation with TiO_2 immobilized on SiC foams. Journal of Photochemistry and Photobiology A: Chemistry 368:1–6
88. Ahmed H, Hashim A, Abdul Jalil HM (2020) Determination of optical parameters of films of PVA/TiO_2/SiC and PVA/MgO/SiC nanocomposites for optoelectronics and UV-detectors. Ukr J Phys 65(6):533–533
89. Karimi-Shamsabadi M, Behpour M (2021) Comparing photocatalytic activity consisting of Sb_2S_3 and Ag_2S on the TiO_2–SiO_2/TiO_2 nanotube arrays-support for improved visible-light-induced photocatalytic degradation of a binary mixture of basic blue 41 and basic red 46 dyes. Int J Hydrogen Energy 46(53):26989–27013
90. Katiyar P, Mishra S, Srivastava A, Prasad NE (2020) Preparation of TiO_2–SiO_2 hybrid nanosols coated flame-retardant polyester fabric possessing dual contradictory characteristics of superhydrophobicity and self-cleaning ability. J Nanosci Nanotechnol 20(3):1780–1789
91. Bak C, Kang HW, Lee C, Jeong DY, Eom H (2019) Laser irradiation synthesis and photocatalytic properties of TiO_2-SiO_2 hybrid thin films. Thin Solid Films 678:32–41
92. Chen H, Zhao L, He X, Huang Z, Wang G, Fang W, Li W (2018) Highly enhanced photocatalytic degradation of dye rhodamine B on the biogenic hierarchical porous TiO_2/SiO_2 with 1D/3D-chain. Clean Technol Environ Policy 20(4):887–897
93. Huang F, Hao H, Sheng W, Lang X (2021) Dye-TiO_2/SiO_2 assembly photocatalysis for blue light-initiated selective aerobic oxidation of organic sulfides. Chem Eng J 423:129419
94. Koli VB, Mavengere S, Kim JS (2019) An efficient one-pot N doped TiO_2-SiO_2 synthesis and its application for photocatalytic concrete. Appl Surf Sci 491:60–66
95. Chen C, Liu X, Long H, Ding F, Liu Q (2019) Preparation and photocatalytic performance of graphene Oxide/WO_3 quantum Dots/TiO_2@ SiO_2 microspheres. Vacuum 164:66–71
96. Zangeneh H, Zinatizadeh AA, Feyzi M, Zinadini S, Bahnemann DW (2018) Photomineralization of recalcitrant wastewaters by a novel magnetically recyclable boron doped-TiO_2-SiO_2 cobalt ferrite nanocomposite as a visible-driven heterogeneous photocatalyst. J Environ Chem Eng 6(5):6370–6381
97. Zhu Q, Pan D, Sun Y, Qi D (2022) Controllable microemulsion synthesis of hybrid TiO_2–SiO_2 hollow spheres and Au-doped hollow spheres with enhanced photocatalytic activity. Langmuir 38(13):4001–4013
98. Parale VG, Kim T, Lee KY, Phadtare VD, Dhavale RP, Park HH (2020) Hydrophobic TiO_2–SiO_2 composite aerogels synthesized via in situ epoxy-ring opening polymerization and sol-gel process for enhanced degradation activity. Ceram Int 46(4):4939–4946

99. Florez DMMG, Ale RBG, Idme AFH, Alarcon LAL, Huallpa EA, Castro Y, Apestegui PGR, Rodriguez Rodriguez JM (2020) SiO_2-TiO_2 films supported on ignimbrite by spray coating for the photocatalytic degradation of NOx gas and methyl orange dye. International Journal of Photoenergy 2020:1–6
100. Wu XQ, Shao ZD, Liu Q, Xie Z, Zhao F, Zheng YM (2019) Flexible and porous TiO_2/SiO_2/carbon composite electrospun nanofiber mat with enhanced interfacial charge separation for photocatalytic degradation of organic pollutants in water. J Colloid Interface Sci 553:156–166
101. Rahman KU, Ferreira-Neto EP, Rahman GU, Parveen R, Monteiro AS, Rahman G, Van Le Q, Domeneguetti RR, Ribeiro SJ, Ullah S (2021) Flexible bacterial cellulose-based BC-SiO_2-TiO_2-Ag membranes with self-cleaning, photocatalytic, antibacterial and UV-shielding properties as a potential multifunctional material for combating infections and environmental applications. J Environ Chem Eng 9(1):104708
102. Suzie VA, Jean-Baptiste T, Elie A, Natacha M, Serge N, Georges KY, Samuel L (2022) Photocatalytic performance of N-TiO_2@SiO_2 composite obtained under gliding arc plasma processing at atmospheric pressure. Results in Engineering 15:100516
103. Ma N, He HY (2021) Synthesis of highly active Si-doped TiO_2 nanotube/SnS hybrids for photocatalytic degradation of toxic dyes
104. Chinh VD, Bavasso I, Di Palma L, Felici AC, Scarsella M, Vilardi G, Bracciale MP, Van NT (2021) Enhancing the photocatalytic activity of TiO_2 and TiO_2–SiO_2 by coupling with graphene–gold nanocomposites. J Mater Sci: Mater Electron 32(4):5082–5093
105. Liu J, Liu J, Shi F, Hu S, Jiang S, Liu S, Liu D, Tian X (2019) F/W co-doped TiO_2-SiO_2 composite aerogels with improved visible light-driven photocatalytic activity. J Solid State Chem 275:8–15
106. Bilgic A (2022) Fabrication of monoBODIPY-functionalized Fe_3O_4@SiO_2@TiO_2 nanoparticles for the photocatalytic degradation of rhodamine B under UV irradiation and the detection and removal of Cu (II) ions in aqueous solutions. J Alloy Compd 899:163360
107. Liao S, Lin L, Huang J, Jing X, Chen S, Li Q (2022) Microorganism-templated nanoarchitectonics of hollow TiO_2-SiO_2 microspheres with enhanced photocatalytic activity for degradation of methyl orange. Nanomaterials 12(9):1606
108. Khannyra S, Luna M, Gil MA, Addou M, Mosquera MJ (2022) Self-cleaning durability assessment of TiO_2/SiO_2 photocatalysts coated concrete: effect of indoor and outdoor conditions on the photocatalytic activity. Build Environ 211:108743
109. Cardoso BN, Kohlrausch EC, Laranjo MT, Benvenutti EV, Balzaretti NM, Arenas LT, Santos MJ, Costa TM (2019) Tuning anatase-rutile phase transition temperature: TiO_2/SiO_2 nanoparticles applied in dye-sensitized solar cells. International Journal of Photoenergy 2019:1–9
110. Diaz C, Valenzuela ML, Cifuentes-Vaca O, Segovia M, Laguna-Bercero MA (2020) Incorporation of NiO into SiO_2, TiO_2, Al_2O_3, and $Na_{4.2}Ca_{2.8}(Si_6O_{18})$ matrices: medium effect on the optical properties and catalytic degradation of methylene blue. Nanomaterials 10(12):2470
111. Akhtar J, Tahir MB, Sagir M, Bamufleh HS (2020) Improved photocatalytic performance of Gd and Nd co-doped ZnO nanorods for the degradation of methylene blue. Ceram Int 46(8):11955–11961
112. Thambiliyagodage C, Mirihana S, Wijesekera R, Madusanka DS, Kandanapitiye M, Bakker M (2021) Fabrication of Fe_2TiO_5/TiO_2 binary nanocomposite from natural ilmenite and their photocatalytic activity under solar energy. Current Research in Green and Sustainable Chemistry 4:100156
113. Ahmad R, Ehsan A, Khan IU, Aslam M, Rout PR (2021) Photocatalytic membrane reactors (PMRs) for wastewater treatment: photodegradation mechanism, types, and optimized factors. In: Urban mining for waste management and resource recovery. CRC Press, Boca, pp 257–272
114. Su CY, Wang LC, Liu WS, Wang CC, Perng TP (2018) Photocatalysis and hydrogen evolution of Al-and Zn-doped TiO_2 nanotubes fabricated by atomic layer deposition. ACS Appl Mater Interfaces 10(39):33287–33295
115. Baran T, Wojtyła S, Minguzzi A, Rondinini S, Vertova A (2019) Achieving efficient H_2O_2 production by a visible-light absorbing, highly stable photosensitized TiO_2. Appl Catal B 244:303–312

116. Abdi G, Shekh MI, Amirian J, Alizadeh A, Zinadini S (2021) Photocatalytic membranes in degradation of organic molecules. Photocatalysis: Advanced Materials and Reaction Engineering 100:1–56
117. Sheng W, Shi JL, Hao H, Li X, Lang X (2020) Polyimide-TiO_2 hybrid photocatalysis: visible light-promoted selective aerobic oxidation of amines. Chem Eng J 379:122399
118. Wu Y, Xu Y, Zhang Y, Feng J, Li Y, Lan J, Cheng X (2022) Fabrication of NiCoP decorated TiO_2/polypyrrole nanocomposites for the effective photocatalytic degradation of tetracycline. Chin Chem Lett 33(5):2741–2746
119. Ma J, Dai J, Duan Y, Zhang J, Qiang L, Xue J (2020) Fabrication of PANI-TiO_2/rGO hybrid composites for enhanced photocatalysis of pollutant removal and hydrogen production. Renewable Energy 156:1008–1018
120. Adnan MAM, Phoon BL, Julkapli NM (2020) Mitigation of pollutants by chitosan/metallic oxide photocatalyst: a review. J Clean Prod 261:121190
121. Adnan MAM, Afzal S, Johan MR, Julkapli NM (2022) A comparative study on the photodegradation efficiency of TiO_2-CS hybrid beads under wet and dry conditions. Int J Mater Prod Technol 65(1):67–79
122. Balakrishnan A, Appunni S, Gopalram K (2020) Immobilized TiO_2/chitosan beads for photocatalytic degradation of 2, 4-dichlorophenoxyacetic acid. Int J Biol Macromol 161:282–291
123. Sugashini S, Gomathi T, Devi RA, Sudha PN, Rambabu K, Banat F (2022) Nanochitosan/carboxymethyl cellulose/TiO_2 biocomposite for visible-light-induced photocatalytic degradation of crystal violet dye. Environ Res 204:112047
124. Feng X, Li X, Su B, Ma J (2022) Solid-phase fabrication of TiO_2/Chitosan-biochar composites with superior UV–Vis light driven photocatalytic degradation performance. Colloids Surf, A 648:129114
125. Nikaido M, Furuya S, Kakui T, Kamiya H (2009) Photocatalytic behavior of TiO_2 nanoparticles supported on porous aluminosilicate surface modified by cationic surfactant. Adv Powder Technol 20(6):598–604
126. Li F, Jiang Y, Xia M, Sun M, Xue B, Ren X (2009) A high-stability silica–clay composite: synthesis, characterization and combination with TiO_2 as a novel photocatalyst for Azo dye. J Hazard Mater 165(1–3):1219–1223
127. Yuan J, Hu H, Chen M, Shi J, Shangguan W (2008) Promotion effect of Al_2O_3–SiO_2 interlayer and Pt loading on TiO_2/nickel-foam photocatalyst for degrading gaseous acetaldehyde. Catal Today 139(1–2):140–145
128. Yuan X, Kobylanski MP, Cui Z, Li J, Beaunier P, Dragoe D, Colbeau-Justin C, Zaleska-Medynska A, Remita H (2020) Highly active composite TiO_2-polypyrrole nanostructures for water and air depollution under visible light irradiation. J Environ Chem Eng 8(5):104178
129. Liu J, Zhou S, Gu P, Zhang T, Chen D, Li N, Xu Q, Lu J (2020) Conjugate polymer-clothed $TiO_2@V_2O_5$ nanobelts and their enhanced visible light photocatalytic performance in water remediation. J Colloid Interface Sci 578:402–411
130. Ardani MR, Pang AL, Pal U, Zheng R, Arsad A, Hamzah AA, Ahmadipour M (2022) Ultrasonic-assisted polyaniline-multiwall carbon nanotube photocatalyst for efficient photodegradation of organic pollutants. Journal of Water Process Engineering 46:102557
131. Mousli F, Chaouchi A, Hocine S, Lamouri A, Vilar MR, Kadri A, Chehimi MM (2019) Diazonium-modified TiO_2/polyaniline core/shell nanoparticles. Structural characterization, interfacial aspects and photocatalytic performances. Appl Surf Sci 465:1078–1095
132. Aminuddin NF, Nawi MA, Bahrudin NN, Jawad AH (2021) Iron ion assisted photocatalytic-adsorptive removal of acid orange 52 by immobilized TiO_2/polyaniline bilayer photocatalyst. Applied Surface Science Advances 6:100180
133. Thakare SR, Mate VR, Urkude K, Gawande SB (2020) Graphene-TiO_2-polyaniline nanocomposite: a new green and efficient catalyst as an alternative for noble metal and $NaBH_4$ induced the reduction of 4-nitro phenol. FlatChem 22:100179

Alginate-Based Hybrid Materials for the Treatment of Textile Dyes

Muhammad Alamzeb, Behramand Khan, and Haroon Subhani

Abstract Uncontrolled anthropogenic activities like rapid industrialization and urbanization are among the most important causes of water pollution. The major constituents of polluted water are several kinds of organic dyes, mainly discharged by textile and dyeing industries. Textile industries convert fiber into fabric and involve many chemical processes and several synthetic and natural chemicals. The fabric produced is then subjected to printing, dying, or combination of both processes to convert it into clothes. During the dying and printing processes, some of the dyes are released, as effluents, into water bodies. Water pollution due to industrial discharge is a serious concern to the environment and human health. The removal of these dyes from wastewater is a major issue nowadays. Many methods have been developed and applied for the removal of dyes from contaminated water, but each one of them have its own merits and demerits in terms of operation, efficiency, design, and total cost. This chapter provides a brief review of the use of alginates and alginate-based hybrid materials for the removal of textile dyes from wastewater.

Keywords Textile dyes · Alginate · Hybrid material · Water pollution

1 Introduction

Alginate, an anionic biopolymer, found on the cell wall of brown algae. The cell wall and inter cellular matrix of seaweeds are mainly composed of alginate. Alginate provides strength and flexibility seaweeds against the power of water where alginates grow and flourish [1]. Alginates have different composition in different seaweeds. Alginate obtained from other than brown algae species consists different amount of dry matter (20–60%); however, the alginate which is obtained from brown species

M. Alamzeb (✉) · H. Subhani
Department of Chemistry, University of Kotli, Kotli 11100, Azad Jammu & Kashmir, Pakistan
e-mail: zebchemist@gmail.com

B. Khan
Department of Chemistry, Islamia College University, Peshawar 25000, KPK, Pakistan

© The Author(s), under exclusive license to Springer Nature Singapore Pte Ltd. 2023

A. Ahmad et al. (eds.), *Nanohybrid Materials for Treatment of Textiles Dyes*,
Smart Nanomaterials Technology, https://doi.org/10.1007/978-981-99-3901-5_19

of algae consists of 40% dry matter. The alginate obtained from brown algae species occurs in the form of a gel. The gel is composed of barium, sodium, magnesium, strontium, and calcium ions [2]. There are also some species of bacteria which can produce alginate, but at present time almost all of the alginate is obtained from algae [3]. The industrial applications of alginates are dependent upon their viscosity, gelation, and stability which they impart to the materials in which they are present. Normally is matrix containing alginic acid bound cations (magnesium, calcium, sodium etc.). The stability of alginic acid is dependent upon the presence of these ions. Divalent cations provide more stability and rigidity alginate as compared to monovalent cations. Alginates are nowadays very popular as pharmaceutical, biomaterial ingredients, and biotechnological material. Due to these potential and wide range properties, alginates are encouraging scientists for more detailed studies about the properties and structures of alginates.

2 History

Alginic acid was discovered in 1881 by a British scientist named E.C.C Stanford. Stanford patented alginic acid in 1881. Stanford explained, in his patent, the chemical structure of alginic acid. Stanford also provided detailed informed about the extraction of alginate. He described that first of all that algal mass need to soak in water or dilute acidic solution. The extract is then extracted with sodium carbonate. After this alginate was precipitated with the addition of dilute acid [4]. The research work on alginic acid continued and in twentieth century it was reported that uronic acid is a constituent of alginic acid. Further, research and studies reported the presence of D-mannuronic in the hydrolyzed samples of alginates. The nature of the bonds (β 1, 4 bond) in uronic acid was identical to that in cellulose. In 1950, the chromatographic screening/analysis of a mixture of uronic acid was carried out which led to the discovery of a new variety of uronic acid, this new variety was named as L-guluronic acid. They also developed a method to determine quantitatively the presence of guluronic acid and mannuronic acid in alginate. Thereafter, it was found that alginate is a binary co-polymer of guluronic and mannuronic acids [5]. The block co-polymer nature of alginate was reported in 1974 [6]. Later on, it was reported that the physical properties of alginates are directly co-related with their block structures types [7].

3 Chemical Structure

Alginate is biopolymer having unbranched structure. Alginates are composed of 1-4 1,4-β-D-mannuronic acid (M) and 1,4 α-L-guluronic acid (G) monomers. The block composition in alginates may either be homogenous (Poly G, Poly-M) or heterogenous (MG). The composition of the blocks can be determined by using the process of partial hydrolysis. Alginates produced by different species possess different composition, sequence, and different ratio of guluronic and mannuronic acids in their blocks. The proportions of these two acids, in every alginate, vary from specie to specie and even if the alginate is obtained from different parts of the same seaweed [6, 8]. Alginates do not consist of repeating regular units. It is almost impossible to describe and elucidate the distribution of various monomers present the polymeric chains of alginate. So, the information about the monomers, present in the polymeric chains of alginates, may not be sufficient and helpful to elucidate and establish the structures of alginates from a wide variety of species. Alginates occur as metallic salts (mainly sodium and potassium), in the inter-cellular regions of cell walls. In seaweeds, the main role and biological functions of alginates are ion exchange and structural support. The young cell wall tissues or intercellular areas are composed of alginates enriched with polymannuronic acid while polyguluronic enriched alginates are the major constituent of cell wall. The polyguluronic acid possess higher affinity for Ca^{2+} ions [9]. Alginates are transported to cytoplasm after their synthesis inside cytoplasm. Alginates can be obtained from both algal and bacterial sources At molecular level, the main difference between algal and bacterial alginate is the presence of O-acetyl groups at C2 and/or C3 in case of bacterial alginates and the absence of O-acetyl groups at the mention carbon atoms in case of algal alginates [10].

4 Sources of Alginates

Alginic acid salts are the most commonly available sources of alginates. Alginates are one of the most abundant families of the naturally occurring anionic polysaccharides [11]. Brown seaweeds like Laminaria japonica, Macrocystis pyreifera, Laminaria hyperborean and Laminaria digitata, etc., are considered as the major sources of alginates [12]. The extracts are then treated with alkali, and finally, the alkaline extracted can be converted into alginic acid by treatment with dilute mineral acids. Alginate and its salt form exist in about 200 different grades. Alginate can be found in neutral, charged, and ultra-pure form in association with a wide range of materials [13].

5 Biodegradability of Alginates

Microbes have alginate lyases enzymes for the degradation of alginates, but human bodies do not have the ability to degrade alginates due to lack of alginate lyases enzymes. Microbes can degrade alginates because they possess alginate lyase enzyme; however, human beings cannot degrade alginates due the lack of alginate lyase enzyme [14]. The alginate lyases, in marine organisms, break the polymeric chains via the process of β-elimination. The cell wall of brown algae metabolizes and degrade alginates into carbon and energy [15]. Naturally occurring alginates are usually non-degradable, however, those alginates which possess cross-linking through monovalent or divalent ions can undergo degradation if the pH conditions are appropriate [16]. Commercially available alginates cannot be easily degraded due to their high molecular masses. The partial oxidation of alginates can speed up its biodegradation because partial oxidation converts the stable chair conformation into open chain adduct. The degradation of alginates depends on degree of oxidation, pH, and temperature [17].

6 Water Pollution by Industries

Many industries like paper, textile, plastics, leather, and cosmetic industries use dyes and colors in their products [18]. All these industries utilize substantial amount of water during dyeing processes, and hence, they generate large volumes of colored wastewaters. These colored wastewaters are released into the environment either directly or indirectly. The toxic dyes present in these industrial effluents results in the ultimate degradation of ecosystem [19]. To prevent the ecosystems and minimize the dangerous effects, these dyes must be removed from water. Several techniques and processes have been devised and developed to remove dyes from contaminated water [20, 21]. These processes include ion exchange, electrochemical destruction, precipitation, irradiation, flocculation, membrane filtration, ozonation, adsorption, advanced oxidation, and aerobic and anaerobic degradation. Among all the mentioned processes, adsorption is comparatively better and widely applicable due to it cost effectiveness, easy experimental setup, less side effects, and eco-friendly nature [22–24].

7 Textile Dyes Pollution, Conti, and Its Treatment

Textile industry, around the world, generates around 1000 billion dollars. 7% of the total world exports are textile industry oriented. Textile industry employs 35 million workers across the globe [25]. Despite the undeniable importance, textile industry is one of the major culprits of several kinds of environmental pollutions including air pollution, solid waste pollution, organic pollution, and heavy metal pollution [26]. However, the main damage is caused from the discharge of textile industry effluents into water bodies [27]. These effluents mainly contain large amount non-biodegradable compounds and chemicals, especially textile dyes [28] (Table 1).

Dyes are synthetic-colored substances and resistant to degradation owing to their complex aromatic structures [29]. Each year an estimated amount of 700,000 tons of around 10,000 different kinds of dyes produced around the world. Most of these dyes are utilized and produced in textile and dyeing related industries. Approximately, 15% of these dyes are lost as effluents by the textile industries [30] (Table 2). These

Table 1 Chemicals and dyes used in textile industries

S. No.	Name of dye or chemical	kg/month
1	Reactive dyes	45
2	Disperse dyes (polyester)	1500
3	Sulphur dyes	302
4	Vat dyes	900
5	Acetic acid	1610
6	Sulfuric acid	678
7	Soap	154
8	Ammonium sulfate	858
9	Oxalic acid	471
10	Softener	856
11	Polyethylene emulsion	1174
12	P V acetate	954
13	Leveling and dispersing agent	546
14	Wetting agent	126
15	Hydrogen peroxide	1038
16	Caustic soda	6212
17	Hydrosulfites	6563
18	Formic acid	1227
19	Organic solvent	247
20	Hydrochloric acid	309
21	Formic acid	1227
22	Organic resin	5115

dyes can harm the environment by blocking of sunlight, slowing photosynthesis, coloring the water, increasing chemical oxygen demand (COD), limiting the growth of aquatic plants, and increasing the toxicity levels of water [31]. In addition to all these, the contamination will also make the ground water unsafe for use and may also cause cancer and mutagenesis [32, 33].

7.1 Major Characteristic of Textile Dye

Some of the major characteristics of textile dyes have been given in Table 3 [34].

Table 2 Types of fabric and percent amount of un-fixed dyes

Fiber type	Dye name	% Un-fixed dye amount
Acrylic	Modified basic	2–3
Nylon and wool	Reactive dyes for wool Metal complexed dyes	7–20
Polyester	Disperse	8–20
Cotton and viscose	Vat dyes	6–21
	Sulfur dyes	31–41
	Direct dyes	6–21
	Pigments	1
	Azoic dyes	6–10
	Reactive dyes	20–50

7.2 Classification of Textile Industries Dyes

The classification of all sorts of commercial textile dyes, by their generic names and chemical compositions, has already been done by American Association of Textile Chemists and Colorists (AATC) and Color Index (C.I). However, on the basis of dye chemistry, textiles dyes can be classified into 12 categories: Acid Dyes, Vat Dyes, Azoic Dyes, Sulfur Dyes, Basic Dyes, Reactive Dyes, Oxidation Dyes, Direct Dyes, Disperse Dyes, Mordant Dyes, Solvent Dyes, and Florescent/Optical Dyes. For the sake of easy understanding, all these dyes have grouped into three categories namely [31]:

- Dyes for synthetic fibers
- Dyes for cellulose fibers
- Dyes for protein fibers.

Textiles dyes contain a number of auxochromic and chromophoric groups (Table 4) [35].

8 Textile Dyes Adverse Effect on Human Health

Textile dyes have been reported to have moderate to severe adverse effects on central nervous system (CNS), reproductive system, enzymes production, skin, kidney, liver, and chromosomes of human beings [36, 37]. Figure 1 provides the brief detail of textile dyes bad effects on human health.

Table 3 Some characteristics of textile industry effluents and their quantities

S. No.	Name of characteristic	Range	
		Minimum	Maximum
1	COD (mg/L)	150	30,000
2	pH	5.5	11.8
3	BOD (mg/L)	80	6000
4	Color (Pt–Co)	50	2500
5	Temperature (°C)	21	62
6	TDS (mg/L)	1500	12,000
7	TS (mg/L)	6000	7000
8	TA (mg/L) as $CaCO_3$	16	801
9	TSS (mg/L)	17	8000
10	Chlorine (mg/L)	1001	5999
11	Chlorides (mg/L)	201	6000
12	Free chlorine (mg/L)		< 9.9
13	Sodium (mg/L)	400	7000
14	Dye (mg/L)	70	–
15	Fluorine (mg/L)	–	< 9.9
16	Zinc (mg/L)	–	< 9.9
17	Silica (mg/L)	–	< 9.9
18	Nickel (mg/L)	–	< 9.9
19	Mercury (mg/L)	–	< 9.9
20	Manganese (mg/L)	–	< 9.9
21	Arsenic (mg/L)	–	< 9.9
22	Iron (mg/L)	–	< 9.9
23	Boron (mg/L)	–	< 9.9
24	Copper (mg/L)	–	< 9.9
25	Oil and grease (mg/L)	10	31
26	Sulfides (mg/L)	5	20
27	Phosphate (mg/L)	–	< 9.9
28	Sulfates (mg/L)	500	1000
29	Free ammonia (mg/L)	–	< 9.9
30	NaCl (mg/L)	–	300
31	NaOH (mg/L)	–	10
32	Na_2CO_3 (mg/L)	–	20
33	NO_3–N (mg/L)	–	< 6

Table 4 Auxochromic and chromophoric groups in textile dyes

Auxochromic groups	Chromophoric groups
Triphenyl methane (–CH(ph)$_3$)	Carboxyl (–CO_2H)
Azo (–N=N–)	Methoxy (–OCH_3)
Phthalocyanine	Electron donor (Br, I, At)
Nitroso (–NO)	Sulfonic acid (–SO_3H)
Anthraquinone	Nitro (–NO_2)
Polymethine (=HC–CH=CH–CH=)	Amino (–NH_2)
Carbonyl (–C=O)	Alkoxyl (–OR)
Sulfide (–C=S)	Methyl-amino (–$NHCH_3$)
Keto-imine (–C=NH)	Hydroxyl (–OH)
Ethylenic (–C=C–)	Dimethyl-amino (–$N(CH_3)_2$

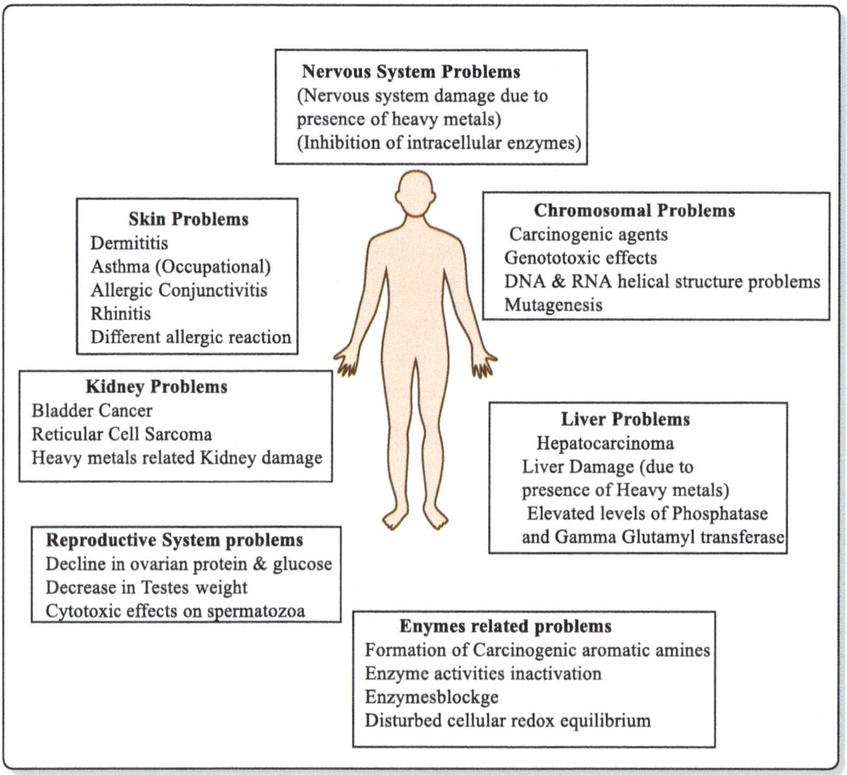

Fig. 1 Adverse effects and human diseases caused by textile dyes

9 Alginate-Based Hybrid Materials for the Treatment of Textile Dyes

Many alginates and alginate-based hybrid materials have been successfully used, for the removal of several types of textile dyes from water, with good and encouraging results (Table 5). These results indicate that alginates and their hybrid materials possess good potential for the removal of textile dyes.

10 Treatment Methods for Textile Dyes Removal Waste Water

Many techniques have been developed to find an economic and efficient way of treating the textile industry effluents (Table 6). These techniques can be classified into the following three classes:

I. Physical treatment
II. Chemical treatment
III. Biological treatment

These techniques are very efficient for the removal of textile dyes from water [65]. Each technique has its own merits and demerits, the most widely used technique is physical treatment but biological treatment is the only green methodology while chemical treatment method is used for special purposes [27].

11 Conclusion and Prospects

Due to textile dyes and chemicals discharge into the water bodies, the textile industries related pollution is going to increase further unless strict measures are taken. Textile industries related pollution is a serious concern, and if it continued with the present rate, then human population will suffer severely. We must be going green and search for relatively safer materials which can be used for the removal of textile dyes from water. In this connection, alginates and their hybrid material can be one of the possible candidates. Alginate is biomaterial which diverse range of applications in the fields of drug delivery, medicines, and textile dyes treatment. Alginates suffer from lack of good mechanical properties and that is why its hybrid materials are prepared to enhance its mechanical potential. Alginate and its hybrid materials have also shown better removal potential for the removal of textile dyes from water. In the future, alginates and alginates-based hybrid materials may become the best materials for the removal of textile dyes.

Table 5 Textile dyes removed by using Alginate and Alginate based hybrid materials

Hybrid material	Name of dye removed	References
Alginate-chitosan beads	Blue 78 and direct yellow 106	[38]
Alginate-hydroxyapatite	Reactive blue 21	[39]
Sodium alginate- magnetic iron oxide-charcoal-β-cyclodextrin	Methylene blue	[40]
Alginate-magnetite nanoparticles	Azo blue dye	[41]
Alginate	Basic violet 16 Basic red 18 Acid black 1	[42] [43]
Glutaraldehyde crosslinked alginate/kaolin hydrogel	Methylene blue Malachite green Methyl violet	[44]
Polyacrylic acid-alginate	Methylene blue Methyl orange	[45]
Lemon peel-sodium alginate	Methylene blue	[46]
Sodium alginate-kaolin	Methylene blue	[47]
Polyethyleneimine crosslinked alginate	Methyl Red	[48]
Calcium alginate-multi-walled carbon nanotubes	Methyl blue Methyl orange	[49]
Fe_3O_4-sodium alginate	Congo red Direct red 23	[49]
Montmorillonite-alginate	Basic red 46	[50]
Alginate-collagen-polyurethanes	Methylene blue Malachite green Indigo carmine	[51]
Algal alginate	Crystal violet	[52]
Bentonite–alginate	Crystal violet Methylene blue	[53, 54]
Sodium Alginate	Congo red Methylene blue Methyl orange	[55, 56]
PVA-sodium alginate-zeolite membrane	Congo red	[57]
Sodium alginate-Poly (N-vinyl-2-pyrrolidone)	Red-120 Brilliant red Remazol brilliant blue R	[58]
Fe_3O_4-Alginate	Methyl orange Methylene blue	[59]
Clinoptilolite-sodium alginate	Methylene blue	[60]
Calcium alginate-organobentonite	Methylene blue	[61]
Calcium alginate beads	Basic black	[62]

(continued)

Table 5 (continued)

Hybrid material	Name of dye removed	References
Magnetic alginate beads	Methylene blue Methyl orange	[63]
Alginate-magnetic chitosan-Zeolite nanocomposite	Methylene blue	[64]

Table 6 Industrial effluents treatment methods

Chemical methods	Biological methods	Physical methods
Ion exchange	Flocculation	Distillation
Neutralization	Stabilization	Adsorption
Oxidation	Activated sludge	Filtration
Electrolysis	Aerated lagoons	Membrane treatment
Reduction	Fungal treatment	Ionization radiation
Catalysis	Trickling filters	Sedimentation
–	Anaerobic digestion	Foam fractionation
–	–	Reverse osmosis
–	–	Coagulation
–	–	Incineration
–	–	Solvent extraction
–	–	Floatation

References

1. Industrial Applications of Marine Biopolymers (2017) 1st edn. CRC Press, Boca Raton
2. Haug A, Smidsrod O (1967) Strontium-calcium selectivity of alginates. Nature 215(5102):757. https://doi.org/10.1038/215757a0
3. Perullini M, Calcabrini M, Jobbágy M, Bilmes SA (2015) Alginate/porous silica matrices for the encapsulation of living organisms: tunable properties for biosensors, modular bioreactors, and bioremediation devices. Open Material Sciences 2(1):3–12. https://doi.org/10.1515/mesbi-2015-0003
4. Hoagland DR, Lieb LL (1915) The complex carbohydrates and forms of sulphur in marine algae of the pacific coast. J Biol Chem 23(1):287–297. https://doi.org/10.1016/S0021-9258(18)87617-7
5. Fischer FG, Dörfel H (1955) Die Polyuronsäuren der Braunalgen (Kohlenhydrate der Algen I). 302 (Jahresband):186–203. https://doi.org/10.1515/bchm2.1955.302.1-2.186
6. Haug A, Larsen B, Smidsrød O (1974) Uronic acid sequence in alginate from different sources. Carbohyd Res 32(2):217–225. https://doi.org/10.1016/S0008-6215(00)82100-X
7. Draget KI, Skjåk Bræk G, Smidsrød O (1994) Alginic acid gels: the effect of alginate chemical composition and molecular weight. Carbohydrate Polymers 25(1):31–38. https://doi.org/10.1016/0144-8617(94)90159-7
8. Szekalska M, Puciłowska A, Szymańska E, Ciosek P, Winnicka K (2016) Alginate: current use and future perspectives in pharmaceutical and biomedical applications. International Journal of Polymer Science 2016:1–17. https://doi.org/10.1155/2016/7697031

9. Rashedy S, Abd El Hafez MSM, Dar M, Cotas J, Pereira L, Raymundo A (2021) Evaluation and characterization of alginate extracted from brown seaweed collected in the Red Sea
10. Skjåk-Braek G, Grasdalen H, Larsen B (1986) Monomer sequence and acetylation pattern in some bacterial alginates. Carbohydr Res 154:239–250. https://doi.org/10.1016/s0008-6215(00)90036-3
11. Xu Z, Wang B, Xu Z (2014) Encyclopedia of biomedical polymers and polymeric biomaterials-processing methods and applications. https://doi.org/10.1081/E-EBPP-120050032
12. Goh CH, Heng PWS, Chan LW (2012) Alginates as a useful natural polymer for microencapsulation and therapeutic applications. Carbohyd Polym 88(1):1–12. https://doi.org/10.1016/j.carbpol.2011.11.012
13. Hecht H, Srebnik S (2016) Structural characterization of sodium alginate and calcium alginate. Biomacromol 17(6):2160–2167. https://doi.org/10.1021/acs.biomac.6b00378
14. Hasnain M, Nayak A, Yadav M, Ahmadi Y, Milivojevic M, Pajić-Lijaković I, Branko B, Manzano V, Pacho N, Tasqué J, Beatriz N, Accorso D, Singh V, Singh A, El-Sherbiny I, Abd M, Aziz A, Abdelsalam E, Garcia N, Dmour I (2019) Alginates versatile polymers in biomedical applications and therapeutics chemically modified alginates for advanced biomedical applications, p 335
15. Zhu B, Yin H (2015) Alginate lyase: review of major sources and classification, properties, structure-function analysis and applications. Bioengineered 6(3):125–131. https://doi.org/10.1080/21655979.2015.1030543
16. Forster RE, Thürmer F, Wallrapp C, Lloyd AW, Macfarlane W, Phillips GJ, Boutrand JP, Lewis AL (2010) Characterisation of physico-mechanical properties and degradation potential of calcium alginate beads for use in embolisation. J Mater Sci Mater Med 21(7):2243–2251. https://doi.org/10.1007/s10856-010-4080-y
17. Kristiansen KA, Potthast A, Christensen BE (2010) Periodate oxidation of polysaccharides for modification of chemical and physical properties. Carbohydr Res 345(10):1264–1271. https://doi.org/10.1016/j.carres.2010.02.011
18. Yaqoob AA, Parveen T, Umar K, Mohamad Ibrahim MNM (2020) Role of nanomaterials in the treatment of wastewater: a review. Water 12(2):495–499
19. Yaqoob AA, Ahmad H, Parveen T, Ahmad A, Oves M, Ismail IM, Qari HA, Umar K, Mohamad Ibrahim MNM (2020) Recent advances in metal decorated nanomaterials and their various biological applications: a review. Front Chem 8:341–366
20. Idris MO, Yaqoob AA, Ibrahim MNM, Ahmad A, Alshammari MB (2023) Introduction of adsorption techniques for heavy metals remediation. Emerging techniques for treatment of toxic metals from wastewater, 1st edn. Elsevier, Amsterdam, pp 1–18
21. Yaqoob AA, Ibrahim MNM, Ahmad A, Reddy AVB (2021) Toxicology and environmental application of carbon nanocomposite. Environmental remediation through carbon based nano composites. Springer, New York, pp 1–18
22. Dawood SA, Sen TK (2013) Review on dye removal from its aqueous solution into alternative cost effective and non-conventional adsorbents
23. Pandey S (2017) A comprehensive review on recent developments in bentonite-based materials used as adsorbents for wastewater treatment. J Mol Liq 241:1091–1113. https://doi.org/10.1016/j.molliq.2017.06.115
24. Rafatullah M, Sulaiman O, Hashim R, Ahmad A (2010) Adsorption of methylene blue on low-cost adsorbents: a review. J Hazard Mater 177(1):70–80. https://doi.org/10.1016/j.jhazmat.2009.12.047
25. Desore A, Narula SA (2018) An overview on corporate response towards sustainability issues in textile industry. Environ Dev Sustain 20(4):1439–1459. https://doi.org/10.1007/s10668-017-9949-1
26. Bhatia SC (2017). In: Devraj S (ed) Pollution Control In Textile Industry, 1st edn. WPI Publishing, New York. https://doi.org/10.1201/9781315148588
27. Kant R (2012) Textile dyeing industry an environmental hazard. Natural Science 04. https://doi.org/10.4236/ns.2012.41004

28. Orts F, del Río AI, Molina J, Bonastre J, Cases F (2018) Electrochemical treatment of real textile wastewater: Trichromy Procion HEXL®. J Electroanal Chem 808:387–394. https://doi.org/10.1016/j.jelechem.2017.06.051
29. Umar K, Yaqoob AA, Ibrahim M, Parveen T, Safian M (2020) Environmental applications of smart polymer composites. Smart Polym Nanocompos Biomed Environ Appl 15:295–320
30. Alamzeb M, Tullah M, Ali S, Ihsanullah KB, Setzer WN, Al-Zaqri N, Ibrahim MNM (2022) Kinetic, thermodynamic and adsorption isotherm studies of detoxification of Eriochrome Black T dye from wastewater by native and washed garlic peel. Water 14(22):3713
31. Samchetshabam G, Hussan A, Gon Choudhury T, Gita S, Soholars P, Hussan A (2017) Impact of textile dyes waste on aquatic environments and its treatment
32. Ejhieh AN, Khorsandi M (2010) Photodecolorization of Eriochrome Black T using NiS-P zeolite as a heterogeneous catalyst. J Hazard Mater 176(1–3):629–637. https://doi.org/10.1016/j.jhazmat.2009.11.077
33. Islam A, Ahmad A, Laskar MA (2012) Characterization of a chelating resin functionalized via azo spacer and its analytical applicability for the determination of trace metal ions in real matrices. Journal of Applied Polymer Science 123(6):3448–3458. https://doi.org/10.1002/app.34844
34. Yaseen DA, Scholz M (2019) Textile dye wastewater characteristics and constituents of synthetic effluents: a critical review. Int J Environ Sci Technol 16(2):1193–1226. https://doi.org/10.1007/s13762-018-2130-z
35. Berradi M, Hsissou R, Khudhair M, Assouag M, Cherkaoui O, El Bachiri A, El Harfi A (2019) Textile finishing dyes and their impact on aquatic environs. Heliyon 5(11):e02711. https://doi.org/10.1016/j.heliyon.2019.e02711
36. Al-Tohamy R, Ali SS, Li F, Okasha KM, Mahmoud YAG, Elsamahy T, Jiao H, Fu Y, Sun J (2022) A critical review on the treatment of dye-containing wastewater: ecotoxicological and health concerns of textile dyes and possible remediation approaches for environmental safety. Ecotoxicol Environ Saf 231:113160. https://doi.org/10.1016/j.ecoenv.2021.113160
37. Safian MT, Umar K, Parveen T, Yaqoob AA, Ibrahim MNM (2021) Biomedical applications of smart polymer composites (Chap. 8). In: Smart polymer nanocomposites: biomedical and environmental applications. Woodhead Publishing Series in Composites Science and Engineering, Elsevier Inc., Cambridge, pp 183–204
38. Semeraro P, Fini P, D'Addabbo M, Rizzi V, Cosma P (2017) Removal from wastewater and recycling of azo textile dyes by alginate-chitosan beads. International Journal of Environment, Agriculture and Biotechnology 2:1835–1850. https://doi.org/10.22161/ijeab/2.4.48
39. Ganesan V, Girija E (2015) Investigations on textile dye adsorption onto hydroxyapatite-alginate nanocomposite prepared by a modified method. Cellul Chem Technol 49:87–91
40. Yadav S, Asthana A, Chakraborty R, Jain B, Singh AK, Carabineiro SAC, Susan M (2020) Cationic dye removal using novel magnetic/activated charcoal/β-cyclodextrin/alginate polymer nanocomposite. Nanomaterials (Basel) 10(1). https://doi.org/10.3390/nano10010170.
41. Lincy A, Jegathambal P, Mkandawire M, MacQuarrie SL (2020) Nano bioremediation of textile dye effluentusing magnetite nanoparticles encapsulated alginate beads
42. Mahmoodi NM (2011) Equilibrium, kinetics, and thermodynamics of dye removal using alginate in binary systems. J Chem Eng Data 56(6):2802–2811. https://doi.org/10.1021/je101276x
43. Vijayaraghavan G, Shanthakumar S (2018) Effective removal of acid black 1 dye in textile effluent using alginate from brown algae as a coagulant. Iranian Journal of Chemistry and Chemical Engineering 37(4):145–151. https://doi.org/10.30492/ijcce.2018.35074
44. Parlayıcı Ş (2022) Natural mineral and biopolymers based adsorbent for cationic dyes removal: glutaraldehyde crosslinked alginate/kaolin bead. Journal of Materials and Environmental Science 13(1):95–114
45. ALSamman MT, Sánchez J (2022) Chitosan- and alginate-based hydrogels for the adsorption of anionic and cationic dyes from water. Polymers 14 (8):1498
46. Amina A, Hassina Z-B, Viseras C, Sánchez-Polo M (2018) Bioadsorbent beads prepared from activated biomass/alginate for enhanced removal of cationic dye from water medium: kinetics,

equilibrium and thermodynamic studies. Journal of Molecular Liquids 256. https://doi.org/10.1016/j.molliq.2018.02.073
47. Marzban N, Moheb A, Filonenko S, Hosseini SH, Nouri MJ, Libra J, Farru G (2021) Intelligent modeling and experimental study on methylene blue adsorption by sodium alginate-kaolin beads. International Journal of Biological Macromolecules 186. https://doi.org/10.1016/j.ijbiomac.2021.07.006
48. Kim H, Purev O, Myung E, Choi N, Cho K (2022) Removal of methyl red from aqueous solution using polyethyleneimine crosslinked alginate beads with waste foundry dust as a magnetic material. International Journal of Environmental Research and Public Health 19(15). https://doi.org/10.3390/ijerph19159030
49. Dutta S, Gupta B, Srivastava SK, Gupta AK (2021) Recent advances on the removal of dyes from wastewater using various adsorbents: a critical review. Materials Advances 2(14):4497–4531. https://doi.org/10.1039/D1MA00354B
50. Hassani A, Soltani RDC, Karaca S, Khataee A (2015) Preparation of montmorillonite–alginate nanobiocomposite for adsorption of a textile dye in aqueous phase: isotherm, kinetic and experimental design approaches. J Ind Eng Chem 21:1197–1207. https://doi.org/10.1016/j.jiec.2014.05.034
51. Carrillo Cortes S, Claudio-Rizo J, Burciaga-Montemayor N, Caldera Villalobos M (2022) Removal of textile dyes from aqueous solutions and wastewaters using biobased interpenetrating networks of alginate, collagen, and polyurethane. Asian Journal of Applied Science and Technology 06:164–174. https://doi.org/10.38177/ajast.2022.6119
52. Vijayaraghavan G, Shanthakumar S (2020) Removal of crystal violet dye in textile effluent by coagulation using algal alginate from brown algae Sargassum sp. Desalination and Water Treatment 196:402–408. https://doi.org/10.5004/dwt.2020.25569
53. Fabryanty R, Valencia C, Soetaredjo FE, Putro JN, Santoso SP, Kurniawan A, Ju Y-H, Ismadji S (2017) Removal of crystal violet dye by adsorption using bentonite—alginate composite. J Environ Chem Eng 5(6):5677–5687. https://doi.org/10.1016/j.jece.2017.10.057
54. Oussalah A, Boukerroui A (2020) Alginate-bentonite beads for efficient adsorption of methylene blue dye. Euro-Mediterranean Journal for Environmental Integration 5(2):31. https://doi.org/10.1007/s41207-020-00165-z
55. Gap R, Ong S-T (2018) Application of experimental design for dyes removal in aqueous environment by using sodium alginate-TiO_2 thin film. Chemical Data Collections 15–16:32–40. https://doi.org/10.1016/j.cdc.2018.03.002
56. Rocher V, Bee A, Siaugue JM, Cabuil V (2010) Dye removal from aqueous solution by magnetic alginate beads crosslinked with epichlorohydrin. J Hazard Mater 178(1–3):434–439. https://doi.org/10.1016/j.jhazmat.2010.01.100
57. Radoor S, Karayil J, Parameswaranpillai J, Siengchin S (2020) Removal of anionic dye Congo red from aqueous environment using polyvinyl alcohol/sodium alginate/ZSM-5 zeolite membrane. Sci Rep 10(1):15452. https://doi.org/10.1038/s41598-020-72398-5
58. İnal M, Erduran N (2015) Removal of various anionic dyes using sodium alginate/poly(N-vinyl-2-pyrrolidone) blend hydrogel beads. Polym Bull 72(7):1735–1752. https://doi.org/10.1007/s00289-015-1367-7
59. Minh VX, Hanh LTM, Lan PT, Van Bien T, Dung NT (2020) Synthesis of Fe_3O_4/Alginate composite for dye removal from aqueous solution. Vietnam Journal of Chemistry 58(2):185–190. https://doi.org/10.1002/vjch.201900151
60. Nigiz FU (2019) Synthesis and characterization of clinoptilolite-alginate beads for dye removal from water. Water Practice and Technology 14(2):311–318. https://doi.org/10.2166/wpt.2019.015
61. Djebri N, Boutahala M, Chelali N-E, Boukhalfa N, Zeroual L (2016) Enhanced removal of cationic dye by calcium alginate/organobentonite beads: modeling, kinetics, equilibriums, thermodynamic and reusability studies. Int J Biol Macromol 92:1277–1287. https://doi.org/10.1016/j.ijbiomac.2016.08.013
62. Aravindhan R, Fathima NN, Rao JR, Nair BU (2007) Equilibrium and thermodynamic studies on the removal of basic black dye using calcium alginate beads. Colloids Surf, A 299(1):232–238. https://doi.org/10.1016/j.colsurfa.2006.11.045

63. Rocher V, Siaugue J-M, Cabuil V, Bee A (2008) Removal of organic dyes by magnetic alginate beads. Water Res 42(4):1290–1298. https://doi.org/10.1016/j.watres.2007.09.024
64. Kazemi J, Javanbakht V (2020) Alginate beads impregnated with magnetic Chitosan@Zeolite nanocomposite for cationic methylene blue dye removal from aqueous solution. Int J Biol Macromol 154:1426–1437. https://doi.org/10.1016/j.ijbiomac.2019.11.024
65. Robinson T, McMullan G, Marchant R, Nigam P (2001) Remediation of dyes in textile effluent: a critical review on current treatment technologies with a proposed alternative. Biores Technol 77(3):247–255. https://doi.org/10.1016/S0960-8524(00)00080-8

The manufacturer's authorised representative in the EU is Springer Nature Customer Service Centre GmbH, Europaplatz 3, 69115 Heidelberg, Germany. If you have any concerns regarding our products, please contact ProductSafety@springernature.com

Printed and bound by CPI Group (UK) Ltd, Croydon, CR0 4YY

01/10/2026

02225864-0017